W9-DAQ-985

Digital Arithmetic

Digital Arithmetic

Miloš D. Ercegovac

COMPUTER SCIENCE DEPARTMENT
UNIVERSITY OF CALIFORNIA, LOS ANGELES

Tomás Lang

DEPARTMENT OF ELECTRICAL AND COMPUTER ENGINEERING
UNIVERSITY OF CALIFORNIA, IRVINE

QA
76.9
. C62
E72
2004

MORGAN KAUFMANN PUBLISHERS

AN IMPRINT OF ELSEVIER SCIENCE

SAN FRANCISCO   SAN DIEGO   NEW YORK   BOSTON
LONDON   SYDNEY   TOKYO

Senior Editor	Denise E. M. Penrose
Publishing Services Manager	Simon Crump
Project Manager	Justin Palmeiro
Project Management	Graphic World Publishing Services
Editorial Coordinator	Alyson Day
Cover Design Manager	Cate Rickard Barr
Cover Design Coordinator	Elisabeth Beller
Cover Art	© Jasper Johns/Licensed by VAGA, New York, NY
Cover Photos	© CNAC/MNAM/Dist. Réunion des Musées Nationaux/Art Resource, NY

Johns, Jasper. Zero of Figures 0–9. 1960–1971. Lithograph, re-worked with paint, fiber collage and Japanese papers. Dimensions: $12\,^3/_4'' \times 10\,^1/_2''$. Inv.: Am 1983–314(1). Photo: Philippe Migeat. Musee National d'Art Moderne, Centre Georges Pompidou, Paris, France.

Johns, Jasper. One of Figures 0–9. 1960–1971. Lithograph, re-worked with paint, fiber collage and Japanese papers. Dimensions: $12\,^7/_8'' \times 10\,^1/_2''$. Inv.: Am 1983–314(2). Photo: Philippe Migeat. Musee National d'Art Moderne, Centre Georges Pompidou, Paris, France.

Text Design	Frances Baca Design
Composition	International Typesetting and Composition
Printer	Maple-Vail, York

IPFW
FEB 16 2005
WITHDRAWN
HELMKE LIBRARY

Designations used by companies to distinguish their products are often claimed as trademarks or registered trademarks. In all instances in which Morgan Kaufmann Publishers is aware of a claim, the product names appear in initial capital or all capital letters. Readers, however, should contact the appropriate companies for more complete information regarding trademarks and registration.

Morgan Kaufmann Publishers
An imprint of Elsevier Science
340 Pine Street, Sixth Floor
San Francisco, CA 94104-3205, USA
www.mkp.com

© 2004 by Elsevier Science (USA)
All rights reserved.
Printed in the United States of America

07 06 05 04 03   5 4 3 2 1

No part of this publication may be reproduced, stored in a retrieval system, or transmitted in any form or by any means—electronic, mechanical, photocopying, or otherwise—without the prior written permission of the publisher.

Library of Congress Control Number: 2002114337

ISBN: 1-55860-798-6

This book is printed on acid-free paper.

About the Authors

Miloš D. Ercegovac is a professor and chair in the UCLA Computer Science Department. He earned an MS (1972) and PhD (1975) in computer science from the University of Illinois, Urbana-Champaign, and a BS in electrical engineering (1965) from the University of Belgrade, Yugoslavia. Dr. Ercegovac specializes in research and teaching in digital arithmetic, digital design, and computer system architecture. His research contributions have been extensively published in journals and conference proceedings. He is a coauthor of two textbooks on digital design and a monograph in the area of digital arithmetic. Dr. Ercegovac has been involved in organizing the IEEE Symposia on Computer Arithmetic since 1978. He served as an editor of the *IEEE Transactions on Computers* and as a subject area editor for the *Journal of Parallel and Distributed Computing*. Dr. Ercegovac is a fellow of the IEEE Computer Society and a member of the ACM.

Tomás Lang is a professor in the Department of Electrical and Computer Engineering at the University of California, Irvine. Previously he was a professor in the Computer Architecture Department of the Polytechnic University of Catalonia, Spain, and a faculty member of the Computer Science Department at the University of California, Los Angeles. He received an electrical engineering degree from the Universidad de Chile in 1965, an MS from the University of California Berkeley in 1966, and the PhD from Stanford University in 1974. Dr. Lang's primary research and teaching interests are in digital design and computer architecture, with current emphasis on high-speed and low-power numerical processors and multiprocessors. He is a coauthor of two textbooks on digital systems, two research monographs, one IEEE Tutorial, and the author or coauthor of numerous research contributions to scholarly publications and technical conferences.

Contents

CHAPTER 1

Review of the Basic Number Representations and Arithmetic Algorithms   3

CHAPTER 2

Two-Operand Addition   51

CHAPTER 5

Division by Digit Recurrence   247

CHAPTER 8

Floating-Point Representation, Algorithms, and Implementations   397

CHAPTER 9

Digit-Serial Arithmetic   489

CHAPTER 10

Function Evaluation　549

Preface

Objectives and Importance

Our main objective in preparing this book is to provide a comprehensive discussion of the main ideas and concepts in digital arithmetic, reflecting both the theory and design aspects, and to help students and practicing engineers develop a good understanding of the "arithmetic style" of algorithms and designs. The research in digital arithmetic continues to be active, and new areas of applications are being introduced, making such a book useful in understanding the state of the art in digital arithmetic in order to develop sound solutions and avoid mistakes and repetitions. Lastly, a thorough exposition of digital arithmetic is likely to stimulate interest in the field.

Digital arithmetic has continued to play an important role in the design of digital processors and application-specific (embedded) systems found in signal processing, graphics, and communications. In spite of a mature body of knowledge, it is not unusual that in each new generation of processors or digital systems new arithmetic design problems need to be solved. A good solution benefits greatly from a comprehensive exposition to digital arithmetic as provided in this book.

Audience

The material covered in this book is intended for graduate students in computer engineering/electrical engineering and computer science who are interested in the design of digital arithmetic for general-purpose processors, application-specific and embedded digital systems, and signal processing systems. It will also be useful to practicing digital design engineers involved in logic and circuit design of arithmetic and floating-point units and in their implementation in VLSI technologies. The background expected consists primarily of college-level mathematics, digital systems and logic design, and, for those interested in applying the material to implementation, a knowledge of VLSI design tools.

Features and Approach

The main feature of our approach has been in providing a unified treatment of digital arithmetic, tying the underlying theory and design practice in a technology-independent manner. We consistently use an algorithmic approach in defining arithmetic operations, illustrate with examples of designs at the logic level, and discuss cost/performance characteristics. To enhance learning, we developed a large set of exercises with solutions and extensive reading lists. These are included in each chapter, and general references (books and compilation of articles) are given in Chapter 1. For instructors we have developed a complete set of lecture viewgraphs.

Ways of Use

The main use of this book is as a text for a graduate course. As such, it can be covered completely in a semester course or alternatively, by eliminating some material, in a quarter course. Many options exist for what is not covered, depending on the emphasis required. For instance, for an emphasis in floating-point units, the most-detailed parts of Chapters 9 and 11 can be skipped; on the other hand, if the emphasis is on other applications, such as signal processing, it might be better to skip parts of Chapter 8. In our opinion, it would be best to cover the chapters in order, to make best use of the knowledge acquired before; however, other sequences are possible. For instance, the chapter on floating point could be covered earlier since it does not depend much on the details of previous chapters. The exercises at the end of the chapters allow for practice and extension of the material and can be used for design and implementation projects. The "Further Readings" sections and extensive bibliography provide material for additional self-study.

The book can also be used as a reference for designers of hardware for numerical applications. In this case, if they have not had a comprehensive course on the topic, the most profitable approach would be to study complete chapters, instead of only particular algorithms or implementations. This approach would provide the basis to experiment with alternative designs to choose the best for the particular requirements and constraints.

Additional Resources

The book is supported with a Web site (http://www.cs.ucla.edu/digital_arithmetic) that contains

- Appendix A: Material for instructors, consisting of solutions to all exercises, sample exams, and source files for lecture viewgraphs. This material will be available to instructors in a password-protected section of the Web site.
- Appendix B: One-third of solutions to exercises.
- Appendix C: Lecture viewgraphs associated with each chapter (in PS and PDF forms).
- Appendix D: Short notes on selected topics.
- Appendix E: Comments and errata.

Overview of Topics

The book begins with a review of basic material in terms of representations and algorithms for the basic operations (Chapter 1) and provides an introduction to the notation and description formats used. It then concentrates on a thorough presentation of alternative algorithms and implementations for addition/subtraction (of two and more than two operands), multiplication, division, and square root (Chapters 2–7). These algorithms and implementations can be directly used for fixed-point applications.

Chapter 8 concentrates on floating-point representation and on the corresponding algorithms and implementations. It contains an extensive discussion of alternative implementations for floating-point addition/subtraction and multiplication and describes the basic approaches to produce correctly rounded results in division and square root.

Chapter 9 presents serial arithmetic, both least-significant-digit first (LSDF) and most-significant-digit first (MSDF). The LSDF approach is effective for algorithms consisting only of additions and multiplications, whereas MSDF can be used for cases that include also division, square root, and comparisons. After considering the basic operations, the chapter illustrates their use in composite operations and in multimodule systems.

Chapters 10 and 11 discuss methods for function generation. Two main approaches are considered: (1) approximations based on multiplications, additions, and table lookup and (2) recurrences with linear convergence. The first approach results in polynomial approximations (also included are methods based only on addition and table lookup) and is applicable to a large variety of functions. On the other hand, the second method is based on multiplicative and additive normalization and is practical only for some important functions, such as logarithm, exponential, sine, cosine, and arctan. In particular, the CORDIC algorithm presented in Chapter 11 is attractive for the multivariable functions rotation by an angle, modulus of a vector, and arctan(y/x). The discussion in that chapter also considers the generalization to hyperbolic and linear coordinates.

Topics Not Covered

Several major areas of digital arithmetic, such as residue number system arithmetic, logarithmic number system arithmetic, modular arithmetic, asynchronous multiplication and division, design for low-power arithmetic, arithmetic error codes, and verification and testing, are not included in this book. This does not imply that the omitted topics are less important than the ones presented; there will be short notes and a bibliography on the Web site.

Acknowledgments

We thank the many people who have influenced us in developing this book and, in particular, our colleague at UCLA, Algirdas Avizienis, for his work in digital arithmetic and contributions to the graduate course CS 252A (Arithmetic Algorithms and Processors). In addition, seminal works of James E. Robertson, Daniel E. Atkins, and Antonin Svoboda had a strong impact on our work.

We have benefited greatly from interactions and collaborations with numerous colleagues from academia and industry including Elisardo Antelo, Jean-Claude Bajard, Javier Bruguera, Neil Burgess, Luigi Dadda, Luigi Ciminiera, Jordi Cortadella, Warren Ferguson, Michael Flynn, David Goldberg, Mary Jane Irwin, Graham Jullien, William Kahan, Simon Knowles, Israel Koren, Peter Kornerup, Willy McAllister, David Matula, Paolo Montuschi, Jean-Michel Muller, Vojin Oklobdzija, Stott Parker, Michael Schulte, Renato Stefanelli, Earl Swartzlander, Naofumi Takagi, George Taylor, Alexandre Tenca, Arnaud Tisserand, Julio Villalba, and Dan Zuras. We thank them all and in particular those that reviewed the manuscript and provided constructive comments.

Our former and current students provided comments that helped us in developing this book: Charles Chien, Raffi Dionysian, John Fernando, Ian Ferguson, Abdolali Gorji-Sinaki, John Harding, Zhijun Huang, Jeong-A Lee, Marianne Louie, Robert McIlhenny, Peter Montgomery, Alberto Nannarelli, Vojin Oklobdzija, John Pipan, Alexandre Tenca, Paul Tu, Dean Tullsen, and Osaaki Watanuki. We thank them all for their suggestions and interest.

We have been very pleased working with our publisher, Morgan Kaufmann. Our thanks to our editor, Denise Penrose; editorial coordinator, Alyson Day; production manager, Jodie Allen; and production editor, Carol O'Connell for their effort and excellent guidance. The secretarial help of Terry Valai at UCLA has been invaluable and enjoyable.

Symbols and Notation

·(+)	logical AND (logical OR)
$(p:$	a column of p bits
$:q]$	a row of q bits (weighted)
[3:2]	reduction of 3 to 2 digit-vectors ([3:2] adder; [3:2] carry-save adder (CSA))
[4:2]	reduction of 4 to 2 digit-vectors ([4:2] adder; [4:2] carry-save adder)
$(p:q]$	p-input, q-output counter
$[p:2]$	reduction of p to 2 digit-vectors ($[p:2]$ adder (compressor))
b	base of floating-point representation $x = M_x \times b^{E_x}$
B	bias in floating-point representation of exponent
CLA	carry-lookahead adder
CLG	carry-lookahead generator
CMPL	complementer
CPA	carry-propagate adder
CRA	carry-ripple adder
CS	carry-save form
CSA	carry-save adder
CSK	carry-skip adder

δ	on-line delay: number of initial cycles in online operation
δ	number of bits of estimate of divisor/argument in division/square root
EOP	effective operation
FA	full-adder
G	guard bit
HA	half-adder
INCR	incrementer
$L_k\ (U_k)$	lower (upper) boundary of selection interval
LOD	leading-one detection
LOP	leading-one prediction
LSDF	least-significant-digit-first mode of computation
LZA	leading-zeros anticipation
MAC	multiply-accumulate
MAF	multiply-add fused
MG	multiple generator
MSDF	most-significant-digit-first mode of computation
MUX	multiplexer (selector)
NAN	not-a-number
ovf	overflow condition
q_j	jth quotient digit
r	radix (base) of number representation
R	round bit
$\rho = a/(r-1)$	redundancy factor for maximum digit value a and radix r

REC	recoder
s_j	jth square root digit
SD	signed digit
T	sticky bit
ulp	unit in the last place
VAND (VOR)	vector AND (OR) gate
$w[j]$	residual
$\overline{w}(\underline{w})$	upper bound (lower bound) of w
WS (WC)	pseudosum (stored-carry) bit-vectors
$X = (x_{n-1}, \ldots, x_0)$	n-digit vector
$X[_j]$	digit-vector X at step (iteration) j
x_i	digit in the ith position of a digit-vector
$\{x\}_t$	x truncated to t fractional bits
$\overline{X}$	bit-complement of vector X
$\hat{y}$	low-precision estimate of the scaled residual $rw[j]$

THE TOPICS REVIEWED IN THIS CHAPTER INCLUDE

- Digital arithmetic and arithmetic units
- Fixed-point number representation systems
- Representation of signed integers (S&M, two's complement, and ones' complement)
- Basic algorithms for signed integers: range extension, arithmetic shifts, addition, subtraction, multiplication, and division

CHAPTER 1

Review of Basic Number Representations and Arithmetic Algorithms

In this chapter we briefly review basic number representations and algorithms used in digital arithmetic. The treatment is very concise; readers that need a more detailed review should consult some of the references listed at the end of the chapter. More advanced algorithms as well as the implementations are the topic of later chapters.

1.1 Digital Arithmetic and Arithmetic Units

Digital arithmetic encompasses the study of number representations, algorithms for operations on numbers, implementations of arithmetic units in hardware, and their use in general-purpose and application-specific systems.

An *arithmetic unit (processor)* is a system that performs operations on numbers. We limit ourselves to the most common cases in which these numbers are

1. fixed-point numbers

   - integers $I = \{-N, \ldots, N\}$
   - rational numbers of the form $x = a/2^f$ ("binary" rationals), $a \in I$ and f positive integer

2. floating-point numbers $x \times b^E$, x rational number, b the integer base, and E integer exponent. The floating-point numbers approximate real numbers and facilitate computations over a wide dynamic range.

Collectively, we refer to these numbers as DA (digital arithmetic) numbers.

An arithmetic processor operates on one, two, or more *operands* depending on the operation. The operands are characterized by a *representation* and a *set* of values as defined in the next section. The *operation* is selected from an allowable set, which usually includes addition, subtraction, multiplication, division, square root, change of sign, comparison, and so on. The *results* can be DA numbers, logical variables (conditions), and/or singularity conditions (exceptions). Logical results occur for operations such as comparison, check for zero, and the like. Singularity conditions correspond to overflow, divide by zero, square root of a negative number, hardware error, and so on.

The *parameters* that describe the processor to the user include the number representation and precision, the operation set, the time required to execute each operation, the cost of the processor, and its energy consumption.

The function (*functional description*) of the arithmetic processor can be given at three levels:

1. **Abstract (mathematical) level.** The domain of operands and results is the set of numbers. The operations are specified as functions (sets of pairs). Also, some abstract properties such as commutativity, associativity, and distributivity can be given. At this level the objective is a functional specification (description). This is also known as high-level description. It has no implementation details.

2. **Arithmetic-algorithm level.** The numbers are represented by vectors of digits (digit-vectors), and the operations are described by algorithms composed of primitive operations (transformations) that are performed on these digit-vectors. This level provides a behavioral description, typically using arithmetic expressions and composition of functions. It introduces constraints affecting implementation.

3. **Implementation level.** The digit-vectors are encoded on bit-vectors. The operations are described by register-transfer algorithms. The description at this level is structural, specifying the modules, their interconnection, and the control flow.

In this text, we discuss arithmetic processors at the arithmetic-algorithm and implementation levels.

In the next section we introduce the basic number systems for fixed-point representation, which are used in the following chapters. Other representations are discussed in later chapters together with their uses. We then present the basic algorithms.

1.2   Basic Fixed-Point Number Representation Systems

To perform operations on fixed-point numbers at the arithmetic-algorithm level, a specific number representation is required. In a *digital representation*, such a number is represented by an ordered n-tuple. Each of the elements of the n-tuple is called a *digit*, and the n-tuple is called a *digit-vector*. The number of digits n is called the *precision* of the representation. We begin with the representation of nonnegative integers, followed by the representation of signed integers, and concluding with an extension to fixed-point numbers.

1.2.1   Representation of Nonnegative Integers

The digit-vector that represents the integer x is denoted by

$$X = (X_{n-1}, X_{n-2}, \ldots, X_1, X_0) \qquad \textbf{1.1}$$

Note that we use a zero-origin, leftward-increasing indexing.

The *number system* to represent x consists of the following elements:

1.   The number of digits n.

2.   A set of (numerical) *values* for the digits. We call D_i the set of values of X_i. The cardinality of set D_i is denoted by $|D_i|$. For example, $\{0, 1, 2, \ldots, 9\}$ is the digit set for the conventional decimal number system with cardinality 10.

3.   A rule of *interpretation*. This rule corresponds to a mapping between the set of digit-vector values and the set of integers.

There are many number systems differing in these elements.

The set of integers, each represented by a digit-vector with n digits, is a *finite set* with at most $K = \prod_{i=0}^{n-1} |D_i|$ different elements since this is the maximum number of different digit-vectors. For example, in a conventional decimal system a digit-vector of six digits can represent a million values. Sets that have been found generally useful to perform basic arithmetic operations include, for example, all integers from 0 to $K - 1$.

A number system is *nonredundant* if each digit-vector represents a different integer; that is, if the representation mapping is one to one. It is *redundant* if there are integers that are represented by more than one digit-vector. Redundant number systems are sometimes used to reduce the complexity of the arithmetic algorithms and increase the speed of execution.

The number systems most frequently used are *weighted systems*. For them the representation mapping is

$$x = \sum_{i=0}^{n-1} X_i W_i \qquad\qquad 1.2$$

where $W = (W_{n-1}, \ldots, W_0)$ is the *weight-vector*.

A *radix number system* is a weighted number system in which the weight-vector is related to the *radix-vector* $R = (R_{n-1}, \ldots, R_0)$ as follows:

$$W_0 = 1; \quad W_i = W_{i-1} \cdot R_{i-1} \quad (1 \le i \le n - 1) \qquad\qquad 1.3$$

This is equivalent to

$$W_0 = 1; \quad W_i = \prod_{j=0}^{i-1} R_j \qquad\qquad 1.4$$

Radix number systems are classified according to the radix-vector into fixed-radix and mixed-radix systems.

In a *fixed-radix* system all elements of the radix-vector have the same value r (the radix). Consequently, the weight vector is

$$W = (r^{n-1}, \ldots, r^2, r, 1) \qquad\qquad 1.5$$

and

$$x = \sum_{i=0}^{n-1} X_i \cdot r^i \qquad\qquad 1.6$$

The most frequently used radices are powers of two, such as 2 (binary), 4 (quaternary), 8 (octal), and 16 (hexadecimal). The corresponding weight-vectors are $W = (\ldots, 16, 8, 4, 2, 1)$ for $r = 2$, $W = (\ldots, 256, 64, 16, 4, 1)$ for $r = 4$, and so on. The other radix that is sometimes used is 10 (decimal); this is done because of our familiarity with this representation and because the interface with humans is more convenient in decimal. Because some arithmetic algorithms are simpler in binary than in decimal, in many systems the input-output is decimal but the internal processing is done in binary. Conversion is therefore required between these representations.

In a *mixed-radix* system the elements of the radix-vector are different. For example, the representation of time in terms of hours, minutes, and seconds in a 24-hour period uses a radix-vector $R = (24, 60, 60)$. The corresponding weight-vector is $W = (3600, 60, 1)$. Consequently, the digit-vector $X = (5, 37, 43)$ represents 20,263 seconds.

According to the *set of digit values*, the radix number systems are classified into canonical and noncanonical systems.

In a *canonical system* the set of values for D_i is $\{0, 1, \ldots, R_i - 1\}$ with $|D_i| = R_i$. For example, the canonical digit sets in the binary, quaternary, octal, and hexadecimal number systems are respectively $\{0, 1\}$, $\{0, 1, 2, 3\}$, $\{0, 1, 2, \ldots, 7\}$, and $\{0, 1, 2, \ldots, 15\}$. The corresponding range of values of x represented with n radix-r digits is

$$0 \leq x \leq r^n - 1 \qquad\qquad\qquad \textbf{1.7}$$

In a *noncanonical system* the set of digit values is not canonical. For example, $D_i = \{-4, -3, -2, -1, 0, 1, 2, 3, 4, 5\}$ is a digit set in a noncanonical decimal system and $\{-1, 0, 1\}$ and $\{0, 1, 2\}$ in noncanonical binary systems.

A noncanonical digit set D_i such that $|D_i| > R_i$ produces a redundant system allowing more than one representation of a value; for example, in the $\{-1, 0, +1\}$ binary system the vectors $(1, 1, 0, 1)$ and $(1, 1, 1, -1)$ both represent the integer "thirteen."[1]

A system with fixed positive radix r and canonical set of digit values is called a *radix-r conventional number system*. These are by far the most commonly

---

1. To distinguish the integer from its radix-10 representation, we give the name of the number as its decimal representation in letters.

Number System	Digit Vector
Conventional radix-2 system (binary)	0011110
Conventional radix-3 system	0001010
Conventional radix-4 system	0000132
Conventional radix-10 system	0000030
Radix-2 system with digit set $\{-1 = \bar{1}, 0, 1\}$	0011110
	$01000\bar{1}0$
Residue system with $P = (17, 13, 11, 7, 5, 3, 2)$	(13)482000

TABLE 1.1 Representations of the integer "thirty."

used number systems. As indicated, the favored radices are powers of 2 and 10 (decimal). In the following sections we discuss algorithms for these systems emphasizing the conventional binary system.

There exist also *nonradix number systems* in which weights are not defined recursively as in (1.3). One example is the *residue number system* (RNS), where for a given set of pairwise relatively prime integers $P = (P_{n-1}, \ldots, P_0)$, a positive integer x (for $0 \leq x < \prod_{i=0}^{n-1} P_i$) is represented by the vector X such that

$$X_i = x \bmod P_i \qquad\qquad 1.8$$

This is a nonredundant system that allows fast implementation of addition and multiplication. In this system there is no notion that the digits on the left are more significant than the digits on the right. In that sense, there is no notion of "weight," and RNS is sometimes classified as a nonweighted system. As an example, we represent in Table 1.1 the integer "thirty" in several number systems using a digit-vector with seven components.

In this text we use fixed radix and mainly radix 2.

Bit-Vector Representation

For the implementation of arithmetic algorithms in (binary) digital systems, it is necessary to represent the digit-vectors by bit-vectors. This is done by defining a *code* for a digit and mapping the digit-vector by mapping each digit according to this code.

In the binary (conventional) number system, the code is direct: the binary-digit values 0 and 1 are represented by the binary-variable values 0 and 1, respectively.

For higher power-of-two radices the most common code is the binary code in which a digit d is represented by a bit vector $(d_{k-1}, \ldots, d_0)$ of $k = \log_2 r$ bits such that

$$d = \sum_{i=0}^{k-1} d_i 2^i \qquad 1.9$$

The use of this code for each digit results in a bit-vector for x that is the same for any power-of-two radix, the only difference being the way the bits are grouped to form a digit. In the binary case, each bit corresponds to a digit, while in the radix-r case, groups of $\log_2 r$ bits form a digit. Therefore, conversion from a bit-vector in a radix-2 representation to a radix-r representation and vice versa is trivial. For example, the bit-vector

$$
\begin{aligned}
X &= (1, 1, 0, 0, 0, 1, 0, 1, 1, 1, 0, 1) \\
&= ((1, 1, 0), (0, 0, 1), (0, 1, 1), (1, 0, 1)) \qquad 1.10 \\
&= ((1, 1, 0, 0), (0, 1, 0, 1), (1, 1, 0, 1))
\end{aligned}
$$

corresponds to the octal digit-vector $(6, 1, 3, 5)$ and the hexadecimal digit-vector $(C, 5, D)$.[2]

The fact that the bit-vectors are identical permits the use of some binary algorithms to perform operations on integers represented in these higher radices.

1.2.2 Representation of Signed Integers

In the previous section we presented the representation of nonnegative integers. We now extend the discussion to the representation of signed integers (positive and negative). Two representations are by far the most common: the sign-and-magnitude representation and the true-and-complement representation; these are the topic of this section.

---

2. The integers 10, 11, ..., 15 are denoted with letters A, B, ..., F, respectively.

Sign-and-Magnitude (SM) System

A signed integer x is represented in the SM system by a pair (x_s, x_m), where x_s is the *sign* and x_m is the *magnitude* (positive integer). The two values of the sign $(+, -)$ are represented by a binary variable, where traditionally 0 corresponds to $+$ and 1 to $-$.

The magnitude can be represented by any system for the representation of positive integers. If a conventional radix-r system is used, the range of signed integers, for n digits in the representation of the magnitude, is

$$0 \leq x_m \leq r^n - 1 \qquad\qquad \textbf{1.11}$$

Note that zero has two representations: $x_s = 0, x_m = 0$ (positive zero) and $x_s = 1, x_m = 0$ (negative zero).

True-and-Complement (TC) System

In the true-and-complement system there is no separation between the representation of the sign and the representation of the magnitude, but the whole signed integer is represented by a positive integer. Consequently, this representation involves an additional mapping as indicated in Figure 1.1.

The signed integer x is represented by a positive integer x_R, which in turn is represented by the digit-vector X. Map 2 defines the mapping between integers

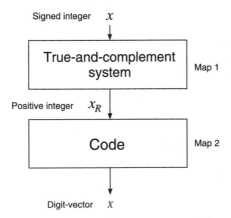

FIGURE 1.1 Signed integer represented by positive integer.

and digit-vectors as discussed in the previous section. We now define the mapping Map 1 for the true-and-complement system.

A signed integer x is represented in the true-and-complement system by a positive integer x_R such that

$$x_R = x \bmod C \qquad\qquad \textbf{1.12}$$

where C is a positive integer, called the *complementation constant*. By the definition of the mod function, for $\max |x| < C$, this is equivalent to

$$x_R = \begin{cases} x & \text{if } x \geq 0 \\ C - |x| = C + x & \text{if } x < 0 \end{cases} \qquad\qquad \textbf{1.13}$$

In order to have an unambiguous representation, the region for $x > 0$ should not overlap with the region for $x < 0$. This requires that

$$\max |x| < C/2 \qquad\qquad \textbf{1.14}$$

The converse mapping is

$$x = \begin{cases} x_R & \text{if } x_R < C/2 \\ x_R - C & \text{if } x_R > C/2 \end{cases} \qquad\qquad \textbf{1.15}$$

When $x_R = C/2$ is representable, it is usually assigned to $x = -C/2$, making the representation asymmetrical.

The representations of positive integers are called *true forms*, and those of negative integers, *complement forms*.

The positive integer x_R can be represented in any system for positive integers. For a digit-vector of n digits, the range is

$$0 \leq x_R \leq r^n - 1 \qquad\qquad \textbf{1.16}$$

The usual choices for the complementation constant are $C = r^n$ (*Range Complement System (RC)*) and $C = r^n - 1$ (*Digit Complement System (DC)*). We now consider a radix-2 representation and leave as an exercise the more general radix-r case.

The choice $C = 2^n$ defines the *two's complement* system. The corresponding mapping is illustrated in Table 1.2. In this system the value $x_R = C$ is outside the range, and, therefore, there is only one representation of $x = 0$. The value $x_R = 2^{n-1}$ could represent either $x = 2^{n-1}$ or $x = -2^{n-1}$, resulting in an asymmetric representation. It is usual to make the second choice in order to

x	x_R	
0	0	
1	1	
2	2	
–	–	True forms
–	–	(positive)
–	–	$x_R = x$
$2^{n-1} - 1$	$2^{n-1} - 1$	
-2^{n-1}	2^{n-1}	
$-(2^{n-1} - 1)$	$2^{n-1} + 1$	
–	–	Complement forms
–	–	(negative)
-2	$2^n - 2$	$x_R = 2^n - |x|$
-1	$2^n - 1$	

TABLE 1.2 Mapping in the two's complement system.

simplify the sign detection, as discussed later in Section 1.2.3. The range of signed integers is

$$-2^{n-1} \leq x \leq 2^{n-1} - 1 \qquad \textbf{1.17}$$

The choice $C = 2^n - 1$ defines the *ones' complement* system. The corresponding mapping is shown in Table 1.3. In this system $x_R = C$ is representable with n digits so that there are two representations of $x = 0$: $x_R = 0$ and $x_R = 2^n - 1$.

The range of signed integers is

$$-(2^{n-1} - 1) \leq x \leq 2^{n-1} - 1 \qquad \textbf{1.18}$$

EXAMPLE 1.1   Represent $-4 \leq x \leq 3$ in the two's complement and ones' complement systems. The mappings

$$x \rightarrow x_R \rightarrow X$$

x	x_R	
0	0	
1	1	
2	2	True forms
–	–	(positive)
–	–	$x_R = x$
–	–	
$2^{n-1} - 1$	$2^{n-1} - 1$	
$-(2^{n-1} - 1)$	2^{n-1}	
–	–	
–	–	Complement forms
-2	$2^n - 3$	(negative)
-1	$2^n - 2$	$x_R = 2^n - 1 - |x|$
0	$2^n - 1$	

TABLE 1.3 Mapping in the ones' complement system.

and

$$X \to x_R \to x$$

are shown in Tables 1.4 and 1.5, respectively.   ■

The following properties of the two's complement and ones' complement systems can be seen from the previous example:

- The representation of zero is unique in the two's complement system since the complementation constant $C = 2^n$ is not representable by an n-bit vector. In the ones' complement system there are two representations of zero since $C = 2^n - 1$ is representable as $(1, 1, \ldots, 1)$.
- The range in the two's complement system is not symmetrical since $x = -2^{n-1}$ is representable but $x = 2^{n-1}$ is not. That is, the range is

$$[-2^{n-1}, 2^{n-1} - 1] \qquad \qquad \textbf{1.19}$$

This means that the system is not closed under the change of sign operation. The range in the ones' complement system is symmetrical:

$$[-(2^{n-1} - 1), 2^{n-1} - 1] \qquad \qquad \textbf{1.20}$$

	Two's Complement $(C = 8)$		Ones' Complement $(C = 7)$	
x	x_R	X	x_R	X
3	3	011	3	011
2	2	010	2	010
1	1	001	1	001
0	0	000	0	000
-0	–	–	7	111
-1	7	111	6	110
-2	6	110	5	101
-3	5	101	4	100
-4	4	100	–	–

TABLE 1.4 Two's and ones' complement representations for $n = 3$: mapping from value to bit-vector.

		Two's Complement $(C = 8)$	Ones' Complement $(C = 7)$
X	x_R	x	x
000	0	0	0
001	1	1	1
010	2	2	2
011	3	3	3
100	4	-4	-3
101	5	-3	-2
110	6	-2	-1
111	7	-1	-0

TABLE 1.5 Two's and ones' complement representations for $n = 3$: converse mapping from bit-vector to value.

1.2.3 Sign Detection

We now present algorithms for sign detection in the sign-and-magnitude and true-and-complement systems. Let

$$\text{sign}(x) = \begin{cases} 0 & \text{if } x \geq 0 \\ 1 & \text{if } x \leq 0 \end{cases} \qquad \textbf{1.21}$$

which allows representation of positive and negative 0s.

In the sign-and-magnitude system, the sign detection is trivial since there is a sign bit.

In the true-and-complement system, since $|x| \leq C/2$, the sign is determined as follows:[3]

$$\text{sign}(x) = \begin{cases} 0 & \text{if } x_R < C/2 \\ 1 & \text{if } x_R \geq C/2 \end{cases} \qquad \textbf{1.22}$$

Consequently, in the two's complement and ones' complement systems the sign is determined from the most-significant bit as follows:

$$\text{sign}(x) = \begin{cases} 0 & \text{if } X_{n-1} = 0 \\ 1 & \text{if } X_{n-1} = 1 \end{cases} \qquad \textbf{1.23}$$

that is,

$$\text{sign}(x) = X_{n-1} \qquad \textbf{1.24}$$

1.2.4 Converse Mapping between Bit-Vectors and Values

Since X_{n-1} corresponds to the sign, it is straightforward to perform the converse mapping using the bit-vector X as follows:

1. $X_{n-1} = 0$ indicates a positive x and, consequently,

$$x = x_R = 0 \times 2^{n-1} + \sum_{i=0}^{n-2} X_i 2^i \qquad \textbf{1.25}$$

2. $X_{n-1} = 1$ indicates a negative x. In this case $x = x_R - C$.

---

3. Assigning $x_R = C/2$ to represent $x = -C/2$ allows a simple sign detection based on the most-significant bit of X.

- For $C = 2^n$, we have

$$x = 1 \times 2^{n-1} + \sum_{i=0}^{n-2} X_i 2^i - 2^n = -1 \times 2^{n-1} + \sum_{i=0}^{n-2} X_i 2^i \quad \textbf{1.26}$$

Combining both cases we get for two's complement

$$x = -X_{n-1} 2^{n-1} + \sum_{i=0}^{n-2} X_i 2^i \quad \textbf{1.27}$$

In other words, in converting a bit-vector to a value, we use the fact that the most-significant bit of X has a negative weight while the remaining bits have positive weights. For example,

$$X = (11011) \rightarrow -16 + 8 + 2 + 1 = -5 = x$$

- For $C = 2^n - 1$, we get

$$x = 1 \times 2^{n-1} + \sum_{i=0}^{n-2} X_i 2^i - (2^n - 1) = -1 \times (2^{n-1} - 1) + \sum_{i=0}^{n-2} X_i 2^i$$

$$\textbf{1.28}$$

Again, after combining both cases we get for ones' complement converse mapping

$$x = -X_{n-1}(2^{n-1} - 1) + \sum_{i=0}^{n-2} X_i 2^i \quad \textbf{1.29}$$

For example,

$$X = (10101) \rightarrow -(16 - 1) + 4 + 1 = -10 = x$$

1.2.5   Extension to Fixed-Point Representations

A fixed-point representation of a number $x = x_{INT} + x_{FR}$ consists of integer and fraction components represented by m and f digits, respectively. Consequently, it is convenient to use the following notation:

$$X = (X_{(m-1)} \ldots X_1 X_0 . X_{-1} \ldots X_{-f}) \quad \textbf{1.30}$$

so that

$$x = \sum_{-f}^{m-1} X_i r^i \qquad \textbf{1.31}$$

For example, $0 \le x \le 7\frac{7}{8}$ is represented in radix-2 as $X = (X_2 X_1 X_0.X_{-1} X_{-2} X_{-3})$.

When representing fractions (no integer part), the convention used sometimes is to assign positive indices to the fractional part. That is,

$$X = (X_1 X_2 \dots X_f) \qquad \textbf{1.32}$$

so that

$$x = \sum_{1}^{f} X_i r^{-i} \qquad \textbf{1.33}$$

1.3   Addition, Change of Sign, and Subtraction

In this section, we discuss addition, subtraction, change of sign, and overflow detection.

1.3.1   Addition and Subtraction of Positive Integers

Consider the operation $z = x + y$ in which the operands and the result are positive integers, represented in a conventional radix-r number system. If the operands are represented by a digit-vector of n digits, the result is represented by a digit-vector of $n + 1$ digits, where the most-significant digit (Z_n) has values in the set $\{0, 1\}$. To limit the number of digits of the result to n digits, an additional binary variable is introduced (the carry-out c_{out}) for the additional digit. Moreover, a carry-in (c_{in}) is included so that

$$c_{out} r^n + z = x + y + c_{in} \qquad \textbf{1.34}$$

resulting in

$$z = (x + y + c_{in}) \bmod r^n \qquad \textbf{1.35}$$

and

$$c_{out} = \begin{cases} 1 & \text{if } (x + y + c_{in}) \ge r^n \\ 0 & \text{otherwise} \end{cases} \qquad \textbf{1.36}$$

In terms of the digit vectors we can write

$$(c_{out}, Z) = ADD(X, Y, c_{in}) \qquad\qquad \textbf{1.37}$$

which is implemented by an adder. As indicated on page 9, for power-of-two radices and using the binary code for the digits, the bit-vectors are the same independent of the radix. As a consequence, at the bit level, the adder is the same for these radices.

The c_{out} signal can be used to indicate an *overflow* (OVF), which indicates that the sum has a value outside the range representable by Z.

Similarly, for subtraction

$$-b_{out}r^n + d = x - y - b_{in} \qquad\qquad \textbf{1.38}$$

and the algorithm is described as

$$(b_{out}, D) = SUB(X, Y, b_{in}) \qquad\qquad \textbf{1.39}$$

1.3.2   Addition, Change of Sign, and Subtraction of Signed Integers

We now describe algorithms for the addition and subtraction of signed integers. Specifically, let x and y be signed integers represented by the digit-vectors X and Y, respectively. The addition algorithm $ADDS$ produces the digit-vector Z representing the signed integer $z = x + y$. That is,

$$Z = ADDS(X, Y) \qquad\qquad \textbf{1.40}$$

For the difference $d = x - y$, since $d = x + (-y)$, it is sufficient to consider the algorithms $ADDS$ and CS (change of sign). That is,

$$D = ADDS(X, CS(Y)) \qquad\qquad \textbf{1.41}$$

If the range of integers represented by Z is the same as that of X and Y, the result of the addition or subtraction might not be representable by Z. In such a case the result of the algorithm cannot be correct and an *overflow (OVF)* signal indicates this situation.

The complexity of implementing the $ADDS$ and CS algorithms in hardware depends on the representation system used for the signed integers.

We now consider these algorithms for the sign-and-magnitude and true-and-complement systems. Although the representation in sign-and-magnitude might seem more natural, and therefore a candidate to be considered first, the algorithm for addition is simpler in the true-and-complement system. Consequently, we consider this first, since the algorithm for sign-and-magnitude makes use of this algorithm.

Addition in the True-and-Complement System

Consider the case where there is no overflow, that is, the result is representable. If this is not the case, the overflow is detected as described later on page 25. In the true-and-complement system $z = x + y$ is obtained by computing

$$z_R = (x_R + y_R) \bmod C \qquad \qquad \textbf{1.42}$$

where x_R, y_R, and z_R are the positive integers representing x, y, and z in this number system, and C is the complementation constant.

To prove the correctness of this algorithm consider

$$(x_R + y_R) \bmod C \qquad \qquad \textbf{1.43}$$

We now show that it corresponds to z_R, the representation of the sum z. By definition of the representation,

$$x_R = x \bmod C; \quad y_R = y \bmod C \qquad \qquad \textbf{1.44}$$

so that

$$(x_R + y_R) \bmod C = (x \bmod C + y \bmod C) \bmod C \qquad \qquad \textbf{1.45}$$

This can be simplified because $(a \bmod C + b \bmod C) \bmod C = (a + b) \bmod C$ and consequently,

$$(x_R + y_R) \bmod C = (x + y) \bmod C = z \bmod C \qquad \qquad \textbf{1.46}$$

x	y	x_R	y_R	z_R	z
13	9	13	9	22	22
13	−9	13	55	68 mod 64 = 4	4
−13	9	51	9	60	−4
−13	−9	51	55	106 mod 64 = 42	−22

TABLE 1.6 Examples of true-and-complement addition ($C = 64$).

and by definition

$$z \bmod C = z_R \qquad\qquad \textbf{1.47}$$

This means that to perform the addition of two signed integers represented in the true-and-complement system, we add the (positive) representations and obtain the residue (mod) of the sum with respect to the complementation constant C. Table 1.6 illustrates several cases of addition for $C = 64$ and $-32 \leq x, y, z \leq 31$.

The algorithm consists of two steps: the addition of the positive representations and the mod operation. The first step is done by the algorithm ADD discussed in Section 1.3.1. We now consider the mod operation.

Let $w_R = x_R + y_R$. Then, since $x_R, y_R < C$, we have that $w_R < 2C$. Therefore, the mod operation results in

$$z_R = w_R \bmod C = \begin{cases} w_R & \text{if } w_R < C \\ w_R - C & \text{if } C \leq w_R < 2C \end{cases} \qquad \textbf{1.48}$$

Consequently, this operation consists of determining if $w_R \geq C$ and, if so, subtracting C from it. The complexity of this operation depends on the value of the complementation constant. We now consider the two's complement and ones' complement systems.

Two's Complement System

In the two's complement system the complementation constant is $C = 2^n$. Since $w_R < 2C$, the representation of w_R is the digit-vector $W = (W_n, W_{n-1}, \ldots, W_0)$ of $n + 1$ digits (bits). Consequently, to determine whether $w_R \geq C$, it is sufficient

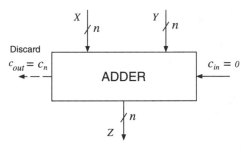

FIGURE 1.2 Two's complement adder.

to check the most significant digit (bit) of W:

$$w_R = \begin{cases} <2^n & \text{if } W_n = 0 \\ \geq 2^n & \text{if } W_n = 1 \end{cases} \qquad \textbf{1.49}$$

In the first case, $w_R \bmod 2^n = w_R$ and its representation is W. In the second situation, it is necessary to subtract 2^n from w_R. This is simple to do since the representation of 2^n is a 1 followed by n 0s. Consequently,

$$w_R \bmod 2^n \iff (1, W_{n-1}, \ldots, W_0) - (1, 0, \ldots, 0) = (W_{n-1}, \ldots, W_0) \quad \textbf{1.50}$$

That is, the mod operation is performed by discarding the most-significant bit. Note that this bit corresponds to the carry-out of the adder that adds x_R and y_R to produce w_R.

In summary, in the two's complement system the result of the addition corresponds to the output of the adder, discarding the carry-out. We describe this by the following bit-level algorithm:

$$ADDS_{2's}(X, Y): \quad Z = ADD(X, Y, 0) \qquad \textbf{1.51}$$

where ADD is a bit-level algorithm for the addition of positive integers and the third operand corresponds to the carry-in. The two's complement addition scheme is shown in Figure 1.2.

EXAMPLE 1.2 We now show two examples of addition of signed integers in the two's complement number system.

$n = 4$	$C = 2^4$	
$X = 1011$	$x_R = 11$	$x = -5$
$Y = 0101$	$y_R = 5$	$y = 5$
$W = 10000$	$w_R = 16$	
$Z = 0000$	$z_R = 0$	$z = 0$

$n = 8$	$C = 2^8$	
$X = 11011010$	$x_R = 218$	$x = -38$
$Y = 11110001$	$y_R = 241$	$y = -15$
$W = 111001011$	$w_R = 459$	
$Z = 11001011$	$z_R = 203$	$z = -53$

■

Ones' Complement System

We now consider the mod operation in the ones' complement system. In this system the complementation constant is $2^n - 1$. To perform $z_R = w_R \bmod (2^n - 1)$ consider the following three cases:

1. If $w_R < 2^n - 1$, then $w_R \bmod (2^n - 1) = w_R$ and $W_n = 0$.
2. If $w_R = 2^n - 1$, then $w_R \bmod (2^n - 1) = 0$ and $W_n = 0$.
3. If $(2^n - 1) < w_R < 2(2^n - 1)$, then $w_R \bmod (2^n - 1) = w_R - (2^n - 1) = (w_R - 2^n) + 1$ and $W_n = 1$.

Consequently,

- if $W_n = 0$, the result is equal to W, and
- if $W_n = 1$ the result is obtained by discarding W_n (subtracting 2^n) and adding 1. Note that this produces a result vector of $(1, 1, \ldots, 1)$ in case 2, which is correct since this is another representation of 0 in the ones' complement system.

Since the bit W_n is produced as the carry-out of the adder, the addition of 1 can be accomplished by an *end-around carry* as shown in Figure 1.3. The effect

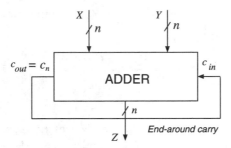

FIGURE 1.3 Ones' complement adder.

of this end-around carry on the implementation is discussed in Chapter 2. The corresponding bit-level algorithm is

$$ADDS_{1s'}(X, Y): \quad Z = ADD(X, Y, c_n) \qquad \textbf{1.52}$$

Change of Sign in the True-and-Complement System

The change of sign operation consists of obtaining the representation of z such that $z = -x$ where z and x are signed integers. Since x and z are represented in the true-and-complement system by x_R and z_R, we have

$$z_R = (-x)_R = (-x) \bmod C = C - x \bmod C = C - x_R \qquad \textbf{1.53}$$

Consequently, the change of sign operation consists of subtracting x_R from the complementation constant C. The complexity of this operation depends on the value of the complementation constant. We now discuss this operation for the the two's complement and ones' complement systems.[4]

Ones' Complement System. In this case the complementation constant is $2^n - 1$, which is represented by the digit-vector $(1, 1, \ldots, 1)$. Therefore the subtraction is performed by complementing each digit of X with respect to 1, obtaining the vector $\overline{X}$. Therefore, the change of sign bit-level algorithm is

$$CS_{1s}(X): \quad Z = \overline{X} \qquad \textbf{1.54}$$

---

4. Note that we discuss first the ones' complement system; this is because the algorithm is simpler and is used as a component in the algorithm for the two's complement system.

EXAMPLE 1.3   The following example illustrates the change of sign in the ones' complement system with $n = 4, C = 2^4 - 1$:

$$
\begin{array}{lll}
X & 1100 & x = -3 \\
\hline
Z = \overline{X} & 0011 & z = 3
\end{array}
$$
■

Two's Complement System. In this case the complementation constant is 2^n. The direct subtraction $2^n - x_R$ requires a complete subtraction, which is complex. Since $2^n = (2^n - 1) + 1$ and the complement with respect to $2^n - 1$ is performed by complementing each digit, the change of sign operation is done in two parts:

1.   Complement each digit with respect to 1.
2.   Add 1.

The addition of 1 can be accomplished by setting $c_{in} = 1$. The corresponding bit-level algorithm is

$$CS_{2's}(X): \quad Z = ADD(\overline{X}, \underline{0}, 1) \qquad\qquad \textbf{1.55}$$

EXAMPLE 1.4   We give an example of the change of sign in the two's complement system. For $n = 4, C = 2^4$, and $x = -3$ we have

$$
\begin{array}{lll}
X & 1101 & x = -3 \\
\hline
\overline{X} & 0010 & \\
0 & 0000 & \\
+ \quad c_0 & 1 & \\
\hline
Z \quad 0011 & z = 3 &
\end{array}
$$
■

Subtraction in the True-and-Complement System

As already indicated, to perform subtraction we combine change of sign and addition since $d = x - y = x + (-y)$. The corresponding bit-level algorithms are

$$
\begin{aligned}
SUB_{2's}: \quad & D = ADD(X, \overline{Y}, 1) \\
SUB_{1's}: \quad & D = ADD(X, \overline{Y}, c_n)
\end{aligned}
\qquad\qquad \textbf{1.56}
$$

Addition/Subtraction in the Sign-and-Magnitude System

The direct algorithm for addition in sign-and-magnitude requires the comparison of the signs, performing an addition if the signs are equal or a subtraction if they are different, and in the latter case a comparison of magnitudes to determine the order of the operands in the subtraction. This is significantly more complex than the algorithm for true-and-complement, in which the operation is always the same, independent of the signs and the relative magnitudes. As a consequence of this, addition in sign-and-magnitude is usually performed by converting the operands to true-and-complement representation, performing the addition in the true-and-complement system, and finally converting the result to sign-and-magnitude. Variations of this algorithm are presented in Section 8.4.

Overflow Detection

An overflow exists whenever the magnitude of result of addition or subtraction exceeds the largest representable magnitude. If this occurs, the result is incorrect and, consequently, it is necessary to detect this situation. Since, as indicated above, even in sign-and-magnitude, the actual addition is performed in the true-and-complement system, we consider only this case.

In the true-and-complement system, an overflow exists when the operands are of the same sign and the result of the addition represents an integer of opposite sign. Since in a ones' complement or two's complement system the sign is determined by the most-significant bit (bit $n - 1$), the overflow detection is specified by the following switching expression:[5]

$$OVF = X'_{n-1} \cdot Y'_{n-1} \cdot Z_{n-1} + X_{n-1} \cdot Y_{n-1} \cdot Z'_{n-1} \qquad \textbf{1.57}$$

Moreover, in the two's complement system, the overflow can also be detected by checking the two most significant carries of the adder (see Exercise 1 . 16).

$$OVF = c_n \oplus c_{n-1} \qquad \textbf{1.58}$$

---

5. We use $'$, $\cdot$ and $+$ to represent the logical NOT, AND and OR functions.

1.4   Range Extension and Arithmetic Shifts

We now present algorithms for range extension and arithmetic shifts for signed integers represented in the radix-2 system. The generalization to radix-r is straightforward. These operations are useful in the implementation of multiplication and division.

1.4.1   Range Extension

The range extension algorithm is performed when it is necessary to represent the value x by a digit-vector of m digits, given its representation by a vector of n digits ($m > n$). That is,

$$z = x \qquad\qquad \textbf{1.59}$$

and

$$Z = (Z_{m-1}, Z_{m-2}, \ldots, Z_0), \quad X = (X_{n-1}, X_{n-2}, \ldots, X_0) \qquad \textbf{1.60}$$

For example, if a single-precision operand is to be added to a double-precision operand, its range must be extended to double precision before the operation. This is also used in multiplication and division algorithms.

In the *sign-and-magnitude system*, for $x = (x_s, X)$ and $z = (z_s, Z)$, the range extension algorithm is

$$\begin{aligned}
z_s &= x_s \quad \text{(sign)} \\
Z_i &= 0, \quad i = m-1, m-2, \ldots, n \\
Z_i &= X_i, \quad i = n-1, \ldots, 0
\end{aligned} \qquad \textbf{1.61}$$

This is illustrated in Figure 1.4(a). The proof is straightforward, resulting directly from the definition of range extension. For example, for $r = 2, n = 3$, and $x_s = 1$, $X = (1, 0, 1)$, the extension to $m = 5$ is $z_s = 1$, $Z = (0, 0, 1, 0, 1)$.

In the *true-and-complement system*, the range extension algorithm is

$$\begin{aligned}
Z_i &= X_{n-1}, \quad i = m-1, \ldots, n \\
Z_i &= X_i, \quad i = n-1, \ldots, 0
\end{aligned} \qquad \textbf{1.62}$$

This is illustrated in Figure 1.4(b). A proof is left as Exercise 1.19. As an example consider the case $r = 2, n = 4, m = 7$, and $X = (1, 0, 1, 1)$. Then in the ones' complement system, $Z = (1, 1, 1, 1, 0, 1, 1)$.

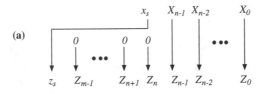

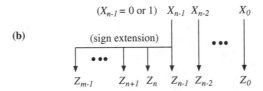

FIGURE 1.4 Range extension. (a) Sign-and-magnitude. (b) True-and-complement.

1.4.2 Arithmetic Shifts

Two elementary arithmetic operations that are used in multiplication and division are the left and right arithmetic shifts. They correspond to scaling operations (multiplying and dividing by the radix).

A *left arithmetic shift* is defined in a conventional radix-2 number system for integers as

$$z = 2x \tag{1.63}$$

and a *right arithmetic shift* as

$$z = 2^{-1}x - \epsilon, \quad |\epsilon| < 1 \tag{1.64}$$

The value of ϵ is such that it makes z an integer. Note that ϵ can be positive or negative. Its sign depends on the representation, as discussed below.

These operations are denoted by $SL(X)$ and $SR(X)$ in the algorithms developed later.

Assuming that *overflow does not occur*, the algorithms to perform these shift operations are as given below.

Arithmetic Shifts in the Sign-and-Magnitude System

The left arithmetic-shift algorithm is

$$\begin{aligned}
z_s &= x_s \quad \text{(sign)} \\
Z_{i+1} &= X_i, \quad i = 0, \ldots, n-2 \\
Z_0 &= 0
\end{aligned} \tag{1.65}$$

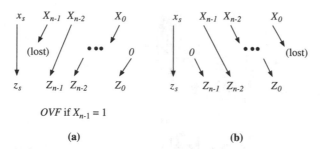

$$OVF \text{ if } X_{n-1} = 1$$

(a)   (b)

FIGURE 1.5 Sign-and-magnitude shift operations. (a) Left shift. (b) Right shift.

The right arithmetic-shift algorithm is

$$z_s = x_s$$
$$Z_{i-1} = X_i, \quad i = 1, \ldots, n-1 \qquad \text{1.66}$$
$$Z_{n-1} = 0$$

Note that in this case, the sign of ϵ is the sign of x.

The proof of these algorithms is straightforward. They are illustrated in Figure 1.5.

EXAMPLE 1.5   For binary SM representation with 8-bit magnitude and $x = -45$, the arithmetic-shift operations result in

$$(x_s, X) = (1,00101101)$$
$$SL(X) = (1,01011010)$$
$$SR(X) = (1,00010110)$$ ■

Arithmetic Shifts in True-and-Complement Systems

In the two's complement system the left arithmetic-shift algorithm is

$$Z_{i+1} = X_i, \quad i = 0, \ldots, n-2$$
$$Z_0 = 0 \qquad \text{1.67}$$

In the ones' complement system, the algorithm is

$$Z_{i+1} = X_i, \quad i = 0, \ldots, n-2$$
$$Z_0 = X_{n-1} \qquad \text{1.68}$$

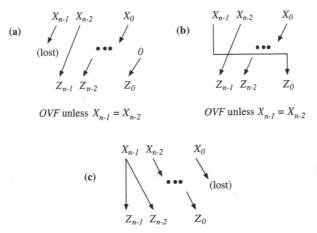

FIGURE 1.6 True-and-complement shift operations for $r = 2$. (a) Two's complement left shift. (b) Ones' complement left shift. (c) TC right shift.

The two's and ones' left arithmetic-shift algorithms are illustrated in Figure 1.6(a) and (b). Proofs are left as Exercise 1.20.

In both the two's and ones' complement systems, the right arithmetic-shift algorithm is

$$Z_{n-1} = X_{n-1}$$
$$Z_{i-1} = X_i, \qquad i = 1, \ldots, n-1$$

1.69

In this case, ϵ is always positive.

The true-and-complement right shift algorithm is illustrated in Figure 1.6(c).

EXAMPLE 1.6   Table 1.7 shows examples of the left and right arithmetic shifts in the true-and-complement system. Signed integers are given in decimal.   ∎

1.5 Basic Multiplication Algorithms

In this section we discuss the basic multiplication algorithms on positive and signed integers. More advanced algorithms and implementations are discussed in Chapter 4.

| | Two's Complement System | | Ones' Complement System | |
	Bit-Vector	Signed Integer	Bit-Vector	Signed Integer
X	001101	13	001111	15
$SL(X)$	011010	26	011110	30
$SR(X)$	000110	6	000111	7
Y	110101	-11	111010	-5
$SL(Y)$	101010	-22	110101	-10
$SR(Y)$	111010	-6	111101	-2

TABLE 1.7 Examples of arithmetic shifts in the true-and-complement system.

1.5.1   Multiplication of Positive Integers

For simplicity we first consider an algorithm for the multiplication of positive integers. Later we extend this algorithm to operate on signed integers. Let x and y be the multiplicand and the multiplier, represented by the n-digit vectors X and Y in the radix-r conventional number system. The multiplication operation produces $p = x \times y$, with p being represented by the digit-vector P of $2n$ digits. The usual method of multiplication is described by the expression

$$p = x \sum_{i=0}^{n-1} Y_i r^i \qquad \text{1.70}$$

$$= \sum_{i=0}^{n-1} x r^i Y_i \qquad \text{1.71}$$

This expression indicates that one first computes the n terms $x r^i Y_i$ and then performs the summation. The computation of the ith term requires an i-position arithmetic left shift of X and a multiplication by the single radix-r digit Y_i. The direct use of this expression leads to a combinational multiplication unit.

If, instead of using $n - 1$ adders, a single adder is reused, the sequential algorithm is

$$
\begin{aligned}
&p[0] = 0 \\
&p[j+1] = r^{-1}(p[j] + x r^n Y_j) \quad \text{for } j = 0, 1, \ldots, n-1 \qquad \text{1.72} \\
&p = p[n]
\end{aligned}
$$

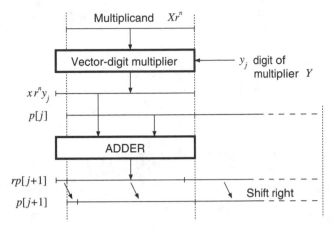

FIGURE 1.7 Relative position of operands in multiplication recurrence.

Since the expansion of this recurrence results in $p[n] = x \times y$, the product is obtained in n steps. Each step consists of the multiplication of x by a radix-r digit to form xY_j, followed by a two-operand addition, and by a one-position arithmetic right shift. The factor r^n multiplying x indicates only that X has to be aligned with the most significant half of the partial product.

This form of recurrence, using a right shift, is chosen so that the multiplication can proceed from the least-significant digit of the multiplier, while the multiplicand retains the same position with respect to the single-precision adder. This is illustrated in Figure 1.7. Note that the adder has $n + 1$ digits because xY_j can have $n + 1$ digits (except for the radix-2 case in which n digits are sufficient).

An example of the execution of the algorithm for radix-2 is given in Figure 1.8. Note that a temporary overflow, indicated by (*) in the figure, may occur in the process of forming the partial products, but it is immediately corrected by the right shift in the following step. Also note that only the significant part of the partial products $((n + j)$ bits of the $2n$ bits) is shown; this is consistent with the implementation of the multiplication algorithm, as described in Chapter 4.

1.5.2 Multiplication of Signed Integers (Radix-2)

The extension of the previous multiplication procedure to signed integers in radix-2 representation is considered next. The operands are represented with n bits (including sign) and the product with $2n - 1$ bits.

$$n = 5 \qquad x = 23 \, (X = 10111) \qquad y = 26 \, (Y = 11010)$$

$p[0]$	00000	
$2^5 x\, Y_0$	00000	
	00000	
$p[1]$	00000	0
$2^5 x\, Y_1$	10111	
	10111	0
$p[2]$	01011	10
$2^5 x\, Y_2$	00000	
	01011	10
$p[3]$	00101	110
$2^5 x\, Y_3$	10111	
	11100	110
$p[4]$	01110	0110
$2^5 x\, Y_4$	10111	
$(*)$	100101	0110
$p[5]$	10010	$10110 = 598$

FIGURE 1.8 Example of magnitude multiplication.

1. **Sign-and-magnitude algorithm.** The algorithm presented before produces the correct magnitude of the product. Therefore, the extension consists only in computing the sign p_s by the common rule of signs: $p_s = x_s \oplus y_s$.

2. **Two's complement algorithm.** The value of the multiplier in the two's complement system can be expressed as in (1.27):

$$y = -Y_{n-1} 2^{n-1} + \sum_{i=0}^{n-2} Y_i 2^i \qquad\qquad \textbf{1.73}$$

Therefore,

$$xy = x \sum_{i=0}^{n-2} Y_i 2^i - x Y_{n-1} 2^{n-1} \qquad\qquad \textbf{1.74}$$

$$n = 5, r = 2 \qquad x = -3 \, (X = 11101) \qquad y = -4 \, (Y = 11100)$$

$p[0]$	0 00000	
$2^5 x Y_0$	0 00000	
	0 00000	
$p[1]$	0 00000	0
$2^5 x Y_1$	0 00000	
	0 00000	0
$p[2]$	0 00000	00
$2^5 x Y_2$	1 11101	
	1 11101	00
$p[3]$	1 11110	100
$2^5 x Y_3$	1 11101	
	1 11011	100
$p[4]$	1 11101	1100
$-2^5 x Y_4$	0 00011	
$p[5]$	0 00000	$1100 = xy = 12$

FIGURE 1.9 Example of two's complement multiplication.

Consequently, the algorithm for multiplication of signed integers in the two's complement system consists of performing the basic recurrence for the first $n - 1$ steps and then subtracting (instead of adding) the multiplicand in the last step.

To avoid losing the sign of the partial product in the case of a temporary overflow, the multiplicand, and the partial product are extended one bit to the left (sign extension). An example is given in Figure 1.9.

3. **Ones' complement algorithm.** The multiplication algorithm for operands in the ones' complement system also requires a corrective step that can be specified in a manner similar to the two's complement case. However, since the change of sign is a simple operation in this system, an alternative approach is to make the negative multiplier positive before applying the

multiplication algorithm, in which case the product should be complemented at the end.

1.6   Basic Division Algorithms

Here we consider integer division.[6] That is, the dividend x, the divisor d, the quotient q, and the remainder w are integers such that

$$x = qd + w \qquad \text{1.75}$$

with the restriction $0 \leq |w| < |d|$ and the sign of the remainder is equal to the sign of the dividend.

We first consider the case of positive integers. These are all represented in a radix-r number system. To obtain a quotient with n digits ($0 \leq q \leq r^n - 1$), the dividend should have $2n$ digits and the divisor n digits. We consider the case $0 < d$ and $x < r^n d$, which precludes division by zero and quotient overflow.

The basic division algorithm consists of n iterations of the following residual recurrence:

$$w[0] = x \qquad \text{1.76}$$

$$w[j + 1] = r w[j] - d^* q_{n-1-j} \quad j = 0, \ldots, n - 1 \qquad \text{1.77}$$

where $q = \sum_{i=0}^{n-1} q_i r^i$ and $d^* = dr^n$; that is, the divisor is aligned with the most-significant half of the residual. In each iteration one digit of the quotient is determined by the quotient-digit selection function

$$q_{j+1} = SEL(w[j], d) \qquad \text{1.78}$$

The value of the quotient digit is such that the next residual $w[j + 1]$ is bounded, such that

$$0 \leq w[j + 1] < d^* \qquad \text{1.79}$$

We now consider the selection function for the restoring and nonrestoring algorithms.

---

6. Non-integer division for floating-point operation is discussed in Chapters 5 and 8.

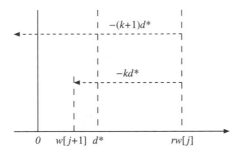

FIGURE 1.10 Selecting quotient digit $q_{n-1-j} = k$.

1.6.1 Restoring Division

In the *restoring algorithm*, the quotient-digit set is the nonredundant set $\{0, 1, 2, \ldots, r-1\}$. In this case, to achieve the residual bound $w[j] < d^*$, it is necessary to use the following quotient-digit selection function:

$$q_{n-1-j} = k \quad \text{if} \quad d^* k \leq rw[j] < d^*(k+1) \qquad (0 \leq k \leq r-1) \qquad \textbf{1.80}$$

This selection is illustrated in Figure 1.10. Its implementation requires comparisons of $rw[j]$ with multiples of d^*. To avoid the need of several comparators, it is possible to subtract the divisor repetitively until the resulting residual is smaller than d^*. This would still need one comparator with d^*. The implementation can be further simplified if the subtraction is continued until the sign of the tentative residual is negative, in which case an addition of d^* produces the correct residual. This addition step is called a *restoring step*. Because of the large number of subtractions required for high radices, this algorithm is only suited for radix-2. In this case, an iteration consists of the following two substeps:

1. A *tentative* residual is calculated as

$$\widetilde{w}[j+1] = 2w[j] - d^* \qquad \textbf{1.81}$$

2. The quotient digit is selected according to the sign of the tentative residual $\widetilde{w}[j+1]$, namely:
If $\widetilde{w}[j+1] \geq 0$ then

$$q_{n-1-j} = 1 \quad \text{and} \quad w[j+1] = \widetilde{w}[j+1] \qquad \textbf{1.82}$$

If $\widetilde{w}[j + 1] < 0$ then

$$q_{n-1-j} = 0 \quad \text{and} \quad w[j + 1] = 2w[j] = \widetilde{w}[j + 1] + d^* \qquad \textbf{1.83}$$

That is, depending on the sign of the tentative residual, the value of the quotient digit is selected. Moreover, when the tentative residual is negative, the new residual is obtained by adding d^*. This restoring division procedure is formalized in Algorithm RD.

Algorithm RD: Restoring Divide

1. [*Initialize*]
    $w[0] = x$
2. [*Recurrence*]
    for $j = 0 \ldots n - 1$
    2.1 $\widetilde{w}[j + 1] = 2w[j] - d^*$
    2.2 if $\widetilde{w}[j + 1] \geq 0$ then
    $\qquad q_{n-1-j} = 1; w[j + 1] = \widetilde{w}[j]$
    else
    $\qquad q_{n-1-j} = 0; w[j + 1] = \widetilde{w}[j] + d^*$
    end for

In an implementation the tentative and the true residuals are stored in the same register. Thus a "restoration" operation $\widetilde{w} + d^*$ is performed whenever the tentative residual is negative.

EXAMPLE 1.7 An example of binary restoring division with $n = 4$ is given in Figure 1.11. Note that the subtractions are performed by adding the two's complement of the divisor d^*. In order to preserve the sign of the shifted residual, the representations of the residuals and divisor are extended by one additional bit to the left. ∎

The restoring division algorithm is simple to implement but is relatively slow. In order to obtain an n-digit quotient, n subtractions, n shifts, and $n/2$ restoration additions (on the average) are required.

Dividend $x = 11_{10} = (00001011)_2$, divisor $d = 2 = (0010)_2$

$w[0] =$	0 0000	1011	
$2w[0] =$	0 0001	0110	
$-d^* =$	1 1110		
$\widetilde{w}[1] =$	1 1111	0110	$q_3 = 0$
$+d^* =$	0 0010		*restore*
$w[1] =$	0 0001	0110	
$2w[1] =$	0 0010	1100	
$-d^* =$	1 1110		
$\widetilde{w}[2] =$	0 0000	1100	$q_2 = 1$
$2w[2] =$	0 0001	1000	
$-d^* =$	1 1110		
$\widetilde{w}[3] =$	1 1111	1000	$q_1 = 0$
$+d^* =$	0 0010		*restore*
$w[3] =$	0 0001	1000	
$2w[3] =$	0 0011	0000	
$-d^* =$	1 1110		
$\widetilde{w}[4] =$	0 0001	0000	$q_0 = 1$
$w[4] =$	$\widetilde{w}[4]$		

Quotient $q = (0101)_2 = 5$, remainder $w = (0001)_2 = 1$. Check: $11 = 2 \times 5 + 1$.

FIGURE 1.11 Example of radix-2 restoring division.

The restoring algorithm can be made faster by not storing the tentative residuals at all and thus avoiding the restoration steps. Such an algorithm is called *nonperforming division*, and it is specified in Algorithm NPD. Now in each iteration, the register has to be loaded either with $2w[j]$ or with the result of the subtraction.

Algorithm NPD: Nonperforming Divide

1. *[Initialize]*
    $w[0] = x$
2. *[Recurrence]*
    for $j = 0 \ldots n - 1$
    if $2w[j] - d^* \geq 0$ **then**
    $q_{n-j-1} = 1; w[j + 1] = 2w[j] - d^*$
    else
    $q_{n-j-1} = 0; w[j + 1] = 2w[j]$
end for

1.6.2   Nonrestoring Division

The speed of the restoring division algorithm can also be improved in the following manner. It is easily observed that the restoration of the jth residual $w[j]$ can be combined with the next subtraction of the divisor. When restoration is required; that is, when $\widetilde{w}[j] < 0$, $q_{n-j} = 0$, then

$$\widetilde{w}[j + 1] = 2w[j] - d^* = 2(\widetilde{w}[j] + d^*) - d^* = 2\widetilde{w}[j] + d^* \qquad 1.84$$

and when no restoration is necessary; that is, when $\widetilde{w}[j] \geq 0$, $q_{n-j} = 1$, then

$$\widetilde{w}[j + 1] = 2w[j] - d^* = 2\widetilde{w}[j] - d^* \qquad 1.85$$

Therefore, an equivalent algorithm can be implemented in which $\widetilde{w}[j]$ is the residual (instead of $w[j]$). These residuals can be positive or negative and are bounded by

$$|\widetilde{w}[j]| < d^* \qquad 1.86$$

Since the residuals can be negative and we want a positive final remainder, the last step of the procedure is modified to assure a positive remainder.

The nonrestoring algorithm is described in Algorithm NRD. To simplify the notation we use $w[j]$ to denote this new residual. This algorithm requires n shifts and n additions/subtractions to obtain an n-digit quotient, and is therefore faster than the restoring one.

Algorithm NRD: Nonrestoring Divide

1. [*Initialize*]
   $w[0] = x$
2. $w[1] = 2w[0] - d^*$
3. [*Recurrence*]
   for $j = 1 \ldots n - 1$
   if $w[j] \geq 0$ then
   $q_{n-1-j} = 1; w[j + 1] = 2w[j] - d^*$
   else
   $q_{n-1-j} = 0; w[j + 1] = 2w[j] + d^*$
4. [*Correct*]
   if $w[n] < 0$ then
   $q_0 = 0; w[n] = w[n] + d^*$
   else
   $q_0 = 1$
   endfor

EXAMPLE 1.8 An example of nonrestoring division of positive fractions is given in Figure 1.12. Note that the subtractions are performed by adding the two's complement of the divisor d^*. In order to preserve the sign of the shifted residual, the representations of the residuals and divisor are extended by an additional bit to the left. ∎

An alternative description of the nonrestoring algorithm consists in defining the digit set for the quotient as $\{-1, +1\}$ instead of the canonical $\{0, 1\}$ and to perform directly the recurrence with the quotient-digit selection

$$q_{n-j-1} = 1 \quad \text{if} \quad w[j] \geq 0 \quad \text{and} \quad -1 \quad \text{otherwise} \qquad \textbf{1.87}$$

For compatibility reasons, usually the quotient eventually has to be transformed to the canonical representation. If this transformation is done digit by digit during the division process, the NRD algorithm results.

Dividend $x = 11_{10} = (00001011)_2$, divisor $d = 2 = (0010)_2$

$w[0] =$	0 0000	1011	
$2w[0] =$	0 0001	0110	
$-d^* =$	1 1110		
$w[1] =$	1 1111	0110	$q_3 = 0$
$2w[1] =$	1 1110	1100	
$+d^* =$	0 0010		
$w[2] =$	0 0000	1100	$q_2 = 1$
$2w[2] =$	0 0001	1000	
$-d^* =$	1 1110		
$w[3] =$	1 1111	1000	$q_1 = 0$
$2w[3] =$	1 1111	0000	
$+d^* =$	0 0010		
$w[4] =$	0 0001	0000	$q_0 = 1$

Quotient $q = (0101)_2 = 5$, remainder $w = (0001)_2 = 1$. Check: $11 = 2 \times 5 + 1$.

FIGURE 1.12 Example of radix-2 nonrestoring division.

1.7 Exercises

Representation of Positive Integers

1.1 (a) Determine how many digits are necessary to represent integers in the range 0 to $(297)_{10}$ using

1. radix-2 conventional system
2. radix-8 conventional system
3. radix-17 conventional system
4. mixed-radix system with radix vector $R = (n + 1, n, \ldots, 3, 2)$ and canonical digit-set

(b) What is the largest integer that can be represented by the digit-vectors of the size determined in each of the cases?

(c) Specify a coding for the digits and determine the number of bits of the bit-vector that represents these integers. Determine the efficiency of each representation, defined as the ratio of the number of bits in the binary representation and the number of bits required by the digit-vector.

1.2 Represent the integers 0, 13, 15, 19, 22, and 127 using a residue number system with $P = (7, 5, 3, 2)$ as the set of moduli. Specify a bit coding for the digits and determine the efficiency of the representation.

1.3 What happens if the moduli used in a residue number system are not relatively prime?

1.4 A processor word has eight bits. Determine the set of positive integers representable with two words for the following representation systems. Determine the efficiency of the representations.

1. Conventional, radix-2

2. Conventional, radix-10, BCD

3. Conventional, radix-16

Representation of Signed Integers

1.5 Given the digit-vector

$$X = (1, 0, 1, 0, 1, 1)_r$$

(a) Determine its representation value x_R in decimal for $r = 2, 8, 10,$ and 16.
(b) What is the greatest value of x_R that a six-component vector X can represent for $r = 2, 10,$ and 16?

1.6 Show the bit-vectors that represent x, $-6 \leq x \leq 6$, in the binary true-and-complement systems with complementation constants $C = 16, 15, 19,$ and 127. Use the minimum number of bits required by the number system.

1.7 Given the digit-vector

$$X = (1, 0, 1, 1)_r$$

(a) determine the representation value x_R in a weighted, radix-r representation system for the radices 2, 7, and 16;

(b) determine the value x in the following cases:

r	Attributes
2	Integer, two's complement
4	Integer, digit complement
8	Integer, digit complement

1.8 Complete the following table, assuming (a) conventional number system, $r = 4$, range complement, $n = 6$ digits (b) conventional number system, $r = 2$, ones' complement, $n = 8$ bits (c) conventional number system, $r = 2$, two's complement, $n = 5$ bits.

	Value x	Value x_R	Digit Vector X
(a)	-39_{10}		
(b)		215_{10}	
(c)			11101

1.9 Complete the following table. All values are given in the decimal system.

Number System	Radix r	Number of Digits n	Value x	Value x_R	Digit-Vector X
SM	10	4	-837		
Two's compl.	2				110010
Range compl.	3	4	-37		
Range compl.	8	3		363	
Ones' compl.	2	8	-83		
Two's compl.	2		$-19/64$		
Digit compl.	8	4			6527
Ones' compl.	2		$-19/64$		

1.10 Given the digit-vector

$$X = (X_6, X_5, \ldots, X_0)$$

and the radix $r = 2$, if the radix point is between bits X_4 and X_3, determine the values of the most positive number x_{max} and the most negative number x_{min} and show their corresponding digit-vector representation in (a) the sign-and-magnitude system (b) the two's complement system (c) the ones' complement system.

1.11   Determine the value x, represented by the digit-vector $X = (1, 0, 1, 0, 1)$ for the following cases:
(a) Values are integers, $r = 2$, and the two's complement system is used.
(b) As in (a), but the ones' complement system is used.
(c) As in (a), but the sign-and-magnitude system is used.
(d) Repeat (a), (b), and (c), assuming that the values are fractions (signed).

1.12   Given the four-component vector $X = (1, 0, 1, 1)_2$:
(a) Assuming that X represents the integer x, represent x with the six-component vector Y using the two's complement and ones' complement systems.
(b) Repeat (a), assuming that the given X represents the fraction x. Do not change the position of the radix point when extending X to the six-component vector.

Algorithms and Implementations for Addition, Subtraction, and Change of Sign

1.13   Perform the operations $x + y, y - x, x - y, -x - y, -y - x$, and $|x - y|$ on digit vectors X and Y that represent the integers $x = -17$ and $y = 9$ in the radix-2 sign-and-magnitude, two's complement, and ones' complement number systems. Determine the minimal number of digits so that no overflow will occur.

1.14   Show an algorithm for the computation

$$z = |x| - |y|$$

where the signed integers x, y, and z are represented in the two's complement system.

1.15   Prove that the following bit-serial algorithm performs the "change of sign" operation in the two's complement system. Let

$$Z = (Z_{n-1}, Z_{n-2}, \ldots, Z_0)_2$$

and

$$X = (X_{n-1}, X_{n-2}, \ldots, X_0)_2$$

represent z and x such that $z = -x$.

Algorithm: If k is the index of the rightmost bit of X that is 1,

$$Z_i = X_i \quad i = 0, 1, \ldots, k$$
$$Z_i = X_i' \quad i = k+1, \ldots, n-1$$

1.16   (a) Show that the overflow in addition in the two's complement system can be detected by the exclusive-or of the carry-in and the carry-out of the most significant bit.

(b) Show that the last expression does not work properly in the ones' complement system.

1.17   In many computers two types of integers are represented: signed integers and unsigned (positive) integers. This is done to use more effectively the available number of bits in a word.

(a) Determine the range of signed and unsigned integers that can be represented by a 16-bit word. In the signed case consider the sign-and-magnitude, two's complement, and ones' complement systems.

(b) Suppose we want to perform the operations of addition and subtraction for both types of integers. The basic module used for these operations is a 16-bit adder and four flags: the zero flag Z is set to 1 when the result is zero, the sign flag SGN is loaded with bit 15 of the result vector, the carry flag CO is loaded with the carry-out of the adder (used for multiprecision operations), and the overflow flag OVF is set to one if there is an overflow in the addition (assuming true-and-complement representation). Indicate whether the same algorithms can be used for addition and subtraction for both types of integers. Consider the two cases for the representation of signed integers: two's complement and ones' complement.

(c) Consider the operation of comparison. The operation is performed by a subtraction (without storing the result) and setting the flags. Determine the values of the flags for greater, equal, and smaller for both types of integers. Consider the three representations in the signed case. Suppose that the computer has conditional branch instructions that branch or not depending on the result of a previous comparison (branch on greater, branch on equal, and branch on smaller). Indicate whether the same instructions can be used if the comparison was done on unsigned integers or on signed integers. If not, determine how the flags would be used by the branch instructions in each case.

Algorithms for Range Extension and Arithmetic Shifts

1.18   Given the digit-vectors $A = (7, 3, 6, 2)$ and $B = (3, 2, 1, 6)$:

(a) Determine the integers represented in a range complement decimal system and in a digit complement decimal system.

(b) Extend the vectors to six digits; that is, represent the same numbers using six digits.

(c) Obtain the representation for the integers $10a$ and $a/10$ using vectors with seven digits.

1.19   Prove the algorithm presented in the text for true-and-complement range extension.

1.20   Prove the algorithms presented in the text for arithmetic shifts in true-and-complement systems.

1.21   For the digit-vectors $X = 00101101$ and $Y = 11010110$ apply the arithmetic-shift algorithms for two's complement and ones' complement radix-2 systems. Check the results.

1.22   Determine conditions for overflow in the right arithmetic-shift algorithm.

1.23   For $A = (1, 1, 0, 1)$, $B = (1, 1, 0)$, $C = (0, 1, 0, 1)$, and $D = (1, 0, 1, 0, 1)$, obtain X representing $x = (a + b) + 8c - 2d$ in the two's complement system.

Multiplication

1.24   Perform the multiplication of $x = 21$ and $y = 14$ using the basic multiplication algorithm for positive integers and $r = 2$.

1.25   Perform multiplication of $x = 21$ and $y = -17$ using algorithms for

- two's complement
- ones' complement

assuming radix 2. What is the minimum number of bits necessary to represent the operands and the result?

1.26   Determine the execution time of the basic multiplication algorithm for n-bit non-negative integers assuming that a partial product of each iteration is stored in a register and

- t_{vd} time to perform vector-digit multiplication
- t_{add} addition time
- t_{reg} register loading time

Propose a modification to the algorithm so that the execution time is reduced to about 50% of the original time.

1.27   Derive a recurrence for multiplication of positive integers assuming that the multiplier digits are used from left to right; that is, the algorithm begins with y_{n-1}. Show a figure indicating relative position of operands, partial products, and the adder. Is there an effect on the execution time compared with the original right-to-left multiplication algorithm?

Division

1.28   Prove the nonrestoring division algorithm.

1.29   Perform nonrestoring division of $x = 14$ by $d = 3$. Use conventional binary representation and perform subtraction by adding the two's complement of the divisor.

1.30   Derive a nonrestoring algorithm for two's complement operands.

1.8   Further Readings

The basic concepts of digital arithmetic are covered in many books on digital design and computer organization such as Wakerly (2001) and Hennessay and Patterson (1995). A broader and more detailed treatment of number systems and arithmetic operations can be found in books on digital arithmetic such as Parhami (2000), Knuth (1998), Omondi (1994), Koren (1993), Scott (1985), Cavanagh (1984), Wasser and Flynn (1982), Kulisch and Miranker (1981), Spaniol (1981), Gosling

(1980), and Hwang (1978). Some of the papers cited in this book as well as other papers are reprinted in a comprehensive two-volume collection (Swartzlander 1990). Oklobdzija (1999) presents an extensive collection of papers on high-performance circuits, logic, and system design, many of them related to digital arithmetic. A survey of digital arithmetic in the 1950s and 1960s, when many of the most important ideas in number representation, algorithms, and implementations were introduced, appears in Garner (1965). A view of the several levels involved in the specification and implementation of arithmetic processors is presented in Avizienis (1971). A theoretical treatment of basic digit sets for radix representation is given in Matula (1982). A classic book on residue arithmetic is Szabo and Tanaka (1967). A tutorial on residue number system appears in Taylor (1984), and a collection of papers on residue number system arithmetic is provided in Soderstrand et al. (1986). Numbers, various representation systems, and their long history are the subject of many books (Gazale 2000; Guedj 1996; McLeish 1991; Ifrah 1985; Dantzig 1954). Sweitz (1987) describes how arithmetic was done in the 15th century. There is also a dictionary of curious and interesting numbers (Wells 1997).

1.9 Bibliography

Avizienis, A. (1971). Digital computer arithmetic: A unified algorithmic specification. In *Proceedings of the Symposium on Computers and Automata*, pages 509–25, April 13–15.

Cavanagh, J. (1984). *Digital Computer Arithmetic*. McGraw-Hill, New York.

Dantzig, T. (1954). *The Number: The Language of Science*. Free Press (Macmillan Publishing Co.), New York.

Ercegovac, M. D. and T. Lang, (1985). *Digital Systems and Hardware/Firmware Algorithms*. John Wiley & Sons, Inc., New York.

Garner, H. L. (1965). Number systems and arithmetic. In *Advances in Computers*, volume 6, pages 131–94. Academic Press, New York.

Gazale, M. (2000). *Number from Ahmes to Cantor*. Princeton University Press, Princeton, New Jersey.

Gosling, J. B. (1980). *Design of Arithmetic Units for Digital Computers*. Springer-Verlag, New York.

Guedj, D. (1996). *Numbers: The Universal Language*. Harry B. Abrams, Inc., New York.

Hennessy, J. L. and D. A. Patterson, (1995). *Computer Architecture: A Quantitative Approach*. Morgan Kaufmann, San Francisco, 2nd edition.

Hwang, K. (1978). *Computer Arithmetic Principles, Architecture and Design*. John Wiley & Sons, Inc., New York.

Ifrah, G. (1985). *From One to Zero: A Universal History of Numbers*. Viking, New York.

Knuth, D. E. (1998). *The Art of Computer Programming: Seminumerical Algorithms*. Addison Wesley, Reading, Massachusetts, 3rd edition.

Koren, I. (1993). *Computer Arithmetic Algorithms*. Prentice Hall, Englewood Cliffs, New Jersey.

Kulisch, U. W., and W. L. Miranker, (1981). *Computer Arithmetic in Theory and Practice*. Academic Press, New York.

Matula, D. W. (1982). Basic digit sets for radix representation. *Journal of the ACM*, 29(4):1131–43.

McLeish, J. (1991). *Number*. Fawcett Columbine, New York.

Oklobdzija, V. G. editor (1999). *High-Performance System Design: Circuits and Logic*. IEEE Press, Piscataway, New Jersey.

Omondi, A. R. (1994). *Computer Arithmetic Systems, Algorithms, Architecture and Implementations*. Prentice Hall International Series in Computer Science, Englewood Cliffs, New Jersey.

Parhami, B. (2000). *Computer Arithmetic: Algorithms and Hardware Designs*. Oxford University Press, New York.

Scott, N. R. (1985). *Computer Number Systems & Arithmetic*. Prentice Hall, Englewood Cliffs, New Jersey.

Soderstrand, M., W. Jenkins, G. Jullien, and F. Taylor (1986). *Residue Number System Arithmetic: Modern Applications in Digital Signal Processing*. IEEE Press, New York.

Spaniol, O. (1981). *Computer Arithmetic: Logic and Design*. John Wiley & Sons, Inc., New York.

Swartzlander, E. E. editor. (1990). *Computer Arithmetic, Vol. 1 and Vol. 2*. IEEE Computer Society Press, Los Alamitos, California.

Sweitz, F. J. (1987). *Capitalism & Arithmetic: The New Math of the 15th Century*. Open Court, La Salle, Illinois.

Szabo, N. S., and R. I. Tanaka (1967). *Residue Arithmetic and Its Applications to Computer Technology.* McGraw-Hill, New York.

Taylor, F. (1984). Residue arithmetic: A tutorial with examples. *IEEE Computer Magazine*, 17(5):50–62.

Wakerly, J. F. (2001). *Digital Design Principles & Practices.* Prentice Hall, Englewood Cliffs, New Jersey.

Wasser, S., and M. J. Flynn (1982). *Introduction to Arithmetic for Digital Computers.* Holt, Rinehart, Winston, New York.

Wells, D. (1997). *Curious and Interesting Numbers.* Penguin Books, New York.

IN THIS CHAPTER, WE PRESENT AND DISCUSS

- Properties of carries
- Carry-ripple adder (CRA) and full-adder
- Approaches to reducing adder delay
- Switched carry-ripple adder
- Carry-skip adder
- Carry-lookahead adder
- Carry-select adder
- Conditional-sum adder
- Prefix adder
- Variable-time adder
- Redundant adders (carry-save and signed-digit)
- Summary of presented schemes

Two-Operand Addition

We begin by considering the addition of two positive fixed-point operands in fixed-radix representation. We first present algorithms and implementations for conventional representation and then consider the case of redundant representations. The adders can then be used for addition of signed operands in alternative representations, such as sign-and-magnitude and two's complement.

The algorithms and implementations we present are for radix 2. However, this includes other power-of-two radices with binary coding of the digits, since in that case the bits of the representation are the same as those of a radix-2 representation.

A binary n-bit adder, shown in Figure 2.1(a), has two operands $0 \leq x, y \leq 2^n - 1$ and carry-in $c_{in} \in \{0, 1\}$ as inputs, and produces as outputs the sum $0 \leq s \leq 2^n - 1$ and carry-out $c_{out} \in \{0, 1\}$ such that

$$x + y + c_{in} = 2^n c_{out} + s \qquad \textbf{2.1}$$

The solution to this equation is

$$s = (x + y + c_{in}) \bmod 2^n$$

$$c_{out} = \begin{cases} 1 & \text{if } (x + y + c_{in}) \geq 2^n \\ 0 & \text{otherwise} \end{cases} \qquad \textbf{2.2}$$

$$= \lfloor (x + y + c_{in})/2^n \rfloor$$

For $n = 1$, the adder reduces to a primitive module called *full-adder* (FA) with three binary inputs x_i, y_i, and c_i and two binary outputs s_i and c_{i+1} indicated in Figure 2.1(b), such that

$$x_i + y_i + c_i = 2c_{i+1} + s_i \qquad \textbf{2.3}$$

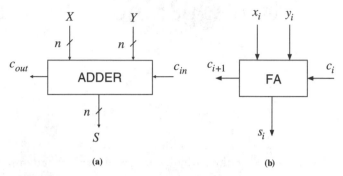

FIGURE 2.1 (a) An n-bit adder. (b) 1-bit adder (full-adder module).

with solution

$$s_i = (x_i + y_i + c_i) \bmod 2$$
$$c_{i+1} = \lfloor (x_i + y_i + c_i)/2 \rfloor$$

2.4

Adder Schemes

In this chapter several addition schemes are presented which provide trade-offs between delay and other characteristics, such as area and energy dissipation. Because of this no scheme can be considered as superior, but they provide alternatives from which to choose in a specific context with specific requirements and constraints.

The most common implementations are of the fixed-time type. That is, the adder has no signal to indicate when the addition is completed, and therefore the worst-case delay has to be considered. On the other hand, variable-time adders have completion signals so that the result of the addition can be used as soon as the completion signal is asserted.

We consider both carry-propagate adders (CPA), which produce the result in a conventional fixed-radix number system, and redundant adders, in which the result is in a redundant number representation. As we discuss later, these redundant adders have a lower delay that is independent of the number of operand bits.

There are many schemes of carry-propagate adders, with the main objective of reducing the delay in obtaining carries. Among them we study the following:

- Switched carry-ripple adder
- Carry-skip adder
- Carry-lookahead adder
- Prefix adder
- Carry-select adder and conditional-sum adder

Redundant adders are characterized by limited carry propagation (independent of the number of bits of the adder). The main types are:

- Carry-save adder
- Signed-digit adder

Different adder schemes are sometimes combined to achieve delay/area constraints, resulting in *hybrid adders*. These adders are not discussed in this book, but references to them are included at the end of this chapter.

2.1　About Carries

The production of the bit s_i $(0 \leq i \leq n - 1)$ in the addition $s = x + y$ can be decomposed into the following two steps, as illustrated in Figure 2.2:

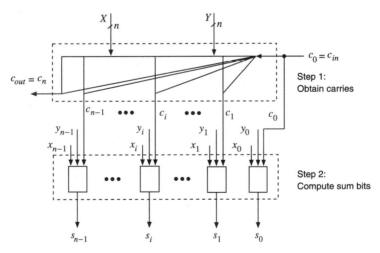

FIGURE 2.2 Steps in addition.

1. Obtaining the carry c_i. This carry represents the influence of bits x_j and y_j for $j < i$ on s_i. That is,

$$c_i = F(x_{i-1}, \ldots, x_0, y_{i-1}, \ldots, y_0, c_{in})$$　　　　**2.5**

More specifically, calling

$$x^{(i)} = \sum_{j=0}^{i} x_j 2^j$$

and

$$y^{(i)} = \sum_{j=0}^{i} y_j 2^j$$

we have

$$c_i = \left\lfloor \frac{x^{(i-1)} + y^{(i-1)} + c_{in}}{2^i} \right\rfloor$$　　　　**2.6**

2. Computing the sum bit s_i from the input bits x_i and y_i and the carry obtained in Step 1. Specifically,

$$s_i = (x_i + y_i + c_i) \bmod 2$$　　　　**2.7**

Since the sum bit s_i at position i depends only on x_i, y_i, and c_i; that is, it is a local function, once the carries are known all sum bits can be computed in parallel.

Consequently, the main objective of all methods for reducing the time of addition for conventional representation is to speed up the process for obtaining all carries. We now discuss several general observations that are commonly used in various approaches for computing carries.

At position i of the addition, consider the relation between the carry-out (c_{i+1}) and the carry-in (c_i). From expression (2.6), we can see that there are three mutually exclusive cases, as summarized in Table 2.1. The determination of the particular case depends only on the local variables x_i and y_i and can be performed in parallel (for all i) by the following switching expressions:

Case Kill:

$$k_i = x_i' y_i' = (x_i + y_i)'$$　　　　**2.8**

Case	x_i	y_i	$x_i + y_i$	c_{i+1}	Comment
Kill ($k_i = 1$)	0	0	0	0	Kill (stop) carry-in
Propagate ($p_i = 1$)	0	1	1	c_i	Propagate carry-in
	1	0	1	c_i	Propagate carry-in
Generate ($g_i = 1$)	1	1	2	1	Generate carry-out

TABLE 2.1 Carry-out cases.

Case Propagate:

$$p_i = x_i \oplus y_i \qquad\qquad \textbf{2.9}$$

Case Generate:

$$g_i = x_i y_i \qquad\qquad \textbf{2.10}$$

Consequently, the carry-out of position i can be expressed in terms of the carry-in to that position as

$$c_{i+1} = g_i + p_i c_i = x_i y_i + (x_i \oplus y_i)c_i \qquad\qquad \textbf{2.11}$$

From the identity $g_i + p_i c_i = g_i + (g_i + p_i)c_i$ and naming $p_i + g_i = a_i$, we get an alternative expression for the carry-out

$$c_{i+1} = g_i + a_i c_i \qquad\qquad \textbf{2.12}$$

which is somewhat simpler to implement than (2.11). The variable a_i corresponds to the combined case when $x_i + y_i$ is 1 or 2. Since $a_i = k_i'$, we call it "alive."
Similarly,

$$c_{i+1}' = k_i + p_i c_i' \qquad\qquad \textbf{2.13}$$

From expressions (2.11) and (2.13) we observe that carries propagate from least-significant to most-significant bit (right to left), forming carry chains of two types: 1-carry chain consisting of carry $= 1$ and 0-carry chain consisting of carry $= 0$. The following example illustrates these chains:

i	9	8	7	6	5	4	3	2	1	0	$c_{in} = 0$
x_i	1	0	1	0	1	1	1	1	0	0	
y_i	0	0	0	1	0	1	0	0	1	0	
	p	k	p	p	p	g	p	p	p	k	
	a		a	a	a	a	a	a	a		
c_{i+1}	0 ←	0	1 ←	1 ←	1 ←	1	0 ←	0 ←	0 ←	0	

Observe that a 1-carry chain begins always with a $g_i = 1$ (or $c_{in} = 1$) and propagates to the left over all consecutive positions $j > i$ where $p_j = 1$. Similarly, a 0-carry chain begins with $k_i = 1$ (or $c_{in} = 0$) and propagates to the left over all consecutive positions $j > i$ where $p_j = 1$. Moreover, the chains are independent.[1]

Expressions (2.11) and (2.12) can be generalized to consider a group of bits, by replacing the bit-generate g_i, the bit-propagate p_i, and the bit-alive a_i by the corresponding group variables. That is,

$$c_{j+1} = g_{(j,i)} + p_{(j,i)}c_i = g_{(j,i)} + a_{(j,i)}c_i \qquad \textbf{2.14}$$

This expression indicates that $c_{j+1} = 1$ if a carry is generated in the group of bits from i to j or if a carry comes in to that group and is propagated (or kept alive) by the group.

From the definition of the carry,

$$g_{(j,i)} = \begin{cases} 1 & \text{if } \sum_{v=i}^{j}(x_v + y_v)2^{v-i} \geq 2^{j+1-i} \\ 0 & \text{otherwise} \end{cases} \qquad \textbf{2.15}$$

$$p_{(j,i)} = \begin{cases} 1 & \text{if } \sum_{v=i}^{j}(x_v + y_v)2^{v-i} = 2^{j+1-i} - 1 \\ 0 & \text{otherwise} \end{cases} \qquad \textbf{2.16}$$

For $a_{(j,i)}$ we observe that $p_{(j,i)} = 1$ iff $p_k = 1$ for all k in $i \leq k \leq j$. That is,

$$p_{(j,i)} = \underset{v=i}{\overset{j}{\text{AND}}}(p_v) \qquad \textbf{2.17}$$

---

1. As discussed later, these carry chains have an effect on the delay of the adder. In some circuit technologies, all carry signals are cleared before the operation. In such a case, 0-carry chains need not be considered for the delay. A similar situation occurs with the 1-carry chains if all carries are preset to 1.

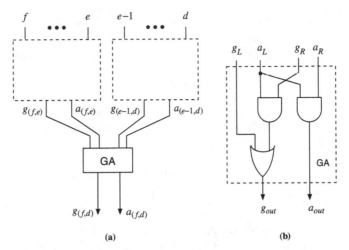

FIGURE 2.3 Computing $(g_{(f,d)}, a_{(f,d)})$.

Therefore, it is easy to verify that the last part of expression (2.14) is satisfied if we define

$$a_{(j,i)} = \mathop{\mathrm{AND}}_{v=i}^{j}(a_v) \qquad \textbf{2.18}$$

By making $i = 0$ in expression 2.14, we obtain

$$c_{j+1} = g_{(j,0)} + p_{(j,0)}c_0 = g_{(j,0)} + a_{(j,0)}c_0 \qquad \textbf{2.19}$$

That is, to compute c_{j+1} it is sufficient to compute the pair $(g_{(j,0)}, p_{(j,0)})$ or the pair $(g_{(j,0)}, a_{(j,0)})$.

Moreover, as shown in Figure 2.3, the computation of the variables for the range of bits (f, d) can use the values of these variables for the subranges (f, e) and $(e - 1, d)$, with $d < e < f$. Specifically, from the definitions we obtain the following switching expressions:

$$g_{(f,d)} = g_{(f,e)} + p_{(f,e)}g_{(e-1,d)} = g_{(f,e)} + a_{(f,e)}g_{(e-1,d)}$$

$$a_{(f,d)} = a_{(f,e)}a_{(e-1,d)} \qquad \textbf{2.20}$$

$$p_{(f,d)} = p_{(f,e)}p_{(e-1,d)}$$

These expressions[2] are the basis for linear and treelike structures to obtain all the carries. These structures are the main topic of this chapter.

Once all the carries are obtained, the sum bits are computed in parallel as:

$$s_i = x_i \oplus y_i \oplus c_i = p_i \oplus c_i \qquad\qquad 2.21$$

EXAMPLE 2.1   We now illustrate the use of these expressions to obtain bit 13 of the sum[3] of the following 16-bit operands ($c_{in} = 0$):

$$\underline{x} = 0110|0010|1100|0011$$

$$\underline{y} = 1011|1101|0001|1110$$

First we need to obtain the carry c_{13}. To do this we divide the operands in groups of four bits, and for each group we obtain (in parallel) the values of the g and p variables (we use here the p variables, although in a practical case the a variables might be preferable). From expressions (2.15) and (2.16)

$$p_{(12,12)} = 1, \quad p_{(11,8)} = 1, \quad k_{(7,4)} = 1, \quad g_{(3,0)} = 1$$

Note that since k, p, and g correspond to mutually exclusive situations, the other variables in the group have value 0.

Now we use expressions (2.20) to combine two adjacent groups. We obtain

$$p_{(12,8)} = 1, \quad k_{(7,0)} = k_{(7,4)} + p_{(7,4)}k_{(3,0)} = 1$$

and finally

$$k_{(12,0)} = 1$$

resulting in

$$c_{13} = g_{(12,0)} + p_{(12,0)}c_{in} = 0$$

---

2. The expressions can be generalized to the case in which the subranges overlap; that is, subranges (f, e) and (h, d), with $h \geq e$.
3. Of course in the addition case all bits have to be obtained; alternative ways of doing this are the topic of this chapter.

Now, the sum bit is

$$s_{13} = x_{13} \oplus y_{13} \oplus c_{13} = 0$$ ∎

2.2 Basic Carry-Ripple Adder (CRA) and FA Implementation

We review here the carry-ripple adder, which corresponds to the basic (carry-propagate) addition algorithm. As shown in Figure 2.4, this adder consists of an array of 1-bit adders (*full-adders* or FAs) defined by expressions (2.4). The correctness of this adder implementation can be shown by induction, using the definition of c_i given by expression (2.6).

Consider now the delay of the addition. Since there is no completion signal, it is necessary to consider the worst-case delay. As shown in Figure 2.4, this worst case corresponds to the delay of the propagation of the carry through $n-1$ bits plus the largest delay between the propagation of the carry through the last bit or the computation of the sum bit s_{n-1}. Consequently, calling t_c the delay from the inputs of the full adder to the carry output and t_s the delay from the inputs to the sum output, the (worst-case) delay of the adder is given by

$$T_{CRA} = (n-1)t_c + \max(t_c, t_s)$$ **2.22**

The largest component of the delay is $(n-1)t_c$. Since this is linearly dependent on n, this adder is slow for large n. The actual value of the delay depends on the technology and on the implementation. The main advantage of this adder is the simplicity of its cells and of the connections among them.

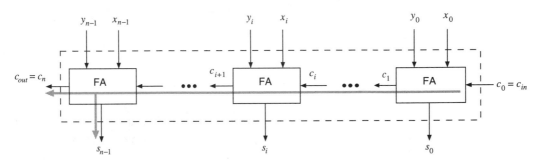

FIGURE 2.4 Carry-ripple adder.

2.2.1 Implementations of Full-Adder

Let us now consider implementations of the full-adder. From expression (2.4), we get the following tabular description:

x_i	y_i	c_i	c_{i+1}	s_i
0	0	0	0	0
0	0	1	0	1
0	1	0	0	1
0	1	1	1	0
1	0	0	0	1
1	0	1	1	0
1	1	0	1	0
1	1	1	1	1

Minimal sum of products expressions for these functions are

$$s_i = x_i y_i' c_i' + x_i' y_i c_i' + x_i' y_i' c_i + x_i y_i c_i$$
$$c_{i+1} = x_i y_i + x_i c_i + y_i c_i$$

2.23

These expressions are the basis for the two-level implementation shown in Figure 2.5(a).

An alternative implementation is based on the use of expression (2.11), namely,

$$c_{i+1} = g_i + p_i c_i$$

2.24

Moreover, $s_i = (x_i + y_i + c_i) \bmod 2$ indicates that

$$s_i = x_i \oplus y_i \oplus c_i = p_i \oplus c_i$$

2.25

These expressions are the basis for the implementation shown in Figure 2.5(b). The submodule producing p_i and g_i (or g_i' for the implementation with a NAND gate shown in Figure 2.5c) is called a *half-adder* (HA) because it performs the addition of two bits (instead of three for a full-adder); the sum bit is p_i and the

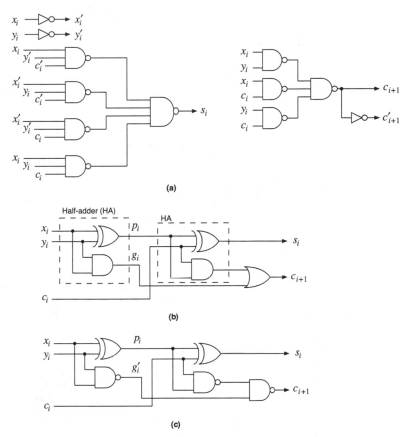

FIGURE 2.5 Implementation of full-adder. (a) Two-level network. (b) Multilevel network with XOR, AND, and OR gates. (c) Multilevel implementation with XOR and NAND gates.

carry bit is g_i, as shown in the following table:

x_i	y_i	g_i	p_i
0	0	0	0
0	1	0	1
1	0	0	1
1	1	1	0

The implementation of a full-adder using two half-adders and one NAND gate requires fewer gates than the two-level network; moreover, although the

Input	(standard loads)
c_i	1.3
x_i	1.1
y_i	1.3
Size: 7 (equivalent gates)	

		Propagation delays		
		t_{pLH}	t_{pHL}	t_p (average)
From	To	(ns)	(ns)	(ns)
c_i	s_i	$0.43 + 0.03L$	$0.49 + 0.02L$	$0.46 + 0.03L$
x_i	s_i	$0.68 + 0.04L$	$0.74 + 0.02L$	$0.71 + 0.03L$
y_i	s_i	$0.68 + 0.04L$	$0.74 + 0.02L$	$0.71 + 0.03L$
c_i	c_{i+1}	$0.36 + 0.04L$	$0.40 + 0.02L$	$0.38 + 0.03L$
x_i	c_{i+1}	$0.73 + 0.04L$	$0.71 + 0.02L$	$0.72 + 0.03L$
y_i	c_{i+1}	$0.37 + 0.04L$	$0.64 + 0.02L$	$0.52 + 0.03L$

L: load on the output

TABLE 2.2 Characteristics of the full-adder in a family of CMOS gates.

two-level implementation has fewer logic levels, the carry delay (from carry-in to carry-out), which is critical for the delay of the carry-ripple adder, is smaller in the two half-adders case than in the implementation of Figure 2.5(a) because it corresponds to the delay of two two-input gates.

Using the implementation with two half-adders, the worst-case delay of the carry-ripple adder is[4]

$$T_{CRA} = t_{XOR} + 2(n - 1)t_{NAND} + \max(2t_{NAND}, t_{XOR}) \qquad 2.26$$

The first t_{XOR} corresponds to p_0. Note that the delay of p_i for $i > 0$ is not in the critical path because all p_i's are computed simultaneously.

Most families of standard cells include a full-adder module. For instance, in the CMOS family we are using as an example in this book, the full-adder module has the characteristics listed in Table 2.2.

---

4. This expression does not include the effect of the load on the gates output (see Exercise 2.1).

2.3 Reducing the Adder Delay

The delay of the carry-ripple adder can be reduced by the following four approaches:

1. Reducing the carry delay t_c. This is achieved in the switched carry-ripple (Manchester) adder.

2. Changing the linear factor n to a "smaller" factor (such as n/k or $\log n$). This is achieved by the carry-skip adder, the carry-lookahead adder, the prefix adder, the carry-select adder, and the conditional-sum adder.

3. Including a completion signal so that the addition time corresponds to the actual addition and not to the worst case.

4. Changing the number representation system. We explore in this chapter the use of redundant representations.

We now consider each of the adder schemes mentioned above.

2.4 Switched Carry-Ripple (Manchester) Adder

The main idea is to use a fast circuit for propagating carry chains. As discussed in Section 2.1, a 1-carry chain starts in position with $g_i = 1$ and propagates to the left over consecutive positions with $p_j = 1$. Similarly, a 0-carry chain begins in position with $k_i = 1$ and propagates to the left over consecutive positions with $p_j = 1$. Since propagate p_i, generate g_i, and kill k_i variables can be obtained in parallel as functions of x_i and y_i only, all chains begin at the same time. By providing a fast circuit path to perform propagation of chains, the total carry delay is reduced. Such a circuit may consist of transmission gates or special transistors.

As discussed in Section 2.1, there are the following mutually exclusive cases:

$x_i + y_i$	g_i	p_i	k_i	c_{i+1}
0	0	0	1	0
1	0	1	0	c_i
2	1	0	0	1

As a consequence of these disjoint situations, the carry-out can be produced by a switch network as shown in Figure 2.6. Since the three situations are disjoint,

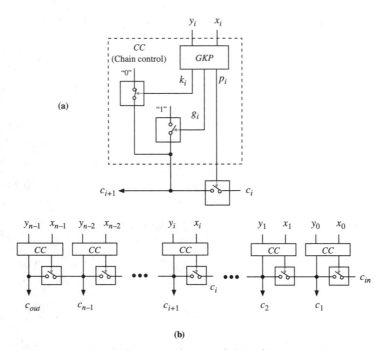

FIGURE 2.6 Switched carry-ripple network (Manchester circuit).

only one of the three switches per bit is closed. The switches for all bits are set simultaneously and then the carry propagates through the closed switches in the horizontal path.

2.4.1   Delay

The delay consists of three components: the setting of the switches (switches of all bits are set simultaneously) with delay t_{sw}, the propagation of the carry through $(n - 1)$ bits (the propagation delay of a switch is t_p), and the production of the sum bit. That is,

$$T_{SRA} = t_{sw} + (n - 1)t_p + (n/m)t_{buf} + t_s \qquad 2.27$$

where the term $(n/m)t_{buf}$ corresponds to the buffers required each m bits to restore the signal. The scheme is effective if t_p is small.

2.5 Carry-Skip Adder

The carry-skip adder is obtained by a modification of the carry-ripple adder. The objective is to reduce the worst-case delay by reducing the number of FA cells through which the carry has to propagate (see expression (2.21)). To achieve this, the adder is divided into groups of m bits and the carry into group $j + 1$ is determined by one of the following two conditions:

1. The carry is propagated by group j. That is, the carry-out of group j is equal to the carry-in of that group. This situation occurs only when the sum of the inputs to that group is equal to $2^m - 1$. Calling $x^{(j)}$ and $y^{(j)}$ the integers corresponding to these inputs, the group propagate signal, defined in Section 2.1, is

$$P^{(j)} = \begin{cases} 1 & \text{if } x^{(j)} + y^{(j)} = 2^m - 1 \\ 0 & \text{otherwise} \end{cases} \qquad 2.28$$

2. The carry is not propagated by the group (that is, it is generated or killed inside the group).

Consequently, to reduce the length of the propagation of the carry, a skip network is provided for each group of m bits so that when a carry is propagated by this group, the skip network makes the carry bypass the group. The m-bit adder is shown in Figure 2.7(a), and a network of these modules implementing an n-bit adder is indicated in Figure 2.7(b). The carry into group $j + 1$ is described by the following expression,[5]

$$c_{in}^{(j+1)} = c_{out}^{(j)} \left(P^{(j)} \right)' + c_{in}^{(j)} P^{(j)} \qquad 2.29$$

At the bit level, the carry is propagated by the group when it is propagated by all the bits in the group. That is,

$$P^{(j)} = \mathop{\text{AND}}_{i=0}^{m-1} p_i \qquad 2.30$$

where the index $0 \leq i \leq m - 1$ is for a generic group.

---

5. In many descriptions of the carry-skip scheme, an AND-OR network is used instead of the multiplexer, resulting in $c_{in}^{(j+1)} = c_{out}^{(j)} + c_{in}^{(j)} P^{(j)}$. However, to obtain the expected speedup with this implementation, it is necessary that all transient $c_{out}^{(j)}$ be 0, as discussed further in Example 2.3.

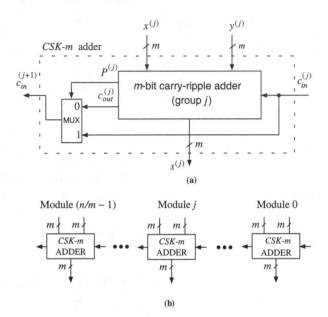

FIGURE 2.7 Carry-skip adder: (a) A group with carry bypass (*CSK-m* adder). (b) *n*-bit carry-skip adder.

2.5.1   Delay

Since there is no completion signal, the worst-case delay has to be considered. To identify this worst-case delay it is important to notice the following:

- As indicated in Section 2.1, the addition process produces several carry-propagation chains. Each of these chains is initiated in a bit with $p = 0$, propagates through consecutive bits with $p = 1$, and terminates in a bit with $p = 0$. These chains can propagate a 0-carry or a 1-carry and therefore can start with $(x_i, y_i) = (0, 0)$ or $(1, 1)$. An example of these propagation chains is shown in Figure 2.8. Note that the carry advances in all chains simultaneously.

- In a carry-skip adder, a chain is initiated in a group and can either terminate in the same group, or skip zero, one, or more groups, and then terminate in another group. That is, the carry in a chain can at most travel inside two groups: the initiating group and the terminating group.

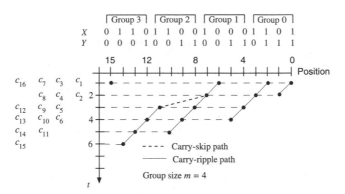

FIGURE 2.8 Carry chains in carry-skip adder. (MUX delay not included.)

As a consequence of the property above, the worst-case delay is produced when a carry is generated in the first bit of the adder (no propagate in the first group), and propagated through all bits up to but not including the most significant bit. That is, it skips all groups except the first and last and terminates in the last bit of the last group (to produce the sum). This critical path is illustrated in Figure 2.9(a) and (b).

The worst-case delay is then

$$T_{CSK} = mt_c + t_{mux} + \left(\frac{n}{m} - 2\right)t_{mux} + (m-1)t_c + t_s$$

$$= (2m-1)t_c + \left(\frac{n}{m} - 1\right)t_{mux} + t_s \qquad 2.31$$

EXAMPLE 2.2   Consider the case $n = 32, m = 4$. To simplify, consider $t_c = t_s = t_{mux} = \delta$. From the expression we obtain

$$T_{CSK} = 15\delta$$

In contrast, the delay of the corresponding carry-ripple adder (expression (2.21)) is 32δ. ■

We now illustrate that the use of an AND-OR network, shown in Figure 2.10, instead of a multiplexer can produce a delay as large as that of the carry-ripple adder.

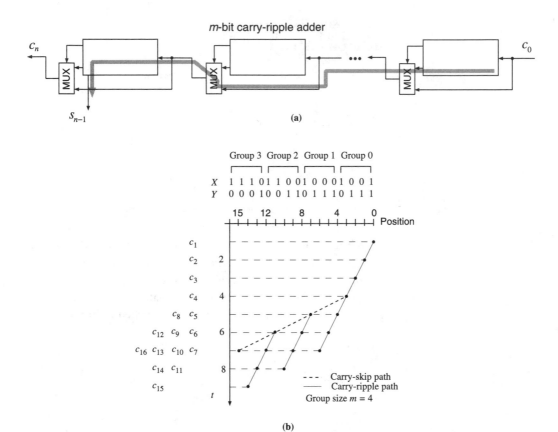

FIGURE 2.9 (a) Critical path in carry-skip adder. (b) Illustration of the worst-case situation for $n = 16$. (MUX delay not included.)

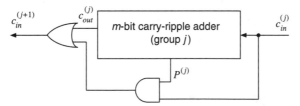

FIGURE 2.10 Carry-skip adder using AND-OR for bypass.

EXAMPLE 2.3   Consider the case in which the carry-skip adder is implemented with an AND-OR network instead of multiplexers and that the following two additions are performed, one after the other without clearing the carries left at the end of the first addition.

Operation 1					
x	0000	1111	0000	1111	$c_{in} = 0$
y	1111	0000	1111	0001	
c	11111	1111	1111	111-	

Operation 2					
x (change last bit)	0000	1111	0000	1110	$c_{in} = 0$
y (change last bit)	1111	0000	1111	0000	
c	00000	0000	0000	000-	

Now consider how this change of carries is produced. We consider the sequence of events, assuming that $t_{p,g} = t_c = t_{AND\text{-}OR} = 1$.

- First (at $t = 0$) the inputs are changed (all bits simultaneously, in the example only bit 0 changes).
- At time $t = 1$ the p and gs of all bits are produced (only bit 0 changes from $g = 1$ to $k = 1$).
- At time $t = 2$, the carry $c_1 = 0$ is produced.
- This carry is propagated through the 4-bit carry-ripple adder, so that at $t = 5$ we obtain $c_4 = 0$. Up to here it is as expected in the carry-skip adder.
- Now, since for all bits in the second group $p = 1$, the carry skips the second group and a 0 appears at the input of the OR gate. However, the other input to the OR gate is still 1, since all the carries in the second group are still 1 (the 0-carry has just entered that group). Consequently, the skip network is not effective in this case. The carry-in to the third group will become 0 only after this 0 has propagated through the second group and appeared at the input of the OR gate.
- This same process will occur for all groups and the delay of the adder will be the same as the worst-case delay of the carry-ripple adder.   ∎

One way to use the AND-OR network and produce the desired speedup is to initialize all carries to 0 and assure that no glitches occur at the carry-out of the groups. This can be accomplished by having a precharge phase, used in dynamic logic.

2.5.2   Group Size

As shown in expression (2.30), the delay of the carry-skip adder depends on the size of the group m. Differentiating this expression with respect to m, we obtain

$$m_{opt} = \sqrt{(t_{mux}/2t_c)n} \quad \text{(minimum delay)}$$

$$T_{opt} \approx \sqrt{8t_{mux}t_c n}$$

2.32

which is proportional to $\sqrt{n}$.

Variable Group Size

The previous analysis assumes that all groups are of the same size. However, this does not produce the minimum delay. This is due to the fact that, for instance, carries generated in the first group have to traverse more skip networks to get to the last group than carries generated in some internal group. So we now consider the case in which the groups can have different sizes. Because of this, to determine the worst case we need to compare the delay of all carry propagation chains. A particular chain is initiated in group i and terminates in group j with $j \geq i$, being propagated by the $j - i - 1$ groups in between. Consequently, if group i has size m_i,

$$T_{CSK} = \max_{i,j}((m_i + m_j - 1)t_c + (j - i - 1)t_{mux}) + t_{mux} + t_s \qquad 2.33$$

with $\sum m_i = n$. Because of the term $j - i - 1$, the worst-case delay can be reduced by reducing the size of the groups close to the beginning and end, as illustrated in Figure 2.11.[6]

EXAMPLE 2.4   The effect of variable group size on the worst-case delay is illustrated for $n = 60$ and $t_c = t_s = t_{mux} = \delta$.

$m = 6$	$T_{CSK} = 21\delta$
$m_i = 4, 5, 6, 7, 8, 8, 7, 6, 5, 4$	$T_{CSK} = 17\delta$

■

---

6. Further details in Guyot et al. (1987) and Chan et al. (1992).

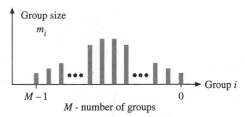

FIGURE 2.11 Optimal distribution of group sizes in carry-skip adder.

Further reduction of the worst-case delay can be obtained by putting several groups into blocks and providing carry-skip around blocks. This process is generalized to multilevel carry-skip adders (see Exercise 2.9).

2.6　Carry-Lookahead Adder (CLA)

The basic idea of this adder is to compute several carries simultaneously. In the extreme, all carries could be computed at the same time. As stated in expression (2.6), if we call $x^{(i)}$ and $y^{(i)}$ the integers represented by the bit-vector from bit 0 to bit i; that is,

$$x^{(i)} = \sum_{v=0}^{i} x_v 2^v \qquad\qquad 2.34$$

and similarly for $y^{(i)}$, the carry is computed by the following expression:

$$c_i = 1 \quad \text{if}\,(x^{(i-1)} + y^{(i-1)} + c_0) \geq 2^i \qquad\qquad 2.35$$

This results in a switching function of $2i + 1$ variables. It is known that any such function can be implemented by a two-level network (for example, NAND-NAND). However, for large i this implementation is impractical because of the large number of gates with large number of inputs. Because of this, in the carry-lookahead adder the input vector is divided into groups and the carries inside a group are computed simultaneously.

2.6.1　One-Level Carry-Lookahead Adder (1-CLA)

Let us consider first the one-level carry-lookahead adder. As shown in Figure 2.12, the input vector is divided into groups of m bits and the groups are connected as in a carry-ripple adder. However, in contrast to the carry-ripple adder, after

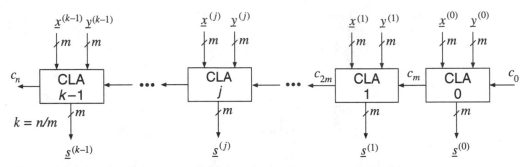

FIGURE 2.12 One-level carry-lookahead adder.

the input carry to the group is known, all carries inside a group as well as the output carry of the group are computed simultaneously. Consequently, if we call t_{group} the delay of this calculation, we obtain the worst-case delay

$$T_{1\text{-}CLA} = \frac{n}{m} t_{group} + t_s \qquad\qquad 2.36$$

Now let us consider the implementation of the group module. To simplify the notation, we index the bits of a generic group from 0 to $m-1$ and call the carry into the group c_0. We could implement directly the switching function resulting from the arithmetic expression (2.34) in a two-level network. However, it can be shown that the number of gates required for $m > 4$ would be too large for typical implementation. Because of this, it is more convenient to add another level producing the variables p_i, g_i, and a_i, which we already introduced in Section 2.1. That is,

$$p_i = x_i \oplus y_i$$
$$g_i = x_i y_i \qquad\qquad 2.37$$
$$a_i = x_i + y_i$$

Consequently, as shown in Figure 2.13, the module consists of three parts: the computation of p_i, g_i, a_i, the computation of the carries in the carry-lookahead generator (CLG), and the computation of the sum $s_i = x_i \oplus y_i \oplus c_i = p_i \oplus c_i$. The outputs A and G are used for the carry-lookahead adder with more than one level.

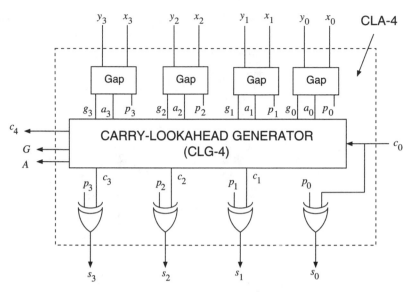

FIGURE 2.13 Carry-lookahead adder module CLA-4 ($m = 4$).

Now we consider the computation of the carries. The switching expression for the carry-out of a 1-bit adder is defined in (2.12) as

$$c_{i+1} = g_i + a_i c_i \qquad \textbf{2.38}$$

We determine expressions for all the carries in the group by substitution. For example, for a group size of 4 ($m = 4$), we get

$$c_1 = g_0 + a_0 c_0$$
$$c_2 = g_1 + a_1 c_1 = g_1 + a_1(g_0 + a_0 c_0) = g_1 + a_1 g_0 + a_1 a_0 c_0 \qquad \textbf{2.39}$$

We see that $c_2 = 1$ if a carry is generated in bit 1, or if a carry is generated in bit 0 and alive (not killed) in bit 1, or if $c_0 = 1$ and it is alive in bits 0 and 1. Similarly then we can write

$$c_3 = g_2 + a_2 g_1 + a_2 a_1 g_0 + a_2 a_1 a_0 c_0$$
$$c_4 = g_3 + a_3 g_2 + a_3 a_2 g_1 + a_3 a_2 a_1 g_0 + a_3 a_2 a_1 a_0 c_0 \qquad \textbf{2.40}$$

An implementation is shown in Figure 2.14.

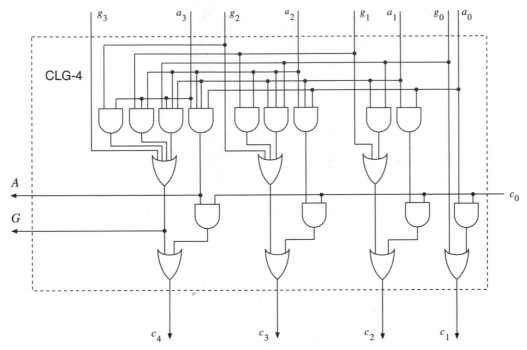

FIGURE 2.14 4-bit carry-lookahead generator CLG-4.

This implementation is easily generalized to any number of inputs: c_{i+1} is 1 if a carry is generated in bit i, or if it is generated in bit $i-1$ and alive in bit i, and so on. The general expression is

$$c_{i+1} = g_i + a_i g_{i-1} + a_i a_{i-1} g_{i-2} + \cdots + (a_i a_{i-1} \cdots a_0) c_0 \qquad \textbf{2.41}$$

or equivalently

$$c_{i+1} = \mathop{\mathrm{OR}}_{j=0}^{i} \left(\mathop{\mathrm{AND}}_{k=j+1}^{i} a_k \right) g_j + \left(\mathop{\mathrm{AND}}_{k=0}^{i} a_k \right) c_0 \qquad \textbf{2.42}$$

The implementation of c_{i+1} requires

- One OR gate of $i+2$ inputs
- $i+1$ AND gates with $2, 3, \ldots, i+2$ inputs

As can be seen, the number of gates and the number of inputs per gate increases with the size of the group. This limits the maximum group size for a practical implementation.

Delay

The delay of this implementation is given by the following expression (see Figures 2.12 and 2.13):

$$T_{1\text{-}CLA} = (\text{Compute } a_i, g_i) + (\text{Ripple between groups}) + (\text{Compute } s_i)$$

Consequently, calling t_{clg} the delay of the carry-lookahead generator,

$$T_{1\text{-}CLA} = t_{a,g} + \frac{n}{m}t_{clg} + t_s \qquad\qquad 2.43$$

As in the carry-skip adder, the dependency on n is now divided by the group size m. The delay of the one-level carry-lookahead adder is smaller than that of the carry-ripple adder as long as t_{clg} is smaller than mt_c. Whether it is faster than the carry-skip adder depends on the relative values of t_c, t_{mux}, $t_{a,g}$, and t_{clg} and on the corresponding group sizes.[7]

2.6.2 Two-Level Carry-Lookahead Adder

For large n the number of groups in a one-level CLA is large, resulting in a slow operation. To reduce the delay, apply the CLA principle among groups. As defined in Section 2.1, for each group we have two signals: $A = 1$ if a carry is alive in the group and $G = 1$ if a carry is generated by the group. Consequently, the carry-out of the group is described by the following switching expression:

$$c_{out} = G + Ac_{in} \qquad\qquad 2.44$$

The switching expressions for A and G are

$$A = \mathop{\text{AND}}_{i=0}^{m-1} a_i \quad (\text{group alive}) \qquad\qquad 2.45$$

$$G = \mathop{\text{OR}}_{j=0}^{m-1} \left(\mathop{\text{AND}}_{i=j+1}^{m-1} a_i \right) g_j \quad (\text{group generate}) \qquad\qquad 2.46$$

The implementation for a group of 4 bits is shown in Figure 2.14.

A variation of the CLA adder, called the Ling adder, which reduces the complexity of producing the group generate, is considered in Exercise 2.18.

---

7. The group for the carry-skip adder might be larger since the number of gates is smaller.

Now we define a section of p groups and determine with a CLG the carry-out of each group in the section. That is,

$$c^{(1)} = G_0 + A_0 c_0$$

$$c^{(2)} = G_1 + A_1 G_0 + A_1 A_0 c_0$$

$$\cdots$$

$$c^{(p)} = G_{p-1} + A_{p-1} G_{p-2} + \cdots + (A_{p-1} A_{p-2} \cdots A_0) c_0$$

2.47

Once the carries out of the groups are produced, these carries are used by the first-level CLA modules to produce the bit carries and the sums. Figure 2.15 shows a 32-bit adder with two-level lookahead (with $p = m = 4$).

Note that the CLA module is used twice: first to compute A and G (which are independent of the carry-in to the group) and then, once the carry-in is known, to compute the internal carries and the sum. Moreover, in this case, the carryout of the CLA module is not used.

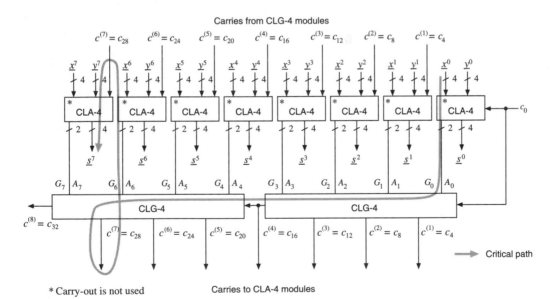

FIGURE 2.15 Two-level carry-lookahead adder ($n = 32$).

Delay

The delay is given by the following expression (see critical path in Figure 2.15):

$$T_{2\text{-}CLA} = t_{a,g} + t_{A,G} + \frac{n}{pm}t_{clg} + t_{clg} + t_s \qquad \textbf{2.48}$$

This delay is smaller than that of the one-level CLA because of the factor n/pm instead of n/m.

2.6.3 ## Three and More Levels

The scheme can be extended to three levels by having lookahead between sections. In general, for L levels of lookahead, the critical path corresponds to the following: (to simplify the notation we consider the case in which the groups at all levels have the same size m)

- a first level to compute a_i, p_i, and g_i (p_i not in the critical path)
- $L - 1$ levels of carry-lookahead generators to compute the As and Gs
- n/m^L carry-lookahead generators connected in a ripple fashion to compute carries of sections at level L
- $L - 1$ levels of carry-lookahead generators to compute the carries of bits (the last of these included in the CLA module)
- one level of exclusive-OR gates to compute the sum

  The corresponding delay is then

$$T_{L\text{-}CLA} = t_{a,g} + (L-1)t_{A,G} + \frac{n}{m^L}t_{clg} + (L-1)t_{clg} + t_s \qquad \textbf{2.49}$$

Moreover, since the same module is used to compute the A, G signals and the corresponding carries, the number of CLG modules is

$$N_{clg} = \sum_{i=1}^{L} \left(\frac{n}{m^i}\right) = n\frac{m^L - 1}{m^L(m-1)} \qquad \textbf{2.50}$$

For $L = 2, m = 4$, and $n = 32$ this results in $N_{clg} = 10$, as shown in Figure 2.15.

The maximum number of levels is obtained when there is only one section at level L. That, is

$$m^L = n$$

or

$$L = \log_m n \qquad \textbf{2.51}$$

The resulting delay is

$$T_{max\text{-}CLA} = t_{a,g} + (\log_m n - 1)t_{A,G} + (\log_m n)t_{clg} + t_s \qquad \textbf{2.52}$$

That is, the delay has a logarithmic dependence on n. Making $t_{A,G} = t_{clg}$, we get that $T_{max\text{-}CLA}$ is proportional to $2 \log_m n$.

Moreover, in this configuration with maximum number of levels, the number of CLG modules is

$$N_{max\text{-}clg} = \frac{n-1}{m-1} \qquad \textbf{2.53}$$

EXAMPLE 2.5 Figure 2.16 illustrates a three-level carry-lookahead adder for $n = 8$ and $m = 2$. The delay is

$$T_{CLA8\text{-}2} = t_{a,g} + 2t_{A,G} + 3t_{clg} + t_s \qquad \textbf{2.54}$$

proportional to $2 \log_2 8$. ∎

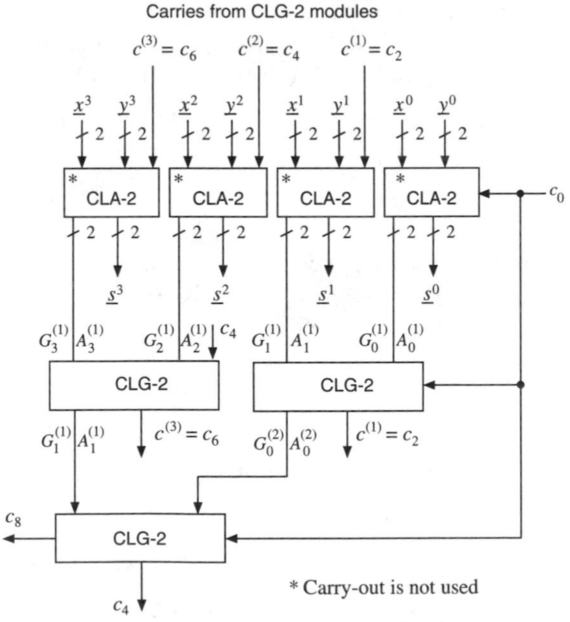

FIGURE 2.16 Three-level carry-lookahead adder ($n = 8, m = 2$).

2.6.4 Choice of Group Size and Number of Levels

As can be seen the group size and the number of levels affect both the delay and the number of modules. A suitable choice depends on the technology and on the adder requirements. With MSI technology, the size of the group was mainly determined by the number of pins in a package. Moreover, the best size was the maximum that could be included in a chip since this reduces the number of chips as well as the number of signal hops between chips. On the other hand, with VLSI technology, in which a whole adder fits in a chip, the constraints are different, and simplicity of cells and regularity of connections become the most critical. Because of this, groups of size two are quite popular.

2.7 Prefix Adder

The prefix adder is a structure that is based on considering the carry computation as a prefix computation. In general, a prefix combinational network of n inputs $x_0, x_2, \ldots, x_{n-1}$ uses the associative (arbitrary) operator $\circ$ to produce the vector of outputs described by

$$z_i = x_i \circ x_{i-1} \circ \cdots \circ x_1 \circ x_0 \qquad \textbf{2.55}$$

As indicated in Section 2.1, for the carry computation we have

$$z_i = (g_{(i,0)}, a_{(i,0)}), \quad x_i = (g_i, a_i) \qquad \textbf{2.56}$$

and the operator (implemented by a cell) has as input two pairs of bits (g_L, g_R) and (a_L, a_R) and as output one pair (g_{out}, a_{out}). It is described by the switching expressions

$$
\begin{aligned}
g_{out} &= g_L + a_L g_R \\
a_{out} &= a_L a_R
\end{aligned}
\qquad \textbf{2.57}
$$

where as before, g and $a = k'$ correspond to generate and to alive signals, respectively.

With this cell, a variety of networks are used to produce the carries. They are all based on the fact that the carry c_i corresponds to the generate signal spanning the bit positions (-1) to $i - 1$. We call this generate signal $g_{(i-1,-1)}$ so that

$$c_i = g_{(i-1,-1)} \qquad \textbf{2.58}$$

where $(g_{-1}, a_{-1}) = (c_0, c_0)$.

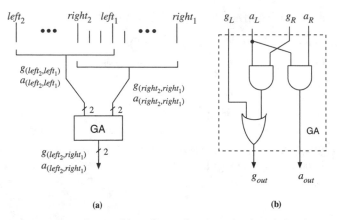

FIGURE 2.17 Composition of spans in computing (g, a) signals.

A prefix adder is then an interconnection of the above-mentioned cells to produce $g_{(i-1,-1)}$ for all i. These carries are then used to obtain the sum bits as

$$s_i = p_i \oplus c_i \qquad\qquad \textbf{2.59}$$

To obtain the carries the cells are connected in a recursive manner to produce the g signals that span an increasing number of bits. That is, beginning with the variables g and a of each bit, the first level of modules produces g and a for groups of two bits, the second level for groups of four bits, and so on. In general, if the right input spans the bits $[right_2, \ right_1]$ and the left input spans the bits $[left_2, \ left_1]$ with $right_2 + 1 \ge left_1$ then the output spans the bits $[left_2, \ right_1]$ as illustrated in Figure 2.17. For instance, for $right = [5, 2]$ and $left = [8, 4]$, the output spans the bits $[8, 2]$.

An array of cells for an 8-bit adder is shown in Figure 2.18. The outputs of the cells are labeled with a pair of integers corresponding to the initial and the final bit that is spanned by the output. Because each level produces a doubling of bits spanned, for n power-of-two, the number of levels is

$$L = \log_2(n) + 1 \qquad\qquad \textbf{2.60}$$

where the additional level is due to the carry-in c_0. In the figure for eight bits there are four levels. Note that the additional level due to c_0 does not increase the

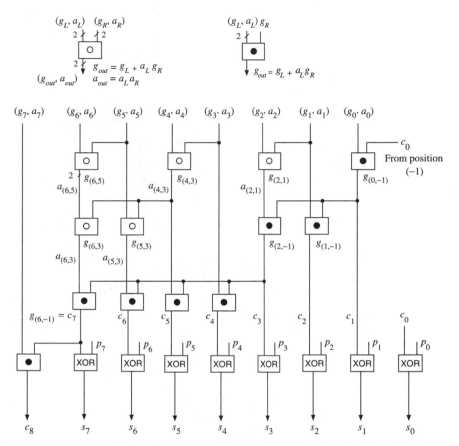

FIGURE 2.18 8-bit prefix adder. (Modules to obtain p_i, g_i, and a_i signals not shown.)

overall delay because the computation of c_8 is in parallel to the calculation of the sum bits. The expression for the delay is

$$T_{PA} = t_{a,g} + \log_2(n)t_{cell} + t_{XOR} \qquad \textbf{2.61}$$

Since each level (except the last) has $n/2$ cells, the number of cells is

$$N = (n/2)\log_2 n + 1 \qquad \textbf{2.62}$$

(not including the gates to produce g_i and a_i nor the XOR gates).

Since the cells are simple, their delay and area are small, resulting in an effective implementation. The main disadvantage of this implementation is the large fanout of some cells (as well as the long interconnection wires). For example, in the 8-bit adder there is a cell with internal fanout of four, so that in general for an adder of n bits the maximum fanout is $n/2$. The large fanout and long interconnections produce an increase in the delay, which can be reduced by including buffers. However, the delay of these buffers might still be significant. In such a case, the large fanout can be eliminated by two approaches, or a combination of both:

1. Increasing the number of levels

2. Increasing the number of cells

We now illustrate an example of each of these approaches.

2.7.1   Increasing the Number of Levels

The fanout can be reduced by increasing the number of levels, as shown in Figure 2.19. This is achieved by reducing the parallelism in the determination of the carries. For instance, the calculation of $g_{(6,3)}$ is obtained from $g_{(5,3)}$ instead of $g_{(4,3)}$. The resulting number of levels in the limit (fanout = 2) is

$$L = 2\log_2(n-1) + 1 \qquad\qquad \textbf{2.63}$$

where the last 1 corresponds again to the stage with one cell, due to c_0. The number of cells is the same as for the basic scheme. Of course, the disadvantage of the scheme is the added delay of the additional levels. To reduce the overall delay, a choice is made between the maximum fanout and the number of levels.

2.7.2   Increasing the Number of Cells

The maximum fanout is reduced to two (without increasing the number of levels) by the structure shown in Figure 2.20. This structure is constructed as follows:

- Level 1 is formed of cells having as inputs neighboring bits. So, groups are formed with bits c_0 and 0, with bits 0 and 1, with bits 1 and 2, and so on. Consequently, for n bits there are n cells, instead of the $n/2$ cells required

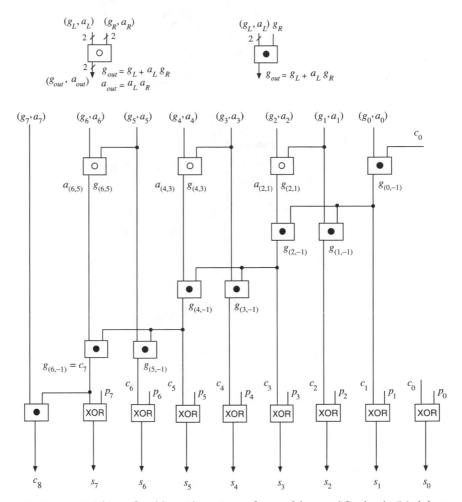

FIGURE 2.19 8-bit prefix adder with maximum fanout of three and five levels. (Modules to obtain p_i, g_i, and a_i signals not shown.)

in this level for the "basic" array. The outputs are labeled as indicated above.

- Level 2 combines outputs of cells of level 1 whose indexes differ by 2. That is, c_0 and 1, 0 and 2, and so on. There are $n - 1$ cells at this level.

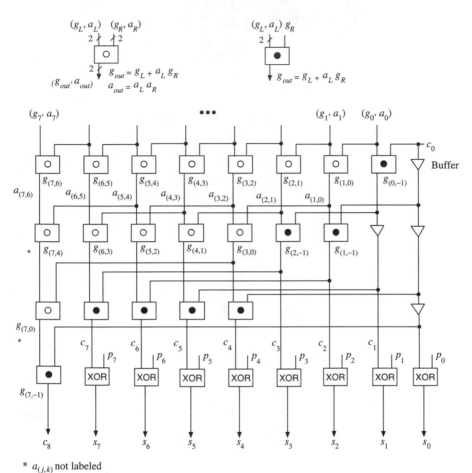

FIGURE 2.20 8-bit prefix adder with minimum number of levels and fanout of two. (Modules to obtain p_i, g_i, and a_i signals not shown.)

- Level 3 combines outputs of cells of level 2 whose indexes differ by 4. That is, c_0 and 3, 0 and 4, and so on. There are $n - 3$ cells.
- In general, level k combines outputs of level $(k - 1)$ whose indexes differ by 2^{k-1}. It has $n - (2^{k-1} - 1)$ cells.

As in the basic scheme there are $\log_2(n) + 1$ levels. Note again the single cell in the last level, because of input c_0. As can be seen, the fanout of all cells is two and the connections are regular.[8] The number of cells is

$$N = n + (n - 1) + (n - 3) + (n - 7) \cdots + (n - (n/2 - 1)) + 1$$

$$= \sum_{i=0}^{\log_2 n - 1} (n - (2^i - 1)) + 1 = (n + 1)(\log_2 n) - \left(\sum_{i=0}^{\log_2 n - 1} 2^i \right) + 1$$

$$= (n + 1)(\log_2 n) - (n - 1) + 1$$

$$= (n)(\log_2 n - 1) + \log_2 n + 2 \qquad\qquad \textbf{2.64}$$

As can be seen from the previous expression, the number of cells of this scheme is about twice that of the basic scheme. If the number of cells is too high, it is possible to use an intermediate scheme, which has an intermediate maximum fanout as well as an intermediate number of cells (see Exercise 2.22).

Prefix Adder with m-Bit Group

The prefix adder can be generalized to use cells that produces the (g, p) pairs of a group of m inputs. This would reduce the (minimum) number of levels to $\log_m n$. However, the cell complexity and delay increases with m. Details are given in the references at the end of the chapter.

2.8   Carry-Select and Conditional-Sum Adders

These two schemes have the same principle and are based on the fact that the main component in the delay of the carry-ripple adders is the propagation of the carry, so that to obtain the sum of bit i it is necessary to wait until the carry has propagated from bit 0 to bit i. Because of this, the idea of these schemes is to compute in parallel two conditional sums: one for a 0-carry and one for a 1-carry, and then select among them when the carry is available. The two schemes differ in the recursive structure; in that sense the carry-select adder is like the one-level lookahead and the conditional-sum adder like the maximum-level case.

The basic principle is to divide the adder into groups of m bits and to compute for each group two conditional sums and carry-outs. If we consider a generic group

---

8. However, the connection span increases with the level, so that buffers might still be needed because of the capacitance of these wires.

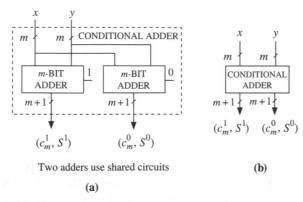

FIGURE 2.21 (a) Obtaining conditional outputs. (b) Combined conditional adder.

in which we label the bits from 0 to $m - 1$, we get

$$\begin{aligned}
\left(c_m^0, S^0\right) &= ADD(X, Y, c_0 = 0) \\
\left(c_m^1, S^1\right) &= ADD(X, Y, c_0 = 1)
\end{aligned}$$

2.65

where X, Y, and S are m-bit vectors.

Then, when the carry-in of the group is known, select from these two forms

$$(c_m, S) = \begin{cases} \left(c_m^0, S^0\right) & \text{if } c_0 = 0 \\ \left(c_m^1, S^1\right) & \text{if } c_0 = 1 \end{cases}$$

2.66

The two m-bit adders for the same group (Figure 2.21(a)) can share components. Because of this, it is better to define a module (Figure 2.21(b)) that has as input the two m-bit operands and produces two $(m + 1)$-bit results. We call this module an m-bit conditional adder (COND-ADDER) and use it in the subsequent structures.

2.8.1 Carry-Select Adder

In the carry-select adder the conditional principle is applied in a linear structure, as shown in Figure 2.22. It consists of an m-bit conditional adder module for each group of m bits and $(n/m) - 1$ multiplexers.

The delay is

$$T_{CSEL} = t_{add,m} + \left(\frac{n}{m} - 1\right) t_{mux}$$

2.67

where $t_{add,m}$ is the delay of the m-bit conditional adder.

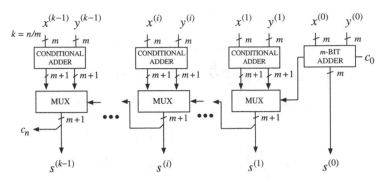

FIGURE 2.22 Carry-select adder.

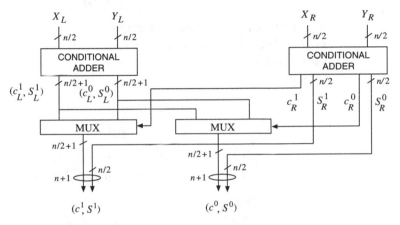

FIGURE 2.23 Doubling the number of bits of the conditional sum.

2.8.2   Conditional-Sum Adder

In the conditional-sum adder, the conditional principle is applied recursively. That is, two groups are combined to form a double-length conditional result as follows (see Figure 2.23):

1.   Decompose.

$$X = (X_L, X_R)$$
$$Y = (Y_L, Y_R)$$

2.68

2. Compute concurrently.

$$\left(c_L^0, S_L^0\right) = ADD(X_L, Y_L, 0) \quad \left(c_R^0, S_R^0\right) = ADD(X_R, Y_R, 0)$$
$$\left(c_L^1, S_L^1\right) = ADD(X_L, Y_L, 1) \quad \left(c_R^1, S_R^1\right) = ADD(X_R, Y_R, 1)$$

2.69

3. Combine to obtain double-length conditional results.

$$(c^0, S^0) = \begin{cases} \left(c_L^0, \left(S_L^0, S_R^0\right)\right) & \text{if } c_R^0 = 0 \\ \left(c_L^1, \left(S_L^1, S_R^0\right)\right) & \text{if } c_R^0 = 1 \end{cases}$$

2.70

$$(c^1, S^1) = \begin{cases} \left(c_L^0, \left(S_L^0, S_R^1\right)\right) & \text{if } c_R^1 = 0 \\ \left(c_L^1, \left(S_L^1, S_R^1\right)\right) & \text{if } c_R^1 = 1 \end{cases}$$

2.71

Note that the right portion of the output comes directly from the corresponding right input, without going through the multiplexer, whereas the left portion is selected by the corresponding carry-out of the right portion. A numerical example is as follows:

$$X_L = 0011 \qquad\qquad X_R = 0111$$
$$Y_L = 1010 \qquad\qquad Y_R = 1001$$
$$\left(c_L^0, S_L^0\right) = (0, 1101) \qquad \left(c_R^0, S_R^0\right) = (1, 0000)$$
$$\left(c_L^1, S_L^1\right) = (0, 1110) \qquad \left(c_R^1, S_R^1\right) = (1, 0001)$$

Combining we obtain

$$(c^0, S^0) = (0, 11100000)$$
$$(c^1, S^1) = (0, 11100001)$$

A 16-bit conditional-sum adder is shown in Figure 2.24.

We observe the following:

- The span of conditional bits doubles each selection level.
- The initial group size is of m bits (limit is $m = 1$).
- For n bits there are $\log_2(n/m)$ selection levels
- If c_{in} is applied in the last selection stage, the number of 2-to-1 multiplexers is roughly n at each level, ignoring multiplexing of the carry-outs. If c_{in} is

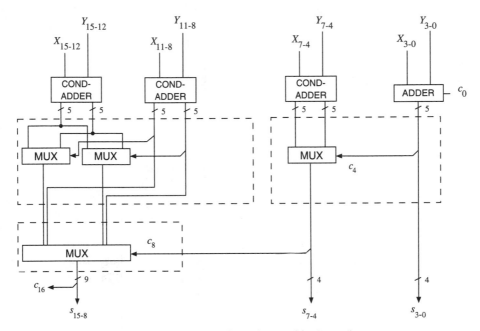

FIGURE 2.24 16-bit conditional-sum adder ($m = 4$).

available in the beginning, each successive selection stage doubles the number of correct least-significant bits of the sum, resulting in a decreased number of MUXes.

- There is a large fanout for mux select signals. For instance, the select signal in the last level goes to $n/2 + 1$ MUX inputs.

An example of the addition process with c_0 available at the beginning is shown in Figure 2.25. The bold carries control the multiplexers.

Delay

The delay is formed by the delay of the m-bit adder plus the multiplexers. That is,

$$T_{cond\text{-}sum} = t_{add\text{-}m} + (\log_2(n/m))t_{mux}$$

2.72

s_3^0	s_2^0	s_1^0	s_0^0	Step 2
c_4^0		c_2^0		
s_3^1	s_2^1	s_1^1	s_0^1	
c_4^1		c_2^1		
s_3^0	s_2^0	s_1^0	s_0^0	Step 3
c_4^0				
s_3^1	s_2^1	s_1^1	s_0^1	
c_4^1				

(a)

	7	6	5	4	3	2	1	0	
x	0	1	0	1	1	0	1	1	$c_0 = 0$
y	0	1	0	1	0	1	0	1	
S^0	0	0	0	0	1	1	1	0	Step 1
c^0	0	**1**	0	**1**	0	**0**	0	**1**	
S^1	1	1	1	1	0	0	0		
c^1	0	**1**	0	**1**	1	**1**	1		
S^0	1	0	1	0	1	1	0	0	Step 2
c^0	0		**0**		0		**1**		
S^1	1	1	1	1	0	0			
c^1	0		**0**		1				
S^0	1	0	1	0	0	0	0	0	Step 3
c^0	0				**1**				
S^1	1	0	1	1					
c^1	0								
S	1	0	1	1	0	0	0	0	

(b)

FIGURE 2.25 Conditional-sum addition for eight bits with $m = 1$: (a) Template. (b) Example.

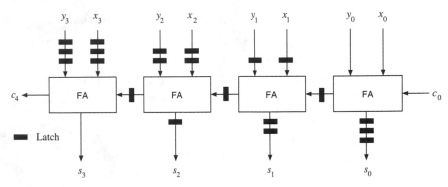

FIGURE 2.26 Pipelined carry-ripple adder (for group size of 1 and $n = 4$).

2.9 Pipelined Adders

The throughput of the adder can be increased by pipelining. To do this, pipeline registers are introduced to shorten the worst-case carry path. For example, a CRA adder is divided into groups of bits and latches are introduced, as shown in Figure 2.26. Most latches are used to synchronize inputs and outputs since different parts (groups) are processed at different cycles. The throughput R is determined by the delay of one group; that is,

$$R = \frac{1}{t_{group}} \qquad\qquad 2.73$$

Adders such as conditional-sum and prefix adders are pipelined by introducing registers between the stages. In this case, no latches are required for synchronization of inputs and outputs.

2.10 Variable-Time Adder

Up to now we have considered fixed-time adders in which, although the actual addition time might be variable, it is necessary to consider the worst-case delay because there is no signal indicating that the operation has terminated. In contrast, the variable-time adders have a completion signal.

In order to make use of the variation in delay, the adder has to be incorporated in a system in which the initiation of the next operation can be triggered by the adder completion signal. Such systems are called *asynchronous* (or *self-timed*).

The addition time is variable because of two factors:

1. Variation in the delay of the components. This can be due to the fabrication process and to environmental factors, such as temperature.

2. The actual input values. As discussed before in the carry-skip adder, the addition time depends on the longest propagation chain, and this length is dependent on the values of the inputs.

Related to these two factors, we consider now two types of variable-time adders. The first is only concerned with the first factor, whereas the second takes both into account.

2.10.1   Type 1: With Self-Timed Carry Circuit

As shown in Figure 2.27, this is a modification of the carry-ripple adder in which there are two carry signals:

$$c_i^0 \quad \text{0-carry}$$

$$c_i^1 \quad \text{1-carry}$$

The coding is as follows:

c_i^0	c_i^1	c_i
0	0	not determined (yet)
0	1	1
1	0	0
1	1	*

* this case does not happen.

Such a coding, known as *double-rail coding*, is typical of asynchronous design.

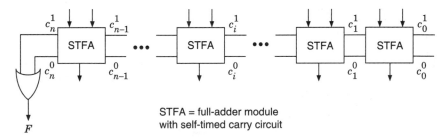

STFA = full-adder module
with self-timed carry circuit

FIGURE 2.27 Variable-time adder: Type 1.

Before each addition, a clearing step is performed that makes

$$c_i^0 = c_i^1 = 0 \quad \text{(not determined yet)} \qquad \textbf{2.74}$$

Then, the operation is started by setting

$$c_0^0 = c_0', \quad c_0^1 = c_0 \qquad \textbf{2.75}$$

The carry signals propagate through all n bits, and the addition finishes when

$$F = c_n^1 + c_n^0 = 1 \qquad \textbf{2.76}$$

This signal indicates the completion of addition only if the delay of the sum of the last 1-bit adder is smaller than the delay of the carry of that adder.

To assure that $F = 1$ is produced by the propagation of the carry signals through the whole adder, the expressions for the carry signals have to be modified to

$$c_{i+1}^0 = k_i\left(c_i^0 + c_i^1\right) + p_i c_i^0 = k_i c_i^1 + (p_i + k_i)c_i^0$$
$$c_{i+1}^1 = g_i\left(c_i^0 + c_i^1\right) + p_i c_i^1 = g_i c_i^0 + (p_i + g_i)c_i^1 \qquad \textbf{2.77}$$

where as before,

$$k_i = x_i' y_i', \quad g_i = x_i y_i, \quad p_i = x_i \oplus y_i \qquad \textbf{2.78}$$

Note that both terms in the expressions depend on the carry-in so that the output carry can only change when the input carry pair is different from (00). This avoids initiating a carry chain at bit $i > 0$ and therefore that F could be 1 before finishing the addition (which depends on the longest carry chain).

The sum is computed as

$$s_i = p_i \oplus c_i^1 \qquad \textbf{2.79}$$

In this scheme, the carry propagates through all bits, independent of the value of the operands. Consequently, the addition time is

$$T_{var\text{-}1} = \sum_{i=0}^{n-1} t_{c,i} \qquad \textbf{2.80}$$

where $t_{c,i}$ is the *actual delay* of the carry network of bit i. This contrasts with the carry-ripple adder, in which the delay corresponds to the worst case, which has to utilize the worst-case delay of the carry network,

$$T_{CRA} = \sum_{i=0}^{n-1} \max(t_c) = n \times \max(t_c) \qquad\qquad \textbf{2.81}$$

where $\max(t_c)$ is the worst-case delay of the carry-out signal.

2.10.2   Type 2: With Parallel Carry Completion Sensing

In this case we want to make use of the fact that the time of addition corresponds to the actual longest carry-propagation chain. The organization is as shown in Figure 2.28.

Note that there are also two carries, but there are two differences with respect to the Type 1 adder:

1. The carry-propagation chains should propagate simultaneously. This requires that the carries be defined by the following expressions:

$$c_{i+1}^0 = k_i + p_i c_i^0$$
$$c_{i+1}^1 = g_i + p_i c_i^1 \qquad\qquad \textbf{2.82}$$

Note that while in the Type 1 adder the carry c_{i+1} always waits for the carry c_i; in this case the carry c_{i+1} is defined right away when either k_i or g_i is 1, initiating the carry-propagation chain.

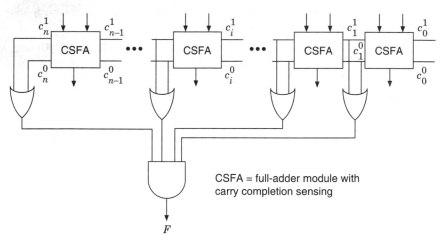

CSFA = full-adder module with carry completion sensing

FIGURE 2.28 Variable-time adder: Type 2.

2. The addition finishes when the carries in all bits are defined. That is, the completion signal is

$$F = \overset{n-1}{\underset{i=0}{\text{AND}}} \left(c_i^0 + c_i^1 \right) \qquad\qquad 2.83$$

The implementation of this signal requires an n-input AND gate. This might be implemented as a tree of $\lceil \log_m n \rceil$ levels of m-input gates.

As in the Type 1 adder, a reset step is required before each addition operation. The addition time is determined by the longest propagation chain. Consequently, the worst-case time is similar to that of the carry-ripple adder. However, the average time depends on the distribution of the operand values. For uniformly distributed operands, it has been shown that the average length of the longest carry-propagation chain is approximately $\log_2(5n/4)$. In a particular situation, it is necessary to determine the average from the specific distribution.

EXAMPLE 2.6   Consider the following operands for an addition. The propagation chains are indicated by the letters a, b, c, d, and e.

```
X  0 1 1 0 0 0 1 1 1 0 0 1 1 0 1 0
Y  1 0 1 0 1 1 0 0 1 1 1 0 0 1 1 0
+  a a a b c c c c c d d d d d d e   Propagation chains
```

Here the longest propagation chain is d. This determines the addition delay.   ∎

We have presented variable-time adders based on the carry-ripple adder structure. However, it is possible to use as a basis any of the other adder structures studied in this chapter (see references at the end).

2.11   Two's Complement and Ones' Complement Adders

We have discussed several adder schemes for addition of positive integers (actually unsigned fixed-point numbers) in a radix-2 representation. We now discuss how these adders are used for addition of signed numbers in two's complement and ones' complement representations.

As shown in Chapter 1, the use for two's complement representation is straightforward: the n-bit sum output of the adder is the result, and the carry-out is discarded. Moreover, the overflow is detected from the most-significant bits of the operands and the result.

For ones' complement addition it is necessary to add the carry-out. We now consider how this addition is done for a carry-ripple adder and for a prefix adder.

For a carry-ripple adder the carry-out is connected to the carry-in (end-around carry) as shown in Figure 1.3. This produces a combinational network with a loop, so that a sequential behavior or oscillations can occur. To study this issue we consider two situations:

1. The operands have at least one position in which $x_i \oplus y_i = 0$. In such a case, the carry loop is initiated and terminates in that position, effectively breaking the loop. In the following example this occurs for position 3.

$$
\begin{array}{llcccccccc}
  & & 7 & 6 & 5 & 4 & 3 & 2 & 1 & 0 \\
X & & 0 & 1 & 0 & 0 & 1 & 1 & 0 & 1 \\
Y & & 1 & 0 & 1 & 1 & 1 & 0 & 1 & 0 \\
  & & & & & & & \text{cout=cin=1} \\
S & & 0 & 0 & 0 & 0 & 1 & 0 & 0 & 0
\end{array}
$$

2. All positions are such that $x_i \oplus y_i = 1$. In this case, the result of the addition corresponds to the value 0. The actual representation of the sum depends on the value of the carry in the loop: if it is 0, then the output is $111\ldots1$, and if it is 1, the output is $000\ldots0$, which both represent 0 in ones' complement.

    However, in this case there might be an oscillation. This occurs if initially (when the operation is initiated) some carries are 1 and others are 0. This pattern of carries goes around the loop producing an oscillation in the output. To avoid this oscillation it is necessary to set the initial carries to a common value (either all 0 or all 1). This requires a "preset" phase in the operation of the adder.

    The fact that the carry chain is effectively broken by a pair for which $x_i \oplus y_i = 0$ indicates that the delay of this modified adder is the same as the delay of the carry-ripple adder, since the worst-case delay is still determined by a carry propagation through $n - 1$ full-adders.

For the prefix adder, the end-around carry approach could also be used. However, if the carry-in is included at the top of the array, as shown for instance in

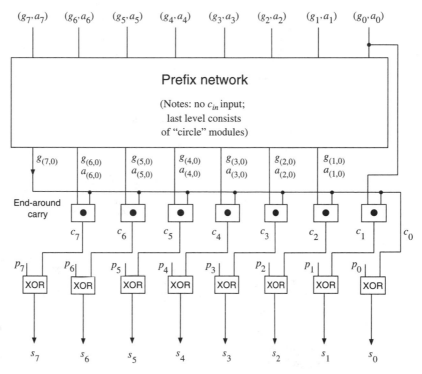

FIGURE 2.29 Implementing ones' complement adder with prefix network. (Modules to obtain p_i, g_i, and a_i signals not shown.)

Figure 2.20, this end-around carry would significantly increase the delay (see exercise 2.31). Consequently, a modification of the adder is more effective, in which the carry-in is added in an additional level. Figure 2.29 shows the resulting adder. Note the high load on the end-around carry signal, which affects the overall delay.

2.12 Adders with Redundant Digit Set

We now consider adders in which the result is represented using a redundant digit set. The operands might be in conventional representation, or one or both also use a redundant set. The objective of having the output in redundant representation is to reduce the addition time by reducing the length of the maximum

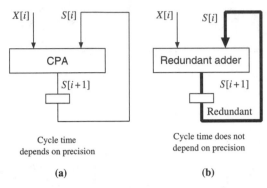

FIGURE 2.30 Accumulation with (a) nonredundant and (b) redundant representation of sum.

carry-propagation chain. We consider the two main redundant digit sets: carry-save and signed-digit.

These adders are used whenever the output in redundant representation is suitable. Typical cases of this are in accumulation (Figure 2.30), multioperand addition, multiplication, division and square root, and other recurrences. These uses are described in the following chapters.

The redundant representation has some disadvantages. One disadvantage is the increase in the number of bits required for the representation, which depends on the degree of redundancy. Another disadvantage is that some operations, such as magnitude comparison and sign detection, are difficult to perform in redundant representation.

2.12.1 Carry-Save Adder (CSA)

The basic idea is to perform an addition of three binary vectors using an array of 1-bit adders (full-adders[9]) but without propagating the carries. As shown in Figure 2.31, the output is represented by two binary vectors called the *carry vector* and the *pseudo-sum vector* (or just *sum*, for short). In terms of the numbers

---

9. In this application FA implementation with 2 HAs might not be effective because of a longer delay for sum than for carry.

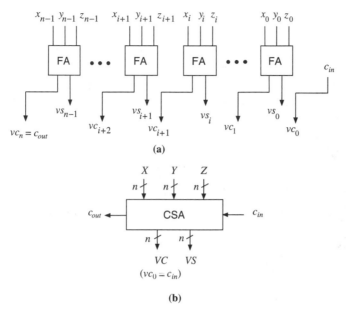

FIGURE 2.31 Carry-save adder: (a) Bit level. (b) Bit-vector level.

represented by the binary vectors, we can write

$$x + y + z = vc + vs = v \qquad \textbf{2.84}$$

Consequently, the sum v of the three numbers x, y, z is represented by the two numbers vc and vs. This representation is redundant since several combinations of the values of vc and vs represent the same number v. Another way of viewing this representation is to consider the corresponding bits of the vectors representing vc and vs as a digit in the radix-2 representation of v. Since these bits are added to obtain the result, they have three possible values: 0, 1, 2. That is, the carry-save representation corresponds to a radix-2 representation with digit set $\{0, 1, 2\}$.

Since the carry output of the full adder of weight i has weight $i + 1$, the carry of bit 0 is $vc_0 = 0$. Consequently, it is possible to include a carry-in c_{in} such that

$$vc_0 = c_{in} \qquad \textbf{2.85}$$

The carry-out c_{out} corresponds to the carry output of the last full-adder. That is,

$$c_{out} = vc_n \qquad \textbf{2.86}$$

EXAMPLE 2.7   The following example shows the carry-save addition of three numbers.

X		0	1	1	1	0	1	**0**	0	
Y		0	0	1	1	1	0	**1**	1	
Z		1	0	1	0	1	0	**1**	0	
VS		1	1	1	0	0	1	**0**	1	
(c_{out}, VC)	0	0	1	1	1	0	1	**0**	1	$* \ vc_0 = c_{in}$
digit value	0	1	2	2	1	0	2	**0**	2	■

The carry-save adder produces a reduction from three binary vectors to two binary vectors. This is called a *3-to-2 reduction*, and the adder a [3:2] adder.

Uses

The following two possibilities exist for the three input vectors:

- Three conventional operands
- One conventional operand and one carry-save operand (this is the case, for example, for the computation of accumulation)

On the other hand, if two carry-save operands are to be added, producing a carry-save result, a 4-to-2 reduction is needed. In this case the adder is called a [4:2] adder. This reduction can be implemented by two CSAs, as shown in Figure 2.32. To reduce the delay, a special network can be designed (see Exercise 2.33).

We discuss in subsequent chapters the cases in which these situations occur.

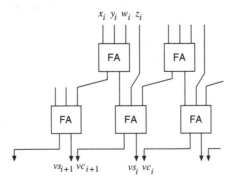

FIGURE 2.32 [4:2] adder.

Addition Time

The time of carry-save addition (3-to-2 reduction) corresponds to the delay of one full-adder, independent of the number of bits. The 4-to-2 reduction implemented as in Figure 2.32 has a delay of two full-adder delays. However, as shown in Exercise 2.33, it can be implemented with a delay of only three XOR gates in the critical path. These delays are significantly smaller than those of carry-propagate adders.

Conversion to Conventional Representation

In some instances, it is necessary to convert a carry-save representation to conventional. This conversion is performed by using a carry-propagate adder with the two operands being the carry vector and the pseudo-sum vector. That is,

$$V = ADD(VS, VC, 0) \qquad\qquad \textbf{2.87}$$

High Radix Carry-Save Representation

One of the disadvantages of the carry-save representation is that the number of bits is doubled. This has an effect on the number of wires and on the number of cells required to store the value. To reduce the number of bits, it is possible to use a high-radix carry-save representation. Calling r the radix of the representation, the (pseudo) sum vs is represented in radix r and the carry vc has one bit per radix-r digit.

EXAMPLE 2.8   The following example illustrates the addition of one carry-save (radix-8) operand and one conventional operand to produce a carry-save (radix-8) result. Note that the addition is performed per radix-8 digit and the carry-out is "saved" in the carry vector.

XS		1	0	1	1	0	1	1	0	0	
XC				1			1			0	
Y		0	1	0	0	0	1	1	1	1	
VS		0	0	0	1	1	1	0	1	1	
(c_{out}, VC)	1			0			1			0	$* vc_0 = c_{in}$

■

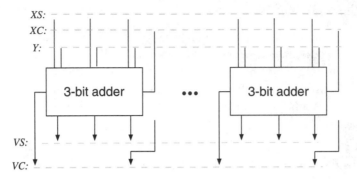

FIGURE 2.33 Radix-8 carry-save adder.

The corresponding implementation is shown in Figure 2.33. Of course, this high-radix implementation results in an increase in delay: from the delay of one full-adder (for the radix-2 case) to the delay of a radix-r adder.

2.12.2 Signed-Digit Adder

In this case the result of the addition uses signed digits, which is a fixed-radix representation with digit values from a signed-integer set. That is,

$$x = \sum_0^{n-1} x_i r^i \qquad \qquad \textbf{2.88}$$

with a digit set

$$D = \{-b, \ldots, -1, 0, 1, \ldots, a\} \qquad \qquad \textbf{2.89}$$

with the restriction $a + b + 1 \geq r$. If $a + b + 1 = r$, the representation is nonredundant, whereas if $a + b + 1 > r$, it is redundant. In most cases, a symmetric digit set is used, that is, $a = b$. For this case, a redundant representation has $a > (r - 1)/2$. In the sequel, we restrict our discussion to symmetric digit sets.

Addition Algorithm

The objective of the signed-digit addition algorithm is to eliminate the carry propagation. To achieve this the following two-step procedure is used:

- **Step 1.** Compute interim sum (w) and transfer (t) such that

$$x + y = w + t \qquad\qquad \textbf{2.90}$$

At the digit level this corresponds to

$$x_i + y_i = w_i + rt_{i+1} \qquad\qquad \textbf{2.91}$$

That is, the transfer digit acts like a carry to the next position.
- **Step 2.** Compute $s = w + t$, which is performed as

$$s_i = w_i + t_i \qquad\qquad \textbf{2.92}$$

This step should be performed without producing a carry. Consequently, as illustrated in Figure 2.34, the transfer digit propagates just one position. As shown, the result has $n + 1$ digits, the most-significant one corresponding to

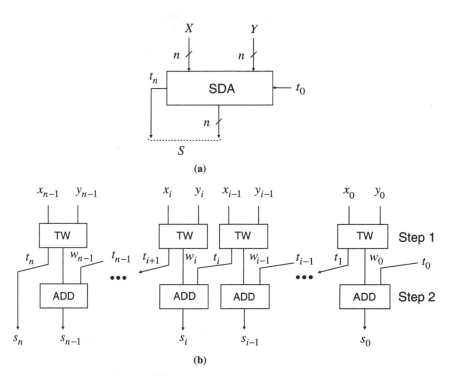

(a)

(b)

FIGURE 2.34 Signed-digit addition.

a "transfer-out." Moreover, a "transfer-in" can be included as the transfer digit to the adder of the first digit.

Since

$$-a \leq s_i \leq a \qquad\qquad 2.93$$

to assure that no carry propagation is produced in the second step, it is necessary that the decomposition in the first step be done so that

$$-a + t^- \leq w_i \leq a - t^+ \qquad\qquad 2.94$$

where

$$-t^- \leq t_{i+1} \leq t^+ \qquad\qquad 2.95$$

The specific addition algorithm depends on the representation of the operands. We now consider the following three cases: (A) both operands in signed-digit representation, (B) one operand conventional and the other signed digit, and (C) both operands in conventional representation.

Case A: Two Signed-Digit Operands

This is the most general case and results in the most complex algorithm. We consider the case in which the digit sets of both operands and of the result are the same, namely, from $-a$ to $+a$. In this case we have

$$-2a \leq x_i + y_i \leq 2a \qquad\qquad 2.96$$

Since $x_i + y_i = r t_{i+1} + w_i$ and $w_i \leq a - t^+$, we obtain for the largest values

$$2a \leq r t^+ + a - t^+ \qquad\qquad 2.97$$

Therefore, $a \leq (r - 1)t^+$, and because $a < r$,

$$-1 \leq t_{i+1} \leq 1 \qquad\qquad 2.98$$

and the algorithm becomes

$$(t_{i+1}, w_i) = \begin{cases} (0, \ x_i + y_i) & \text{if } -a + 1 \leq x_i + y_i \leq a - 1 \\ (1, \ x_i + y_i - r) & \text{if } x_i + y_i \geq a \\ (-1, \ x_i + y_i + r) & \text{if } x_i + y_i \leq -a \end{cases} \qquad 2.99$$

We now determine a bound on a that has to be satisfied for the above algorithm. Consider the case $x_i + y_i = a$. Since

$$w_i = (x_i + y_i - rt_{i+1}) \leq a - 1 \qquad\qquad \textbf{2.100}$$

for this case it is necessary to have $t_{i+1} = 1$, resulting in

$$w_i = a - r \qquad\qquad \textbf{2.101}$$

This is negative, so it is necessary to verify that $w_i \geq -a + 1$. Therefore,

$$a - r \geq -a + 1 \qquad\qquad \textbf{2.102}$$

so that

$$2a \geq r + 1 \qquad\qquad \textbf{2.103}$$

or

$$a \geq (r + 1)/2 \qquad\qquad \textbf{2.104}$$

Note that this value is larger than the minimum value for redundant representation.

EXAMPLE 2.9 Consider an addition operation with operands and results in a signed-digit radix $r = 10$ representation with $a = 6$.

This pair (r, a) satisfies the bound above. The first step requires $-5 \leq w_i \leq 5$ and $-1 \leq t_i \leq 1$.

X		$\bar{2}$	1	$\bar{6}$	3	4
Y		1	3	$\bar{2}$	6	2
W		$\bar{1}$	4	2	$\bar{1}$	$\bar{4}$
T	0	0	$\bar{1}$	1	1	
S		$\bar{1}$	3	3	0	$\bar{4}$

Note that (2.104) is not satisfied for radix-2 with digit set $\{-1, 0, 1\}$. Since this case is important, a modification is developed as follows.

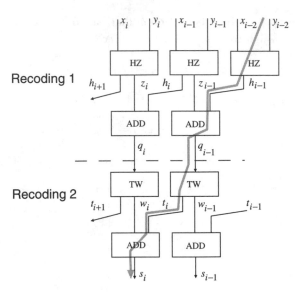

FIGURE 2.35 Double recoding method for signed-bit addition.

Modified Signed-Digit Addition for r=2. We consider two approaches: the double-recoding approach and the approach that uses information from the previous digit position.

Method 1: Double recoding. The signed-digit addition can be viewed as a recoding from the digit set of $x_i + y_i$ (which is $\{-2, \ldots, 2\}$), into the digit set of the result $\{-1, 0, 1\}$. In this method, two recodings (or signed-digit additions) are performed, as follows:

1. The first recoding is from the digit set of $x_i + y_i$, namely, $\{-2, -1, 0, 1, 2\}$ to an intermediate set, for instance $\{-2, -1, 0, 1\}$.[10]

2. The second recoding transforms this intermediate digit set into the radix-2 signed-bit set $\{-1, 0, 1\}$.

Both recodings are accomplished by applying the signed-digit addition algorithm. Figure 2.35 illustrates this method; as can be seen, the critical path is increased with respect to single signed-digit addition.

---

10. The alternative intermediate digit set $\{-1, 0, 1, 2\}$ is also possible.

The two recodings are defined as follows:

Recoding 1

$$x_i + y_i = 2h_{i+1} + z_i \in \{-2, -1, 0, 1, 2\} \qquad \textbf{2.105}$$

such that $h_i \in \{0, 1\}$ and $z_i \in \{-2, -1, 0\}$.

Consequently, the result of this recoding (addition) is

$$q_i = z_i + h_i \qquad \textbf{2.106}$$

with $q_i \in \{-2, -1, 0, 1\}$. Note that the computation of q_i does not have to be done explicitly; that is, it can be kept as z_i and h_i.

Recoding 2

$$q_i = 2t_{i+1} + w_i \in \{-2, -1, 0, 1\} \qquad \textbf{2.107}$$

such that $t_i \in \{-1, 0\}$ and $w_i \in \{0, 1\}$.

Consequently,

$$s_i = w_i + t_i \in \{-1, 0, 1\} \qquad \textbf{2.108}$$

Method 2: Using information from previous digit position. As before,

1. $x_i + y_i = 2t_{i+1} + w_i$

2. $s_i = w_i + t_i$

The critical values in step 1 are $|x_i + y_i| = 1$ because in this case it is not possible to satisfy the condition for t_{i+1} and w_i so as to produce a carry-free second step. However, in each of these cases two combinations of (t_{i+1}, w_i) are possible, and we use information from the previous digit position to choose between them, as follows.

Consider first the case $x_i + y_i = 1$. The two possible combinations are $(t_{i+1}, w_i) = (0, 1)$ and $(t_{i+1}, w_i) = (1, -1)$. If we know that the transfer digit t_i from the previous digit position will not be 1, then we can choose the first combination. Similarly, if we know that it cannot be -1, then we can choose the second combination. Consequently, we define the condition P_i, which gives us the required information, as follows:

$$P_i = \begin{cases} 0 & \text{if } (x_i, y_i) \text{ both nonnegative} \\ & \quad (\text{which implies } t_{i+1} \geq 0) \\ 1 & \text{otherwise } (t_{i+1} \leq 0) \end{cases} \qquad \textbf{2.109}$$

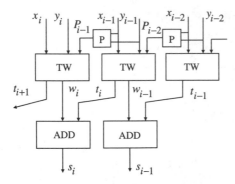

FIGURE 2.36 Signed-bit addition using the information from previous digit.

By symmetry, this information is also useful for the case $x_i + y_i = -1$. Using P_i, the first step results in the following table:

$x_i + y_i$	P_{i-1}	t_{i+1}	w_i
2	—	1	0
1	$0(t_i \geq 0)$	1	−1
1	$1(t_i \leq 0)$	0	1
0	—	0	0
−1	$0(t_i \geq 0)$	0	−1
−1	$1(t_i \leq 0)$	−1	1
−2	—	−1	0

The module-level implementation of this algorithm is shown in Figure 2.36.

EXAMPLE 2.10

$$X \qquad 0\,1\,1\,1\,\bar{1}\,1\,0\,\bar{1}\,1$$
$$Y \qquad 0\,1\,1\,0\,\bar{1}\,0\,1\,0\,1$$

$$P \qquad 0\,0\,0\,0\,1\,0\,0\,1\,0$$

$$W \qquad 0\,0\,0\,1\,0\,\bar{1}\,1\,\bar{1}\,0$$
$$T \qquad 0\,1\,1\,0\,\bar{1}\,1\,0\,0\,1\,0$$

$$S \qquad 1\,1\,0\,0\,1\,\bar{1}\,1\,0\,0 \qquad\qquad ∎$$

Case B: Both Operands Conventional

This is the other extreme case, resulting in the simplest algorithm. In this case the digit set of the operands is $0 \leq x_i, y_i \leq r - 1$. Consequently, $0 \leq x_i + y_i \leq 2r - 2$. We now determine the bounds $t_{i+1} \in [t^-, t^+]$ and $w_i \in [w^-, w^+]$. Since the sum is always positive, we get $t^- = 0$. To determine t^+ and w^+, consider the maximum value of $x_i + y_i = 2r - 2$. That is,

$$r t^+ + w^+ \geq 2r - 2 \qquad\qquad \textbf{2.110}$$

To satisfy the condition (2.93), we make $w^+ = a - t^+$, resulting in

$$t^+ \geq 2 - \frac{a}{r - 1} \qquad\qquad \textbf{2.111}$$

That is,

$$t^+ = \begin{cases} 1 & \text{if } a = r - 1 \\ 2 & \text{otherwise} \end{cases} \qquad\qquad \textbf{2.112}$$

In summary, t_{i+1} and w_i are bounded as follows:

$$0 \leq t_{i+1} \leq t^+ \quad -a \leq w_i \leq a - t^+ \qquad\qquad \textbf{2.113}$$

The algorithm for $a = r - 1$ is

$$(t_{i+1}, \ w_i) = \begin{cases} (0, \ x_i + y_i) & \text{if } x_i + y_i \leq r - 2 \\ (1, \ x_i + y_i - r) & \text{if } x_i + y_i \geq r - 1 \end{cases} \qquad\qquad \textbf{2.114}$$

On the other hand, the algorithm for $\frac{r+1}{2} \leq a < r - 1$ is

$$(t_{i+1}, \ w_i) = \begin{cases} (0, \ x_i + y_i) & \text{if } x_i + y_i \leq a - 2 \\ (1, \ x_i + y_i - r) & \text{if } a - 1 \leq x_i + y_i \leq r + a - 2 \\ (2, \ x_i + y_i - 2r) & \text{if } r + a - 1 \leq x_i + y_i \end{cases} \qquad \textbf{2.115}$$

EXAMPLE 2.11   Operands: conventional; result: signed-digit; $r = 10$, $a = 6$.

X		2	1	9	0	4	1
Y		4	3	9	9	3	4
W		$\bar{4}$	4	$\bar{2}$	$\bar{1}$	3	$\bar{5}$
T	1	0	2	1	1	1	
S	1	$\bar{4}$	6	$\bar{1}$	0	$\bar{2}$	$\bar{5}$

    ■

In this case, the algorithm is valid also for $r = 2$. The algorithm is

$$w_i = \begin{cases} 0 & \text{if } x_i + y_i \neq 1 \\ -1 & \text{if } x_i + y_i = 1 \end{cases} \qquad \text{2.116}$$

$$t_{i+1} = \begin{cases} 0 & \text{if } x_i + y_i = 0 \\ 1 & \text{if } x_i + y_i \geq 1 \end{cases} \qquad \text{2.117}$$

Case C: One Conventional Operand and One Signed Digit

This is an important situation that appears in the implementation of, for example, multiplication and division, discussed in Chapters 4 and 5, respectively. In this case

$$-a \leq x_i + y_i \leq a + r - 1 \qquad \text{2.118}$$

Because of this range, the first step of the procedure can be performed with $0 \leq t_i \leq 1$, so that the procedure is a straightforward extension of Case B. That is,

$$(t_{i+1}, w_i) = \begin{cases} (0, \ x_i + y_i) & \text{if } -a \leq (x_i + y_i) \leq a - 1 \\ (1, \ x_i + y_i - r) & \text{if } (x_i + y_i) \geq a \end{cases} \qquad \text{2.119}$$

For radix-2 the algorithm reduces to

$$w_i = \begin{cases} 0 & \text{if } |x_i + y_i| \neq 1 \\ -1 & \text{if } |x_i + y_i| = 1 \end{cases} \qquad \text{2.120}$$

$$t_{i+1} = \begin{cases} 0 & \text{if } x_i + y_i \leq 0 \\ 1 & \text{if } x_i + y_i \geq 1 \end{cases} \qquad \text{2.121}$$

As usual, this is followed by the second step:

$$s_i = w_i + t_i \qquad \text{2.122}$$

Bit-Level Implementation of the Radix-2 Algorithms

We now present bit-level implementations of the radix-2 cases. These implementations depend on the coding used for the signed digits. Since in many cases, the result is used as the operand in the subsequent operation, the coding for both the operands and result should be the same. If we call d the signed digit and d^+ and d^- the two bits of the representation, a suitable coding is described by the following arithmetic expression:

$$d = d^+ - d^- \qquad\qquad 2.123$$

This corresponds to the following table (note the two representations of 0):

d^+	d^-	d
0	0	0
0	1	−1
1	0	1
1	1	0

First consider the Case C algorithm (one conventional operand, one signed digit). The conventional operand x has digit set $x_i \in \{0, 1\}$, while the signed-digit operand y has digit y_i represented by $y_i^+ \in \{0, 1\}$ and $y_i^- \in \{0, 1\}$.

The first step of the algorithm (expression (2.119)) maps onto the following switching expressions[11] (w_i corresponds to w_i^- and t_i to t_i^+):

$$w_i = x_i \oplus y_i^+ \oplus y_i^-$$
$$t_{i+1} = x_i y_i^+ + x_i (y_i^-)' + y_i^+ (y_i^-)' \qquad 2.124$$

Since $w_i = w_i^- \leq 0$ and $t_i = t_i^+ \geq 0$, the second step is trivial, so that

$$s_i^+ = t_i, \quad s_i^- = w_i \qquad\qquad 2.125$$

As shown in Figure 2.37, the implementation consists of a full adder with one of the inputs and one of the outputs complemented.

Case B is a straightforward simplification of the implementation of Case C, making $y_i^- = 0$.

Now consider implementation of Case A (both operands signed digit). We implement the double recoding approach. It consists of two levels of full adders as shown in Figure 2.38. Note that the variable z_i in the algorithm is mapped into the two signals v_i and y_i^- in the implementation. Note also that the value $x_i + y_i = 0$ is recoded to either $(h_{i+1}, z_i) = (0, 0)$ or $(h_{i+1}, z_i) = (1, -2)$.

---

11. This can be determined from a table of the switching functions.

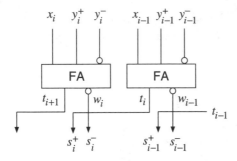

FIGURE 2.37 Radix-2 signed-digit adder: one operand conventional, one operand redundant, result redundant.

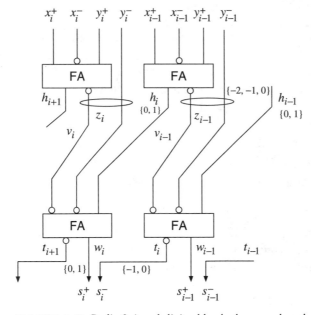

FIGURE 2.38 Radix-2 signed-digit adder: both operands and result redundant.

2.13   Concluding Remarks

We have presented several adder schemes, which differ in characteristics such as delay, area, and energy dissipation. Although the structures are different, these adders are unified by the properties of the carries, as outlined in Section 2.1.

Because of this, in many cases a structure can be converted into another by manipulation of these carry properties, and other similar structures can easily be developed. We have also illustrated the use of completion signals to have variable-time adders; these take advantage of the fact that not all additions have the same delay and their implementations correspond to typical asynchronous combinational systems. Finally, we have considered redundant adders, which have carry-propagation chains of very limited length (one or two digits), independent of the adder width; these produce fast adders with low area and are used in many algorithms that contain the very common addition operation and where conversion to conventional representation does not eliminate the advantage of using the redundant adder. Examples are multioperand addition, multiplication, division, and square root, as discussed in the following chapters. These redundant adders also point to the fact that the addition algorithm is strongly dependent on the representation of operands and result.

As characterization measures we have used delay and number of modules. For delay, we considered expressions in terms of delays of modules and, in some cases, in terms of delays of primitive gates. These measures are rough first-order estimates, which are somewhat independent of the specific implementation and do not include important considerations, such as the effect of interconnection wires and the load on the signals.

None of the adder structures is superior in all aspects since this is an example of the usual trade-off between delay and area. Moreover, the specific characteristics depend on many factors, such as the technology used, the primitive cells available, and the design tools. As a consequence, for a particular application, it is necessary to explore the design space to obtain a suitable implementation.

Although detailed analysis, depending on the physical implementation, is required to compare the schemes accurately, it is informative to classify them according to the complexity with respect to the number of bits. A summary of this is given in Table 2.3. With respect to the delay, we can distinguish the usual (iterative) linear structures, with delay proportional to n/m, and the treelike structures, with delay proportional to $\log_m n$, where m is the number of bits handled by each module. This is typical of switching functions of n variables. The particular properties of the adder are used to share substructures to produce the complete sum vector.

The basic linear structures are the carry-ripple (with $m = 1$) and the one-level carry-lookahead, which corresponds to a carry-ripple with radix 2^m. A variation

Scheme	Delay proportional to	Area proportional to
Linear structures:		
Carry ripple	n	n
Carry lookahead (one level)	n/m	$(k_m m)(n/m) = k_m n$
Carry select (one level)	n/m	$(k_m m)(n/m) = k_m n$
Carry skip (one level)	$\sqrt{n}$	n
Logarithmic structures:		
Carry lookahead (maximum levels)	$2\log_m n$	$(k_m m)(n/m) = k_m n$
Prefix	$\log_m n$	$((k_m m)\log_m n)n$
Conditional sum	$\log_2(n/m)$	$(k_m + \log_2(n/m))n$
Completion signal (average delay)	$(\log_2 n)/m$	$k_m m(n/m) = k_m n$
Redundant	constant	n

TABLE 2.3 Summary of delay and area complexities for adder schemes.

of this radix-2^m case is the carry-select, in which the carry between digits is used to select among two conditional sums. The carry-skip structure is an alternative that makes use of the characteristics of the longest carry chain.

The carry-lookahead concept can be extended to produce multilevel lookahead structures of varying depths, up to a tree-type structure of depth proportional to $\log_m(n)$. Actually, the number of levels is $2\log_m(n)$, since a second pass is required to obtain the carries inside the modules. Various tree-type structures can also be obtained by considering the carry computation as a prefix computation; these structures are very regular, and it is easy to develop alternatives that trade off the signal load, the number of levels, and the number of cells.

The carry-select concept can also be extended to a treelike structure. Note, however, that in this case, the number of levels is proportional to $\log_2(n/m)$ (not $\log_m n$) since the selection is done by considering the two possible values of the carry between consecutive blocks.

In the variable-time case, a suitable measure of delay is the average delay. This delay depends on the distribution of input values. Usually the value is given for a uniform distribution, although other distributions might better represent some applications. In the text we have considered the case of a carry-ripple adder;

in practice other adder structures can be used as the basis for the variable-time case.

The redundant adders have a constant delay, independent of the adder width. Since this constant is quite low (between one and two full-adders, for the radix-2 case), these adders are significantly faster than those for conventional representation.

With respect to the area, we use as measure the number of cells. To take into account the effect of the added complexity introduced by using a fast radix-2^m module, we consider that the area of an m-bit module has a complexity $k_m m$ times the complexity of the radix-2 module. For the case of the multiplexers used in the carry-select and conditional-sum adders we use $k_m = 1$. For the linear structures, we see an increase in area as the group size increases. In the logarithmic structures we observe that the lookahead adder has an area proportional to $(k_m m)n$, whereas for the prefix adder it is proportional to $(k_m m)n \log_m n$. The additional area of the prefix case is due to a reduction of the delay by a factor of two; this is an example of the trade-off between delay and area.

We do not discuss the complexity with respect to energy, since this is highly dependent on circuit technology. Several studies on the energy of adders have been reported in the literature, but at this time there is no general model that can be used to compare adder schemes in a way that is relatively independent of the technology.

2.14   Exercises

Carry-Ripple Adder

2.1   In the full-adder implementation with two half-adders, the load on the carry-in is larger than the load on other signals. Since the delay of the carry-out signal is affected by this load, it is convenient to reduce it. One possibility is to include an inverter in the carry-in input to the XOR gate producing s_i and to change the XOR to an XNOR. Determine the effect of this modification on the delay of the carry-out signal, using the characteristics of Table 2.4 (average delay).

2.2   Determine the delay of a 32-bit adder using the full-adder characteristics of Table 2.4 (average delays).

Gate Type	Fanin	Propagation Delays			Load Factor (standard loads)	Size (equivalent gates)
		t_{pLH} (ns)	t_{pHL} (ns)	t_p (average) (ns)		
AND	2	$0.15 + 0.037L$	$0.16 + 0.017L$	$0.16 + 0.027L$	1.0	2
AND	3	$0.20 + 0.038L$	$0.18 + 0.018L$	$0.19 + 0.028L$	1.0	2
AND	4	$0.28 + 0.039L$	$0.21 + 0.019L$	$0.25 + 0.029L$	1.0	3
OR	2	$0.12 + 0.037L$	$0.20 + 0.019L$	$0.16 + 0.028L$	1.0	2
OR	3	$0.12 + 0.038L$	$0.34 + 0.022L$	$0.23 + 0.025L$	1.0	2
OR	4	$0.13 + 0.038L$	$0.45 + 0.025L$	$0.29 + 0.032L$	1.0	3
NOT	1	$0.02 + 0.038L$	$0.05 + 0.017L$	$0.04 + 0.028L$	1.0	1
NAND	2	$0.05 + 0.038L$	$0.08 + 0.027L$	$0.07 + 0.033L$	1.0	1
NAND	3	$0.07 + 0.038L$	$0.09 + 0.039L$	$0.08 + 0.039L$	1.0	2
NAND	4	$0.10 + 0.037L$	$0.12 + 0.051L$	$0.11 + 0.045L$	1.0	2
NAND	5	$0.21 + 0.038L$	$0.34 + 0.019L$	$0.28 + 0.029L$	1.0	4
NAND	6	$0.24 + 0.037L$	$0.36 + 0.019L$	$0.30 + 0.028L$	1.0	5
NAND	8	$0.24 + 0.038L$	$0.42 + 0.019L$	$0.33 + 0.029L$	1.0	6
NOR	2	$0.06 + 0.075L$	$0.07 + 0.016L$	$0.07 + 0.046L$	1.0	1
NOR	3	$0.16 + 0.111L$	$0.08 + 0.017L$	$0.12 + 0.059L$	1.0	2
NOR	4	$0.23 + 0.149L$	$0.08 + 0.017L$	$0.16 + 0.083L$	1.0	4
NOR	5	$0.38 + 0.038L$	$0.23 + 0.018L$	$0.32 + 0.028L$	1.0	4
NOR	6	$0.46 + 0.037L$	$0.24 + 0.018L$	$0.35 + 0.028L$	1.0	5
NOR	8	$0.54 + 0.038L$	$0.23 + 0.018L$	$0.39 + 0.028L$	1.0	6
XOR[+]	2*	$0.30 + 0.036L$	$0.30 + 0.021L$	$0.30 + 0.029L$	1.1	3
		$0.16 + 0.036L$	$0.15 + 0.020L$	$0.15 + 0.028L$	2.0	
XOR[+]	3*	$0.50 + 0.038L$	$0.49 + 0.027L$	$0.50 + 0.033L$	1.1	6
		$0.28 + 0.039L$	$0.27 + 0.027L$	$0.28 + 0.033L$	2.4	
		$0.19 + 0.036L$	$0.17 + 0.025L$	$0.18 + 0.032L$	2.1	
2-OR/NAND 2	4	$0.17 + 0.075L$	$0.10 + 0.028L$	$0.14 + 0.052L$	1.0	2
2-AND/NOR 2	4	$0.17 + 0.075L$	$0.10 + 0.028L$	$0.14 + 0.052L$	1.0	2
2-MUX	2	$0.20 + 0.050L$	$0.22 + 0.050L$	$0.21 + 0.050L$	0.5	2

L: load on the gate output
* different characteristics for each input
[+] XNOR same characteristics as XOR; for full-adder characteristics see Table 2.2

TABLE 2.4 Characteristics of a family of CMOS gates.

2.3 Design a radix-4 full adder using the CMOS family of gates shown in Table 2.4. Compare delay and size with a 2-bit carry-ripple adder implemented with (radix-2) full-adders (use average delays).

Switched Carry-Ripple Adder

2.4 Compare the delay of a 32-bit switched carry-ripple adder with that of a 32-bit standard carry-ripple adder (using the two-half-adders implementation and the characteristics of gates shown in Table 2.4). To model the delay of the switched carry-ripple case, assume that the delay of the switch (setting or propagation) is equal to that of one 2-input NAND gate (with load of 2) and that to restore the signal a buffer (of delay equal to 1.5 of the delay of a 2-input NAND gate) has to be introduced at 4-bit intervals.

Carry-Skip Adders

2.5 Draw a diagram of carry chains similar to that of Figure 2.8 for the addition of the following bit-vectors:

$$X \quad 0100110001110110$$
$$Y \quad 1011011010001100$$

2.6 As in Example 2.3, determine the delay of the second addition for the following two consecutive additions:

Operation 1				
x	0000	0111	0000	1111
y	1111	0000	1111	0001

Operation 2				
x (same)	0000	0111	0000	1111
y (change last bit)	1111	0000	1111	0000

2.7 Derive the expressions for the optimal group size and optimal delay for a carry-skip adder with fixed-size groups.

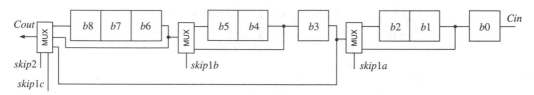

Delays:
    all bi have the same carry delay
    t_MUX = delay for propagating carry through MUXes
    t_skip1 = delay for generating $skip1a$, $skip1b$, or $skip1c$ signals
    t_skip2 = delay for generating $skip2$ signal

FIGURE 2.39 Two-level carry-skip adder.

2.8   Determine the delay of a 64-bit carry-skip adder for the following cases:

(a)   fixed-size groups of eight bits

(b)   optimal fixed-size groups according to the characteristics of Tables 2.2 (for the full-adder) and 2.4 (for the MUX)

(c)   an implementation with variable-size groups that produces a smaller worst-case delay than (b)

2.9   Consider the two-level carry-skip network shown in Figure 2.39 (Turrini 1989). Determine the worst-case delay for an 9-bit adder and for a 36-bit adder (using four of these modules) and assuming that $t_c = t_{mux} = \delta$.

2.10   (a)   Determine the worst-case delay for an n-bit two-level carry-skip adder. Assume fixed-size groups of m bits/group and p fixed-size sections of $s\,m$ bits/section; that is, $n = p s m$. Assume that $t_c = t_{mux}$.

(b)   Using the expression obtained in (a) determine the optimal group size m for $m = s$.

Carry-Lookahead Adder

2.11   Determine the number of equivalent gates, the maximum gate fanin and fanout, and the critical delays of a carry-lookahead generator (CLG) for $m = 4$ and $m = 8$. Use NAND and NOR gates and the gate characteristics of the family in Table 2.4.

2.12   A block carry-lookahead module BCLA generates only the MS carry bit in a group as shown in Figure 2.40. Show a one-level structure (similar to a one-level CLA)

Bit-level generate and propagate signals

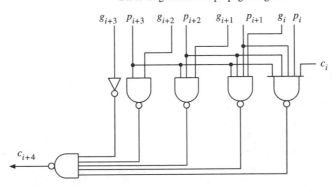

FIGURE 2.40 BCLA module.

for a 32-bit adder constructed only from these BCLA modules, half-adders, and full-adders. Determine its worst-case carry delay using the characteristics of the gates in Table 2.4. Compare with a carry-skip adder with group size of four.

2.13 Using expression (2.40), determine the four carries in a CLG-4 for the following bit-vectors: $X = 0101, Y = 1001, c_0 = 1$.

2.14 Using expressions in Section 2.7.2, determine the carries for a two-level lookahead adder with 4-bit groups for the following input vectors:

$$X \quad 0110111010111001$$
$$Y \quad 1001000011100101$$

2.15 Draw a diagram similar to Figure 2.15 for a 64-bit three-level carry-lookahead adder.

2.16 Derive the expression for the number of CLG modules for an n-bit CLA adder with L levels and groups of size m.

2.17 For a 128-bit adder and using groups and sections of four bits, compare the delay for one-level, two-level, three-level, and four-level carry-lookahead adders for the case $t_{clg} = t_{A,G} = 6t_{a,g} = 3t_s$.

2.18 Ling's adder (Doran 1988; Ling 1981) uses a more efficient recurrence for carries compared with the recurrence used in the carry-lookahead adders discussed in

Section 2.1. The expressions used there are

$$p_i = x_i \oplus y_i, \quad g_i = x_i y_i, \quad c_{i+1} = g_i + p_i c_i, \quad s_i = c_i \oplus p_i$$

Ling defines a new "carry" function $h_i = c_{i+1} + c_i$, resulting in the following adder expressions:

$$t_i = x_i + y_i, \quad g_i = x_i y_i, \quad h_i = g_i + t_{i-1} h_{i-1}, \quad s_i = t_i \oplus h_i + g_i t_{i-1} h_{i-1}$$

(a)   Show that Ling's expressions produce the correct sum.
(b)   Consider the expressions for a group of four bits and show that Ling's approach is more efficient than the conventional one with respect to the number of gates and fanin.

Prefix Adder

2.19   The prefix adder shown in Figure 2.18 includes a carry-in. Draw the modified structures for the case in which there is no carry-in.

2.20   Using the prefix adder of Figure 2.20, perform the addition of the bit-vectors $X = 01010111$, $Y = 11100111$, and $c_0 = 1$.

2.21   Using a prefix adder as a basis, design a network that produces simultaneously $s = x + y$ and $z = x + y + 1$. This network is useful in rounding for floating-point addition.

2.22   Give a diagram like Figure 2.20 for a 16-bit prefix adder with the minimum number of levels and a maximum fanout of four. To do this, you might want to begin with two schemes: one with a maximum fanout of eight and the other with a maximum fanout of two, and then interpolate.

Carry-Select Adder

2.23   Design a conditional-adder module of four bits (see Figure 2.22) to be used in the first stage of a carry-select adder. In your design, share the logic between the adder with the carry-in 0 and the adder with the carry-in 1. Specifically, base your design on the carry-ripple scheme.

2.24   Using a carry-select adder with 4-bit groups, perform the addition of the following bit-vectors: $X = 0111100010101010$, $Y = 1010101110110010$, and $c_0 = 0$.

2.25 A variation of the carry-select adder computes the carries into each group separately from the sums, using a faster network, such as lookahead. This reduces the the worst-case delay and also allows a more effective use of variable-size groups. To evaluate this approach determine the optimal size of groups in a carry-select adder and the corresponding worst-case delay for the following two cases:

(a) Carries between groups are generated by a linear (iterative) network.

(b) Carries between groups are obtained by a tree (lookahead) network.

Define first the relevant delays.

Conditional-Sum Adder

2.26 Using a conditional-sum adder with 2-bit groups, perform the addition for the following bit-vectors: $X = 01010111, Y = 10101111, c_0 = 0$.

2.27 Consider the following two schemes for adding two binary operands of length $n = 2^p$ bits. Scheme A forms conditional sums over groups of $n/4$ bits using carry-ripple adders and then applies the conditional-sum method to obtain the final sum. Scheme B forms $n/4$-bit sums using the conditional-sum method within groups and ripples the carries between the groups. That is, the ith group requires the carry from the group $i - 1$ in order to produce the correct sum and carry-out. If the delay of a full-adder is 2δ and that of a 2-to-1 multiplexer is δ, determine the value of p for which scheme A is faster than scheme B.

2.28 Consider a variation of the conditional-sum adder, called the *conditional-carry adder*. In this variant, only the carries are computed using the conditional approach. These carries are then used to compute the sums as $s_i = x_i \oplus y_i \oplus c_i$.

(a) Design a 16-bit conditional-carry adder, initial group size of 1, and the incoming carry c_0. Show the necessary logic details (which you can abstract as modules). Label all signals. Indicate the critical path.

(b) Compare the design with a 16-bit conditional-sum adder with respect to the cost (type and number of modules) and the delay in the critical path.

Variable-Time Adders

2.29 Using variable-time adders of both types discussed in this chapter, perform the addition of the following bit-vectors: $X = 1000100111, Y = 0111000110$. The adder cells have an actual delay 15% smaller than the worst-case delay, and the delay of the AND gate to determine the completion signal (for an 8-bit adder) is equal to the delay of one adder cell.

Determine the actual delay in both cases and compare with the delay of the carry-ripple adder.

2.30 (a) Consider a 32-bit carry-select adder consisting of four groups of size 8. Suppose that each group is implemented as an 8-bit Type 1 variable-time adder. Design the rest of the network so that the whole adder behaves as a Type 1 adder.

   (b) Compare the design in (a) with a 32-bit Type 1 adder that is based on the carry-ripple adder scheme.

Ones' Complement Adder

2.31 Determine the delay of an n-bit ones' complement adder implemented connecting the carry-out with the carry-in in the scheme of Figure 2.20. Compare with the adder that includes an additional stage to add the carry-out (Figure 2.29).

Redundant Adder

2.32 Perform the addition of the following 4-bit vectors using

   (a) a [4:2] adder composed of two levels of [3:2] adders
   (b) a [4:2] adder composed of the cells of Figure 2.41:

$$0111110000110011$$
$$1110001001101101$$
$$1010101010101010$$
$$1010111101110111$$

2.33 (a) Consider implementing a cell for the [4:2] adder using full-adders. Determine the connections between the full-adders and inputs so that the delay in the critical path is the smallest. Use the average delays of a full-adder given in Table 2.2 and assume the load of 1 for the outputs of the [4:2] adder.

   (b) Show that the scheme of Figure 2.41 implements a cell for the [4:2] adder. Determine the average propagation delays using Table 2.4 (average delay) and compare with the results obtained in (a).

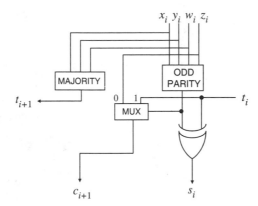

FIGURE 2.41 Alternative implementation of a cell for the [4:2] adder.

2.34 Design a radix-8 carry-save adder using a radix-8 cell composed of three full-adders.

2.35 Perform the radix-8 carry-save addition of the following two operands (one carry-save and one conventional):

$$101110110011$$
$$1 \quad 1 \quad 0 \quad 1$$
$$011100111011$$

2.36 Describe an algorithm and an implementation for the addition of two radix-8 carry-save operands.

2.37 Perform a radix-4 signed-digit addition for the following two operands: $X = 0\bar{2}1\bar{3}101\bar{1}$, $Y = 10\bar{1}\bar{3}\bar{1}110$.

2.38 Give arithmetic expressions for a radix-4 signed-digit adder with inputs $a_i, b_i \in \{-2, -1, 0, 1, 2\}$ and output $s_i \in \{-2, -1, 0, 1, 2\}$—all minimally redundant digit sets. Define all intermediate variables and their digit sets. Specify the corresponding blocks without going into design details at the binary level. Compare this adder with a radix-4 signed-digit adder with the maximally redundant digit set $\{-3, -2, -1, 0, 1, 2, 3\}$.

2.39 Perform a radix-2 signed-digit addition for the following two operands: $X = 01\bar{1}1\bar{1}01\bar{1}$, $Y = 10101\bar{1}\bar{1}1$. Use both approaches described in the chapter.

2.40   Derive high-level and binary-level expressions for the double recoding algorithm resulting in the redundant adder implementation shown in Figure 2.38.

2.41   Design a three-operand adder with two operands x and y in the two's complement form with the digit set $\{0, 1\}$ (MS bit is in the set $\{-1, 0\}$), and the third operand z and the sum s in the signed-digit form with the digit set $\{-2, -1, 0, 1, 2\}$.

Design the adder using the recoding approach. Show the recoding equations and all intermediate digit sets. Show the block diagram of your design. Discuss the critical path.

2.42   Develop an addition algorithm for conventional radix-4 operands with the digit set $\{0, 1, 2, 3\}$ and the sum with $a = 2$ using

(a)   Method 1 (double recoding)
(b)   Method 2 (recoding with information from the previous position)

For each design estimate the delay (logic levels) and compare with the addition using $a = 3$ in the digit set for the sum.

2.43   Perform the radix-2 signed-digit addition of one signed-digit operand and one conventional operand using the implementation of Figure 2.37 for the following operands: $X = 01110110$, $Y = 1\bar{1}100\bar{1}11$.

2.44   Perform the radix-2 signed-digit addition of two signed-digit operands using the implementation of Figure 2.38 for the following operands: $X = 01\bar{1}\bar{1}\bar{1}100$, $Y = 111\bar{1}\bar{1}111$.

Complexity of Adders

2.45   Consider an implementation of an n-input adder using modules of no more than m inputs. Show that the minimum number of levels of these modules is $O(\log_m n)$. *Hint*: Consider the implementation of a switching function of n variables.

2.15   Further Readings

As is apparent from the text of this chapter, there is a variety of adders that have been developed in the last 50 years. The literature on these adders is very extensive; we give here a list of some of the most relevant papers, because of their historical significance and/or because they provide additional insight as well as more detailed information on implementations.

Overviews and comparisons of different adder structures are given in Sklansky (1960b), Lehman (1962), Gosling (1971), Nagendra et al. (1996) and Zimmerman (1998).

Switched-Ripple Adder

The switched carry-ripple adder, also called the Manchester adder, was initially described in Kilburn et al. (1959). In today's technology it relies on the efficient implementation of transmission gates, as described in Fenwick (1987), Rabey (1996), and Weste and Eshragian (1993).

Carry-Skip Adder

The concept of the carry-skip adder is presented in Morgan and Jarvis (1959) and Lehman and Burla (1961) and was analyzed in Majerski (1967). It has been extended in a variety of ways; the main aspects considered are the determination of optimal group sizes for a variety of delay models and the extension to variable group-size and multilevel schemes (Oklobdzija and Burnes 1985; Guyot et al. 1987; Turrini 1989; Kantabutra 1991; Chan et al. 1992; Kantabutra 1993a). The concept of carry-skip has been applied to switched carry-ripple adders in Chan and Schlag (1990).

Carry-Lookahead Adder

The carry-lookahead adder has been the most popular adder with logarithmic delay. Introduced in Weinberger and Smith (1958), it has led to numerous variations (mainly in number of levels and group size) and implementations. A variant that simplifies the implementation for some technologies, called the Ling adder, was presented in Ling (1981). A general class, of which the Ling adder is a member, is discussed in Doran (1988). A CMOS implementation of the Ling adders is presented in Quach and Flynn (1992).

Carry-Select and Conditional-Sum Adders

The carry-select adder was introduced in Bedrij (1962) and its simplification in VLSI implementation is shown in Tyagi (1993). The conditional-sum adder was first presented in Sklansky (1960a).

Prefix Adder

Prefix adders have become very popular because of their regularity and suitability for VLSI implementations. The initial paper describing addition as a prefix computation is Kogge and Stone (1973). A variation with larger fanout but fewer cells is presented in Ladner and Fisher (1980). In Brent and Kung (1982) a scheme is proposed with the minimal fanout of one. In Han and Carlson (1987) a good overview is given and higher radix prefix schemes are proposed. A design of area-time optimal adder is discussed in Wei and Thompson (1990). An analysis of the whole class of prefix adders and a comparison of different implementations is given in Knowles (1999). In Lynch and Swartzlander (1992) a variation is proposed for efficient adders with non-power-of-two width.

Reverse Carry Adder

An approach proposing reverse carries to overlap levels in a multilevel adder is presented and evaluated in Bruguera and Lang (2000).

Variable-Time Adder

Variable-time adders are an example of asynchronous and self-timed combinational networks, so the literature on these types of networks is relevant. Particular to adders, the carry-completion adder was described in Gilchrist et al. (1955), with recent VLSI realizations presented in Salomon (1987) and Ramachandran and Lu (1996). A self-timed carry-lookahead adder is presented in Cheng et al. (2000). A conditional-sum adder with completion detection is described in Martin and Hufnagel (1980). Asynchronous adders are evaluated in Franklin and Pan (1994) and Kinniment (1996). The sequential and indeterminate behavior of an adder with end-around-carry is examined in Shedletsky (1977).

Redundant Adder

Redundant adders, because of the nonconventional representation of the output, are used as building blocks for more complex operations, such as multioperand addition, multiplication, and division. Consequently, most references are found in the corresponding chapters. Carry-save addition was introduced in Estrin et al. (1956) in the context of sequential multiplication, following an observation of Burks, Goldstine, and Von Neumann. Apparently, Babbage articulated the idea of "postponed" carries in the design of the calculating engine Randell (1975). Signed-digit representation, addition, and other basic operations were investigated in Avizienis (1960, 1961, 1962, 1964, 1966); the first extensive use of this representation was in the Illiac III computer as described in Atkins (1970). The relationship between radix-2 signed-digit adder and carry-save adder is discussed in Duprat and Muller (1991), where the term *borrow save* is used for the signed-digit case. The borrow-save coding was discussed as early as 1967 in Robertson (1967), where a deterministic procedure for the design of carry-save adders and borrow-save subtracters was proposed. Related work on the set transformations and design of adders/subtracters appears in Rohatsch (1967), Borovec (1968), and Chow and Robertson (1978). Later work on systematic procedures and operand coding for the design of redundant adders is presented in Parhami (1988), Carter and Robertson (1990), Bajard et al. (1994), Ercegovac and Lang (1997), and Phatak et al. (2001). Zero, sign, and overflow detection in signed-digit addition are discussed in Parhami (1993). Issues of carry-save addition, such as overflow detection and correction and saturation control, are presented in Noll (1991). Recoding (conversion) between digit sets provides another view in the design of redundant adders. General aspects of recoding are discussed in Kornerup (1994), Ercegovac and Lang (1996), and Kornerup (1999).

Implementation of Adders

The literature describing design and implementation of various types of adders is very extensive (for example, MacSorley 1961; Anderson et al. 1967; Bayoumi et al. 1983; Ngai et al. 1986; Oklobdzija 1988; Naffziger 1996; Knowles 1999; Flynn and Oberman 2001). Application of the logical effort model in the design of adders is presented in Dao and Oklobdzija (2001). An energy-efficient adder design is described in Parhi (1999). Oklobdzija (1999) presents an extensive collection of papers on high-performance circuits, logic, and system design, many of them related to implementation of digital arithmetic schemes.

Incrementer

Incrementers, a special case of adders, are typically used in implementing counters. Schemes that achieve constant cycle time independent of the length are presented in Ercegovac and Lang (1989), Vuillemin (1991), Lutz and Jayasimha (1996), and Stan et al. (1998).

Hybrid Adder

Hybrid adders combine several addition schemes to achieve implementation delay/area constraints. Hybrid adders using a carry-lookahead and carry-select schemes are described in Dobberpuhl et al. (1992), Lynch and Swartzlander (1992), and Kantabutra (1993b). A hybrid adder using carry-skip and carry-select schemes is discussed in Burgess (2001). Hybrid adders are also appropriate when the operand bits to the final adder in tree multipliers (discussed in Chapter 4) do not arrive simultaneously. In such a situation, a hybrid adder provides an efficient implementation as presented in Oklobdzija and Villeger (1995).

Pipelined Adder

A good discussion of general approaches to pipelined adders is presented in Dadda and Piuri (1996). Pipelined designs of several adder schemes are described in Unwala and Swartzlander (1993). Advanced design techniques using asynchronous circuits and wave pipelining are described in Singh and Nowick (2000) and Wong et al. (1993).

Condition Detection Using Adder

Adders are often used to detect conditions such as zero-sum. Such conditions are trivially obtained when the result is computed by full-precision carry-propagate addition. Schemes discussed in Weinberger (1978), Cortadella and Llaberia (1992), Vassiliadis et al. (1993), and Lutz and Jayasimha (1997) present various solutions to obtain conditions without using carry propagation.

Serial Adder

Digit-serial addition schemes and related literature are discussed in Chapter 9.

Bounds on Delay

Theoretical bounds on the delay of addition are presented in Winograd (1965), Spira (1973), and Brent (1970).

2.16   Bibliography

Anderson, S. F., J. G. Earle, R. E. Goldschmidt, and D. M. Powers (1967). The IBM 360/370 model 91: floating-point execution unit. *IBM Journal of Research and Development*, pages 34–53.

Atkins, D. E. (1970). Design of the arithmetic units of ILLIAC III: Use of redundancy and higher radix methods. *IEEE Transactions on Computers*, C-19(8):720–33.

Avizienis, A. (1960). *A Study of Redundant Number Representations for Parallel Digital Computers*. PhD thesis, University of Illinois, Urbana.

Avizienis, A. (1961). Signed digit number representations for fast parallel arithmetic. *IRE Transactions on Electronic Computers*, EC-10(9):389–400.

Avizienis, A. (1962). On flexible implementation of digital computer arithmetic. In *Proc. IFIP Congress*, pages 664–70.

Avizienis, A. (1964). Binary-compatible signed-digit arithmetic. In *Proc. Fall Joint Computer Conference*, pages 663–72.

Avizienis, A. (1966). Arithmetic microsystems for the synthesis of function generators. *Proceedings of the IEEE*, 54(12):1910–19.

Bajard, J. C., J. Duprat, S. Kla, and J.-M. Muller (1994). Some operators for on-line radix 2 computations. *Journal of Parallel and Distributed Computing*, 22(2):336–45.

Bayoumi, M. A., G. A. Jullien, and W. C. Miller (1983). An area-time efficient NMOS adder. *Integration*, 1:317–34.

Bedrij, O. J. (1962). Carry-select adder. *IRE Transactions on Electronic Computers*, EC-11(6):340–46.

Borovec, R. T. (1968). The logical design of a class of limited carry-borrow propagation adders. Technical report no. 275, Dept. of Computer Science, University of Illinois.

Brent, R. P. (1970). On the addition of binary numbers. *IEEE Transactions on Computers*, C-19(8):758–59.

Brent, R. P., and H. T. Kung (1982). A regular layout for parallel adders. *IEEE Transactions on Computers*, C-31(3):260–64.

Bruguera, J. D., and T. Lang (2000). Multilevel reverse-carry adder. In *Proceedings of the IEEE International Conference on Computer Design: VLSI in Computers and Processors (ICCD'00)*, pages 155–62.

Burgess, N. (2001). Accelerated carry-skip adders with low hardware cost. In *Proceedings of the 35th Asilomar Conference on Signals, Systems and Computers*, pages 852–56.

Carter, T. M., and J. E. Robertson (1990). The set theory of arithmetic decomposition. *IEEE Transactions on Computers*, C-39(8):993–1005.

Chan, P. K., M. D. Schlag, C. D. Thomborson, and V. G. Oklobdzija (1992). Delay optimization of carry-skip adders and block carry-lookahead adders using multidimensional dynamic programming. *IEEE Transactions on Computers*, 41(8):920–30.

Chan, P. K., and M. D. F. Schlag (1990). Analysis and design of CMOS Manchester adder with variable carry-skip. *IEEE Transactions on Computers*, C-39(8): 983–92.

Cheng, F.-C., S. H. Unger, and M. Theobald (2000). Self-timed carry-lookahead adders. *IEEE Transactions on Computers*, 49(7):659–72.

Chow, C. Y., and J. E. Robertson (1978). Logical design of a redundant binary adder. In *Proceedings of the 4th IEEE Symposium on Computer Arithmetic*, pages 109–15.

Cortadella, J., and J. M. Llaberia (1992). Evaluation of $A + B = K$ conditions without carry propagation. *IEEE Transactions on Computers*, 41(11):1484–88.

Dadda, L., and V. Piuri (1996). Pipelined adders. *IEEE Transactions on Computers*, 45(3):348–56.

Dao, H., and V. G. Oklobdzija (2001). Application of logical effort for speed optimization and analysis of representative adders. In *Proceedings of the 35th Asilomar Conference on Signals, Systems and Computers*, pages 1666–69.

Dobberpuhl, D. W., et al. (1992). A 200-MHz 64-b dual-issue CMOS microprocessor. *IEEE Journal of Solid-State Circuits*, 27(11):1555–64.

Doran, R. W. (1988). Variants of an improved carry-lookahead adder. *IEEE Transactions on Computers*, C-37(9):1110–13.

Duprat, J., and J.-M. Muller (1991). Writing numbers differently for faster calculation. *Technique et Science Informatiques*, 10(3):211–24.

Ercegovac, M. D., and T. Lang (1989). Binary counter with counting period of one half adder independent of counter size. *IEEE Transactions on Circuits and Systems*, 36(6):924–26.

Ercegovac, M. D., and T. Lang (1996). On recoding in arithmetic algorithms. *Journal of VLSI Signal Processing*, 14:283–94.

Ercegovac, M. D., and T. Lang (1997). Effective coding for fast redundant adders using radix-2 digit set $\{0, 1, 2, 3\}$. In *Proceedings of the 31st Asilomar Conference on Signals, Systems and Computers*, pages 1163–67.

Estrin, G., B. Gilchrist, and J. H. Pomerane (1956). A note on high-speed digital multiplication. *IRE Transactions on Electronic Computers*, page 140.

Fenwick, P. M. (1987). A fast-carry adder with CMOS transmission gates. *Computer Journal*, 30(1):77–79.

Flynn, M. J., and S. F. Oberman (2001). *Advanced Computer Arithmetic Design*. John Wiley & Sons, Inc., New York.

Franklin, M. A., and T. Pan (1994). Performance comparison of asynchronous adders. In *Proceedings of the International Symposium on Advanced Research in Asynchronous Circuits and Systems*, pages 117–25.

Gilchrist, B., J. H. Pomerene, and S. Y. Wong (1955). Fast carry logic for digital computers. *IRE Transactions on Electronic Computers*, EC-4:133–36.

Gosling, J. B. (1971). Review of high-speed addition techniques. *Proceedings of IEE*, 118(1):29–35.

Guyot, A., B. Hochet, and J.-M. Muller (1987). A way to build efficient carry-skip adders. *IEEE Transactions on Computers*, C-36(10).

Han, T., and D. A. Carlson (1987). Fast area-efficient VLSI adders. In *Proceedings of the 8th IEEE Symposium on Computer Arithmetic*, pages 49–56.

Kantabutra, V. (1991). Designing optimum carry-skip adders. In *Proceedings of the 10th IEEE Symposium on Computer Arithmetic*, pages 146–55.

Kantabutra, V. (1993a). Accelerated two-level carry-skip adders—a type of very fast adder. *IEEE Transactions on Computers*, C-42(11):1389–93.

Kantabutra, V. (1993b). A recursive carry-look-ahead/carry-select hybrid adder. *IEEE Transactions on Computers*, C-42(12):1495–99.

Kilburn, T., D. B. G. Edwards, and D. Aspinall (1959). Parallel addition in a digital computer—a new fast carry. *Proceedings of the IEE*, 106B:460–64.

Kinniment, D. J. (1996). An evaluation of asynchronous addition. *IEEE Transactions on VLSI Systems*, 4(1):137–40.

Knowles, S. (1999). A family of adders. In *Proceedings of the 14th IEEE Symposium on Computer Arithmetic*, pages 30–34.

Kogge, P. M., and H. S. Stone (1973). A parallel algorithm for the efficient solution of a general class of recurrence equations. *IEEE Transactions on Computers*, C-22(8):783–91.

Kornerup, P. (1994). Digit-set conversions: Generalizations and applications. *IEEE Transactions on Computers*, 43(5):622–29.

Kornerup, P. (1999). Necessary and sufficient conditions for parallel, constant time conversion and addition. In *Proceedings of the 14th IEEE Symposium on Computer Arithmetic*, pages 152–56.

Ladner, R., and M. Fisher (1980). Parallel prefix computation. *Journal of the ACM*, 27(4):831–38.

Lehman, M. (1962). A comparative study of propagation speed-up circuits in binary arithmetic units. *Information Processing*, pages 671–77.

Lehman, M., and N. Burla (1961). Skip techniques for high-speed carry propagation in binary arithmetic units. *IRE Transactions on Electronic Computers*, EC-10:691–98.

Ling, H. (1981). High-speed binary adder. *IBM Journal Research and Development*, 25(3):156–66.

Lutz, D. R., and D. N. Jayasimha (1996). Programmable modulo-k counters. *IEEE Transactions on Circuits and Systems I: Fundamental Theory and Applications*, 43(11):939–41.

Lutz, D. R., and D. N. Jayasimha (1997). The half-adder form and early branch condition resolution. In *Proceedings of the 13th Symposium on Computer Arithmetic*, pages 266–73.

Lynch, T., and E. E. Swartzlander (1992). A spanning tree carry lookahead adder. *IEEE Transactions on Computers*, C-41(8):931–39.

MacSorley, O. L. (1961). High-speed arithmetic in binary computers. *IRE Proceedings*, 49:67–91.

Majerski, S. (1967). On determination of optimal distributions of carry skips in adders. *IEEE Transactions on Electronic Computers*, EC-16(1):45–58.

Martin, N. M., and S. P. Hufnagel (1980). Conditional-sum early completion adder logic. *IEEE Transactions on Computers*, C-29:753–56.

Morgan, C. P., and D. B. Jarvis (1959). Transistor logic using current switching routing techniques and its application to a fast carry-propagation adder. *Proceedings of the IEE*, 106B:467–68.

Naffziger, S. (1996). A sub-nanosecond 0.5 micron 64b adder design. *Digest of IEEE International Solid-State Circuits Conference*, pages 362–63.

Nagendra, C., M. J. Irwin, and R. M. Owens (1996). Area-time-power tradeoffs in parallel adders. *IEEE Transactions Circuits and Systems II: Analog and Digital Signal Processing*, 43(10):689–702.

Ngai, T. F., M. J. Irwin, and S. Rawat (1986). Regular, area-time efficient carry-lookahead adders. *Journal of Parallel and Distributed Computing*, 3(1):92–105.

Noll, T. (1991). Carry-save architectures for high-speed digital signal processing. *Journal of VLSI Signal Processing*, 3(1-2):121–40.

Oklobdzija, V. G. (1988). Simple and efficient CMOS circuit for fast VLSI adder realization. In *Proceedings of the IEEE Symposium on Circuits and Systems*, pages 235–38.

Oklobdzija, V. G., editor (1999). *High-Performance System Design: Circuits and Logic*. IEEE Press, Piscataway, New Jersey.

Oklobdzija, V. G., and E. R. Burnes (1985). Some optimal shemes for ALU implementation in VLSI technology. In *Proceedings of the 7th IEEE Symposium on Computer Arithmetic*, pages 2–8.

Oklobdzija, V. G., and D. Villeger (1995). Improving multiplier design by using improved column compression tree and optimized final adder in CMOS technology. *IEEE Transactions on VLSI*, 3(2):292–301.

Parhami, B. (1988). Carry-free addition of recoded binary signed-digit numbers. *IEEE Transactions on Computers*, C-37(11):1470–76.

Parhami, B. (1993). On the implementation of arithmetic support functions for generalized signed-digit number systems. *IEEE Transactions on Computers*, 42(3):379–84.

Parhi, K. K. (1999). Low-energy CSMT carry generators and binary adders. *IEEE Transactions on VLSI Systems*, 7(12):450–62.

Phatak, D. S., T. Geoff, and I. Koren (2001). Constant-time addition and simultaneous format conversion based on redundant binary representation. *IEEE Transactions on Computers*, 50(11):1267–87.

Quach, N. T., and M. J. Flynn (1992). High-speed addition in CMOS. *IEEE Transactions on Computers*, 41(12):1612–15.

Rabaey, J.-M., A. Chandrakasan, and B. Nikolić (2003). *Digital Integrated Circuits: A Design Perspective*. Prentice Hall, Englewood Cliffs, New Jersey, 2 edition.

Ramachandran, R., and S.-L. Lu (1996). Efficient arithmetic using self-timing. *IEEE Transactions on VLSI*, 4(4):445–54.

Randell, B., editor (1975). *On the Mathematical Powers of the Calculating Engine (C. Babbage)*. Springer-Verlag, New York, 2nd edition.

Robertson, J. E. (1967). A deterministic procedure for the design of carry-save adders and borrow-save subtracters. Technical report No. 235, Dept. of Computer Science, University of Illinois, Urbana-Champaign.

Rohatsch, F. A. (1967). *A Study of Transformations Applicable to the Development of Limited Carry-Borrow Propagation Adders*. PhD thesis, Department of Computer Science, University of Illinois, Urbana-Champaign.

Salomon, D. (1987). A design for an efficient NOR-gate only, binary ripple adder with carry-completion detection logic. *Computer Journal*, 30(3):283–85.

Shedletsky, J. J. (1977). Comment on the sequential and indeterminate behaviour of an end-around-carry adder. *IEEE Transactions on Computers*, C-26(3):271–72.

Singh, M., and S. M. Nowick (2000). Fine-grain pipelined asynchronous adders for high-speed DSP applications. In *Proceedings of the IEEE Computer Society Workshop on VLSI 2000, System Design for a System-on-Chip Era*, pages 111–18.

Sklansky, J. (1960a). Conditional-sum addition logic. *IRE Transactions on Electronic Computers*, EC-9:226–31.

Sklansky, J. (1960b). An evaluation of several two-summand binary adders. *IRE Transactions on Electronic Computers*, EC-9:213–26.

Spira, P. M. (1973). Computation times of arithmetic and Boolean functions in (d,r) circuits. *IEEE Transactions on Computers*, C-22(6):552–55.

Stan, M. R., A. F. Tenca, and M. D. Ercegovac (1998). Long and fast up/down counters. *IEEE Transactions on Computers*, 47(7):722–35.

Turrini, S. (1989). Optimal group distribution in carry-skip adders. In *Proceedings of the 9th IEEE Symposium on Computer Arithmetic*, pages 96–103.

Tyagi, A. (1993). A reduced area scheme for carry-select adders. *IEEE Transactions on Computers*, C-42(10):1163–70.

Unwala, I. H., and E. E. Swartzlander (1993). Superpipelined adder designs. In *Proceedings of the International Symposium on Circuits and Systems (ISCAS)*, volume 3, pages 1841–44.

Vassiliadis, S., J. Philips, and B. Blaner (1993). Condition code predictor for fixed-point arithemtic units. *IEEE Transactions on Computers*, 42(7):825–39.

Vuillemin, J. E. (1991). Constant time arbitrary length synchronous binary counters. In *Proceedings of the 10th IEEE Symposium on Computer Arithmetic*, pages 180–83.

Wei, B. W. Y., and C. D. Thompson (1990). Area-time optimal adder design. *IEEE Transactions on Computers*, 39(5):666–75.

Weinberger, A. (1978). High-speed zero-sum detection. In *Proceedings of the 4th IEEE Symposium on Computer Arithmetic*, pages 200–207.

Weinberger, A., and J. L. Smith (1958). A logic for high-speed addition. *Nat. Bur. Stand. Circ.*, 591:3–12.

Weste, N. H. E., and K. Eshragian (1993). *Principles of CMOS VLSI Design: A System Perspective*. Addison-Wesley Publishing Co., Reading, Massachusetts, 2nd edition.

Winograd, S. (1965). On the time required to perform addition. *Journal of the ACM*, 12(2):277–85.

Wong, D. C., G. De Micheli, and M. J. Flynn (1993). Designing high-performance digital circuits using wave pipelining: Algorithms and practical experiences. *IEEE Transactions Computer-Aided Design of Integrated Circuits and Systems*, 12(1):25–46.

Zimmerman, R. (1998). *Binary Adder Architectures for Cell-Based VLSI and Their Synthesis (Ph.D. dissertation)*. Series in Microelectronics, Vol. 37. Hartung-Gore, Konstanz, Switzerland.

IN THIS CHAPTER WE PRESENT AND DISCUSS THE FOLLOWING TOPICS:

- Bit-arrays for unsigned and signed operands
- Reduction by rows and by columns: $|p{:}2|$ modules and $|p{:}2|$ adders for reduction by rows, and $(p{:}q|$ counters and multicolumn counters for reduction by columns[1]
- Sequential implementation
- Combinational implementation with reduction by rows (arrays of adders) and reduction by columns $((p{:}q|$ counters)
- Pipelined adder arrays
- Partially combinational implementation

---

1. We use the notation "$(p{:}$" for a column of p bits, and "$q|$" for a row of q bits (weighted).

CHAPTER **3** | Multioperand Addition

In this chapter we consider algorithms and implementations for addition of more than two operands. That is, for m operands we want to obtain s such that

$$s = \sum_{i=1}^{m} x(i)$$

3.1

This operation is used in several algorithms. Examples are multiplication, recurrences, transforms, and filters. The implementations can be classified into sequential and combinational and, in the latter, into adder arrays and column reduction schemes. It is also possible to perform the operation partly combinational and partly sequential. Moreover, the combinational part can be pipelined for higher throughput.

We consider here both the case in which the operands are magnitudes (positive values) and signed values. For the latter, we consider two's complement representation, since this is the simplest and most frequently used. Moreover, the range of the result can be such that no overflow is possible, or this range might be restricted, in which case an overflow detection should be included. We discuss only the first case.

The input operands are represented using bit-vectors, and the set of input bit-vectors forms a *bit-array*. We now discuss the bit-arrays for unsigned (magnitudes only) and signed (two's complement) operands.

3.1 Bit-Arrays for Unsigned and Signed Operands

Before considering the addition, we determine the bit-array to be added. In general, the range of values of each operand can be different, resulting in a nonrectangular bit-array. To simplify the notation, in this chapter we consider the case in which all operands have the same range of values and illustrate the

$$
\begin{array}{cccccccc}
a_0 & a_0 & a_0 & a_0 . & a_1 & a_2 \cdots a_n \\
b_0 & b_0 & b_0 & b_0 . & b_1 & b_2 \cdots b_n \\
c_0 & c_0 & c_0 & c_0 . & c_1 & c_2 \cdots c_n \\
d_0 & d_0 & d_0 & d_0 . & d_1 & d_2 \cdots d_n \\
e_0 & e_0 & e_0 & e_0 . & e_1 & e_2 \cdots e_n
\end{array}
$$

Sign extension

FIGURE 3.1 Sign-extended array for $m = 5$.

more general case only later in Example 3.2. However, the methods and techniques discussed are applicable to nonrectangular arrays, as presented in Chapter 4 for the important application of multiplication.

Consider the case in which each of the m operands is represented by an n-bit vector. The bit-array to be added is then an n by m rectangular array, and the sum bit-vector has $n + p$ bits with $p = \lceil \log_2 m \rceil$. To perform the addition it is necessary to extend the range of the operands to $n + p$ bits. For magnitudes this is trivial since the extension is done by adding most significant 0s.

For two's complement representation, the extension consists of replicating the sign, as shown for $m = 5$ in Figure 3.1. To simplify the description that follows, we place the binary point after the "sign" bit and index as for fractions. That is, the operands are in the range $-1 \leq x \leq 1 - 2^{-n}$, and the two's complement representation is

$$
x_0 . x_1 x_2 \ldots x_n
$$

with value

$$
x = -x_0 + \sum_{i=1}^{n} x_i 2^{-i} \tag{3.2}
$$

The sign-extended operands are then

$$
x_{-p} x_{-(p-1)} \ldots x_{-1} x_0 . x_1 x_2 \ldots x_n \tag{3.3}
$$

with $x_{-i} = x_0$ for $1 \leq i \leq p$.

To avoid the additional adder bits required by these sign extensions, we now present a way of reducing these extended bits. Since the sign position has a negative weight, apply the following identity:[2]

$$(-x_0) + 1 - 1 = (1 - x_0) - 1 = x_0' - 1 \qquad \textbf{3.4}$$

which transforms a signed operand as follows:

$$x_0. \quad x_1 \quad x_2 \quad x_3 \quad \cdots \quad x_n$$

is replaced by

$$\begin{array}{cccccc} x_0'. & x_1 & x_2 & x_3 & \cdots & x_n \\ -1 & & & & & \end{array}$$

The resulting bit-array is shown in Figure 3.2(a). Now we can add the array of -1s. Since we placed the integer point after the sign bits, the value of this array of -1s is $-m$, which is represented by $yyy \ldots y$ in Figure 3.2(a). This bit-vector can be combined with the last row so that the total number of rows in the array remains m. An example is shown for $m = 5$ in Figure 3.2(b). In this case, $yyyy = 1011$. Calling e_0 the sign bit of the fifth operand, we get

$$\begin{array}{ccc} & e_0 = 0 & e_0 = 1 \\ 1011 + e_0' & 1100 & 1011 \end{array}$$

and both cases are included in the bit-vector $1e_0'e_0e_0$.

3.2 Reduction

The inputs in multioperand addition are bit-vectors forming a bit-array. The primitive operation performed on the input bit-array is a *reduction*, which produces an output bit-array with a smaller number of bits, by adding the input bits.

Two main reduction approaches are used: *reduction by rows* and *reduction by columns*. The modules used for reduction by rows are called adders, and those used for reduction by columns are called counters. We now discuss these modules and then use them for multioperand addition.

---

2. Note the use of the bit inversion operation (denoted by x_0') in an arithmetic expression. This should be interpreted in the intuitive way as converting a value 1 to 0 and vice versa.

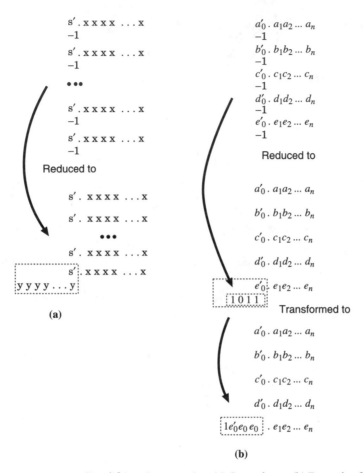

FIGURE 3.2 Simplifying sign extension: (a) General case. (b) Example of simplifying array for $m = 5$.

3.2.1   [p:2] Adders for Reduction by Rows

The adders in the reduction by rows can be either carry-propagate adders, which produce the output in conventional representation, or redundant adders, with redundant output either in carry-save or signed-digit form. Since the redundant adders have a smaller delay because of the limited carry propagation, we consider only the latter. In Chapter 2 we considered two-operand redundant adders.

In fact, the adder with one operand in carry-save and one operand in conventional representation can be used to add three operands in conventional representation and produces a result as the sum of two vectors. Therefore, it performs a 3-to-2 reduction and is called a [3:2] adder. Similarly, the carry-save adder for two carry-save operands is a [4:2] adder. We now generalize this to a [p:2] adder.

A [p:2] adder reduces p bit-vectors to 2 bit-vectors, as shown in Figure 3.3(a). The implementation consists of modules that have p rows of k bits as input and produce two rows of k bits as output. To achieve this the module also produces

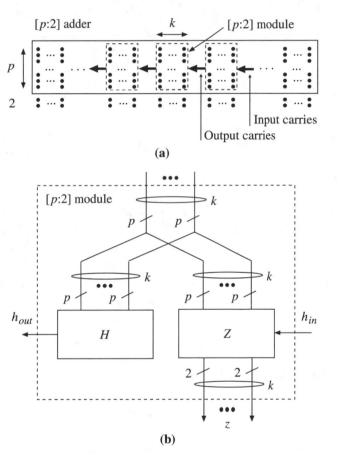

(a)

(b)

FIGURE 3.3 A [p:2] adder: (a) Input-output bit-matrix. (b) k-column [p:2] module decomposition.

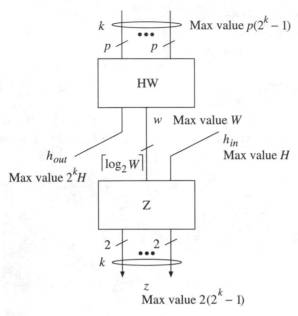

FIGURE 3.4 A model of a [p:2] module.

carries, which are added in the next module. Consequently, a module also has carries as input. In order to have a limited carry propagation, the output carries of a module should not depend on the input carries of that module (Figure 3.3(b)).

The complexity and delay of the module is determined by the number of columns k. Consequently, the number of columns of a group should be minimized. To determine the minimum number of columns k we consider the model of Figure 3.4, consisting of the following two modules:

- Module HW, which computes the output carries h_{out} and an intermediate sum w only in terms of the inputs
- Module Z, which adds w and the incoming carries h_{in} to produce the output z

Since the output signals (including the output carry) have to be able to represent at least the maximum input (including the carry), if we call H the maximum value of the carry (both in and out), we get

$$p(2^k - 1) + H \leq 2^k H + 2(2^k - 1) \qquad \textbf{3.5}$$

so that

$$H \geq p - 2 \qquad\qquad \textbf{3.6}$$

Moreover, calling W the maximum value of w, the following three conditions have to be satisfied:

1. Considering module HW, the outputs have to be able to represent the input

$$p(2^k - 1) \leq W + 2^k H \qquad\qquad \textbf{3.7}$$

which results in

$$W \geq 2^k(p - H) - p \qquad\qquad \textbf{3.8}$$

2. In the same module, since w is the residual after subtracting the carry-out (which has weight 2^k),

$$W \geq 2^k - 1 \qquad\qquad \textbf{3.9}$$

3. For module Z,

$$W + H \leq 2(2^k - 1) \qquad\qquad \textbf{3.10}$$

which results in

$$W \leq 2(2^k) - (H + 2) \qquad\qquad \textbf{3.11}$$

These three conditions are summarized in

$$\max(2^k - 1, 2^k(p - H) - p) \leq W \leq 2(2^k) - (H + 2) \qquad\qquad \textbf{3.12}$$

If we now use the minimum value of H (to minimize the number of carries), that is, $H = p - 2$, this is reduced to

$$\max(2^k - 1, 2(2^k) - p) \leq W \leq 2(2^k) - p \qquad\qquad \textbf{3.13}$$

Consequently,

$$2(2^k) - p \geq 2^k - 1 \qquad\qquad \textbf{3.14}$$

resulting in

$$2^k \geq p - 1 \qquad\qquad \textbf{3.15}$$

So, for example, $p = 4$ results in $H = 2, k = 2$, and $W = 4$ (from (3.13)).

Table 3.1 gives the values of H, k, and W for typical modules. Notice that as the number of bits p increases, so do also H, k, and W, resulting in a more complex module with a larger delay.

p	H	k	W
3	1	1	1
4	2	2	4
5	3	2	3
6	4	3	10
7	5	3	9
9	7	3	7
11	9	4	21

TABLE 3.1 Values of H, k, and W for typical modules.

An implementation requires the coding of the variables h and w. Then the modules can be implemented using gate networks. Figure 3.5 shows the implementation of a [4:2] module. In this case, $h = h_{i,1} + h_{i,2}$ (unary code) and $w = 2c + 2b + a$, as shown in the figure.

Another possible implementation uses a network of full-adders. Such an implementation for a [4:2] module is given in Chapter 2; Figure 3.6 illustrates this implementation for [5:2] and [7:2] modules. The internal carries are coded in a unary code, and w is represented by the signals with a dot.

3.2.2 $(p:q]$ Counters for Reduction by Columns

The reduction by columns is done by modules that add a column of p bits of the same weight and produce q bits of adjacent weights. That is,

$$\sum_{i=0}^{p-1} x_i = \sum_{j=0}^{q-1} y_j 2^j \qquad\qquad \textbf{3.16}$$

Consequently, the relation between p and q is

$$2^q - 1 \geq p \qquad\qquad \textbf{3.17}$$

that is,

$$q = \lceil \log_2(p + 1) \rceil$$

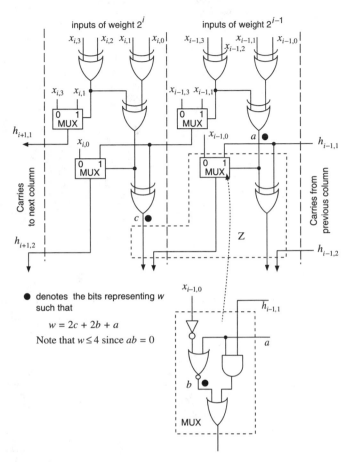

FIGURE 3.5 Gate network implementation of [4:2] module.

The module to perform this reduction is called a $(p:q]$ counter. Typical examples of these counters are $(3:2]$, $(7:3]$, and $(15:4]$. A representation of a $(p:q]$ counter is given in Figure 3.7.

Implementation of $(p:q]$ Counters

A $(p:q]$ counter is a module with p inputs and q outputs. It can be implemented by a $2^p \times q$ ROM, a network of full-adders, or a specialized gate network. While flexible, ROMs are relatively slow; thus the other two approaches are usually preferred.

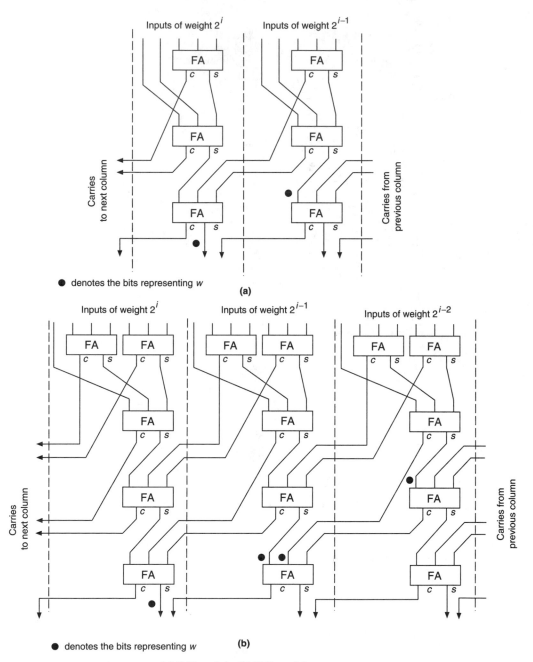

FIGURE 3.6 (a) [5:2] module. (b) [7:2] module.

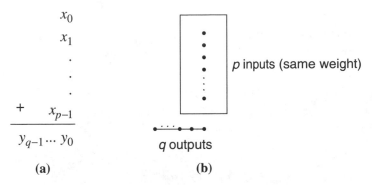

FIGURE 3.7 (a) $(p{:}q]$ reduction. (b) Counter representation.

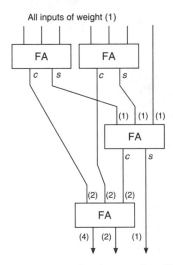

FIGURE 3.8 Implementation of $(7{:}3]$ counter by an array of full-adders.

The network of full-adders systematically uses the 3-to-2 reduction property of the full-adder. Consequently, a $(3{:}2]$ counter is implemented by one full-adder and a $(7{:}3]$ counter by a network of four full-adders, as shown in Figure 3.8.

The delay of the implementation of a $(p{:}q]$ counter with full-adders can be reduced by making use of the different delays for different input-output pairs. A systematic approach to do this might be to use half-adders as basic building blocks.

Specialized gate networks are suitable for intermediate values of p, such as 7 or 15, but become too complicated for larger p. A gate network implementation of a (7:3] counter is described in the following example.

EXAMPLE 3.1   We derive expressions and show a gate network for a (7:3] counter. The inputs are the seven binary variables $X = (x_6, x_5, x_4, x_3, x_2, x_1, x_0)$, and the output is

$$q = \sum_{i=0}^{6} x_i = 4q_2 + 2q_1 + q_0 \qquad \text{3.18}$$

We partition the input vector X into two subvectors $X_B = (x_6, x_5, x_4, x_3)$ and $X_A = (x_2, x_1, x_0)$. The partial sums corresponding to the subvectors are

$$q_A = \sum_{i=0}^{2} x_i = 2q_{A1} + q_{A0}$$

$$\qquad \text{3.19}$$

$$q_B = \sum_{i=3}^{6} x_i = 4q_{B2} + 2q_{B1} + q_{B0}$$

For the sum q_A the switching expressions, corresponding to the sum and the carry outputs of a full-adder, are

$$q_{A0} = x_2 \oplus x_1 \oplus x_0$$

$$\qquad \text{3.20}$$

$$q_{A1} = x_2 x_1 + x_2 x_0 + x_1 x_0$$

For the sum q_B we have

$$q_{B0} = x_6 \oplus x_5 \oplus x_4 \oplus x_3$$

$$q_{B1} = [x_6 x_5 + x_6 x_4 + x_6 x_3 + x_5 x_4 + x_5 x_3 + x_4 x_3] \cdot (x_6 x_5 x_4 x_3)'$$

$$= [x_6 x_5 + x_4 x_3 + (x_6 + x_5)(x_4 + x_3)] \cdot (x_6 x_5 x_4 x_3)' \qquad \text{3.21}$$

$$= a \cdot (x_6 x_5 x_4 x_3)' = a q'_{B2}$$

$$q_{B2} = x_6 x_5 x_4 x_3$$

Finally, $q = q_A + q_B$ so that

$$q_0 = q_{B0} \oplus q_{A0}$$

$$q_1 = (q_{B1} \oplus q_{A1}) \oplus (q_{B0}q_{A0})$$

$$q_2 = q_{B2} + q_{B1}q_{A1} + (q_{B1} \oplus q_{A1})(q_{B0}q_{A0})$$

$$= q_{B2} + a\,q_{A1} + (q_{B1} \oplus q_{A1})(q_{B0}q_{A0})$$

$\qquad\qquad$ **3.22**

An implementation after transforming the expressions to allow use of faster gates such as NANDs, NORs, AND-OR-INVERT, and OR-AND-INVERT is shown in Figure 3.9.

Comparison of the delay of the critical path and the cost of the network shown in Figure 3.9 with that of the network of full-adders shown in Figure 3.8 is left as Exercise 3.7. ∎

Multicolumn Counter

It is possible to generalize the counter concept to the reduction of several columns. That is, counter

$$(p_{k-1}, p_{k-2}, \ldots, p_0 : q]$$

$\qquad\qquad$ **3.23**

reduces k columns with p_i bits in column of weight 2^i into q output bits. If we call a_{ij} the bit of column i and row j and v the value represented by the q-bit output, we get

$$v = \sum_{i=0}^{k-1} \sum_{j=1}^{p_i} a_{ij} 2^i \le 2^q - 1$$

$\qquad\qquad$ **3.24**

An example is the (5, 5:4] counter shown in Figure 3.10(a). For this case,

$$v \le 5 \times 2 + 5 \times 1 = 15 = 2^4 - 1$$

$\qquad\qquad$ **3.25**

A second example is the (1, 2, 3:4] counter shown in Figure 3.10(b). In this case,

$$v \le 1 \times 4 + 2 \times 2 + 3 \times 1 = 11 < 2^4 - 1$$

$\qquad\qquad$ **3.26**

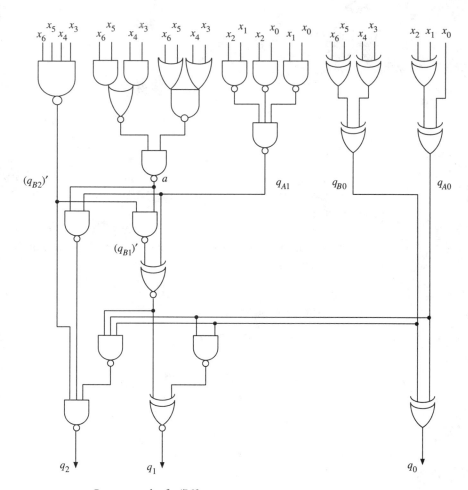

FIGURE 3.9 Gate network of a (7:3] counter.

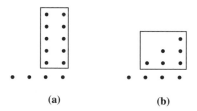

FIGURE 3.10 (a) (5, 5:4] counter. (b) (1, 2, 3:4] counter.

3.3 Sequential Implementation

This implementation consists of one adder and a register. To have the cycle time independent of precision and usually equal to the delay of a few full-adders, redundant adders are preferred. Using a $[p{:}2]$ adder each iteration adds $p - 2$ operands for a total of $\lceil m/(p - 2)\rceil$ iterations. The algorithm is

$$S[0] = 0$$
$$\textbf{for } i = 1 \textbf{ to } \lceil m/(p - 2)\rceil \textbf{ do:}$$
$$S[i] = S[i - 1] + \sum_{j=(i-1)(p-2)+1}^{i(p-2)} x(j);$$

and the result is $S[\lceil m/(p - 2)\rceil]$. This result is in carry-save representation. If the result is required in conventional representation, it is converted at the end using a CPA. Figure 3.11(b) shows an example using a $[p{:}2]$ adder. The case with $p = 3$ (carry-save adder) is shown in Figure 3.11(c).

3.3.1 Unsigned and Signed Operands

Depending on the type of operands (unsigned or signed in two's complement), the corresponding addition algorithm has to be used. In both cases, the range has to be extended by $\lceil \log_2 m\rceil$ bits to accommodate the range of the result. For two's complement, the extension is done as discussed in Section 3.1.

3.4 Combinational Implementation

The whole multioperand addition can be performed by a combinational network. Two alternatives exist: reduction by rows, performed by an array of adders, and reduction by columns, performed by an array of counters.

3.4.1 Reduction by Rows: Array of Adders

The organization of the array of adders can be classified into two extreme classes: linear array and tree array.

Linear Array

This corresponds to an unfolding of the sequential algorithm. If $[p{:}2]$ adders are used, for an addition of m operands the array consists of $\lceil (m - 2)/(p - 2)\rceil$ adders, since the first adder now receives p operands and the rest receive $p - 2$.

Cycle time dependent on precision

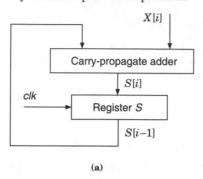

(a)

Cycle time not dependent on precision

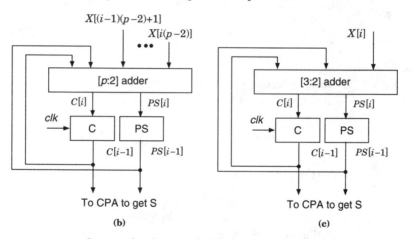

To CPA to get S   To CPA to get S

(b)   **(c)**

FIGURE 3.11 Sequential multioperand addition: (a) With conventional adder. (b) With $[p:2]$ adder. (c) With [3:2] adder.

This scheme is shown in Figure 3.12. The width of the adders increases to adapt to the width of the partial sum. As indicated before, the number of bits of the final sum is $n + p$, where $p = \lceil \log_2 m \rceil$. Moreover, for the two's complement case, the last adder has to include the additional extension, as explained in the previous section. The delay is equal to $\lceil (m - 2)/(p - 2) \rceil t_{[p:2]}$.

A comparison of linear arrays using carry-ripple adders and [3:2] adders is left as Exercise 3.19.

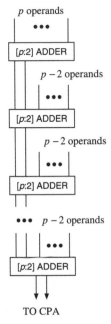

p operands

$\bullet\bullet\bullet$

[p:2] ADDER

$p - 2$ operands

$\bullet\bullet\bullet$

[p:2] ADDER

$p - 2$ operands

$\bullet\bullet\bullet$

[p:2] ADDER

$\bullet\bullet\bullet$ $p - 2$ operands

$\bullet\bullet\bullet$

[p:2] ADDER

TO CPA

FIGURE 3.12 Linear array of [p:2] adders for multioperand addition.

Adder Tree

Since addition is associative, it is possible to organize the array of adders as a tree which has fewer levels than the linear array. Again, the use of redundant adders is preferable for lower delay.

The number of adders required for the tree is the same as that for the linear array. This is shown by an argument on the total number of inputs to the adders and the use of these inputs to accept the operands and connect to the outputs of other adders. We now show this for the case of using [p:2] adders.

Calling k the number of adders, the total number of adder inputs is kp. These inputs are used for the m operands and for the $2(k - 1)$ adder outputs, since each adder has two outputs (and the outputs of one adder are used as the array output or the input to the conversion adder). That is,

$$pk = m + 2(k - 1) \qquad\qquad \textbf{3.27}$$

l	1	2	3	4	5	6	7	8	9
m_l	3	4	6	9	13	19	28	42	63

TABLE 3.2 [3:2] reduction sequence.

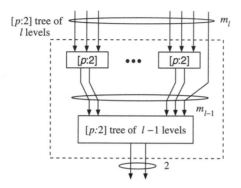

FIGURE 3.13 Construction of a $[p{:}2]$ carry-save adder tree.

and

$$k = \left\lceil \frac{m-2}{p-2} \right\rceil \quad [p{:}2] \text{ carry-save adders} \qquad \textbf{3.28}$$

Now we consider the number of adder levels. For this, we develop a recurrence for m_l, the number of adder operands that can be added with a tree of l levels. As shown in Figure 3.13, the m_{l-1} operands are divided in groups of two, so that each group corresponds to the outputs of one adder at level l. Since each adder has p inputs, we get

$$m_l = p \left\lfloor \frac{m_{l-1}}{2} \right\rfloor + m_{l-1} \bmod 2 \qquad \textbf{3.29}$$

where $m_1 = p$. For instance, for $p = 3$ the resulting sequence is shown in Table 3.2 and for $p = 4$ we get $m_l = 2^{l+1}$.

In particular, calling L the total number of levels to add m operands, this number of levels is obtained from the recurrence (3.29) by making $m_L = m$. For example, Figure 3.14 shows a [3:2] adder tree for $m = 9$, and Figure 3.15 a [4:2] adder tree for $m = 16$.

An approximation of the number of levels, which is good for large m_l, is obtained by considering that for all levels the number of operands is even.

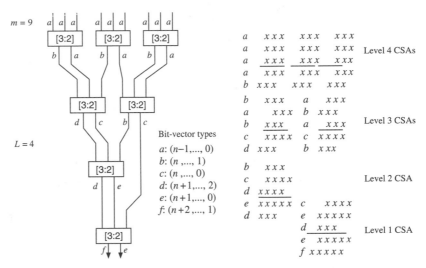

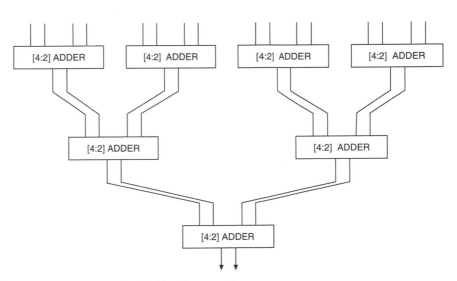

FIGURE 3.14 [3:2] adder tree for 9 operands (magnitudes with $n = 3$).

FIGURE 3.15 Tree of [4:2] adders for $m = 16$.

Since the best-case reduction per level is $p/2$, for l levels we get

$$m_l \approx \frac{p^l}{2^{l-1}} \qquad\qquad \textbf{3.30}$$

and

$$l \approx \log_{p/2}(m_l/2) \qquad\qquad \textbf{3.31}$$

Selecting the Value p

A larger p results in a smaller number of adder levels, in both the linear array and in the tree. However, the delay and complexity of the adder increase. Moreover, the connections between levels are more regular for values of p that are powers of two, as illustrated in Figure 3.15. Consequently, the best value of p depends on the requirements for the multioperand adder.

As an example, we compare the use of trees of [3:2] and [4:2] adders for a multioperand addition for $m = 16$, with the reduction of delay as the main requirement. In general, calling $T_{[4:2]}$ and $T_{[3:2]}$ the delays using [4:2] and [3:2] adders, respectively, we get

$$T_{[4:2]} < T_{[3:2]} \quad \text{if } L_{[4:2]}\, t_{[4:2]} < L_{[3:2]}\, t_{[3:2]} \qquad\qquad \textbf{3.32}$$

where $t_{[4:2]}$ and $t_{[3:2]}$ are the delays of the corresponding adders.

For $m = 16$ the number of levels of [4:2] adders is three, whereas for [3:2] adders it is six. Consequently, the [4:2] case has a smaller delay if $T_{[4:2]} < 2T_{[3:2]}$.

3.4.2 Reduction by Columns with $(p{:}q]$ Counters

In this method the bit array is reduced by using several levels of $(p{:}q]$ counters. As discussed before, a $(p{:}q]$ counter reduces a column of p bits to a bit-vector of q bits. Since this is done to every column of the bit matrix, the result is the reduction of a matrix of p rows into a matrix of q rows. This is illustrated in Figure 3.16 for a (7:3] counter.

Number of Counter Levels

To reduce the whole bit-array to two rows, it might be necessary to use several levels of counters. If only $(p{:}q]$ counters are used, then the whole array of m rows can be reduced to q rows by L levels of counters. The number of levels

$$
\begin{array}{cccc}
1 & 0 & 1 & 1 \\
0 & 0 & 1 & 0 \\
1 & 0 & 0 & 1 \\
0 & 1 & 1 & 0 \\
1 & 0 & 1 & 0 \\
1 & 1 & 1 & 1 \\
0 & 1 & 1 & 0 \\
\hline
0 & 1 & 0 & 1 \\
\end{array}
$$

$$
\begin{array}{ccc}
0 & 1 & 1 & 1
\end{array}
$$

$$
\begin{array}{cccc}
1 & 0 & 1 & 0
\end{array}
$$

FIGURE 3.16 Example of reduction using (7:3] counters.

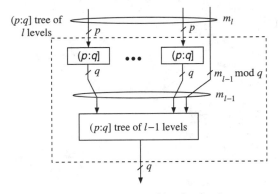

FIGURE 3.17 Construction of $(p{:}q]$ reduction tree.

required is determined in a similar way as discussed before for the $[p{:}2]$ adder. As before, calling m_l the number of bits in a column that can be reduced using l levels, we obtain

$$
m_1 = p
$$

$$
m_l = p \left\lfloor \frac{m_{l-1}}{q} \right\rfloor + m_{l-1} \bmod q \tag{3.33}
$$

As shown in Figure 3.17, this is obtained by grouping the m_{l-1} operands into groups of q operands and then using a $(p{:}q]$ counter for each group. For large m_l this results in

$$
l \approx \log_{p/q}(m_l/q) \tag{3.34}
$$

Number of Levels	1	2	3	4	...
Maximum number of rows	7	15	35	79	...

TABLE 3.3 Sequence for (7:3] counters.

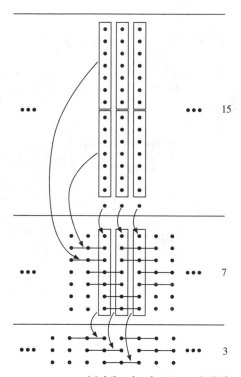

FIGURE 3.18 Multilevel reduction with (7:3] counters.

For instance, the sequence for (3:2] counters is the same as for [3:2] adders, given in Table 3.2, and for (7:3] counters the sequence is shown in Table 3.3.

Figure 3.18 shows a multilevel reduction using (7:3] counters.

Systematic Design Method

If the number of bits in each column of the array is the same, the method of reduction by columns is similar to the reduction by rows. However, this is not always the case; in particular it is not the case in multiplication (discussed in

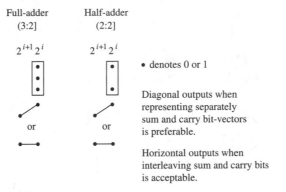

Full-adder
(3:2]

Half-adder
(2:2]

$2^{i+1}\ 2^i$ $2^{i+1}\ 2^i$

• denotes 0 or 1

Diagonal outputs when
representing separately
sum and carry bit-vectors
is preferable.

or   or

Horizontal outputs when
interleaving sum and carry bits
is acceptable.

FIGURE 3.19 Full-adder and half-adder as (3:2] and (2:2] counters.

the next chapter) and for the lower levels in a multilevel reduction. In the latter case, this is because the least-significant columns do not receive bits from the reduction of other columns of lower weight. As a consequence of this, the number of counters required for those columns is smaller. We now discuss a systematic design approach that uses the minimum number of counters. The basic idea is to place the counters in such a way that the reduction at level l of the reduction produces columns of m_{l-1} bits.

Although the method can be used for any counter, we present it for (3:2] counters (full-adders). Moreover, we show that in some places it is advantageous to use also (2:2] counters (half-adders), instead of the more complex full-adders. The notation for using these counters is described in Figure 3.19. For this case the reduction sequence is as shown in Table 3.2. Therefore, the optimal reduction sequence is (3, 4, 6, 9, 13, 19, 28, 42, ...).

The first step in the process is to determine the number of levels and the corresponding reduction sequence. For instance, for $m = 35$ operands the number of levels is $L = 8$ and the sequence is 35, 28, 19, 13, 9, 6, 4, 3.

Since, because of the carries, there might be a different number of bits in each column, the reduction is performed separately for each column. Consider that at a level l in the reduction process column i has e_i bits, and in the reduction sequence the next reduction corresponds to m_{l-1} bits. These m_{l-1} bits are formed by the sum outputs of the full-adders and half-adders of column i plus the carries produced by the adders of column $i - 1$, plus the bits of column i that are not reduced but transferred to the next level. This is illustrated in Figure 3.20.

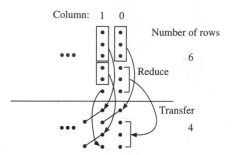

Column: 1   0

Number of rows

6

Reduce

Transfer

4

FIGURE 3.20 Reduction process.

Since a full-adder uses as inputs three bits in column i and produces one bit in that column, the reduction per full-adder is two bits. Similarly, the reduction per half-adder is one bit. Consequently, calling f_i and h_i the number of full-adders and half-adders of column i, respectively, we have the following relation at level l:

$$e_i - 2f_i - h_i + f_{i-1} + h_{i-1} = m_{l-1} \qquad \textbf{3.35}$$

resulting in

$$2f_i + h_i = e_i - m_{l-1} + f_{i-1} + h_{i-1} = p_i \qquad \textbf{3.36}$$

As can be seen from this expression, the determination of f_i and h_i is a sequential process, with the initial conditions $f_{-1} = h_{-1} = 0$. For instance if $e_0 = 19$, then $m_{l-1} = 13$ (next level in the reduction sequence) and $2f_0 + h_0 = p_0 = 6$.

Expression (3.36) is used to determine the number of full-adders and half-adders. Clearly, the solution that produces the minimum number of carries to the next column is

$$f_i = \lfloor p_i/2 \rfloor \qquad h_i = p_i \bmod 2 \qquad \textbf{3.37}$$

In the example above $f_0 = 3, h_0 = 0$.

This reduction process is described by a table, as illustrated in the following example.

EXAMPLE 3.2   The reduction by columns for $m = 8$ magnitudes of $n = 5$ bits is shown in Table 3.4.

		i					
	6	5	4	3	2	1	0
$l = 4$							
e_i			8	8	8	8	8
m_3			6	6	6	6	6
h_i			0	0	0	1	0
f_i			2	2	2	1	1
$l = 3$							
e_i		2	6	6	6	6	6
m_2		4	4	4	4	4	4
h_i		0	0	0	0	1	0
f_i		0	2	2	2	1	1
$l = 2$							
e_i		4	4	4	4	4	4
m_1		3	3	3	3	3	3
h_i		0	0	0	0	0	1
f_i		1	1	1	1	1	0
$l = 1$							
e_i	1	3	3	3	3	3	3
m_0	2	2	2	2	2	2	2
h_i	0	0	0	0	0	0	1
f_i	0	1	1	1	1	1	0

TABLE 3.4 Example of reduction process.

The resulting array of full-adders and half-adders is shown in Figure 3.21. It has 26 full-adders and 4 half-adders. For the final 2-to-1 reduction, a 7-bit CPA is needed.

The delay in the critical path is roughly

$$T = t_{csa.tree} + t_{CPA}$$
$$= 4t_{fa} + t_{CPA(7)} \qquad \textbf{3.38}$$

∎

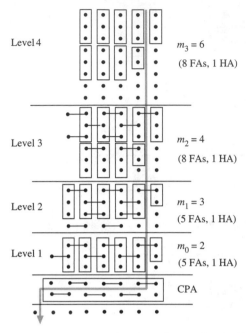

Level 4 $m_3 = 6$
           (8 FAs, 1 HA)

Level 3 $m_2 = 4$
           (8 FAs, 1 HA)

Level 2 $m_1 = 3$
           (5 FAs, 1 HA)

Level 1 $m_0 = 2$
           (5 FAs, 1 HA)

CPA

FIGURE 3.21 Reduction by columns of eight 5-bit magnitudes. Cost of reduction: 26 FAs and 4 HAs.

The total delay of the scheme consists of the delay of the reduction array and the delay of the final CPA. Since the delay of the CPA depends on the number of bits, a more aggressive reduction might be applied to reduce the precision of the final adder at the expense of additional counters (Exercise 3 . 22).

Example with Nonrectangular Array and Operands with Specific Relations

The discussion up to now only considers rectangular bit-arrays and does not take into account special relations among the operands. We here present a first example in which the operands are related, resulting in a nonrectangular array; in the next chapter we discuss the important case of multiplication.

EXAMPLE 3.3   Design an array of full-adders and half-adders to compute

$$f = a + 3b + 3c + d$$

where the operands a, b, c, d are integers in the range -4 to 3, in two's complement representation.

We first determine the range of f. Since each operand is in the range -4 to $+3$, we obtain

$$-4 + (-12) + (-12) - 4 = -32 \le f \le 3 + 9 + 9 + 3 = 24$$

Consequently f requires 6 bits. To perform the operation as an array of adders, we decompose $3b$ and $3c$ into $2b + b$ and $2c + c$, respectively.

We construct the bit-matrix, extended to the left to preserve the sign:

a	a_2	a_2	a_2	a_2	a_1	a_0
b	b_2	b_2	b_2	b_2	b_1	b_0
$2b$	b_2	b_2	b_2	b_1	b_0	0
c	c_2	c_2	c_2	c_2	c_1	c_0
$2c$	c_2	c_2	c_2	c_1	c_0	0
d	d_2	d_2	d_2	d_2	d_1	d_0

which can be transformed into

a		a_2'	a_1	a_0
		-1		
b		b_2'	b_1	b_0
		-1		
$2b$	b_2'	b_1	b_0	
	-1			
c		c_2'	c_1	c_0
		-1		
$2c$	c_2'	c_1	c_0	
	-1			
d		d_2'	d_1	d_0
		-1		

and, finally, reduced to the following bit-matrix by noting that the sum of (-1) entries is -8×2^2, represented in two's complement by 100000.

$$
\begin{array}{cccccc}
1 & 0 & b_2' & a_2' & a_1 & a_0 \\
  &   & c_2' & b_2' & b_1 & b_0 \\
  &   &   & b_1 & b_0 &   \\
  &   &   & c_2' & c_1 & c_0 \\
  &   &   &   & c_1 & c_0 \\
  &   &   & d_2' & d_1 & d_0 \\
\end{array}
$$

The resulting bit-matrix is reduced to two rows by an array of full- and half-adders. The method described before produces Table 3.5.

	\(i\)					
	5	4	3	2	1	0
$l = 3$						
e_i	1	0	2	6	6	4
m_2	4	4	4	4	4	4
h_i	0	0	0	1	0	0
f_i	0	0	0	1	1	0
$l = 2$						
e_i	1	0	4	4	4	4
m_1	3	3	3	3	3	3
h_i	0	0	0	0	0	1
f_i	0	0	1	1	1	0
$l = 1$						
e_i	1	1	3	3	3	3
m_0	2	2	2	2	2	1*
h_i	0	0	0	0	0	0
f_i	0	0	1	1	1	1

* To reduce by one bit the width of the CPA.

TABLE 3.5 Reduction for Example 3.3.

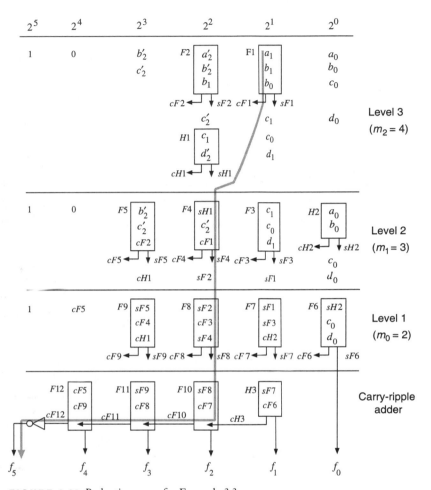

FIGURE 3.22 Reduction array for Example 3.3.

The corresponding network of adders is shown in Figure 3.22. The final stage consists of a carry-propagate adder of only four bits because the last output bit is produced in the previous level (in the figure a carry-ripple adder is used). The most significant bit of the result is $f_5 = 1 \oplus cF12 = (cF12)'$ so that an inverter can be used instead of a half-adder.

The delay of the array is roughly

$$T = 3t_{fa} + t_{CPA}(3)$$

and the number of modules is

$$N = 12FA + 3HA + 1INV$$ ■

3.4.3   Pipelined Adder Arrays

The adder arrays can be pipelined to increase the throughput for the case of many (independent) multioperand additions. This is done by defining stages and separating them by latches as illustrated in Figure 3.23. The stage delay determines

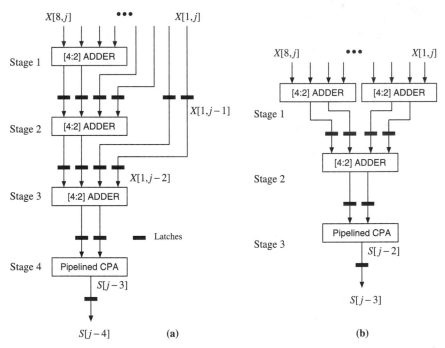

FIGURE 3.23 Pipelined arrays with [4:2] adders for computing $S[j] = \sum_{i=1}^{8} X[i, j]$, $j = 1, \ldots, N$: (a) Linear array. (b) Tree array.

the throughput R. That is,

$$R = \frac{1}{t_{stage}} \qquad \qquad \textbf{3.39}$$

The delay of the operation (also called the *latency*) corresponds to the sum of the delays of the stages. Consequently, the organization of the adders as a tree reduces the latency.

If a conventional output is required, a conversion is needed. The conversion should be pipelined into stages of the same delay as the addition stages.

3.5   Partially Combinational Implementation

The combinational implementation is faster than the sequential because of the following two reasons:

- In the combinational case it is possible to organize the adders in a tree structure and to organize this structure so as to reduce the critical path.
- In the sequential case, the delay of each cycle has to include the delay of loading the partial result in registers.

On the other hand, the combinational implementation requires a larger area. As a compromise, a mixed implementation can be used in which a set of k operands are added per iteration, so that m/k iterations are required. Figure 3.24(a) shows the case in which an adder tree is used to add four operands per iteration. This

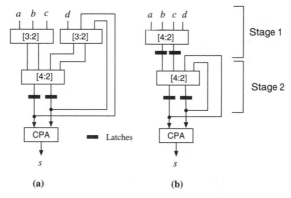

FIGURE 3.24 Partially combinational scheme for summation of four operands per iteration: (a) Nonpipelined. (b) Pipelined.

requires a tree that adds six operands; we show the case in which the first level uses [3:2] adders and the second a [4:2] adder.

The implementation can be pipelined for faster addition of m operands. A problem exists with the accumulation of the partial sum. If it is added at the top of the tree as indicated in Figure 3.24(a), pipelining is not possible. However, the partial sum can be added after the tree, as shown in Figure 3.24(b), resulting in an implementation with two levels of [4:2] adders. Although the nonpipelined implementation could also be done with two levels of [4:2] adders, this would increase the delay.

A generalized network of [3:2] adders for reduction of q inputs at a time followed by accumulation is illustrated in Figure 3.25. Of course, adders with a higher reduction ratio, such as [4:2] adders, can also be used.

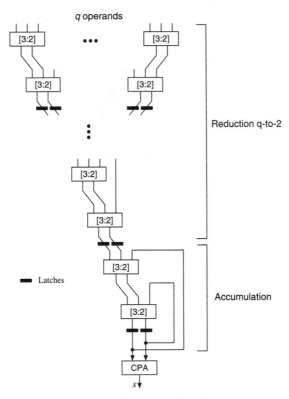

FIGURE 3.25 Scheme for summation of q operands per iteration.

3.6 Exercises

Bit-Arrays for Two's Complement

3.1 Determine the bit-matrix (as in Figure 3.2) for $m = 7$.

Reduction by [p:2] Modules

3.2 (a) For the [4:2] module of Figure 3.5, show that the value of $w \leq 4$. Show the input values for which $w = 4$.

(b) Since the network for both columns is the same, why is it necessary to consider two columns as the basic module when using the model of Figure 3.4?

(c) Show that the module is a [4:2] module.

(d) Compare the implementations of Figures 3.5 and 2.41, in terms of delay and number of equivalent gates.

3.3 Show a linear array of [5:2] modules, implemented as in Figure 3.6(a), to reduce five 8-bit operands to two bit-vectors. Determine the critical path. Is there a carry-propagation chain? Compare with an array of [3:2] adders (carry-save adders).

3.4 Design a [6:2] module with full-adders and determine the critical path.

3.5 Design a [9:2] module with full-adders and determine the critical path.

3.6 Using suitable CAD tools, synthesize [p:2] modules for $p = 4$ and 7 at the gate level and compare with implementations using full-adder modules.

Reduction by (p:q] Counters

3.7 (a) Determine the delay (average) of the critical path of the gate network implementing the (7:3) counter shown in Figure 3.9 using Table 2.4.

(b) Determine the cost of the gate network in (a) in equivalent gates.

(c) Determine the delay (average) of the critical path of the network of FAs implementing the (7:3) counter shown in Figure 3.8 using Table 2.2.

(d) Determine the cost of the network in (c) in equivalent gates.

(e) Compare and discuss your findings in (a), (b), (c), and (d).

3.8   Show a network of full-adders implementing a (15:4] counter.

3.9   Determine how many levels of (15:4] counters are necessary to add 127 operands.

Sequential Implementation

3.10   Show a design of the sequential multioperand addition scheme with the [3:2] adder of Figure 3.11(c) at the binary level using full-adders and registers. The operands $X[i]$ are in the range [0, 127], and the maximum number of operands is 32. The CPA adder is a carry-ripple adder. On the logic diagram indicate all modules used and the precision in bits of all connections. Using a reasonable delay model, estimate the delay in the critical path. If a CRA is used instead of the carry-save adder, what is the increase in the delay in the critical path? Discuss change in cost.

3.11   Repeat Exercise 3.10 for the operands $X[i]$ in the range $[-31, 31]$.

Linear Arrays and Tree of Adders for Reduction by Rows

3.12   Draw the linear arrays as in Figures 3.12 with $p = 3$ and $p = 4$ for $m = 7$ (magnitudes).

3.13   (a)   Design a network consisting of [5:2] and [4:2] adders to reduce 10 4-bit operands to two operands.
          (b)   Compare the network obtained in (a) with a [3:2] adder array in number of full-adders and delay.

3.14   Estimate the delay of a linear array of adders to produce the sum of eight positive integers in the range [0, 255]. Make reasonable assumptions on the delay model in each case (see Chapter 2). The adders are of the following types:
          (a)   Carry-ripple adder (CRA)
          (b)   Single-level carry-skip adder with a fixed group size of 4 (CSK4)
          (c)   Parallel prefix adder with minimum number of levels and fanout of two
          (d)   Carry-select adder
          (e)   [4:2] adder followed by a parallel prefix adder.

3.15 Design a linear array of [3:2] carry-save adders for $m = 8$ and $n = 6$ (two's complement) using full-adders and half-adders. The CPA is of a carry-ripple type. Estimate the delay in the critical path using a reasonable delay model. What percentage of the total delay is in the CPA? Using other modules as needed, design a faster CPA and discuss its effect on the overall design.

3.16 (a) Show a bit-level design of a tree of carry-save adders followed by a CPA to add $m = 6$ operands with $n = 4$ bits each in the two's complement form. Determine the precision of each carry-save adder and the final CPA so that the correct sign and the range of the result is obtained. Label all modules and interconnections.

(b) Estimate the delay in the critical path using a reasonable delay model of the modules.

(c) Give the values on all input and output lines for the following set of input operands:

$$a = 1001$$
$$b = 0010$$
$$c = 1110$$
$$d = 0101$$
$$e = 0011$$
$$f = 1010$$

3.17 Two schemes are considered for reducing four n-bit operands to one. Scheme A uses three carry-propagate adders. Scheme B uses a [4:2] adder and a CPA. Determine under which conditions scheme B is not faster than scheme A.

3.18 Add the following set of integers using [4:2] adders and a carry-propagate adder: $+73, -52, +22, -127, -31, +17, +47, -80$. Use two's complement representation.

3.19 Bit-level linear arrays with carry-ripple adders and carry-save adders are shown in Figure 3.26.

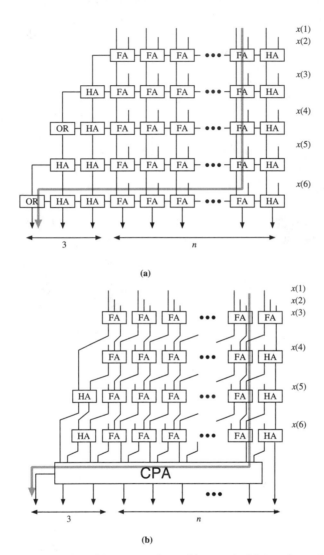

FIGURE 3.26 Linear array for multioperand addition of magnitudes: (a) With carry-ripple adders. (b) With [3:2] adders.

(a)   Determine the delay in the critical path for each scheme.

(b)   Under what circumstances is the scheme with [3:2] adders faster than the scheme with CRAs?

Reduction by Columns

3.20   Modify Figure 3.21 for two's complement representation.

3.21   Show a table of the reduction by columns process for $m = 9$ and $n = 6$ for magnitudes and for two's complement.

3.22   Compare the arrays in Figure 3.21 and in Figure 3.27 with respect to

(a)   the precision of the CPAs

(b)   the delay in the critical path if CRAs are used as final-adder

(c)   the number of FAs and HAs

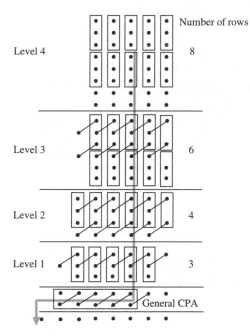

FIGURE 3.27   Reduction by row scheme for adding eight 5-bit magnitudes.

3.23   Determine the critical path in each of the schemes shown in Figures 3.21 and 3.27 using the following delay models for FA and HA modules:

$$t_{FA}(a, b \rightarrow s) = 4\tau$$
$$t_{FA}(c_{in} \rightarrow s) = 2\tau$$
$$t_{FA}(a, b \rightarrow c_{out}) = 3\tau$$
$$t_{FA}(c_{in} \rightarrow c_{out}) = 2\tau$$
$$t_{HA}(a, b \rightarrow s) = 2\tau$$
$$t_{HA}(a, b \rightarrow c_{out}) = 1\tau$$

where $\tau = t_{2-NAND}$.

3.24   Design a network consisting of full-adders and half-adders to compute

$$z = a - 3b + 5c$$

where a, b, c are integers in the range $[-4, 3]$, represented in the two's complement system.

(a)   What is the least number of bits necessary to represent z?
(b)   Show the bit-matrix before and after simplification.
(c)   Show your final network. Minimize the delay and the number of FA/HA modules in the reduction to two operands.
(d)   What is the minimum precision of the carry-propagate adder needed to produce the final result? Which type of CPA would be best suited?

3.25   Design a network using reduction by columns to compute

$$G = |(P0 + 2 \times P1 + P2) - (P6 + 2 \times P4 + P5)| + |(P2 + 2 \times P3 + P4)$$
$$- (P0 + 2 \times P7 + P6)|$$

where the inputs Pi are positive integers in the range $[0, 255]$. If there is an overflow, the output is set to 255 (saturated). This function, known as the *Sobel filter*, is used in processing grayscale images consisting of 8-bit pixels.

Pipelined and Partially Combinational Implementations

3.26   Design a pipelined linear array for addition of eight operands in the range $[0, 63]$ similar to the scheme shown in Figure 3.23(a).

3.27 Show a pipelined implementation as in Figure 3.25 for six operands per iteration. Determine the time required to add 24 operands and compare with a nonpipelined scheme.

3.7 Further Readings

The literature on the basic idea of carry-save addition with [3:2] and [4:2] adders, used in multioperand addition, is discussed in Chapter 2. Multioperand addition schemes for magnitudes and signed operands, [p:2] modules, and column counters are frequently discussed in the literature on multiplication (Chapter 4).

A simplification of the sign extension in multioperand addition of operands in two's complement form, equivalent to the approach discussed in this chapter, is presented in Agrawal and Rao (1978).

Reduction by Rows

Early schemes for reduction by rows using [3:2] (carry-save) adders are described in MacSorley (1961), Bucholz (1962), and in Wallace (1964). The former scheme uses a tree of carry-save adders to add six summands per iteration in a radix-8 sequential multiplier, while in the latter, frequently referred to as the Wallace tree, all summands are applied in parallel. A [4:2] adder with carry-save representation was discussed in Weinberger (1981). It is generalized in Lim (1978) to [p:2] adders, which are also called parallel compressors in Gajski (1980).

Reduction by Columns

A scheme for reduction by columns and the concept of parallel counters were introduced in Dadda (1965, 1976). In particular, a reduction scheme using a full-adder as a (3:2] and a half-adder as a (2:2] counter was developed. This method has been frequently used in reduction arrays in multipliers called Dadda multipliers. In Stenzel et al. (1977) reduction by columns using (p:q] counters is discussed. A tree of full-adders proposed in Foster and Stockton (1971) implements (p:q] counters. Implementation of parallel counters with partially analog counters is discussed in Swartzlander (1973). An implementation of (p:q] counters using several tables and threshold switching functions is developed in Ho and Chen (1973). Another approach for implementing (p:q] counters is presented in

Svoboda (1970): a column of p entries consisting of 1s and 0s is sorted so that a unique transition from 1s to 0s produces a signal indicating the number of 1s, which after encoding, produces q.

Generalized Parallel Counters

Generalized parallel counters and methods of synthesis of large counters from small ones are discussed in Meo (1975), Kobayashi and Ohara (1978), and Dormido and Canto (1981, 1982).

Implementation

A variety of circuit-level implementations have been developed. Gate networks with a minimal number of gates and interconnections for (3:2] counters (carry-save adders) are described in Lai and Muroga (1982). [p:2] modules at the transistor level for different p are developed in Song and Micheli (1991). Gate networks for [7:3] modules and (7:3) parallel counters are presented in Mehta et al. (1991). Implementations for [4:2] modules are presented, among others, in Nagamatsu et al. (1990), Kanie et al. (1994), and Makino et al. (1996) and for [5:2] modules in Kwon et al. (2000). A good discussion of VLSI cell designs for (3:2] and [4:2] modules is presented in Zimmerman (1998). Power-efficient design of [4:2] and [5:2] modules is discussed in Prasad and Parhi (2001).

Bounds on Delay

Bounds on delays and optimization techniques for networks of (3:2] and (2:2] counters are presented in Paterson and Zwick (1993). The complexity of multioperand addition is presented in Atkins and Ong (1979).

Miscellaneous Schemes

Multioperand addition with conditional-sum adders is considered in Efe (1981). Variations on multioperand addition using different digit sets are explored in Parhami (1996). Pipelined multioperand adders are described in Yeh and Parhami (1996).

3.8 Bibliography

Agrawal, D. P., and T. R. N. Rao (1978). On multiple operand addition of signed binary numbers. *IEEE Transactions on Computers*, C-27(11):1068–70.

Atkins, D. E., and S. Ong (1979). Time-component complexity of two approaches to multioperand binary addition. *IEEE Transactions on Computers*, C-28(12):918–26.

Bucholz, W. (1962). *Planning a New Computer System: Project STRETCH*, Chapter 14, p. 210. Wiley and Sons, Inc., New York.

Dadda, L. (1965). Some schemes for parallel multipliers. *Alta Frequenza*, 34:349–56.

Dadda, L. (1976). On parallel digital multipliers. *Alta Frequenza*, 45:574–80.

Dormido, S., and M. A. Canto (1981). Synthesis of generalized parallel counters. *IEEE Transactions on Computers*, C-30(9):699–703.

Dormido, S., and M. A. Canto (1982). An upper bound for the synthesis of generalized parallel counters. *IEEE Transactions on Computers*, C-31(8): 802–5.

Efe, K. (1981). Multi-operand addition with conditional sum logic. In *Proceedings of the 5th IEEE Symposium on Computer Arithmetic*, pages 251–55.

Foster, C. C., and F. D. Stockton (1971). Counting responders in an associative memory. *IEEE Transactions on Computers*, C-20:1580–83.

Gajski, D. D. (1980). Parallel compressors. *IEEE Transactions on Computers*, C-29(5):393–98.

Ho, I. T., and T. C. Chen (1973). Multiple addition by residue threshold functions and their representation by array logic. *IEEE Transactions on Computers*, C-22:762–67.

Kanie, Y., Y. Kubota, S. Toyoyama, Y. Iwase, and S. Suchimoto (1994). 4-2 compressor with complementary pass-transistor logic. *IEICE Transactions on Electronics*, E77-C(4):647–49.

Kobayashi, H., and H. Ohara (1978). A synthesizing method for large parallel counters with a network of smaller ones. *IEEE Transactions on Computers*, C-27(8):753–57.

Kwon, O., K. Nowka, and E. E. Swartzlander (2000). A 16-bit by 16-bit MAC design using fast 5:2 compressors. In *Proceedings of the IEEE International Conference on Application-Specific Systems, Architectures, and Processors*, pages 235–43.

Lai, H. C., and S. Muroga (1982). Logic networks of carry-save adders. *IEEE Transactions on Computers*, C-31:870–82.

Lim, R. S. (1978). High-speed multiplication and multiple summand addition. In *Proceedings of the 4th IEEE Symposium on Computer Arithmetic*, pages 149–53.

MacSorley, O. L. (1961). High-speed arithmetic in binary computers. *IRE Proceedings*, 49:67–91.

Makino, H., H. Suzuki, H. Morinaka, Y. Nakase, H. Shinohara, K. Mashiko, T. Sumi, and Y. Horiba (1996). A design of high-speed 4-2 compressor for fast multiplier. *IEICE Transactions on Electronics*, E79-C(4):538–48.

Mehta, M., V. Parmar, and E. E. Swartzlander (1991). High-speed multiplier design using multi-input counter and compressor circuits. In *Proceedings of the 10th IEEE Symposium on Computer Arithmetic*, pages 43–50.

Meo, A. R. (1975). Arithmetic networks and their minimization using a new line of elementary units. *IEEE Transactions on Computers*, C-24(3):258–80.

Nagamatsu, M., S. Tanaka, J. Mori, K. Hirano, T. Noguchi, and K. Hatanaka (1990). A 15-ns 32 × 32-b CMOS multiplier with an improved parallel structure. *IEEE Journal of Solid-State Circuits*, 25(2):494–97.

Parhami, B. (1996). Variations on multioperand addition for faster logarithmic-time tree multiplier. In *Proceedings of the 30th Asilomar Conference on Signals, Systems and Computers*, pages 899–903.

Paterson, M., and U. Zwick (1993). Shallow circuits and concise formulae for multiple addition and multiplication. *Computational Complexity*, 3(3): 262–91.

Prasad, K., and K. K. Parhi (2001). Low-power 4-2 and 5-2 compressors. In *Proceedings of the 35th Asilomar Conference on Signals, Systems and Computers*, pages 129–33.

Song, P. J., and G. D. Micheli (1991). Circuit and architecture trade-offs for high-speed multiplication. *IEEE Journal of Solid-State Circuits*, 26(9):1184–98.

Stenzel, W. J., W. J. Kubitz, and G. H. Garcia (1977). A compact high-speed parallel multiplication scheme. *IEEE Transactions on Computers*, C-26(10):948–57.

Svoboda, A. (1970). Adder with distributed control. *IEEE Transactions on Computers*, C-19(8):749–51.

Swartzlander, E. E. (1973). Parallel counters. *IEEE Transactions on Computers*, C-22:1021–24.

Wallace, C. S. (1964). A suggestion for a fast multiplier. *IEEE Transactions on Electronic Computers*, EC-13(2):14–17.

Weinberger, A. (1981). 4:2 carry-save adder module. *IBM Technical Disclosure Bulletin*, 23.

Yeh, C.-H., and B. Parhami (1996). Efficient pipelined multi-operand adders with high throughput and low latency: Designs and applications. In *Proceedings of the 30th Asilomar Conference on Signals, Systems and Computers*, pages 894–98.

Zimmerman, R. (1998). *Binary Adder Architectures for Cell-Based VLSI and Their Synthesis*. Ph.D. dissertation. Series in Microelectronics, Vol. 37. Hartung-Gore, Konstanz, Switzerland.

CHAPTER **4** | Multiplication

In this chapter we consider algorithms and implementations of multiplication of signed integers in constant-radix representation (sign-and-magnitude and two's complement). These units are used for fixed-point multiplication (applying appropriate scaling factors) and are part of a floating-point unit, as discussed in Chapter 8.

The multiplication operation is

$$p = x \times y \qquad\qquad\textbf{4.1}$$

where x (multiplicand), y (multiplier), and p (product) are signed integers. High-level descriptions of the algorithms for sign-and-magnitude and two's complement are as follows:

- **Sign-and-magnitude:** Each operand is represented by a sign, with value $+1$ and -1, and an n-digit magnitude, and the result by a sign and a $2n$-digit magnitude. The high-level algorithm is

$$sign(p) = sign(x) \cdot sign(y) \qquad\qquad\textbf{4.2}$$

$$|p| = |x| \cdot |y| \qquad\qquad\textbf{4.3}$$

The representations of the magnitudes are

$$X = (x_{n-1}, x_{n-2}, \ldots, x_0) \qquad |x| = \sum_{i=0}^{n-1} x_i r^i \quad \text{(multiplicand)}$$
$$(0 \le x \le r^n - 1)$$

$$Y = (y_{n-1}, y_{n-2}, \ldots, y_0) \qquad |y| = \sum_{i=0}^{n-1} y_i r^i \quad \text{(multiplier)}$$
$$(0 \le y \le r^n - 1)$$

$$P = (p_{2n-1}, p_{2n-2}, \ldots, p_0) \quad |p| = \sum_{i=0}^{2n-1} p_i r^i \quad \text{(product)}$$
$$(0 \le p \le r^{2n} - 2r^n + 1)$$

- **Two's complement:** We consider here only the radix-2 case. Each operand is represented by an n-bit vector, and the result by an $2n$-bit vector.

181

This $2n$-bit result is required because the range is

$$-(2^{n-1})(2^{n-1}-1) \le p \le (-2^{n-1})(-2^{n-1}) = 2^{2n-2} \qquad \textbf{4.4}$$

so that the most positive value is represented by a vector of $2n$ bits.

If x_R, y_R, and p_R are the corresponding positive integer representations of x, y, and p, respectively, the high-level algorithm is

$$p_R = \begin{cases} x_R y_R & \text{if } x \ge 0,\ y \ge 0 \\ 2^{2n} - (2^n - x_R)y_R & \text{if } x < 0,\ y \ge 0 \\ 2^{2n} - x_R(2^n - y_R) & \text{if } x \ge 0,\ y < 0 \\ (2^n - x_R)(2^n - y_R) & \text{if } x < 0,\ y < 0 \end{cases} \qquad \textbf{4.5}$$

As will be seen, this algorithm can be simplified when using the corresponding digit-vectors.

In the next sections two types of algorithms are considered:

1. Add-and-shift algorithm. For magnitudes, this algorithm is based on the following identity:

$$x \times y = \sum_{i=0}^{n-1} x y_i r^i \qquad \textbf{4.6}$$

which is implemented by digit-by-integer multiplications ($x y_i$), arithmetic shifts by i positions, and a multioperand addition. We consider the sequential and combinational variants as well as the adaptation to two's complement representation.

2. Composition of smaller multiplications.

4.1 Sequential Multiplication with Recoding

This basic algorithm was reviewed in Chapter 1. We here extend it to include the recoding of the multiplier and consider radix-4 pipelined and higher-radix implementations. We consider first the sign-and-magnitude representation and then introduce the modifications for the two's complement representation.

4.1.1   Sign-and-Magnitude

As reviewed in Chapter 1, the basic algorithm for magnitudes is

$$p[0] = 0$$
$$p[j+1] = r^{-1}(p[j] + (xr^n)y_j) \quad \text{for } j = 0, 1, \dots, n-1 \qquad \textbf{4.7}$$
$$p = p[n]$$

The relative position of operands in the recurrence is illustrated in Figure 4.1.

The execution takes n cycles, and each cycle corresponds to the delay of a digit multiplication (one digit of the multiplier times the whole multiplicand), the delay of addition, plus register delay. The delay of shift (constant by one digit position) is negligible since it is implemented by wiring. That is,

$$T = n(t_{digmult} + t_{add} + t_{reg}) \qquad \textbf{4.8}$$

This time is reduced if a redundant adder is used. As shown in Figure 4.2, this adder has one redundant operand and one conventional operand. If the result is required in conventional representation, it has to be converted. The conversion of the least-significant half of the product can be done during the shifting.

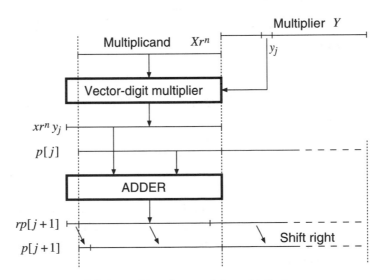

FIGURE 4.1 Relative position of operands in multiplication recurrence.

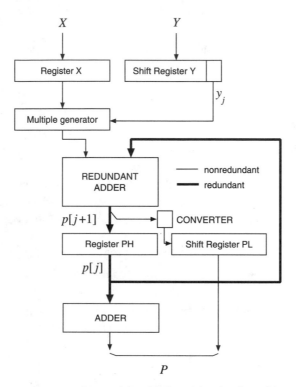

FIGURE 4.2 Sequential multiplier with redundant adder.

Radix 2 and Radix 4

The simplest implementation is obtained if the multiplier is represented in radix 2, since the multiple of the multiplicand is either x or zero. However, the number of iterations is reduced by using a larger radix. When the radix is 2^k, this is equivalent to considering k bits of the radix-2 multiplier per iteration. The main problem with this approach is the digit multiplication, since now the digit of the multiplier has 2^k values.

For radix 4, this digit multiplication can be simplified by recoding the multiplier into radix-4 digits with values $(-2, -1, 0, 1, 2)$ since the multiplication by these digit values is simple (complementation and shifting of the multiplicand). The recoding produces z such that

$$z = y, \quad z_i \in \{-2, -1, 0, 1, 2\} \qquad \textbf{4.9}$$

Since the sequential multiplication algorithm uses the digits of the multiplier from least significant to most significant, the recoding algorithm can also be sequential. In this case, the nonredundant digit set $\{-1, 0, 1, 2\}$ can be used.[1] This set allows a simpler implementation (fewer multiples) than the redundant set $\{-2, -1, 0, 1, 2, \}$. Calling v_i the radix-4 digit of the multiplier (corresponding to two bits of the radix-2 multiplier), the recoding uses a carry bit c_i and is performed by the recurrence

$$z_i = v_i + c_i - 4c_{i+1} \qquad\qquad \textbf{4.10}$$

The carry c_{i+1} is selected so that the value $z_i = 3$ is avoided. Consequently, when $v_i + c_i \geq 3$ we produce $c_{i+1} = 1$ and $z_i = v_i + c_i - 4$. This recoding is described by the following table:

$v_i + c_i$	z_i	c_{i+1}
0	0	0
1	1	0
2	2	0
3	−1	1
4	0	1

Note that the recoding (of a magnitude) produces a final carry. This carry has to be considered as an additional digit. The carry is avoided if the number of bits of the multiplier is odd, so that the most-significant radix-4 digit of the nonrecoded multiplier has values 0 or 1 only. That is, for an n-bit magnitude multiplier the number of radix-4 digits of the recoded version is $\lceil (n + 1)/2 \rceil$.

A radix-4 multiplier using this recoding is shown in Figure 4.3. This implementation is pipelined into three stages as follows:

- Stage 1: multiplier recoding
- Stage 2: generating the multiple of the multiplicand
- Stage 3: addition (using a redundant adder, illustrated with a carry-save adder) and shift (with conversion of the shifted-out bits).

The number of bits of the carry-save adder and of register SCH is $n + 3$ since the n bits of the multiplicand are extended by one bit because of the multiple 2 and

---

1. We later describe a parallel recoding algorithm, which requires the redundant digit set $\{-2, -1, 0, 1, 2\}$.

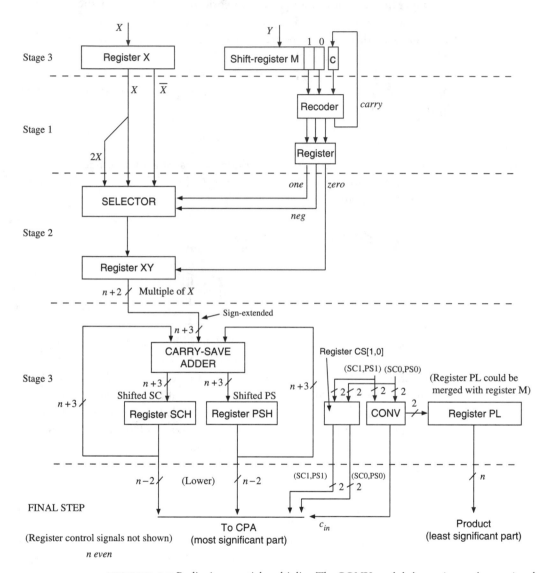

FIGURE 4.3 Radix-4 sequential multiplier. The CONV module has an internal carry signal, which in the last step is used as a c_{in} for the CPA.

Cycle	0	1	2	3	4	5	...	$m+1$	$m+2$	
	LOAD X									
	LOAD Y									
Stage 1	0	z_0	z_1	z_2	z_3	z_4				
Stage 2	0	0	Xz_0	Xz_1	Xz_2	Xz_3		Xz_{m-1}		
Stage 3	0	0	0	PS[1]	PS[2]	PS[3]		PS[$m-1$]	PS[m]	
				SC[1]	SC[2]	SC[3]		SC[$m-1$]	SC[m]	
CPA									Final product	

FIGURE 4.4 Timing diagram.

by another bit because the multiple is signed (in two's complement representation); finally it is necessary to extend one more bit to accommodate the range of the result of the addition.

The number of iterations is equal to the number of digits of the recoded multiplier, that is, $m = \lceil (n+1)/2 \rceil$. The execution is illustrated in the timing diagram shown in Figure 4.4.

The recoding uses the two rightmost bits of the multiplier register (M_1, M_0) and the carry flag C. For this, the M register is shifted two bits per iteration. The signals are described as follows:

$$one = M_0 \oplus C = \begin{cases} 0 & \text{select } 2x \\ 1 & \text{select } x \end{cases} \qquad \textbf{4.11}$$

$$neg = M_1 C + M_1 M_0 = \begin{cases} 0 & \text{select direct} \\ 1 & \text{select complement} \end{cases} \qquad \textbf{4.12}$$

$$zero = M_1 M_0 C + M_1' M_0' C' = \begin{cases} 0 & \text{load nonzero multiple} \\ 1 & \text{load zero multiple (clear)} \end{cases} \qquad \textbf{4.13}$$

$$C_{next} = M_1 M_0 + M_1 C = neg \qquad \textbf{4.14}$$

Note that when $zero = 1$ the selection made by neg is irrelevant. This allows a simpler expression for neg and C_{next}. Figure 4.5 illustrates a recoder implementation.

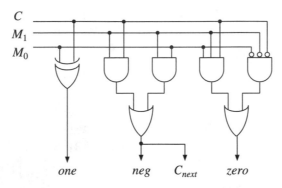

C
M_1
M_0

one   neg C_{next} zero

FIGURE 4.5 Sequential recoder implementation.

The generation of $(-1)x$ is performed by a bit complement, and the added 1 is placed in the least-significant bit of the carry vector, as follows:

$PS[j]$	PS_{n+2}	PS_{n+1}	PS_n	$\cdots$	PS_1	PS_0
$SC[j]$	SC_{n+2}	SC_{n+1}	SC_n	$\cdots$	SC_1	SC_0
$-x$	X'_{n+2}	X'_{n+1}	X'_n	$\cdots$	X'_1	X'_0
CSA	s_{n+2}	s_{n+1}	s_n	$\cdots$	s_1	s_0
	c_{n+2}	c_{n+1}	c_n	$\cdots$	c_1	1^*

* For two's complement of x.

EXAMPLE 4.1 $n = 5$ bits $m = 3$ radix-4 digits

$$x = 29 \qquad X = 11101$$
$$y = 27 \qquad Y = 11011$$
$$Z = 2\overline{1}\overline{1} \quad (z = y) \quad (-1 = \overline{1})$$

As discussed before, the carry-save adder has 8 bits (see Figure 4.6). The least-significant bits $(PS_1[j], PS_0[j])$ and $(SC_1[j], SC_0[j])$ are added and shifted out, producing two final product bits per recurrence step.[2] The final product consists of the lower 6 bits produced during the three

---

2. See Exercise 4.3 for details. Note that in this example the conversion is simple because the digit to be converted is not larger than three; in general this might not be the case.

	CSA	Shifted out		
$PS[0]$	00000000			
$SC[0]$	00000000			
$x\,Z_0$	11100010			
$4PS[1]$	11100010			
$4SC[1]$	0000000**1**			
$PS[1]$	11111000	11		
$SC[1]$	00000000			
$x\,Z_1$	11100010			
$4PS[2]$	00011010			
$4SC[2]$	1100000**1**			
$PS[2]$	00000110	1111		
$SC[2]$	11110000			
$x\,Z_2$	00111010			
$4PS[3]$	11001100			
$4SC[3]$	0110010**0**			
$PS[3]$	11110011	001111		
$SC[3]$	00011001			
P	1100	001111	=	783

FIGURE 4.6 Example of radix-4 sequential multiplication with carry-save adder.

recurrence steps and the 4 upper bits obtained by a 4-bit CPA. Note that the least-significant bit of $4SC[j]$ (shown in boldface) is 1 when the multiplier digit is negative so that the two's complement of the multiplicand is required.   ∎

Higher Radices

Sequential multiplication can be done using a multiplier representation with a radix higher than 4 to further reduce the number of iterations. The algorithm is a direct extension of the radix-4 case. For instance, for radix 8 the multiplier can be recoded into the digit set $\{-3, -2, -1, 0, 1, 2, 3, 4\}$ with a direct extension of the algorithm presented for radix 4. The main problem with the implementation of this multiplication is the generation of $3x$. This can be done as a preprocessing step by addition of $2x$ plus x.

The extension to even higher radices requires the preprocessing of more multiples. An alternative is to use several radix-4 and/or radix-2 stages in one iteration. For instance, Figure 4.7 shows a radix-16 multiplication unit in which the multiplier is recoded into a radix-16 signed digit v_j in the set $\{-10, \ldots, 0, \ldots, 10\}$. This recoding is actually performed by recoding into two redundant radix-4 digits u_j and w_j such that[3]

$$v_j = 4u_j + w_j \quad u_j, w_j \in \{-2, -1, 0, 1, 2\} \qquad \textbf{4.15}$$

The recurrence is

$$
\begin{aligned}
q[j] &= p[j] + xw_j \\
p[j+1] &= \frac{1}{16}(q[j] + (4x)u_j)
\end{aligned}
\qquad \textbf{4.16}
$$

where $p[j]$ is the partial product, $q[j]$ is its intermediate value, and xw_j and $4xu_j$ are multiples of the multiplicand and shifted multiplicand, respectively.

The use of a higher radix $r = 2^k$ reduces the number of iterations to n/k. Due to a more complex iteration step for higher radices, the reduction of the total delay is less than k times with respect to radix-2 multiplication (see Exercise 4.8).

Use of [p:2] Adder

As discussed in Chapter 3, in a multioperand addition it is possible to use $[p:2]$ adders. Since multiplication is a multioperand addition, this can also be done for multiplication. In such a case, $p - 2$ multiples of the multiplicand are used per iteration. For instance, in the radix-16 example of Figure 4.7, two multiples are used per iteration, so that a [4:2] adder can be used.

---

3. The sequential radix-4 recoding algorithm can be used without increasing the cycle delay if the delay of recoding is not larger than the delay of a full-adder.

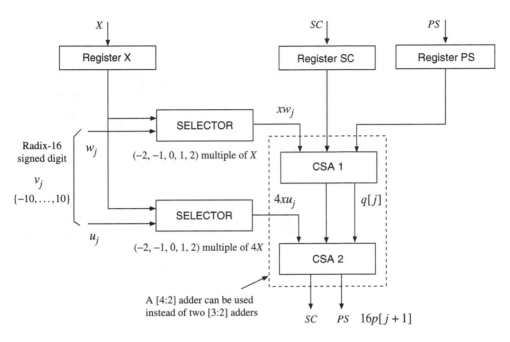

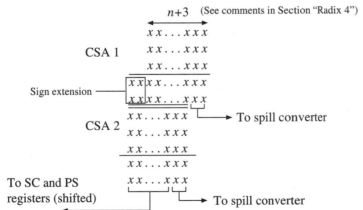

FIGURE 4.7 Radix-16 multiplication datapath (partial).

4.1.2   Two's Complement

We now describe the modifications required when the operands and the result are in two's complement representation. As indicated by the algorithm shown at the beginning of the chapter, one possibility is to transform the operands into sign-and-magnitude, perform the multiplication, and then transform the result. However, this is not necessary when using the add-and-shift algorithm. In this case, the following modifications to the sign-and-magnitude algorithm are suitable.

First, for operands of n bits (in two's complement representation) the range of the product is $-(2^{n-1} - 1)(2^{n-1}) \leq p \leq (2^{2n-2})$, requiring $2n$ bits in two's complement representation.

Second, since the multiplicand is represented in two's complement, the addition and shift operations are performed in this system, as discussed in Chapter 1.

Third, the effect of the two's complement multiplier can be taken into account in the following two ways, both based on the relation of the multiplier value and its two's complement representation, namely,

$$y = -y_{n-1}2^{n-1} + \sum_{i=0}^{n-2} y_i 2^i \qquad\qquad 4.17$$

which in radix 4 corresponds to

$$y = (-2y_{n-1} + y_{n-2})2^{n-2} + \sum_{i=0}^{n-3} y_i 2^i$$

$$= v_{m-1}4^{m-1} + \sum_{i=0}^{m-2} v_i 4^i \qquad\qquad 4.18$$

where $m = n/2$ (n even) and the radix-4 digit values are $v_{m-1} \in \{-2, -1, 0, 1\}$ and $v_i \in \{0, 1, 2, 3\}$.

Consequently the two alternatives are

1. Subtracting instead of adding in the last iteration when the multiplier digit is negative. The subtraction is done by addition of the two's complement of the multiple of the multiplicand.

2. Recoding the multiplier into a signed-digit set.

For radix 2, this can be done by a modification of the recoding for sign-and-magnitude (extending the sign); however, in this case it seems preferable to use the first approach. Consequently, we concentrate on the radix-4 case, in which the recoding is done anyhow to eliminate the multiple $3x$.

The sequential radix-4 recoding for sign-and-magnitude presented before has to be modified for the two's complement case. One possible modification is to extend the sign: one bit if n is odd and a whole radix-4 digit if it is even. Then, as in two's complement addition, the carry-out of the extended digit vector is discarded. However, this increases the number of cycles when n is even.

Two variations are possible to eliminate the need for the additional cycle:

1. Consider the most-significant radix-4 digit as described by expression (4.18). Then, the recoding of the last digit is

$v_{m-1} + c_{m-1}$	z_{m-1}
-2	-2
-1	-1
0	0
1	1
2	2

   This requires a special recoder for the last digit. However, in terms of the bits of y and the carry, this recoder differs from the one for sign-and-magnitude only in the case $v_{m-1} + c_{m-1} = -2$ (that is, $y_{n-1} y_{n-2} = 10$, $c_{m-1} = 0$). Consequently, the recoder is quite easy to modify for this case.

2. Use the parallel recoder discussed for combinational radix-4 multiplication (see page 286).

4.2 Combinational Multiplication with Recoding

Instead of performing the multiplication in several cycles (iterations), reusing the hardware, in the combinational case the operation is performed in a single cycle. The combinational add-and-shift algorithm (actually shift-and-add) is based on

$$p = \sum_{i=0}^{n-1} x y_i r^i \qquad\qquad 4.19$$

In this case, the multiplication is done in two steps:

1. Generation of the (shifted) multiples of the multiplicand $(x \times y_i) r^i$
2. Multioperand addition of the multiples generated in step 1

We now consider each of these steps and then show the implementation of complete multipliers.

4.2.1   Generation of Multiples and Bit-Array

The multiples are

$$m[i] = x y_i r^i \qquad\qquad 4.20$$

This corresponds to a multiplication of the multiplicand by one digit of the multiplier and an arithmetic shift left of i positions. Usually, the multiplicand is in radix-2 representation so that the result of this digit multiplication is a bit-vector and the shift is of $i \times \log_2 r$ bit positions. The set of these bit vectors, adequately range extended, forms a bit matrix that is added in the second step.

Consider now the generation of the bit-matrix for the multiplier in radix 2 and in radix 4.

Radix 2

For the multiplier in radix 2, the digit multiplication is especially simple since y_i has only values 0 and 1. Consequently, each multiple is produced by a set of AND gates, as shown in Figure 4.8(a). The resulting bit-matrix for multiplication of magnitudes is shown in Figure 4.8(b).

For multiplication in two's complement representation, two modifications are required:

1. The range extension is done by replicating the sign bit of each of the multiples. Since the largest negative operand value is -2^{n-1}, the maximum product positive value is $2^{2(n-1)}$. Consequently, to avoid an overflow, the array should produce a product represented by $2n$ bits, so that the extension should be performed accordingly.

2. The multiple $x y_{n-1} 2^{n-1}$ is subtracted instead of added. This is because, as already used in the sequential implementation, in two's complement

$$y = -y_{n-1} 2^{n-1} + \sum_{i=0}^{n-2} y_i 2^i \qquad\qquad 4.21$$

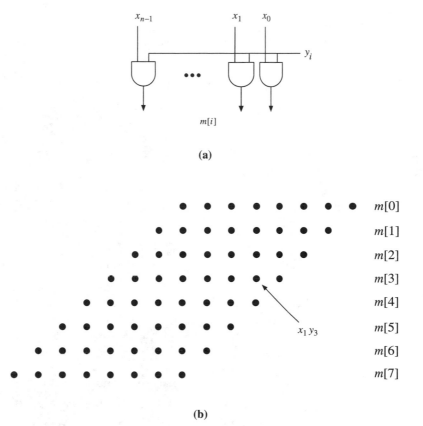

FIGURE 4.8 (a) Radix-2 multiple generation. (b) Bit-matrix for multiplication of magnitudes ($n = 8$).

The subtraction is done by complementation and addition. The complementation is performed by a bit-complement plus the addition of 1.[4]

We now show how to construct a bit-matrix for radix-2 multiplication of two's complement operands $x = (x_{n-1}, x_{n-2}, \ldots, x_0)$ and $y = (y_{n-1}, y_{n-2}, \ldots, y_0)$.

---

4. Instead of doing this subtraction step, it has been proposed to recode the (two's complement) multiplier into the digit set $\{-1, 0, 1\}$. However, there seems to be no advantage in following this approach.

7	6	5	4	3	2	1	0
x_3y_0	x_3y_0	x_3y_0	x_3y_0	x_3y_0	x_2y_0	x_1y_0	x_0y_0
x_3y_1	x_3y_1	x_3y_1	x_3y_1	x_2y_1	x_1y_1	x_0y_1	
x_3y_2	x_3y_2	x_3y_2	x_2y_2	x_1y_2	x_0y_2		
$x_3'y_3$	$x_3'y_3$	$x_2'y_3$	$x_1'y_3$	$x_0'y_3$			
				y_3			

(a)

7	6	5	4	3	2	1	0
				$(x_3y_0)'$	x_2y_0	x_1y_0	x_0y_0
			$(x_3y_1)'$	x_2y_1	x_1y_1	x_0y_1	
		$(x_3y_2)'$	x_2y_2	x_1y_2	x_0y_2		
	$(x_3'y_3)'$	$x_2'y_3$	$x_1'y_3$	$x_0'y_3$			
				y_3			
0	-1	-1	-1	-1			

(b)

7	6	5	4	3	2	1	0
			y_3	$(x_3y_0)'$	x_2y_0	x_1y_0	x_0y_0
			$(x_3y_1)'$	x_2y_1	x_1y_1	x_0y_1	
		$(x_3y_2)'$	x_2y_2	x_1y_2	x_0y_2		
1	$(x_3'y_3)'$	$x_2'y_3$	$x_1'y_3$	$(x_0y_3)'$			

(c)

FIGURE 4.9 Constructing bit-matrix for two's complement multiplier ($n = 4$). (a) Basic bit-matrix with each row sign-extended. (b) Bit-matrix after initial transformation. (c) Bit-matrix after final transformation.

As mentioned above, to handle the largest positive product $(-2^{n-1})(-2^{n-1})$, the bit-matrix has $2n$ columns.

The bit-matrix after sign extension is shown in Figure 4.9(a). The increase in the number of bits due to the sign extension would complicate the addition step. As discussed in Chapter 3, the effect of this sign extension can be reduced by applying the following identity in the sign position (which is of negative weight):

$$(-s) + 1 - 1 = (1 - s) - 1 = s' - 1 \qquad \textbf{4.22}$$

Consequently,

$$x_{n-1}y_i \quad x_{n-2}y_i \quad \cdots \quad x_0 y_i$$

is replaced by

$$(x_{n-1}y_i)' \quad x_{n-2}y_i \quad \cdots \quad x_0 y_i$$
$$-1$$

The bit-matrix resulting from this transformation has $n + 2$ rows as shown in Figure 4.9(b). The two extra rows can be eliminated as follows:

1. The digit-vector in row 6

$$(0, (-1), (-1), (-1), (-1))$$

    has value -15×2^3, which can be replaced by its two's complement representation, namely, 10001000.

2. Precompute

$$x_0' y_3 + y_3 + 1$$

    in column 3, rows 4, 5, and 6, resulting in a carry y_3 and a sum $(x_0 y_3)'$. The carry is placed in row 1, column 4, and the sum in row 4, column 3.

The application of this modification results in the final bit-matrix shown in Figure 4.9(c). Consequently, with respect to the array for magnitudes, only two additional bits are required.

Radix 4

To reduce the number of multiples and, therefore, the complexity of the multioperand addition, it is convenient to consider the multiplier represented in a radix higher than 2. For radix 4 with conventional representation of the multiplier digit, the values of the digit are 0, 1, 2, and 3. As discussed for the sequential case, the implementation of the multiple generation consists of an AND-OR network for each bit to select among the three possible multiples different from 0. The generation of the multiples x and $2x$ is simple, but the multiple $3x$ requires an addition. As also done for the sequential multiplication, to avoid this multiple it is possible to recode the multiplier into a signed-digit set. To optimize the recoding we consider two cases:

1. The bit-array is added by a linear array of adders. As discussed further in the next section, in this case each adder in the array has as operands the partial sum (of the previous additions) and one multiple. Consequently, the recoding can be done as in the sequential multiplication case (digit set $\{-1, 0, 1, 2\}$). This has the advantage of requiring only four values per digit.[5]

2. The bit-array is added by a tree of adders. In this case all the multiples are obtained simultaneously and applied as operands in the first level of the tree. Therefore, the recoding has to be done in a parallel fashion; that is, all digits of the recoded multiplier should be obtained simultaneously. In this case, the digit set is $\{-2, -1, 0, 1, 2\}$ and the recoding algorithm is as follows.

Parallel Radix-4 Recoding

We present the high-level arithmetic algorithm using the same technique as for signed-digit addition in Chapter 2. Then we consider a bit-level implementation. We show that the recoding is correct for two's complement representation of the multiplier. If the multiplier is a magnitude, it should be extended by the sign bit equal to zero.

Let us call v_j a radix-4 digit of the multiplier (obtained by pairing consecutive bits of y). That is,

$$v_j = 2y_{2j+1} + y_{2j} \quad j = (m - 1, \ldots, 0) \quad m = n/2 \qquad \textbf{4.23}$$

and

$$y_{n-1}, y_{n-2}, \ldots, y_1, y_0$$

is the radix-2 representation of the multiplier.

At the high level the algorithm has two steps:

1. Obtain w_j and t_{j+1} such that

$$v_j = w_j + 4t_{j+1} \qquad \textbf{4.24}$$

---

5. If a $[p{:}2]$ adder is used, with $p > 3$, then the parallel recoding is needed.

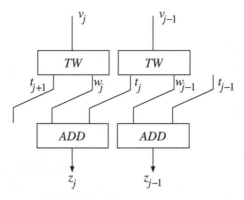

FIGURE 4.10 Radix-4 parallel recoding from $\{0, 1, 2, 3\}$ into $\{-2, -1, 0, 1, 2\}$.

2. Obtain

$$z_j = w_j + t_j \qquad\qquad \textbf{4.25}$$

For a parallel algorithm, the second step should be performed without carry propagation. This is achieved if

$$-2 \le w_j \le 1 \quad 0 \le t_{j+1} \le 1 \qquad\qquad \textbf{4.26}$$

Consequently, the algorithm is

$$(t_{j+1}, w_j) = \begin{cases} (0, v_j) & \text{if } v_j \le 1 \\ (1, v_j - 4) & \text{if } v_j \ge 2 \end{cases} \qquad\qquad \textbf{4.27}$$
$$z_j = w_j + t_j$$

As shown in Figure 4.10, digits $j - 1$ and j of the multiplier are involved in the generation of z_j.

We now show that the recoding algorithm is correct for two's complement representation. For this, consider the most-significant digit. The value of this digit is $u_{m-1} = -2y_{n-1} + y_{n-2}$. On the other hand, the recoding algorithm uses $v_{m-1} = 2y_{n-1} + y_{n-2}$ and produces $w_{m-1} = v_{m-1} - 4t_m$. Since $t_m = 1$ if $v_{m-1} \ge 2$, we get

$y_{n-1}y_{n-2}$	v_{m-1}	w_{m-1}	u_{m-1}
00	0	0	0
01	1	1	1
10	2	-2	-2
11	3	-1	-1

Consequently, the algorithm is correct if we discard the transfer digit t_m.

We now consider a bit-level implementation. The radix-2 multiplier is

$$Y = (y_{n-1}, y_{n-2}, \ldots, y_0) \quad y_i \in \{0, 1\} \qquad 4.28$$

and the recoded radix-4 multiplier

$$Z = (z_{m-1}, z_{m-2}, \ldots, z_0) \quad z_i \in \{-2, -1, 0, 1, 2\} \qquad 4.29$$

From the high-level algorithm, bits $y_{2j+1}, y_{2j}, y_{2j-1}, y_{2j-2}$ are involved in the generation of z_j. However, since $t_{j+1} = 1$ only when $v_j \geq 2$, bit y_{2j-2} has no effect. Specifically,

$$z_j = w_j + t_j = (v_j - 4t_{j+1}) + t_j \qquad 4.30$$

Since $v_j = 2y_{2j+1} + y_{2j}$ and $t_j = y_{2j-1}$, we get

$$z_j = (2y_{2j+1} + y_{2j} - 4y_{2j+1}) + y_{2j-1} = -2y_{2j+1} + y_{2j} + y_{2j-1} \quad 4.31$$

This bit-level recoding can be described by the following table:

y_{2j+1}	y_{2j}	y_{2j-1}	z_j
0	0	0	0
0	0	1	1
0	1	0	1
0	1	1	2
1	0	0	-2
1	0	1	-1
1	1	0	-1
1	1	1	0

EXAMPLE 4.2　Following are two examples of the bit-level recoding:

$$y = 01011110 \qquad y = 10001101$$
$$z = 1 \ 2 \ 0 \ \bar{2} \qquad z = \bar{2} \ 1 \ \bar{1} \ 1$$

∎

We implement the recoder with a representation of z_j by the triple (*sign, one, two*), as follows:

- *sign* $= 1$ if z_j is negative
- *one* $= 1$ if z_j is either 1 or -1
- *two* $= 1$ if z_j is either 2 or -2

From the table we obtain the following switching expressions:

$$\begin{aligned} sign &= y_{2j+1} \\ one &= y_{2j} \oplus y_{2j-1} \\ two &= y_{2j+1} y'_{2j} y'_{2j-1} + y'_{2j+1} y_{2j} y_{2j-1} \end{aligned} \qquad \textbf{4.32}$$

The bit-level implementation of this recoder and the multiple generator are shown in Figure 4.11. The generation of the two's complement for negative multiples is produced by the signal *sign*, which controls the bit-inverter, and the signal c, which completes the two's complement operation. Note that the 0 for $(y_{2j+1}, y_{2j}, y_{2j-1}) = 111$ is obtained by a bit-vector $\underline{1}$ at the output of the bit-inverter and $c = 1$.

The algorithm can be extended to higher radices in a straightforward manner. That is, for $z_j \in \{-(r/2), \ldots, +(r/2)\}$ the recoding is described by

$$(t_{j+1}, \ w_j) = \begin{cases} (0, \ v_j) & \text{if } v_j \le (r/2) - 1 \\ (1, \ v_j - r) & \text{if } v_j \ge r/2 \end{cases} \qquad \textbf{4.33}$$
$$z_j = w_j + t_j$$

The bit-level implementation uses $\log_2(r) + 1$ bits of Y to produce z_j.

Bit-Array

Because of the recoding, the multiples of the multiplicand are signed; consequently, they are represented in the two's complement system (even for multiplication of magnitudes). For multiplication of n-bit operands the result has $2n$ bits so that the rows of the bit matrix have to be extended to $2n$ bits. The negative

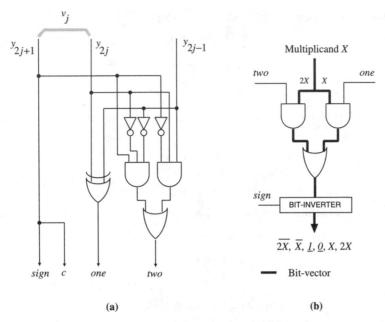

FIGURE 4.11 (a) Implementation of parallel recoder. (b) Implementation of multiple generator.

multiples are obtained by bit complement and addition of 1 (these additional bits are called c in what follows). The bit array is formed by the multiples (and the c bits). Because of the radix 4, consecutive multiples are shifted two positions.

There are some differences in the array, depending on whether the operands are magnitudes or signed in two's complement representation. We now discuss both cases.

For magnitudes, the multiplier is extended with a 0. Consequently, the number of radix-4 digits (rows in the array) is $\lceil (n + 1)/2 \rceil$. Moreover, the most-significant radix-4 digit has to be positive (with value 0, 1, or 2), so that no sign bit nor c bit is needed for that row.

As an example, the bit array for multiplication of 7-bit magnitudes is shown in Figure 4.12(a). To reduce elements of the bit array, we use similar modifications to those discussed for the radix-2 case and for multioperand addition. The resulting array is shown in Figure 4.12(c).

	13	12	11	10	9	8	7	6	5	4	3	2	1	0
xz_0:	s_e	s_e	s_e	s_e	s_e	s_e	e	e	e	e	e	e	e	e
xz_1:	s_f	s_f	s_f	s_f	f	f	f	f	f	f	f	f		c_e
xz_2:	s_g	s_g	g	g	g	g	g	g	g	g		c_f		
xz_3:	h	h	h	h	h	h	h	h		c_g				

(a)

	13	12	11	10	9	8	7	6	5	4	3	2	1	0
xz_0:						s'_e	e	e	e	e	e	e	e	e
xz_1:				s'_f	f	f	f	f	f	f	f	f		c_e
xz_2:		s'_g	g	g	g	g	g	g	g	g		c_f		
xz_3:	h	h	h	h	h	h	h	h		c_g				
		-1		-1		-1								

(b)

	13	12	11	10	9	8	7	6	5	4	3	2	1	0
xz_0:	1		1	s'_e	s_e	s_e	e	e	e	e	e	e	e	e
xz_1:				s'_f	f	f	f	f	f	f	f	f		c_e
xz_2:		s'_g	g	g	g	g	g	g	g	g		c_f		
xz_3:	h	h	h	h	h	h	h	h		c_g				

(c)

FIGURE 4.12 Radix-4 bit-matrix for multiplication of magnitudes ($n = 7$). There are 8 bits plus sign for each row because of the possible multiple 2. The result is a magnitude, no sign included. The -1's of the last row of (b) are combined with s'_e to form the first row of (c).

A similar array is used for multiplication of signed values in two's complement representation since the multiplier recoding is applicable for this representation. In this case, $\lceil n/2 \rceil$ rows are required, and all digits of the recoded multiplier can be negative. The maximum positive value of the product is 2^{2n-2}, requiring $2n$ bits for the two's complement representation. Figure 4.13 illustrates multiplication of two's complement 8-bit operands. Note that one additional row is required (formed of bit c_h).

(a)

	15	14	13	12	11	10	9	8	7	6	5	4	3	2	1	0
xz_0:	s_e	s_e	s_e	s_e	s_e	s_e	s_e	s_e	e	e	e	e	e	e	e	e
xz_1:	s_f	s_f	s_f	s_f	s_f	s_f	f	f	f	f	f	f	f	f		c_e
xz_2:	s_g	s_g	s_g	s_g	g	g	g	g	g	g	g	g		c_f		
xz_3:	s_h	s_h	h	h	h	h	h	h	h	h		c_g				
										c_h						

(b)

	15	14	13	12	11	10	9	8	7	6	5	4	3	2	1	0
xz_0:								s'_e	e	e	e	e	e	e	e	e
xz_1:						s'_f	f	f	f	f	f	f	f	f		c_e
xz_2:				s'_g	g	g	g	g	g	g	g	g		c_f		
xz_3:		s'_h	h	h	h	h	h	h	h	h		c_g				
		-1			-1			-1		c_h	-1					

(c)

	15	14	13	12	11	10	9	8	7	6	5	4	3	2	1	0
xz_0:	1		1		1	s'_e	s_e	s_e	e	e	e	e	e	e	e	e
xz_1:						s'_f	f	f	f	f	f	f	f	f		c_e
xz_2:				s'_g	g	g	g	g	g	g	g	g		c_f		
xz_3:		s'_h	h	h	h	h	h	h	h	h		c_g				
										c_h						

FIGURE 4.13 Radix-4 bit-matrix for two's complement multiplication ($n = 8$). See comments in caption of previous figure.

Radix 8

Considering a radix-8 representation of the multiplier (three bits per digit), reduces the number of multiples by a factor of 3. Since the generation of the multiples of x (up to $7x$) is complicated, the multiplier can be recoded to signed-digit representation. The algorithms for this are direct extensions of the radix-4 cases. For sequential recoding it is possible to use a nonredundant digit set (for instance

−3 to 4), whereas for the parallel recoding the redundant digit set −4 to 4 is appropriate. In any case, the multiple $3x$ has to be generated, usually as $2x + x$, which requires an addition. In principle, it is possible to keep this $3x$ in carry-save representation (two vectors), in which case no addition would be needed; however, this would effectively double the number of vectors that have to be added in the array, which eliminates the advantage of recoding. The addition for $3x$ can be done in parallel with the recoding of the multiplier.

Whether radix-8 recoding reduces the overall delay depends on the organization of the adder array. In any case, it reduces the number of adders required in the array.

4.2.2 Addition of the Bit-Array

For the addition of the bit-array we consider the same approaches discussed for multioperand addition; namely, reduction by rows using adder arrays (linear and tree) and reduction by columns using $(p:q]$ counters. We therefore concentrate here on the effects of the particular shape of the bit-array in multiplication. Since redundant addition is used to limit the carry propagation, the bit-array addition has two steps: reduction to two rows and conversion to conventional representation. In this section we concentrate on the reduction to two rows and discuss further the conversion in the next section.

Reduction by Rows: Linear Adder Array

We consider now the addition by a linear adder array. The difference with respect to the multioperand addition of Chapter 3 is the shape of the bit array. Figure 4.14 shows linear adder arrays for radix 2 and radix 4 using [3:2] and [4:2] adders (for the case with signed-digit adders, see Exercise 4.6).

In the radix-2 case, one bit of the result is produced at each level so that the final adder has $n + 1$ bits. In the radix-4 case, the carry bits do not make 2-bit product digits directly obtainable. One possibility is to carry out conversion of the least-significant outputs of the CSA stage so that 2 bits of the product are obtained per stage (see Exercise 4.13). In this case the final adder has n bits. The final conversion is optional (although almost always done) and can use any of the fast carry-propagate adders discussed in Chapter 2.

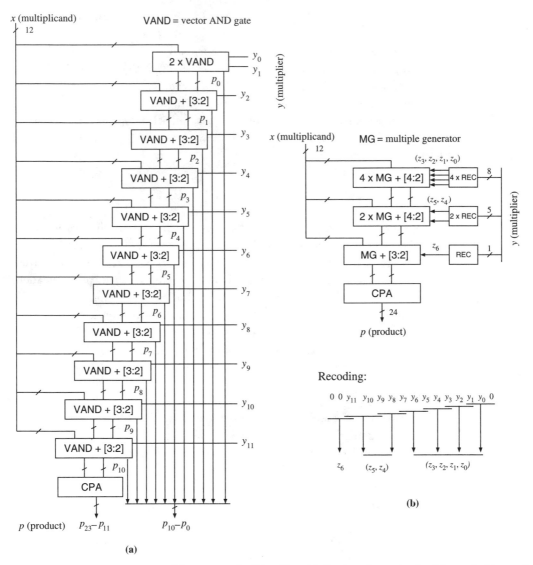

FIGURE 4.14 Linear array for 12×12 multiplication of magnitudes: (a) $r = 2$. (b) $r = 4$. Also included are the modules to produce the multiples.

The delays of the 12×12 multiplication implementations shown in Figure 4.14, with conversion of the four least-significant bits in the radix-4 case, are

- for radix 2,

$$T = t_{AND} + 10t_{fa} + t_{CPA(13)} \qquad \textbf{4.34}$$

- for radix 4

$$T = t_{REC} + t_{MG} + 2t_{[4:2]} + t_{fa} + t_{CPA(21)} \qquad \textbf{4.35}$$

Adder Tree

To reduce the number of levels of adders, a tree can be used, as discussed for multi-operand addition. This is applicable for both radix-2 and radix-4 bit-arrays. If m is the number of digits of the multiplier (n for radix 2 and $n/2$ for radix 4), the number of levels is $\lceil \log_2(m/2) \rceil$ for 4-to-2 adders, and approximately $\lceil \log_{3/2}(m/2) \rceil$ for an array of 3-to-2 adders.

Note that from radix 2 to radix 4 the reduction in the number of levels is just one. Moreover, for the radix-4 case it is necessary to add the delay of the recoder. Consequently, it is not clear that there is a reduction in the overall delay. However, the radix-4 case might be considered because of possible area reduction due to a reduced number of partial products. Figure 4.15 shows [3:2] and [4:2] adder trees. The carry-propagate adder used for conversion is now wider than in the case of a linear array. In the case of using [3:2] adders, the number of full adders can be reduced by considering the shape of the bit-array. This is illustrated in Figure 4.16.

Pipelining

The adder arrays can be pipelined to increase the throughput of multiplication. An implementation for a linear array is shown in Figure 4.17.

Reduction by Columns Using (p :q] Counters

This is an application of the method discussed for multioperand addition. That is, each column of the bit-array is reduced individually until two rows are obtained. These are then reduced to one row by a carry-propagate adder. The method is

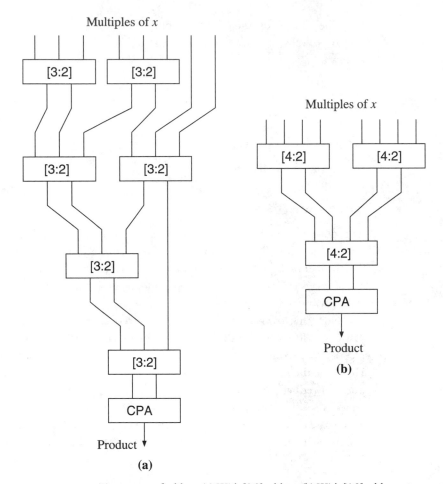

FIGURE 4.15 Tree arrays of adders: (a) With [3:2] adders. (b) With [4:2] adders.

especially suited for multiplication because the different column height is used to reduce the number of counters. The height of each column is better shown in the bit-array of Figure 4.18.

One possibility is to use (3:2] and (2:2] counters. For this case, the design method presented in Chapter 3 produces Table 4.1 and Figure 4.19. As discussed also in Chapter 3, larger $(p:q]$ counters can be used for the reduction.

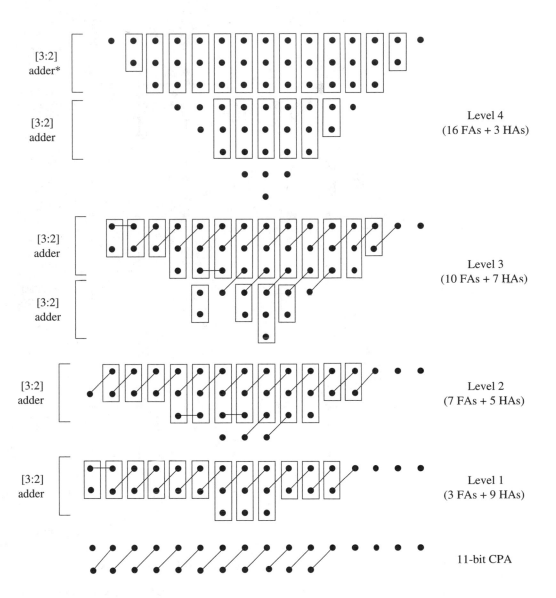

* [3:2] adder uses HAs when possible.

FIGURE 4.16 Reduction by rows with FAs and HAs ($n = 8$): cost 36 FAs, 24 HAs, and 11-bit CPA.

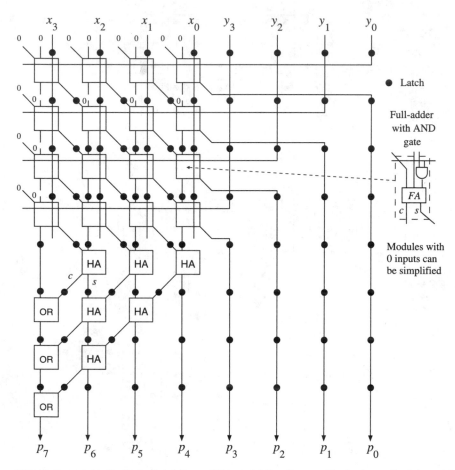

FIGURE 4.17 Radix-2 pipelined linear CSA multiplier for magnitudes ($n = 4$). Adapted from Noll et al. (1986).

4.2.3   Final Adder for Converting Product to Conventional Form

In most multipliers the product is required in conventional representation. In such cases the two-row result of the reduction are inputs to a carry-propagate adder, which produces the product. Since the delay of this addition contributes to the overall multiplication delay, it is convenient to use a fast adder (see Chapter 2).

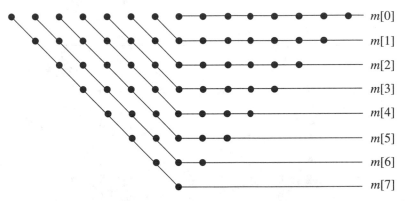

FIGURE 4.18 Bits of (shifted) multiples organized as bit-triangle (for magnitudes and radix-2 multiplier).

However, typical fast adders are designed under the assumption that all operand bits arrive at the same time. This is not the case in the multiplier,[6] especially when using column reduction since the length of the columns (including carries from previous columns) is not uniform. Consequently, it might be beneficial to design an adder that takes into account the characteristics of the arrival time of the operand bits. Consider, for instance, the idealized arrival profile of Figure 4.20(a). We see three regions: at the least-significant side the arrival time increases when moving to the left, in the middle region the time is large and constant, and at the most-significant side the time decreases. For this case, a hybrid adder of the following characteristics might be convenient:[7]

- For the least-significant region a carry-ripple adder seems appropriate, since any faster structure would need to wait for the higher bits.
- For the middle region a fast adder is required.
- For the most-significant region, a carry-select adder would be suitable since the sum of the most-significant bits can be computed earlier.

The corresponding implementation is illustrated in Figure 4.20(b).

---

6. This effect is lost if the multiplier is pipelined and the final adder corresponds to another stage of the pipeline.
7. See Stelling and Oklobdzija (1996) for further details.

	14	13	12	11	10	9	8	7	6	5	4	3	2	1	0
$l=4$															
e_i	1	2	3	4	5	6	7	8	7	6	5	4	3	2	1
m_3	6	6	6	6	6	6	6	6	6	6	6	6	6	6	6
h_i	0	0	0	0	0	0	1	1	1	0	0	0	0	0	0
f_i	0	0	0	0	0	1	1	1	0	0	0	0	0	0	0
$l=3$															
e_i	1	2	3	4	6	6	6	6	6	6	5	4	3	2	1
m_2	4	4	4	4	4	4	4	4	4	4	4	4	4	4	4
h_i	0	0	0	0	0	0	0	0	0	1	1	0	0	0	0
f_i	0	0	0	1	2	2	2	2	2	1	0	0	0	0	0
$l=2$															
e_i	1	2	4	4	4	4	4	4	4	4	4	4	3	2	1
m_1	3	3	3	3	3	3	3	3	3	3	3	3	3	3	3
h_i	0	0	0	0	0	0	0	0	0	0	0	1	0	0	0
f_i	0	0	1	1	1	1	1	1	1	1	1	0	0	0	0
$l=1$															
e_i	1	3	3	3	3	3	3	3	3	3	3	3	3	2	1
m_0	2	2	2	2	2	2	2	2	2	2	2	2	2	2	2
h_i	0	0	0	0	0	0	0	0	0	0	0	1	0	0	0
f_i	0	1	1	1	1	1	1	1	1	1	1	1	0	0	0
CPA	2	2	2	2	2	2	2	2	2	2	2	2	2	2	1

Note: e_i is the number of inputs in column i; f_i is the number of FAs; h_i is the number of HAs; m_j is the number of operands in the next level in the reduction sequence.

TABLE 4.1 Reduction by columns using FAs and HAs for 8×8 radix-2 magnitude multiplier.

4.3   Partially Combinational Implementation

To reduce the area required for a fully combinational multiplier, it is possible to use a mixed implementation, which essentially corresponds to a sequential implementation in which a high radix is used to represent the multiplier. Then, this high-radix digit of the multiplier is represented in radix 2 or radix 4 and

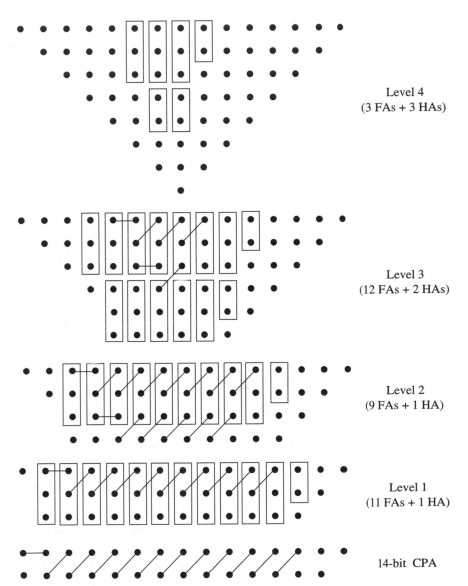

Level 4
(3 FAs + 3 HAs)

Level 3
(12 FAs + 2 HAs)

Level 2
(9 FAs + 1 HA)

Level 1
(11 FAs + 1 HA)

14-bit CPA

FIGURE 4.19 Reduction by columns using FAs and HAs ($n = 8$): cost 35 FAs, 7 HAs, and 14-bit CPA.

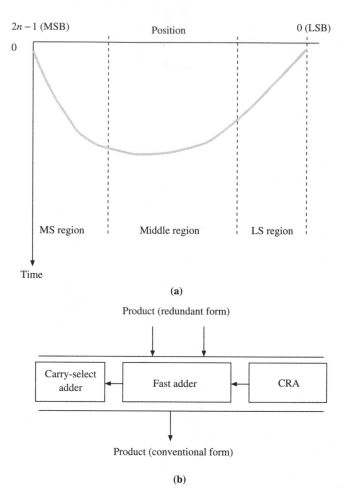

FIGURE 4.20 Final adder: (a) Arrival time of the inputs to the final adder. (b) Hybrid final adder.

a combinational implementation is done for this partial multiplication. Figure 4.21 shows such an implementation in which 12 bits of the multiplier are used each iteration[8] (radix 2^{12}). Note that to reduce the overall delay, the iteration is pipelined in the same way as was discussed for multioperand addition.

---

8. The additional bit of the multiplier is used for the recoding.

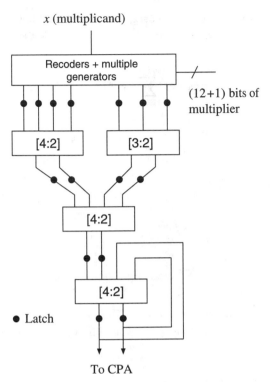

FIGURE 4.21 Radix-2^{12} pipelined sequential multiplier using CSA tree.

4.4   Arrays of Smaller Multipliers

We now consider the multiplication of two n-bit magnitudes using modules that perform the multiplication of a k-bit magnitude by an l-bit magnitude (a k by l multiplication). The module performs

$$p = a \times b \qquad\qquad 4.36$$

in which the bit vectors representing a, b, and p are

$$
\begin{aligned}
A &= (a_{k-1}, a_{k-2}, \dots, a_0) \\
B &= (b_{l-1}, b_{l-2}, \dots, b_0) \qquad\qquad 4.37 \\
P &= (p_{k+l-1}, p_{k+l-2}, \dots, p_0)
\end{aligned}
$$

To perform the $n \times n$ multiplication using these $k \times l$ modules, the operands are decomposed into digits of radix 2^k and 2^l, respectively. That is,

$$x = \sum_{i=0}^{(n/k)-1} x^{(i)} 2^{ki}$$

$$y = \sum_{j=0}^{(n/l)-1} y^{(j)} 2^{lj}$$

4.38

Then the multiplication is

$$p = x \times y = \sum_{i=0}^{(n/k)-1} x^{(i)} 2^{ki} \times \sum_{j=0}^{(n/l)-1} y^{(j)} 2^{lj}$$

$$= \sum x^{(i)} y^{(j)} 2^{ki+lj} = \sum p^{(i,j)} 2^{ki+lj}$$

4.39

That is, $(n/k) \times (n/l)$ modules are needed. The outputs of these modules, suitably aligned, produce a bit-matrix that can be added by any of the methods discussed before.

EXAMPLE 4.3   Consider the 12×12 multiplication of magnitudes using 4×4 multiplication modules. The decomposition of the operands is

$$x = x^{(2)} 2^8 + x^{(1)} 2^4 + x^{(0)}$$
$$y = y^{(2)} 2^8 + y^{(1)} 2^4 + y^{(0)}$$

4.40

The multiplication is then

$$x \times y = x^{(2)} y^{(2)} 2^{16} + x^{(2)} y^{(1)} 2^{12} + x^{(1)} y^{(2)} 2^{12} + x^{(2)} y^{(0)} 2^8$$
$$+ x^{(1)} y^{(1)} 2^8 + x^{(0)} y^{(2)} 2^8 + x^{(1)} y^{(0)} 2^4 + x^{(0)} y^{(1)} 2^4 + x^0 y^0$$ **4.41**

The corresponding bit-matrix is shown in Figure 4.22. This can be reduced by any of the methods discussed before. ■

FIGURE 4.22 12×12 multiplication using 4×4 multipliers: bit-matrix.

	13	12	11	10	9	8	7	6	5	4	3	2	1	0
xz_0:				s'_e	s_e	s_e	e	e	e	e	e	e	e	e
xz_1:			1	s'_f	f	f	f	f	f	f	f	f		c_e
xz_2:		1	s'_g	g	g	g	g	g	g	g		c_f		
xz_3:	h	h	h	h	h	h	h	h		c_g				
w:								w	w	w	w	w	w	w

FIGURE 4.23 Radix-4 bit-matrix for multiply-add of magnitudes ($n = 7$). z_i's are radix-4 digits obtained by multiplier recoding.

4.5 Multiply-Add and Multiply-Accumulate (MAC)

In many applications the operation of multiplication is followed by an addition to perform

$$s = x \times y + w \qquad \qquad 4.42$$

This can be implemented efficiently by including the operand W as part of the bit-array, as illustrated in Figure 4.23. A block-diagram of a multiply-add unit is shown in Figure 24(a).

A variation of the multiply-add is the multiply-accumulate, which is useful to perform operations such as a sum of products of the form

$$s = \sum_{i=1}^{m} x[i] \times y[i] \qquad \qquad 4.43$$

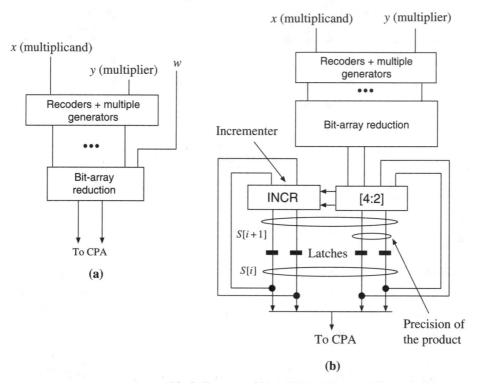

FIGURE 4.24 Block-diagrams of (a) multiply-add unit and (b) multiply-accumulate unit.

This can be accomplished effectively by repeatedly using a multiplier-accumulator (MAC) that performs the operation

$$s[i+1] = x[i] \times y[i] + s[i] \qquad \textbf{4.44}$$

where $s[1] = 0$ and $s = s[m+1]$.

After the carry-save product is produced, the accumulation is performed in two parts: the least-significant (LS) part is obtained in redundant form using a [4:2] adder of the precision required by the multiplication, and the most-significant (MS) part using a carry-save incrementer. The number of bits of the MS and LS parts is determined by the number of bits of the result s. Figure 4.24(b) shows a block diagram of the implementation. The carry-save result is converted to the final result using a carry-propagate adder.

4.6   Saturating Multiplier

Integer multiplication of n-bit operands produces an integer of $2n$ bits. Some applications in signal processing and graphics keep only n bits of the result and, in case of a result larger than what can be represented with n bits (overflow), saturate the result to the maximum representable value ($2^n - 1$ for magnitudes and $2^{n-1} - 1$ and -2^{n-1} for two's complement). These are called *saturating multipliers*.

The direct implementation incorporates a standard multiplier, a detection of the overflow, and setting the result. Detection has two forms:

- For magnitudes, overflow is detected by one or more bits with value 1 in the n most-significant bits. This detection is implemented by an n-input OR gate.
- For two's complement representation, the detection is different for positive and for negative results: if positive (bit in position $2n - 1$ is zero), detect as for magnitudes; if negative, detect by one or more bits with value 0 in the n most-significant bits.

Setting the result also has two forms:

- For magnitudes, the n-bit result is set to $2^n - 1$ (all ones). This can be implemented with a 2-input OR gate for each result bit.
- For two's complement representation, the n bits are set to $011 \ldots 1$ for positive result and to $100 \ldots 0$ for negative.

Figure 4.25 illustrates detection and result setting in the case of multiplication of magnitudes. Exercise 4.22 discusses an implementation that reduces the required hardware.

4.7   Truncating Multiplier

Multiplication of n-bit fractions produces a fraction of $2n$ bits. Some applications in signal processing keep only the n most-significant bits of the result and dispose of the least-significant bits after performing rounding. If a larger roundoff error is allowed, not all LSB bits of the final result are generated, which leads to a simpler implementation as indicated in Figure 4.26.

The error in the result consists of E_{red} due to simplified reduction, and E_{rnd} due to rounding of the $n + k$ computed bits to the n most-significant bits. To achieve a specific total error smaller than one ulp (unit in the last significant place),

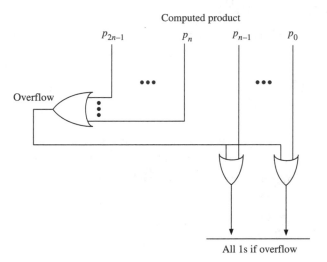

FIGURE 4.25 Detection and result setting for multiplication of magnitudes.

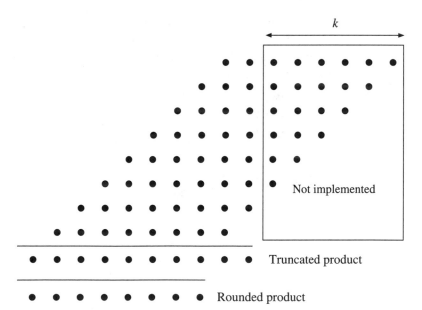

FIGURE 4.26 Bit-matrix of a truncated magnitude multiplier.

the errors E_{red} and E_{rnd} can be reduced by choosing k, the number of positions not implemented, and by adding a suitable constant to the reduction array. This approach and several others are discussed in detail in the literature mentioned at the end of the chapter.

4.8 Rectangular Multipliers

The multipliers discussed up to now had both operands with the same number of bits (square multipliers). In many applications the multiplication operands have different number of bits (say, k and n). These multipliers are called *rectangular multipliers*.

The implementation of these rectangular multipliers follows the same approach as for the square case: construction of the bit array and reduction to two rows. For the two's complement case, the same type of analysis as for the square case has to be performed to reduce the required sign extensions (Exercise 4.23).

4.9 Squarers

A squaring operation $s = x^2$ is frequently used in signal processing and multimedia applications. For example, computing the Euclidean distance between two points a and b in the three-dimensional space

$$\left(\sum_{i=1}^{3}(a_i - b_i)^2\right)^{1/2}$$

where $a_i (b_i)$ is the coordinate of point a (b) in dimension i, is frequently used in graphics applications. It requires computing a large number of squares, making a dedicated implementation desirable. While a general multiplier can be used to compute squares, a dedicated implementation is attractive because of a simplified partial-product bit-array.

A bit-array for x^2 (magnitudes, radix 2) consists of the diagonal with entries $x_i x_i = x_i$ (Identity 1) and two bit-array regions: A above the diagonal and B below the diagonal. Since $x_i x_j = x_j x_i$ (Identity 2), the sum of entries in regions A and B are the same ($A + B = 2A = 2B$). Therefore, an equivalent bit-array consists of the diagonal entries and the A (or B) bit-array moved one position to the left. The transformed bit-array has a reduced number of entries and a reduced number

11	10	9	8	7	6	5	4	3	2	1	0
						x_5x_0	x_4x_0	x_3x_0	x_2x_0	x_1x_0	x_0x_0
					x_5x_1	x_4x_1	x_3x_1	x_2x_1	x_1x_1	x_0x_1	
				x_5x_2	x_4x_2	x_3x_2	x_2x_2	x_1x_2	x_0x_2		
			x_5x_3	x_4x_3	x_3x_3	x_2x_3	x_1x_3	x_0x_3			
		x_5x_4	x_4x_4	x_3x_4	x_2x_4	x_1x_4	x_0x_4				
	x_5x_5	x_4x_5	x_3x_5	x_2x_5	x_1x_5	x_0x_5					
	x_5x_4	x_5x_3	x_5x_2	x_5x_1	x_5x_0	x_4x_0	x_3x_0	x_2x_0	x_1x_0		x_0
	x_5		x_4x_3	x_4x_2	x_4x_1	x_3x_1	x_2x_1		x_1		
			x_4		x_3x_2		x_2				
					x_3						

(a)

11	10	9	8	7	6	5	4	3	2	1	0
	x_5x_4	x_5x_3	x_5x_2	x_5x_1	x_5x_0	x_4x_0	x_3x_0	x_2x_0	x_1x_0		x_0
	x_5		x_4x_3	x_4x_2	x_4x_1	x_3x_1	x_2x_1		x_1		
			x_4	x_3x_2	x_3x_2'		x_2				

(b)

FIGURE 4.27 Bit-array simplification in squaring of magnitudes ($n = 6$). (a) Bit-array after using Identities 1 and 2. (b) Further reduction in number of rows after using Identity 3.

of rows as illustrated in Figure 4.27(a). Moreover, $x_i + x_i x_j = 2x_i x_j + x_i x_j'$ (Identity 3) can be used to achieve further reduction in the number of rows at the expense of extra inverters and AND gates (Figure 4.27(b)). It can also be used to reduce the number of bits of the final adder.

For n-bit magnitudes the number of rows in the simplified bit-array is no larger than $\lceil (n/2) \rceil$ and the number of inputs to the reduction network is

$$\sum_{i=1}^{n} i = (n^2 + n)/2$$

Any of the previously discussed bit-array reduction methods is applicable. Although the bit-array resulting from a multiplier recoding to radix 4 also has $\approx n/2$ rows, the bit-array obtained by the simplification discussed here is preferable to radix-4 recoding since it does not require recoding and multiple-generation networks. Squarers for two's complement operand are obtained in a similar manner.

4.10   Constant and Multiple-Constant Multipliers

If one of the operands of a multiplication is constant, the multiplication $P = X \times C$ can be described by

$$P = \sum_{\{j \mid C_j = 1\}} X \times C_j 2^j \qquad \textbf{4.45}$$

where the set $\{j\}$ corresponds to all the ones' in the binary representation of C. Consequently, the number of rows of the bit-matrix is reduced since no row is necessary when the bit $C_i = 0$. The number of rows can be further reduced by recoding, as follows:

1. Radix-4 recoding reduces n bits to no more than $(n+1)/2$ nonzero digits from the set $\{-2, -1, 0, 1, 2\}$.

2. Canonical recoding into the digit set $\{-1, 0, 1\}$ is a sequential recoding that minimizes the number of nonzero binary digits. It transforms the constant n-bit operand, with $n/2$ nonzero bits on the average, to a representation with $n/3$ nonzero bits on the average. For example, $C = 0101111001$, which has six nonzero digits, is recoded into the canonical form $10\bar{1}000\bar{1}001$, with four nonzero digits. Moreover, since two consecutive digits in the recoded form cannot both be nonzero, the canonical form has at most $n/2$ nonzero digits.

The resulting bit-matrix is then reduced using any of the techniques presented before.

In addition to reducing the nonzero digits, it is possible to factor the constant into the product of smaller constants. The implementation, which consists of the connection of smaller multipliers, might be simpler than the direct implementation. This is illustrated in the following example.

EXAMPLE 4.4   Consider the computation of $45X$ using implementations that utilize only carry-propagate adders.

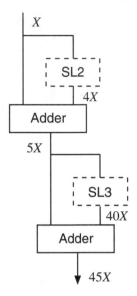

SLk = shift left k positions

FIGURE 4.28 Implementation of $P = X \times C$ for $C = 45$ using common subexpressions.

The binary representation of 45 has four 1s. Consequently, a direct implementation would use three adders. The canonical recoding has also four nonzero digits, requiring also three adders. On the other hand, we can factor as follows:

$$45X = 5X \times 9 = X(2^2 + 1)(2^3 + 1)$$

The implementation of this decomposition requires two adders, as shown in Figure 4.28.   ∎

Of course, the implementation can use redundant adders with more than two inputs. The best decomposition depends on the type of adders used. Exercises 4.26 and 4.27 illustrate designs with [3:2] carry-save adders. In some instances it is beneficial to perform recoding followed by factoring. Several heuristic techniques have been developed for this task, as described in the literature.

If several multiplications by constants are required simultaneously; that is, $P_k = X \times C_k, k = 1, 2, \ldots, K$ (known as *multiple-constants multiplication*),

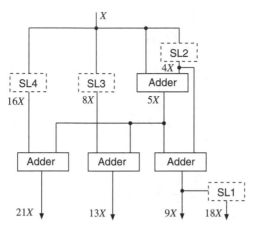

SLk = shift left k positions

FIGURE 4.29 An example of multiple-constants multipliers.

the use of common subexpressions leads to further reductions in the number of adders, compared to implementation using separate constant multipliers.[9] The following example illustrates such a case.

EXAMPLE 4.5 Consider the simultaneous computation of $P_1 = 9X$, $P_2 = 13X$, $P_3 = 18X$, and $P_4 = 21X$.

An implementation using separate constant multipiers requires six adders. By decomposing the products as follows: $P_1 = 5X + 4X$, $P_2 = 8X + 5X$, $P_3 = 2 \times 9X$, and $P_4 = 16X + 5X$ and sharing subexpressions, the implementation can be done with four adders (Figure 4.29). ∎

4.11 Concluding Remarks

We have presented a variety of implementation schemes for the multiplication operation. We have considered the case of magnitude and signed operands represented in a conventional representation; it is important to stress that, as for

---

9. See Potkonjak et al. (1996) for further details.

any operation, the algorithms and implementations for multiplication depend heavily on the representation of operands and result. Since multiplication for conventional representation is performed as multioperand addition, the techniques presented in Chapter 3 are applicable, the main difference being the shifts required in multiplication, which change the shape of the bit-array to be added.

Let us now consider some design choices. The main choice for the delay/area trade-off is between sequential and combinational implementation. In the sequential case, a higher radix reduces the number of cycles at the expense of a longer cycle and a larger area. For high radices, each iteration actually corresponds to a rectangular multiplier; therefore, this scheme is also called a *partially combinational multiplier*, and there is a continuum from a sequential implementation to a combinational one.

Since multiplication is performed by addition of multiples of the multiplicand, it is convenient to have few multiples and fast addition. The number of multiples is reduced by a larger radix; however, this complicates the generation of multiples. This can be simplified somewhat (especially for radix 4 and radix 8) by recoding the multiplier. As a consequence of this, at the implementation level, the basic radices used are 2, 4, and 8, and higher radices are implemented by arrays using these lower radices.

With respect to the additions, as in multioperand addition, the reduction of the bit-array can be done by rows or by columns. In the reduction by rows redundant adders are used because of their small delay and area. Because of the shape of the array the adders have a variable number of bits. The adder array can be linear or a tree. For large multipliers, the tree is advantageous because of the much smaller number of adder levels. However, the area of the tree multiplier might be larger because of the irregularity and length of the interconnections. This length might also affect the delay. Although we have presented only the linear array and the complete tree, there are intermediate solutions that are obtained by partitioning the linear array and merging the partial results.

The reduction by column uses several levels of counters to reduce the columns to size two (two rows). A systematic method is presented to reduce the number of counters. The reduction by columns makes better use of the shape of the array to reduce the number of cells. However, the interconnection is more irregular than that of reduction by rows.

The delay analysis we have done is quite rough since it has considered only the number of cells (mainly for the case of full-adders and half-adders) in the critical path. More detailed studies have been done considering actual delays of sums and carries and restructuring the array to reduce the critical delay. Moreover, since not all outputs of the adder array are produced at the same time, it is possible to reduce the overall delay by taking this into account in the design of the adder for conversion to conventional representation.

Multipliers are sometimes pipelined to increase the throughput. A variety of possibilities exist on the number of stages; this is very dependent on the technology and the requirements. In Chapter 8 we consider these issues for floating-point multiplication.

Large multipliers can also be implemented by the interconnection of smaller multiplier modules. This was particularly interesting when only small multipliers could fit in a chip, which is not the case anymore. However, it might still be convenient to implement multipliers in this form to reduce the number of long interconnections and to provide multiplications of several sizes; that is, partitionable multipliers (this is becoming important because of the variety of data widths for different applications; for instance, the wide data for floating-point applications and the much narrower data for multimedia).

In this chapter, we also consider more specialized units such as multiplier-accumulators, squarers, saturating multipliers, truncating multipliers, and multiplication by a constant. In all these cases, the main operation is a multiplication, but the unit is adapted because of the particular characteristics of the operands and/or result. In addition to providing some design ideas for these cases, the lesson here is that the designer should take into account the particular characteristics of the operation, instead of using always a standard multiplier.

4.12 Exercises

For the design exercises use the circuit data shown on the inside cover pages.

Sequential Multiplication with Recoding

4.1 **[Sequential radix-4 multiplication example]** Show the multiplication of $x = 30$ by $y = -25$ as in Example 4.1 (use two's complement representation with 6 bits). Determine the number of cycles in the execution of this multiplication in the unit of Figure 4.3. Show the values in registers during the last pass through the pipeline.

4.2 **[Sequential recoding]** Determine the table for sequential recoding from a conventional radix-8 digit set to the set $\{-4, -3, -2, -1, 0, 1, 2, 3\}$.

4.3 **[Converter carry-save to conventional]** Design a sequential converter for the radix-4 sequential multiplier shown in Figure 4.3. The converter has as external inputs the least-significant bits produced by the carry-save adder: $(PS_1[j], PS_0[j])$ and $(SC_1[j], SC_0[j])$, one internal state variable $w_0[j-1]$ (the present state of the converter) and produces two product bits p_{2j+1}, p_{2j} and the next state $w_0[j]$ such that

$$2(PS_1[j] + SC_1[j]) + (PS_0[j] + SC_0[j] + w_0[j-1])$$
$$= 4w_0[j] + 2p_{2j+1} + p_{2j}$$

Show the network and determine the minimum cycle time for the converter.

4.4 **[Converter carry-save to conventional]** Design an alternative sequential converter for the radix-4 sequential multiplier shown in Figure 4.3. The external inputs are in this case the least-significant bits produced by the multiple generator and the PS and SC registers. Determine the arithmetic expression relating the inputs, the state variables, and the outputs of the converter, similar to the expression given in Exercise 4.3. Show the network and determine the minimum cycle time for the converter.

4.5 **[Gate-level design]** Design at the gate level a 16×16 radix-4 sequential multiplier for operands in the two's complement form following the scheme given in Figure 4.3. To reduce the critical path it might be convenient to place the CONV module after the register; that is, first load the bits to be converted in a register and then convert (at the beginning of next cycle). Determine the critical path and the total cost in terms of equivalent gates.

4.6 **[Use of signed-digit adder]** Repeat Exercise 4.5 using a signed-digit adder for accumulation of partial products. If you are doing both exercises, compare the cycle time and cost of both implementations.

4.7 **[Generation of multiples for radix 16]** Design a scheme to generate multiples of a positive multiplicand for a radix-16 sequential multiplication unit with the digit set $\{-8, \ldots, -1, 0, 1, \ldots, 8\}$. Consider the following alternatives:

(a) Precompute the multiples.

(b) Generate the multiples during the recurrence execution.

Compare the cost and the effect on the cycle time and the total delay of multiplication.

4.8 [Speedup of radix-16 multiplier] Determine the delay of the critical path of the radix-16 multiplier of Exercise 4.7 and the speedup $S = T_2/T_{16}$ where T_2 and T_{16} are the latencies of the radix-2 multiplier and radix-16 multiplier, respectively. Does the speedup depend on n, the precision of the operands? Discuss.

Combinational Multiplication with Recoding

4.9 [Generation of bit-array for radix 2, two's complement] Generate the bit-array for the multiplication of $x = 125$ by $y = -122$. Use the techniques discussed in the text to reduce the sign extension bits.

4.10 [Parallel radix-4 recoding] Recode the binary representation (two's complement) $y = 10101111$ into radix 4 with digit set $\{-2, -1, 0, 1, 2\}$.

4.11 [Generation of bit-array for radix 4: magnitudes and two's complement] Show the bit-array for radix-4 multiplication of

(a) $x = 67$ by $y = 76$ (magnitudes)

(b) $x = -67$ by $y = -76$ (two's complement representation)

4.12 [Linear adder array] Design a gate network for a radix-4 linear CSA array for two's complement multiplication $(n = 16)$ following the scheme in Figure 4.14(b). Use the design of the radix-4 recoder given in the text. Determine the delay of the critical path.

4.13 [Converter for linear array] In the linear CSA array multiplier shown in Figure 4.14(b) the precision of the final CPA can be shortened by converting in an iterative manner the least-significant radix-4 (redundant) digits produced by each CSA stage. (See Exercise 4.3 for the definition of the converter.) Design a network to perform this conversion in parallel with the CSA reduction and determine the precision of the final CPA. (Note: If the delay of the converter is larger than one full-adder, try to reduce the delay of producing the carry in the converter.)

4.14   **[Reduction by columns]** Modify the bit-triangle of Figure 4.18 for the following cases:

(a)   Radix 2 with operands in two's complement representation
(b)   Radix 4 with operands in magnitude representation and multiplier recoding
(c)   Radix 4 with operands in two's complement representation and multiplier recoding

4.15   **[Reduction by columns]** Develop a table to determine the number of full- and half-adders required for the addition of the cases in the previous exercise.

Partially Combinational Multiplication

4.16   For the multiplier shown in Figure 4.21, determine the number of cycles required for a multiplication with a 47-bit multiplier.

Arrays of Smaller Multipliers

4.17   Show the bit-matrix corresponding to a 16×16 magnitude multiplication using 4×4 multiplication modules.

4.18   Consider the implementation of 12×12 multiplication with operands and product in two's complement representation. Use 5×5 multiplication modules (two's complement representation).

(a)   Determine how many modules are required.
(b)   Show the bit-matrix to be added, identifying the output bits of each multiplication module.
(c)   Determine the network of full-adders and half-adders required to reduce the bit-matrix to two rows, using the column reduction approach.

Multiplier-Accumulator

4.19   For the computation of

$$s = \sum_{i=1}^{32} x[i]y[i]$$

with $x[i]$ and $y[i]$ represented by 16 bits in two's complement representation, compare the execution time of the following two alternatives:

(a) Using a combinational multiplier with delay t_M (this includes the conversion of the product to conventional representation) and an adder with delay t_A

(b) Using a multiplier-accumulator (carry-save result) with delay $t_M - t_A + t_{[4:2]}$ and a final adder of delay t_A

4.20 Consider evaluating the inner product of 16-element vectors A and B:

$$S = \sum_{i=1}^{16} A[i] \times B[i]$$

where each element $A[i]$ $(B[i])$ is a positive integer in the range $[0, 127]$.

(a) Determine the precision of the result S to avoid overflows.

(b) Design a pipelined linear array of [3:2] carry-save adders with radix-4 multiplier recoding (digit set $\{-2, -1, 0, 1, 2\}$). Assume one pair of vector elements is available each clock cycle. The intermediate values of the sum are in carry-save form. The final result is obtained by a carry-propagate adder that requires one cycle.

(c) Determine the critical path in the network in terms of the delay of basic modules and determine the cycle time.

(d) Give a timing diagram of the inner product computation in terms of clock cycles and determine the latency in terms of clock cycles.

(e) Show a modified network if [4:2] adders are used instead of [3:2] adders and compare with the network in (b) with respect to the cycle time.

Saturating Multiplier

4.21 Design a network for obtaining the saturated product for a two's complement multiplier with $n = 8$.

4.22 Consider an $n \times n$-bit magnitude multiplier (Schulte et al. 2000). The saturation is performed when the most-significant half of the product $(p_{2n-2}, \ldots, p_n)$ is nonzero. This condition V happens if any of the bit products $x_j y_i$ in columns n to $2n - 2$ of the multiplication bit-matrix is one or if any of the carries co_k into column n is one.

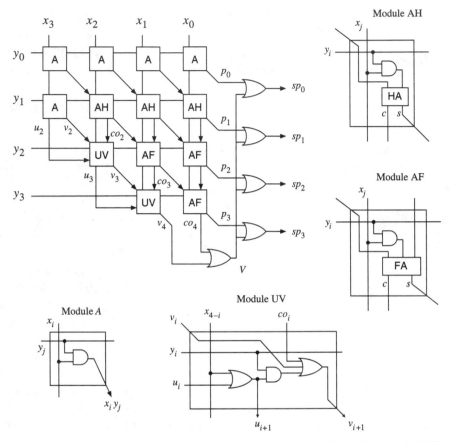

FIGURE 4.30 4×4 magnitude multiplier with saturation. Adapted from Schulte et al. (2000).

(a)   Derive a switching expression for V for $n = 4$.

(b)   Show that the following expressions, implemented in Figure 4.30, compute V:

$$u_{i+1} = u_i + x_{4-i}$$
$$v_{i+1} = v_i + co_i + u_{i+1} \cdot y_i$$
$$V = v_4 + co_4$$

where $i = 2, 3$, $u_2 = x_3$, and $v_2 = x_3 y_1$.

    (c)   Determine the delay in the critical path of the scheme in Figure 4.30.

    (d)   Determine the delay in the critical path for the general case of an $n \times n$ magnitude multiplier. Give the value for $n = 24$.

Rectangular Multiplier

4.23   Determine the bit-array for a 16 by 6 bits multiplication unit with operands in two's complement representation. Reduce the bits required for sign extension.

Squarers

4.24   Apply Identity 3 to the reduced bit-array shown in Figure 4.27(b) to reduce the precision of the final adder.

4.25   Design a squarer for a 6-bit two's complement operand.

Constant Multiplier

4.26   Let $C = 2925$. Design adder networks to compute $P = C \times X$ for

    (a)   carry-ripple adders (CRA)

    (b)   [3:2] CSAs and prefix adder

    For each network determine the delay in terms of the full-adder delay t_{FA} and the number of FAs, and compare the solutions.

4.27   Design adder networks to compute $P_1 = 27X$, $P_2 = 36X$, $P_3 = 41X$, and $P_4 = 67X$ using

    (a)   carry-ripple adders (CRA)

    (b)   [3:2] CSAs and prefix adder

    For each network determine the delay in terms of the full-adder delay t_{FA} and the number of FAs, and compare the solutions.

4.13 Further Readings

The implementation of multiplication has been important since the introduction of the digital computer. Because of its significance in engineering and scientific computations, it has received much attention since. Multiplication is also used in application-specific systems for signal processing, communications, and so on.

Sequential Multiplier and Sequential Recoding

Sequential multiplication was popular when hardware was expensive and bulky; it is still of use in some applications and might be of interest for highly parallel systems on a chip. The use of carry-save addition in sequential multiplication is first mentioned in Burks et al. (1946), described in Estrin et al. (1956), and an early implementation presented in Kilburn et al. (1956). Sequential recoding to radix 4 with the digit set $\{-1, 0, 1, 2\}$ is described in Ware et al. (1982).

Combinational-Sequential Multiplier

Partially combinational multipliers were used in the 1960s to achieve higher performance but still with limited hardware. Examples are described in MacSorley (1961), Anderson et al. (1967), and Gosling (1980). A more recent multiplier of this type is presented in Santoro and Horowitz (1989).

Combinational Multiplier and Parallel Recoding

Today most multipliers are combinational because of the higher speed and the available area. Radix-4 recoding of the multiplier is frequently used. This recoding, called *radix-4 Booth recoding* or *modified Booth recoding*, is an extension of the radix-2 recoding (called *Booth recoding*) introduced in Booth (1951) and its use first described in MacSorley (1961). Proofs of its correctness are given in Rubinfield (1975), Vassiliadis et al. (1989), and Sam and Gupta (1990), and the presentation based on ideas from signed-digit addition is given in Ercegovac and Lang (1996). Radix-4 recoding of a redundant multiplier is discussed in Lyu and Matula (1995) and Ercegovac et al. (1994). Extensions to radix 8 and radix 16 are described in Zurawski and Gosling (1987), Sam and Gupta (1990), Kornerup (1994), and Ercegovac and Lang (1996). Several multipliers in recent floating-point units use radix-8 recoding (Schwarz et al. 1997). Seidel et al. (2001) present multipliers using radix 32 and radix 256.

Two's Complement Multiplier

Although many multipliers are for magnitudes, since the sign-and-magnitude is the standard representation for floating point, two's complement is preferred for fixed-point multiplication. Special cells for the two's complement case are used in Pezaris (1971), and variations of the scheme presented in this chapter are discussed in Robertson (1955) and Baugh and Wooley (1973).

Linear Array Multiplier

Linear adder arrays for multiplication have been frequently considered because of their regularity (Braun 1963; Guild 1969; Agrawal 1979). The idea of separate routing of partial products between odd rows and even rows to reduce the delay to roughly one half while maintaining the regularity of the linear array is presented in Iwamura et al. (1982). The odd/even scheme is generalizable to several partitions followed by an additional reduction array. In the limit, such a scheme is equivalent to a tree of adders.

Tree Array Multiplier

The design of tree adder arrays using row reduction with [3:2] carry-save adders is described in MacSorley (1961), Bucholz (1962), and Anderson et al. (1967), with a general description of the method given in Wallace (1964). The [4:2] adder is described in Weinberger (1981) and used for multiplication in Luk and Vuillemin (1983) and Santoro and Horowitz (1989). The use of signed-digit adders is presented in Takagi et al. (1985), Harata et al. (1987), Briggs and Matula (1993), and Makino et al. (1996). Column reduction with full-adders in a carry-save approach is discussed in Dadda (1965). A comparison of row reduction and column reduction using full-adders and half-adders is reported in Bickerstaff et al. (2001). The use of counters is described in Dadda (1976) and of (7:3] counters in Montoye et al. (1990) and Mehta et al. (1991). In Song and De Micheli (1991), several counters are used and the resulting multipliers compared; there is also a comparison with several commercial multipliers. In Wang et al. (1995), a column reduction scheme without predetermined column height is presented. In Oklobdzija et al. (1996) and Stelling et al. (1998), an algorithmic method is given to minimize the delay of the array reduction taking into account the cell delays.

Pipelined Multiplier

Pipelining of multipliers is a standard technique used to achieve high throughput. An example of a pipelined linear array multiplier is presented in Noll et al. (1986).

Multiply-Accumulate Unit

Multiply-accumulate designs are discussed in Lu and Samueli (1993), Huang et al. (1994), and Stelling and Oklobdzija (1997).

Integer Multiplier

Integer multiplication schemes and designs are the subject of Magenheimer et al. (1988), Zuras (1993, 1994), and Owens et al. (1995).

Hybrid Final Adder

The effect of the nonsimultaneous production of the output bits on the final addition and design of a hybrid adder are discussed in Oklobdzija and Villeger (1995) and Stelling and Oklobdzija (1996).

Left-to-Right Multiplier

Left-to-right multiplication to allow for the on-the-fly conversion of the redundant result without using a carry-propagate adder is presented in Ercegovac and Lang (1990) and improved in Ciminiera and Montuschi (1996) and Takagi and Horiyama (1999). A VLSI implementation of a left-to-right multiplier without a CPA is described in Kolagotla et al. (1997). Left-to-right multipliers have been used in implementing recursive filters (Knowles et al. 1989).

Multiplier with Operands in Redundant Form

Multipliers with operands in redundant form allow preceding arithmetic operations to be performed without using a CPA to produce results in conventional forms. Design of this type of multiplier is discussed in Flynn and Oberman (2001) and Ferguson and Ercegovac (1999).

Miscellaneous Multiplier Schemes

Multiplication by constants is discussed in Dempster and Macleod (1994, 1995) and Potkonjak et al. (1996). Squarers are presented in Chen (1971), Strandberg et al. (1996), and Wires et al. (1999) and multiplication with saturation in Schulte et al. (2000). Finally, there are many discussions of various schemes for truncated multipliers (Yoshida et al. 1991; Lim 1992; Schulte and Swartzlander 1993; King and Swartzlander 1998; Jou and Kuang 1999; Swartzlander 1999; Schulte et al. 1999; Wires et al. 2000, 2001; Van et al. 2000).

Several Ph.D. dissertations have been recently devoted to the design of multipliers (Santoro 1989; Bewick 1994; Stelling 1995; Callaway 1996; Al-Twaijry 1997; Meier 1999).

Low-Power Multiplier

A comparison of the energy dissipation of several combinational multipliers is given in Callaway (1996) and Callaway and Swartzlander (1997). The effects of sign extension techniques and recoder design on energy dissipation are analyzed in de Angel and Swartzlander (1996) and Fried (1997). A comprehensive treatment of analysis and design of low-power multipliers is the subject of Meier (1999). A methodology for analyzing the effect of physical layout on the design of low-power multipliers is presented in Meier et al. (1996). Circuit techniques for low-power multipliers are discussed in Abu-Khater et al. (1996) and Mahanti-Shetti et al. (1999). Cherkauer and Friedman (1997) discuss a hybrid radix-4/radix-8 multiplier design for low-power applications. Schulte et al. (1999) discuss reduction of power dissipation in truncated multipliers.

Delay/Area Bounds

Theoretical bounds on the area and time to perform multiplication are discussed in Winograd (1967) and Brent and Kung (1981). An obvious decomposition is to perform $n \times n$-bit multiplication with four $n/2 \times n/2$-bit multiplications. Karatsuba and Ofman (1962) show that three $n/2 \times n/2$-bit multiplications and a few extra additions are sufficient. Optimal VLSI layouts are presented in Cappello and Steiglitz (1983) and Luk and Vuillemin (1983).

4.14 Bibliography

Abu-Khater, I. S., A. Bellaouar, and M. I. Elmasry (1996). Circuit techniques for CMOS low-power high-performance multipliers. *IEEE Journal of Solid-State Circuits*, 31(10):1535–46.

Agrawal, D. P. (1979). High-speed arithmetic arrays. *IEEE Transactions on Computers*, C-28(3):215–24.

Al-Twaijry, H. A. (1997). *Area and Performance Optimized CMOS Multipliers*. PhD thesis, Stanford University.

Anderson, S. F., J. G. Earle, R. E. Goldschmidt, and D. M. Powers (1967). The IBM 360/370 model 91: floating-point execution unit. *IBM Journal of Research and Development*, pages 34–53.

Baugh, C. R., and B. A. Wooley (1973). A two's complement parallel array multiplication algorithm. *IEEE Transactions on Computers*, C-22:1045–47.

Bewick, G. W. (1994). *Fast Multiplication: Algorithms and Implementation*. PhD thesis, Stanford University.

Bickerstaff, K. C., E. E. Swartzlander, and M. J. Schulte (2001). Analysis of column compression multipliers. In *Proceedings of the 15th IEEE Symposium on Computer Arithmetic*, pages 33–39.

Booth, A. D. (1951). A signed binary multiplication technique. *Quarterly Journal of Mechanics and Applied Mathematics*, 4(2):236–40.

Braun, E. L. (1963). *Digital Computer Design*. Academic Press, New York.

Brent, R. P., and H. T. Kung (1981). The area-time complexity of binary multiplication. *Journal of the ACM*, 28(3).

Briggs, W. S., and D. W. Matula (1993). A 17 × 69 bit multiply and add unit with redundant binary feedback and single cycle latency. In *Proceedings of the 11th IEEE Symposium on Computer Arithmetic*, pages 163–71.

Bucholz, W. (1962). *Planning a New Computer System: Project STRETCH*, Chapter 14, page 210. Wiley and Sons, Inc., New York.

Burks, A., H. H. Goldstine, and J. Von Neumann (1946). Preliminary discussion of the logic design of an electronic computing instrument. Technical report, Institute for Advanced Study, Princeton. Reprinted in C. G. Bell, *Computer Structures, Readings and Examples*, McGraw-Hill, New York, 1971.

Callaway, T. (1996). *Area, Delay, and Power Modeling of CMOS Adders and Multipliers*. PhD thesis, The University of Texas at Austin.

Callaway, T. K., and E. E. Swartzlander (1997). Power-delay characteristics of CMOS multipliers. In *Proceedings of the 13th IEEE Symposium on Computer Arithmetic*, pages 26–32.

Cappello, P. R., and K. Steiglitz (1983). A VLSI layout for a pipelined Dadda multiplier. *ACM Transactions on Computer Systems*, 1(2):157–74.

Chen, T. C. (1971). A binary multiplication scheme based on squaring. *IEEE Transactions on Computers*, C-20:678–80.

Cherkauer, B. S., and E. G. Friedman (1997). A hybrid radix-4/radix-8 low power signed multiplier architecture. *IEEE Transactions on Circuits and Systems—II: Analog and Digital Signal Processing*, 44(8):656–59.

Ciminiera, L., and P. Montuschi (1996). Carry-save multiplication schemes without final addition. *IEEE Transactions on Computers*, 45(9):1050–55.

Dadda, L. (1965). Some schemes for parallel multipliers. *Alta Frequenza*, 34:349–56.

Dadda, L. (1976). On parallel digital multipliers. *Alta Frequenza*, 45:574–80.

de Angel, E., and E. Swartzlander (1996). Low power parallel multipliers. In *Proceedings of the IEEE Workshop on VLSI Signal Processing*, pages 199–208.

Dempster, A. G., and M. D. Macleod (1994). Constant integer multiplication using minimum adders. *IEE Proceedings Circuits Devices Systems*, 141(5):407–13.

Dempster, A. G., and M. D. Macleod (1995). Use of minimum-adder multiplier blocks in FIR digital filters. *IEEE Transactions on Circuits and Systems—II: Analog and Digital Signal Processing*, 42(9):569–77.

Ercegovac, M. D., and T. Lang (1990). Fast multiplication without carry-propagate addition. *IEEE Transactions on Computers*, 39(11):1385–90.

Ercegovac, M. D., and T. Lang (1996). On recoding in arithmetic algorithms. *Journal of VLSI Signal Processing*, 14:283–94.

Ercegovac, M. D., T. Lang, and P. Montuschi (1994). Very-high radix division with prescaling and rounding. *IEEE Transactions on Computers*, 43(8):909–18.

Estrin, G., B. Gilchrist, and J. H. Pomerane (1956). A note on high-speed digital multiplication. *IRE Transactions on Electronic Computers*, page 140.

Ferguson, M. I., and M. D. Ercegovac (1999). A multiplier with redundant operands. In *Proceedings of the 33rd Asilomar Conference on Signals, Systems and Computers*, volume 2, pages 1322–26.

Flynn, M. J., and S. F. Oberman (2001). *Advanced Computer Arithmetic Design*. John Wiley and Sons, Inc., New York.

Fried, R. (1997). Minimizing energy dissipation in high-speed multipliers. In *Proceedings of 1997 International Symposium on Low Power Electronics and Design*, pages 214–19.

Gosling, J. B. (1980). *Design of Arithmetic Units for Digital Computers*. Springer-Verlag, New York.

Guild, H. H. (1969). Fully iterative fast arrays for binary multiplication and addition. *Electronic Letters*, 5:263.

Harata, Y., Y. Nakamura, H. Nagase, M. Takigawa, and N. Takagi (1987). A high-speed multiplier using a redundant binary adder tree. *IEEE Journal of Solid-State Circuits*, SC-22(1):28–34.

Huang, X., W.-J. Liu, and B. W. Y. Wei (1994). A high-performance CMOS redundant binary multiplication-and-accumulation (MAC) unit. *IEEE Transactions on Circuits and Systems I: Fundamental Theory and Applications*, 41(1):33–39.

Iwamura, J., K. Suganama, S. Taguchi, M. Kimura, and K. Maeguchi (1982). A 16-bit CMOS/SOS multiplier-accumulator. In *Proceedings of the ICCD '82 Conference*, pages 151–54, New York.

Jou, J. M., and S. R. Kuang (1999). Design of low-error fixed-width multiplier for DSP applications. *IEEE Transactions on Circuits and Systems—II: Analog and Digital Signal Processing*, 46(6):836–42.

Karatsuba, A., and Y. Ofman (1962). Multiplication of multidigit numbers on automata. *Soviet Physics-Doklaty*, 7(7):595–96.

Kilburn, T., D. B. G. Edwards, and G. F. Thomas (1956). The Manchester University Mark II Computing Machine. *Proceedings of the IEE*, pt. 103B, Suppl. 2:247–68.

King, E. J., and E. E. Swartzlander (1998). Data-dependent truncation scheme for parallel multipliers. In *Proceedings of the 31st Asilomar Conference on Signals, Systems and Computers*, volume 2, pages 1178–82.

Knowles, S. C., R. F. Woods, J. McWirther, and J. McCanny (1989). Bit-level systolic architectures for high-performance IIR filtering. *Journal of VLSI Signal Processing*, 1(1):9–24.

Kolagotla, R. K., H. R. Srinivas, and G. F. Burns (1997). VLSI implementation of a 200-MHz 16*16 left-to-right carry-free multiplier in 0.35 μm CMOS technology for next-generation DSPs. In *Proceedings of the IEEE 1997 Custom Integrated Circuits Conference*, pages 469–72.

Kornerup, P. (1994). Digit-set conversions: Generalizations and applications. *IEEE Transactions on Computers*, 43(5):622–29.

Lim, Y. C. (1992). Single-precision multiplier with reduced circuit complexity for signal processing applications. *IEEE Transactions on Computers*, 41(10):1333–36.

Lu, F., and H. Samueli (1993). A 200-Mhz CMOS pipelined multiplier-accumulator using a quasi-domino dynamic full-adder cell design. *IEEE Journal of Solid-State Circuits*, 28:123–32.

Luk, W. K., and J. E. Vuillemin (1983). Recursive implementation of optimal time VLSI integer multipliers. In *VLSI '83. Proceedings of the IFIP TC WG 10.5 International Conference on Very Large Scale Integration*, pages 155–68. Elsevier Science Publishers (North-Holland).

Lyu, C. N., and D. W. Matula (1995). Redundant binary Booth recoding. In *Proceedings of the 12th IEEE Symposium on Computer Arithmetic*, pages 50–57.

MacSorley, O. L. (1961). High-speed arithmetic in binary computers. *IRE Proceedings*, 49:67–91.

Magenheimer, D. J., L. Peters, K. W. Pettis, and D. Zuras (1988). Integer multiplication and division on the HP precision architecture. *IEEE Transactions on Computers*, C-37:980–90.

Mahant-Shetti, S. S., P. T. Balsara, and C. Lemonds (1999). High performance low power array multiplier using temporal tiling. *IEEE Transactions on Very Large Scale Integration (VLSI) Systems*, 7(1):121–24.

Makino, H., Y. Nakase, H. Suzuki, H. Morinaka, H. Shinohara, and K. Mashiko (1996). An 8.8-ns 54 × 54-bit multiplier with high speed redundant binary architecture. *IEEE Journal of Solid-State Circuits*, 31(6):773–83.

Mehta, M., V. Parmar, and E. E. Swartzlander (1991). High-speed multiplier design using multi-input counter and compressor circuits. In *Proceedings of the 10th IEEE Symposium on Computer Arithmetic*, pages 43–50.

Meier, P. C. H. (1999). *Analysis and Design of Low Power Digital Multipliers*. PhD thesis, Carnegie Mellon University.

Meier, P. C. H., R. A. Rutenbar, and L. R. Carley (1996). Exploring multiplier architecture and layout for low power. In *Proceedings of the IEEE 1996 Custom Integrated Circuits Conference*, pages 513–16.

Montoye, R. K., E. Hokonek, and S. L. Runyan (1990). Design of the floating-point execution unit of the IBM RISC System/6000. *IBM Journal of Research and Development*, 34(1):59–70.

Noll, T., D. Schmitt-Landsiedel, H. Klar, and G. Enders (1986). A pipelined 330-MHz multiplier. *IEEE Journal of Solid-State Circuits*, SC-21(6):411–16.

Oklobdzija, V. G., and D. Villeger (1995). Improving multiplier design by using improved column compression tree and optimized final adder in CMOS technology. *IEEE Transactions on VLSI*, 3(2):292–301.

Oklobdzija, V. G., D. Villeger, and S. S. Liu (1996). A method for speed optimized partial product reduction and generation of fast parallel multipliers using an algorithmic approach. *IEEE Transactions on Computers*, 45(3):294–306.

Owens, R. M., R. S. Bajwa, and M. J. Irwin (1995). Reducing the number of counters needed for integer multiplication. In *Proceedings of the 12th IEEE Symposium on Computer Arithmetic*, pages 38–41.

Pezaris, S. D. (1971). A 40 ns 17 bit by 17 bit array multiplier. *IEEE Transactions on Computers*, C-20(4):442–47.

Potkonjak, M., M. B. Srivastava, and A. P. Chandrakasan (1996). Multiple constant multiplications: Efficient and versatile framework and algorithms for exploring common subexpression elimination. *IEEE Transaction on Computer-Aided Design of Integrated Circuits and Systems*, 15(2):151–65.

Robertson, J. E. (1955). Two's complement multiplication in binary parallel computers. *IEEE Transactions on Electronic Computers*, EC-34(3):118–19.

Rubinfield, L. P. (1975). A proof of the modified Booth's algorithm for multiplication. *IEEE Transactions on Computers*, C-24(4):1014–15.

Sam, H., and A. Gupta (1990). A generalized multibit recoding of the two's complement binary numbers and its proof with application in multiplier implementations. *IEEE Transactions on Computers*, C-39(8):1006–15.

Santoro, M. R. (1989). *Design and Clocking of VLSI Multipliers*. PhD thesis, Stanford University.

Santoro, M. R., and M. A. Horowitz (1989). SPIM: a pipelined 64 × 64-bit iterative multiplier. *IEEE Journal of Solid-State Circuits*, 24:487–93.

Schulte, M. J., and E. Swartzlander (1993). Truncated multiplication with correction constant (for DSP). In *Proceedings of the IEEE Workshop on VLSI Signal Processing*, pages 388–96.

Schulte, M. J., J. E. Stine, and J. G. Jansen (1999). Reduced power dissipation through truncated multiplication. In *Proceedings of the IEEE Alessandro Volta Memorial Workshop on Low-Power Design*, pages 61–69.

Schulte, M. J., P. I. Balzola, A. Akkas, and R. W. Brocato (2000). Integer multiplication with overflow detection or saturation. *IEEE Transactions on Computers*, 49(7):681–91.

Schwarz, E. M., R. Averill, III, and L. Sigal (1997). A radix-8 CMOS S/390 multiplier. In *Proceedings of the 13th IEEE Symposium on Computer Arithmetic*, pages 2–9.

Seidel, P.-M., L. D. McFearin, and D. W. Matula (2001). Binary multiplication radix-32 and radix-256. In *Proceedings of the 15th IEEE Symposium on Computer Arithmetic (Arith-15)*, pages 23 32.

Song, P. J., and G. De Micheli (1991). Circuit and architecture trade-offs for high-speed multiplication. *IEEE Journal of Solid-State Circuits*, 26(9):1184–98.

Stelling, P. F. (1995). *Application of Combinatorics to Parallel Multiplier Design, Tree Reconstruction, and the Analysis of Strings*. PhD thesis, University of California, Davis.

Stelling, P. F., and V. G. Oklobdzija (1996). Design strategies for optimal hybrid final adders in a parallel multiplier. *Journal of VLSI Signal Processing*, 14(3):321–31.

Stelling, P. F., and V. G. Oklobdzija (1997). Implementing multiply-accumulate operation in multiplication time. In *Proceedings of the 13th IEEE Symposium on Computer Arithmetic*, pages 99–106.

Stelling, P. F., C. U. Martel, V. G. Oklobdžija, and R. Ravi (1998). Optimal circuits for parallel multipliers. *IEEE Transactions on Computers*, 47(3):273–85.

Strandberg, R. H., L. G. Bustamante, V. G. Oklobdzija, M. Soderstrand, and J. C. LeDuc (1996). Efficient realizations of squaring and reciprocal used in adaptive sample rate notch filter. *Journal of VLSI Signal Processing Systems*, 14(3):303–9.

Swartzlander, E. (1999). Truncated multiplication with approximate rounding. In *Proceedings of the 33rd Asilomar Conference on Signals, Systems, and Computers*, volume 2, pages 1480–83.

Takagi, N., and T. Horiyama (1999). A high-speed reduced-size adder under left-to-right input arrival. *IEEE Transactions on Computers*, 48(1):76–80.

Takagi, N., H. Yasukura, and S. Yajima (1985). High speed multiplication algorithm with a redundant binary addition tree. *IEEE Transactions on Computers*, C-34(9):789–96.

Van, L. D., S.-S. Wang, and W.-S. Feng (2000). Design of the lower error fixed-width multiplier and its application. *IEEE Transactions Circuits and Systems II: Analog and Digital Signal Processing*, 47(10):1112–18.

Vassiliadis, S., E. M. Schwarz, and D. J. Harahan (1989). A general proof for overlapped multiple-bit scanning multiplication. *IEEE Transactions on Computers*, 38:172–73.

Wallace, C. S. (1964). A suggestion for a fast multiplier. *IEEE Transactions on Electronic Computers*, EC-13(2):14–17.

Wang, Z., G. A. Jullien, and W. C. Miller (1995). A new design technique for column compression multipliers. *IEEE Transactions on Computers*, 44(8):962–70.

Ware, F. A., W. McAllister, J. R. Carlson, D. K. Sun, and R. J. Vlach (1982). 64 bit monolithic floating-point processors. *IEEE Journal of Solid-State Circuits*, SC-17(5):898–907.

Weinberger, A. (1981). 4:2 carry-save adder module. *IBM Technical Disclosure Bulletin*, 23.

Winograd, S. (1967). On the time required to perform multiplication. *Journal of the ACM*, 14(4):793–802.

Wires, K. E., M. J. Schulte, L. P. Marquette, and P. I. Balzola (1999). Combined unsigned and 2's complement squarers. In *Proceedings of the 33rd Asilomar Conference on Signals, Systems, and Computers*, pages 1215–19.

Wires, K. E., M. J. Schulte, and J. E. Stine (2000). Variable-correction truncated floating point multipliers. In *Proceedings of the 34th Asilomar Conference on Signals, Systems, and Computers*, pages 1344–48.

Wires, K. E., M. J. Schulte, and J. E. Stine (2001). Combined IEEE compliant and truncated floating point multipliers for reduced power dissipation. In *Proceedings of the IEEE International Conference on Computer Design: VLSI in Computers and Processors (ICCD'01)*, pages 497–500.

Yoshida, N., E. Goto, and S. Ichikawa (1991). Pseudorandom rounding for truncated multipliers. *IEEE Transactions on Computers*, 40(9):1065–67.

Zuras, D. (1993). On squaring and multiplying large integers. In *Proceedings of the 11th IEEE Symposium on Computer Arithmetic*, pages 260–71.

Zuras, D. (1994). More on squaring and multiplying large integers. *IEEE Transactions on Computers*, 43(8):899–908.

Zurawski, J. H. P., and J. B. Gosling (1987). Design of a high-speed square root, multiply and divide unit. *IEEE Transactions on Computers*, C-36(1):13–23.

IN THIS CHAPTER WE PRESENT AND DISCUSS THE FOLLOWING MAIN TOPICS:

- Fractional division: algorithm and implementation

CHAPTER 5

Division by Digit Recurrence

This chapter describes algorithms and implementations for the division operation. Several classes of algorithms exist for this operation, the most used being the digit recurrence method, the multiplicative method, various approximation methods, and special methods such as the CORDIC and continued product methods. The algorithms and implementations of the type discussed here are based on a *digit recurrence*. In this method, the quotient is represented in a radix-r form and one digit of it is obtained per iteration. Many of the techniques presented here are applicable to other digit recurrences, such as square root and reciprocal square root, as well as to the class of online algorithms.

The implementations of the algorithms presented can be either sequential, combinational, or a combination of both; moreover, in the combinational case the implementation can be pipelined or nonpipelined. The design space is large since many parameters are involved and the best solution depends on the particular requirements. Thus, it is impractical to describe the set of good designs. In this chapter we concentrate on the method of design and give examples of implementations that are representatives of the different approaches.

First, we present the algorithm for *fractional* operands and result, which relates directly to the requirements of floating-point processors (see Chapter 8); we then discuss the modifications required for *integer* division. We concentrate on algorithms that use *redundant quotient-digit sets* since these have significant speed and cost advantages. The most difficult problem in the digit recurrence division algorithms is selection of quotient digits. We discuss a general theory useful in developing quotient-digit selection functions and describe several instances of selection functions and their implementations.

5.1   Definition and Notation

The division operation is defined by the following expressions:

$$x = q \cdot d + rem \qquad\qquad \textbf{5.1}$$

and

$$|rem| < |d| \cdot ulp \quad \text{and} \quad sign(rem) = sign(x) \qquad\qquad \textbf{5.2}$$

where the *dividend x* and the *divisor d* are the operands and the results are the *quotient q* and, optionally, the *remainder rem*. The unit in the last position (*ulp*) defines the granularity of the quotient. The two most typical cases produce a fractional quotient or an integer quotient. For these cases we have

- fractional quotient: $ulp = r^{-n}$ (for radix-r representation and n-digit quotient)
- integer quotient: $ulp = 1$

Correspondingly two types of division operation are defined:

1. Fractional division, in which operands and result are fractions. This case is directly related to floating-point division.[1]

2. Integer division, with integer operands and result.

The bulk of this chapter considers the fractional case. We then, in Section 5.4, describe integer division and discuss its integration with fractional division. Moreover, the most-frequently used representation of operands/result is sign-and-magnitude, so we consider only magnitudes.

Normalized Divisor

As will be apparent in the next sections, the preferred division algorithms require that the divisor be in normalized form. For the fractional case, this corresponds to

$$1/2 \le d < 1 \qquad\qquad \textbf{5.3}$$

---

1. Although the IEEE Floating Point Standard 754 uses significands in the range [1,2), the adaptation to this format is straightforward (see Chapter 8).

This is usually the case for floating-point representations. On the other hand, when this restriction is not part of the representation of the operand, a prenormalization step has to be included. This is achieved by a left shift, as is described in Section 5.4 for integer division.

Range of Quotient

For normalized fractional divisor and fractional dividend (not necessarily normalized), the quotient is in the range $0 < q < 2$. If a normalized fraction is required, a normalization step should be included.

5.2   Algorithm and Implementation of Fractional Division

The digit recurrence algorithm consists of n iterations of a recurrence, in which each iteration (step) produces one digit of the quotient. This is preceded by an initialization step and followed by a termination step. We now consider these steps, beginning with the recurrence step, which is the core of the algorithm.

5.2.1   Recurrence Step

Let us call $q[j]$ the value of the quotient after j steps; that is,

$$q[j] = q[0] + \sum_{i=1}^{j} q_i r^{-i} \qquad 5.4$$

where $q[0]$ is determined by the initialization. The n-digit final quotient is then

$$q = q[n] = q[0] + \sum_{i=1}^{n} q_i r^{-i} \qquad 5.5$$

The quotient-digit set plays a crucial role in the characteristics of the algorithm. The most-direct choice is to use the canonical digit set such that $0 \leq q_j \leq r - 1$. This leads to the basic restoring division, which is not convenient because of an "expensive" quotient-digit selection. For radix 2, the situation is somewhat improved by using the digit set $\{-1, 1\}$ (no 0), resulting in a nonrestoring algorithm. However, because of the nonredundant nature of the digit set, the quotient-digit selection is still complex.[2] In the rest of the chapter, to obtain a simpler selection function (as is shown later), we use a *redundant digit set*. In particular we use the symmetric signed-digit set of consecutive integers.[3]

$$q_j \in \mathcal{D}_a = \{-a, -a+1, \ldots, -1, 0, 1, \ldots, a-1, a\} \qquad \textbf{5.6}$$

Since for redundant representation more than r consecutive integer values including zero are needed (exactly r values produce a nonredundant representation), a has to satisfy

$$a \geq \lceil r/2 \rceil \qquad \textbf{5.7}$$

The redundancy factor ρ is defined as

$$\rho = \frac{a}{r-1}, \quad \frac{1}{2} < \rho \leq 1 \qquad \textbf{5.8}$$

From the definition, a correct division algorithm must produce a quotient q with a positive error (remainder) of less than one *ulp*, with respect to the infinite precision value.[4] That is, for fractional division the error is bounded by

$$0 \leq \epsilon_q = \frac{x}{d} - q < r^{-n} \qquad \textbf{5.9}$$

Moreover, the recurrence has to converge to this error. Calling $\epsilon[j]$ the error after iteration j, we have

$$\epsilon[j] = \frac{x}{d} - q[j] \leq \epsilon[n] + \sum_{i=j+1}^{n} \max(q_i) r^{-i} = \epsilon[n] + \frac{a}{r-1}(r^{-j} - r^{-n}) \quad \textbf{5.10}$$

This equation has as solution[5]

$$\epsilon[j] \leq \rho r^{-j} \qquad \textbf{5.11}$$

As will become clear in Section 5.5, to make use of the negative values of the quotient digit it is necessary to have also negative errors after iteration j,

---

2. Restoring and nonrestoring algorithms are reviewed in Chapter 1.
3. See Chapter 2. There are some instances in which a nonsymmetric set might be preferable; we do not consider this generalization here.
4. This is consistent with the requirement that the remainder be bounded by $|d| \cdot ulp$.
5. The equal condition in the next expression is not applicable for the case $\rho = 1$; in this case it is necessary to use the $<$ condition to assure that the final error is less than r^{-n}.

so that

$$|\epsilon[j]| = \left| \frac{x}{d} - q[j] \right| \le \rho r^{-j} \qquad \textbf{5.12}$$

This assures that the magnitude of the error produced after n iterations is bounded by r^{-n}. However, this error can be negative; in such a case, a correction step is required, as discussed in Section 5.2.2.

From expression (5.12) the recurrence is obtained as follows. First, multiply by d, to eliminate the division operation. We get

$$|x - dq[j]| \le \rho d r^{-j} \qquad \textbf{5.13}$$

The bound of (5.13) decreases with j. To have a variable whose bound is independent of j, we define the residual (or partial remainder) w so that

$$w[j] = r^j (x - dq[j]) \qquad \textbf{5.14}$$

with bound

$$|w[j]| \le \rho d \qquad \textbf{5.15}$$

To obtain the recurrence we compute

$$w[j+1] - rw[j] = r^{j+1}(q[j] - q[j+1])d$$

Since, from (5.4), $q[j+1] = q[j] + q_{j+1} r^{-(j+1)}$, the recurrence is

$$w[j+1] = rw[j] - dq_{j+1} \qquad \textbf{5.16}$$

with the initial value obtained from (5.14) by setting $j = 0$:

$$w[0] = x - dq[0] \qquad \textbf{5.17}$$

Expression (5.16) is the basic recurrence on which the division algorithms are based.

The recurrence is performed so that $w[j+1]$ is bounded by (5.15). This is accomplished by selecting a suitable value of q_{j+1} by means of the *quotient-digit selection function*

$$q_{j+1} = SEL(rw[j], d) \qquad \textbf{5.18}$$

Actually, the use of a redundant digit set for q_j allows that the selection function uses $\widehat{y}$, a truncated $rw[j]$, and $\widehat{d}$, a truncated d. That is,

$$q_{j+1} = SEL(\widehat{y}, \widehat{d}) \qquad \textbf{5.19}$$

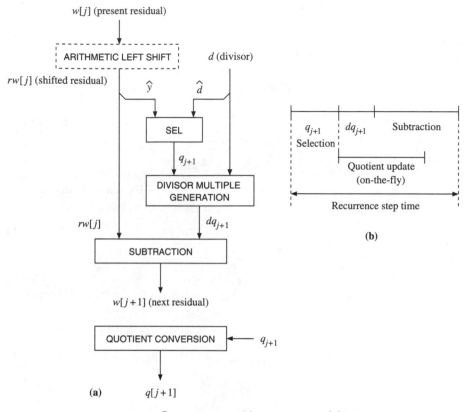

FIGURE 5.1 Recurrence step: (a) components and (b) timing.

The number of bits of $\widehat{y}$ and $\widehat{d}$ depends on the radix and on the quotient-digit set, as discussed in Section 5.5

Implementation of a Recurrence Step

As indicated by the recurrence, each iteration consists of five subcomputations (Figure 5.1(a)):

1. One digit arithmetic left shift of $w[j]$ to produce $rw[j]$

2. Determination of the quotient digit q_{j+1} by the quotient-digit selection function

3. Generation of the divisor multiple $d \times q_{j+1}$

4. Subtraction of dq_{j+1} from $rw[j]$

5. Update of the quotient $q[j]$ to $q[j + 1]$ by the on-the-fly conversion

$$q[j + 1] = CONV(q[j], q_{j+1}) \qquad\qquad \textbf{5.20}$$

discussed in Section 5.2.3, producing at each recurrence step the corresponding quotient value in a nonredundant form

The five subcomputations are executed as indicated in the timing diagram of Figure 5.1(b). Note that no time has been allocated for the arithmetic shift since it is implemented by suitable wiring. Moreover, the relative magnitudes of the delay of each of the components depend on the specific implementation. The quotient update is not in the critical path.

This general description of the recurrence step can result in different specific versions depending on several interrelated factors. We now list the most important factors and mention their effect on the overall execution time and cost of the implementation; the reasons for these effects are clarified later.

1. **Radix r.** For the same quotient precision, the number of iterations of the algorithm is reduced by a factor k when going from a radix r to a radix r^k. However, this increase in radix produces a more complex implementation because of the quotient-digit selection and the generation of the divisor multiples. This additional complexity increases the time of each iteration.

2. **Quotient-digit set.** As indicated, a redundant signed-digit set is used to simplify the quotient-digit. The value of the redundancy factor influences the complexity of the quotient-digit selection and of the generation of the divisor multiples in an opposite manner: a higher ρ reduces complexity of the selection function but increases complexity of the generation of the divisor multiples. Consequently, the choice of ρ is an important design decision.

3. **Representation of the residual.** In particular, it can be represented in *nonredundant* form (for example, conventional two's complement representation) or *redundant* form (for example, carry-save two's complement representation or signed-digit representation). The redundant form has the big advantage that the addition/subtraction part of the

iteration is done using a *carry-free adder* (and is, therefore, fast).[6] Its disadvantages are that it complicates somewhat the quotient-digit selection and that it increases the number of register bits required to store the residual. Moreover, if the remainder is needed, the final residual has to be converted to conventional representation. Even if the remainder is not needed, a sign detection has to be implemented for the correction step, indicated in Section 5.2.2.

4. **Quotient-digit selection function.** The complexity of the implementation of this function depends on all the previous factors. Its delay is an important contributor to the iteration time, especially when a carry-free adder is used.

5.2.2   Initialization, Number of Iterations, and Termination

We now consider the steps of the whole algorithm, which consists of an initialization step, a number of iterations of the recurrence step, and a termination step.

The initialization should satisfy the initial value of the residual and assure convergence. As indicated, $w[0] = x - dq[0]$ and the residual bound is $|w[0]| \leq \rho d$. The following options for the initializations are possible:

*   Make $q[0] = 0$. For $\rho = 1$, we make $w[0] = x/2$; consequently, $w[0] < \frac{1}{2}$ for $x < 1$.
    For $\frac{1}{2} < \rho < 1$, we make $w[0] = x/4$; this produces $w[0] < \frac{1}{4} < \rho d$.
    This initialization produces a scaled quotient (divided by two or by four) in the range $0 \leq q \leq \rho$. To obtain the correct quotient ($0 \leq q < 2$), a scaling by two or four is required during the termination step. Because of this scaling, to obtain a final quotient of n bits, one or two additional bits have to be computed.
*   For $\rho = 1$ and a normalized dividend, make $q[0] = 1$ and $w[0] = x - d$, which results in $|w[0]| < \frac{1}{2}$. In this case the algorithm converges only if the quotient is $q \leq 1 + \rho$ (that is why it can be used only for $\rho = 1$). Moreover, this method requires additional hardware to compute $x - d$.

---

6. We use the term "carry-free adder" to denote adders, such as carry-save adders and signed-digit adders, characterized by carry chains of fixed (small) length (see Chapter 2).

The number of iterations N of the recurrence step is dependent on the number of bits of the final quotient, the scaling of the dividend introduced by the initialization, the guard bit required for rounding (see Chapter 8), and the radix. For example, for a 53-bit quotient, using a radix 8 with $\rho = 1$, and rounding (one guard bit) the number of iterations is

$$N = \left\lceil \frac{53 + 1 + 1}{3} \right\rceil = 19$$

The termination step has to account for the following:

- The algorithm can produce a negative final residual $w[N]$. On the other hand, the definition of division requires a nonnegative remainder (for positive dividend).[7] Consequently, it is necessary to have a correction step that adjusts the quotient as follows:

$$q = \begin{cases} q[N] & \text{if } w[N] \geq 0 \\ q[N] - r^{-N} & \text{if } w[N] < 0 \end{cases} \qquad \textbf{5.21}$$

- If the dividend has been shifted for initialization (divided by two or by four), this is compensated by shifting the quotient correspondingly.[8]
- For most floating-point implementations it is required to detect the zero-remainder condition, to determine exact quotient and for rounding. This condition is determined from $w[N] = 0$ and the bits of $q[N]$ after digit n.

Implementation of Initialization and Termination

The initialization is implemented by a fixed shift of one or two positions or by a subtraction $(x - d)$. When the residual is in carry-save format, no actual subtraction might be required.

The termination is implemented by a sign detection of the residual, by the conditional decrement of the quotient (when the residual is negative), and by a fixed shift (for the case in which the initialization is done with a shift).

---

7. In some applications the sign of the remainder can be arbitrary, as long as the remainder is bounded; in such cases the correction is not necessary.
8. This produces a quotient in the range $(\frac{1}{2}, 2)$. In a floating-point unit the quotient is normalized and rounded (see Chapter 8), so that the shifting can be included as part of this process.

5.2.3   On-the-Fly Conversion

The quotient has to be converted from signed-digit representation to conventional representation. This can be done with an addition step after the quotient is completely computed. However, this addition would increase the overall execution time. To avoid this, we now discuss an algorithm that performs the conversion in a digit-serial manner as the digits of the quotient are produced.

A possibility is the following algorithm. Let $Q[j]$ be the digit vector of the converted quotient consisting of the j most-significant digits, that is,

$$Q[j] = \sum_{i=1}^{j} q_i r^{-i} \qquad \textbf{5.22}$$

Then we have

$$Q[j + 1] = Q[j] + q_{j+1} r^{-(j+1)} \qquad \textbf{5.23}$$

Since q_{j+1} can be negative, we can use the following algorithm for this addition:

$$Q[j + 1] = \begin{cases} Q[j] + q_{j+1} r^{-(j+1)} & \text{if } q_{j+1} \geq 0 \\ Q[j] - r^{-j} + (r - |q_{j+1}|) r^{-(j+1)} & \text{if } q_{j+1} < 0 \end{cases} \qquad \textbf{5.24}$$

This algorithm has the disadvantage that the subtraction $Q[j] - r^{-j}$ requires the propagation of a borrow and, therefore, is slow. To avoid this propagation we keep another form, $QM[j]$, with value

$$QM[j] = Q[j] - r^{-j} \qquad \textbf{5.25}$$

Using this second form, the conversion algorithm is

$$Q[j + 1] = \begin{cases} Q[j] + q_{j+1} r^{-(j+1)} & \text{if } q_{j+1} \geq 0 \\ QM[j] + (r - |q_{j+1}|) r^{-(j+1)} & \text{if } q_{j+1} < 0 \end{cases} \qquad \textbf{5.26}$$

so that the subtraction is replaced by loading the form $QM[j]$. It is necessary to update also the form $QM[j]$, as follows:

$$QM[j + 1] = Q[j + 1] - r^{-(j+1)}$$

$$= \begin{cases} Q[j] + (q_{j+1} - 1) r^{-(j+1)} & \text{if } q_{j+1} > 0 \\ QM[j] + ((r - 1) - |q_{j+1}|) r^{-(j+1)} & \text{if } q_{j+1} \leq 0 \end{cases} \qquad \textbf{5.27}$$

j	q_j	$Q[j]$	$QM[j]$
0		0	0
1	1	0.1	0.0
2	1	0.11	0.10
3	0	0.110	0.101
4	1	0.1101	0.1100
5	-1	0.11001	0.11000
6	0	0.110010	0.110001
7	0	0.1100100	0.1100011
8	-1	0.11000111	0.11000110
9	1	0.110001111	0.110001110
10	0	0.1100011110	0.1100011101
11	1	0.11000111101	0.11000111100
12	0	0.110001111010	0.110001111001

TABLE 5.1 Example of conversion.

Now all additions are concatenations, so no carry/borrow is propagated. We call this an *on-the-fly conversion algorithm*. In terms of concatenations the algorithm is

$$Q[j+1] = \begin{cases} (Q[j], q_{j+1}) & \text{if } q_{j+1} \geq 0 \\ (QM[j], (r - |q_{j+1}|)) & \text{if } q_{j+1} < 0 \end{cases} \qquad \textbf{5.28}$$

$$QM[j+1] = \begin{cases} (Q[j], q_{j+1} - 1) & \text{if } q_{j+1} > 0 \\ (QM[j], ((r-1) - |q_{j+1}|)) & \text{if } q_{j+1} \leq 0 \end{cases} \qquad \textbf{5.29}$$

with the initial conditions $Q[0] = QM[0] = 0$ (for a positive quotient).

As an example consider the radix-2 case in Table 5.1.

Implementation of the Conversion

The implementation of the algorithm requires two registers to contain $Q[j]$ and $QM[j]$, respectively. These registers are shifted one digit left with insertion in the least-significant digit, depending on the value of q_{j+1}. They also require parallel loading to load $Q[j]$ with $QM[j]$ and vice versa. This implementation is shown

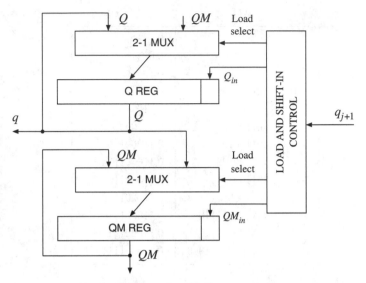

FIGURE 5.2 Implementation of on-the-fly conversion.

in Figure 5.2. The operations on these registers are

$$Q \leftarrow \begin{cases} \textit{shift Q with insert } (Q_{in}) & \textit{if } C_{shiftQ} = 1 \\ \textit{shift QM with insert } (Q_{in}) & \textit{if } C_{loadQ} = 1 \end{cases} \qquad \textbf{5.30}$$

$$QM \leftarrow \begin{cases} \textit{shift QM with insert } (QM_{in}) & \textit{if } C_{shiftQM} = 1 \\ \textit{shift Q with insert } (QM_{in}) & \textit{if } C_{loadQM} = 1 \end{cases} \qquad \textbf{5.31}$$

where

$$Q_{in} = \begin{cases} q_{j+1} & \textit{if } q_{j+1} \geq 0 \\ r - |q_{j+1}| & \textit{if } q_{j+1} < 0 \end{cases} \qquad \textbf{5.32}$$

$$QM_{in} = \begin{cases} q_{j+1} - 1 & \textit{if } q_{j+1} > 0 \\ (r-1) - |q_{j+1}| & \textit{if } q_{j+1} \leq 0 \end{cases} \qquad \textbf{5.33}$$

and the register control signals $C_{loadQ} = C'_{shiftQ}$ and $C_{loadQM} = C'_{shiftQM}$. Table 5.2 describes the operation for the radix-4 case.

q_{j+1}	Q_{in}	C_{shiftQ}	$Q[j+1]$	QM_{in}	$C_{shiftQM}$	$QM[j+1]$
3	3	1	$(Q[j], 3)$	2	0	$(Q[j], 2)$
2	2	1	$(Q[j], 2)$	1	0	$(Q[j], 1)$
1	1	1	$(Q[j], 1)$	0	0	$(Q[j], 0)$
0	0	1	$(Q[j], 0)$	3	1	$(QM[j], 3)$
−1	3	0	$(QM[j], 3)$	2	1	$(QM[j], 2)$
−2	2	0	$(QM[j], 2)$	1	1	$(QM[j], 1)$
−3	1	0	$(QM[j], 1)$	0	1	$(QM[j], 0)$

TABLE 5.2 Control signals and operations for radix-4 on-the-fly conversion.

Quotient Rounding

When division is performed in a floating-point unit, usually the result has to be rounded. When the on-the-fly conversion is used, this rounding can be incorporated as part of the conversion. This is discussed further in Chapter 8.

5.3   Implementations of the Division Algorithm

As indicated, the core of the division algorithm consists of N iterations of the recurrence. The implementation of this core can be *totally sequential*, where the hardware of the recurrence step is reused for all the iterations and the residual is updated in a register (Figure 5.3(a)); *totally combinational*, where the hardware for the recurrence step is replicated (Figure 5.3(b)); or a combination of both, where the step hardware is replicated k times and this superstep is reused N/k times (Figure 5.3(c)). The combinational implementations can be pipelining so that several division operations can use the hardware at the same time, with the corresponding increase in throughput. The selection of one of these alternatives is influenced by cost, speed, and throughput considerations.

    The alternative implementations above are the same as discussed for multiplication in Chapter 4. However, there is a significant difference in the combinational implementations. While in multiplication, because of the associativity of addition, the sequence of additions can be performed either in a linear array or in a tree, for division only the linear structure is possible because of the dependency introduced by the quotient-digit selection. This, together with the fact that the

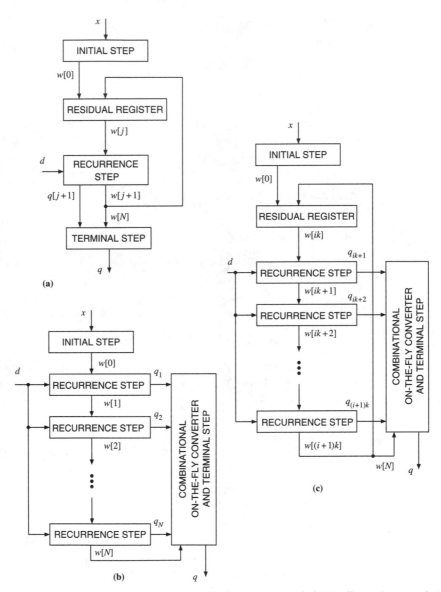

FIGURE 5.3 Division implementation: (a) Totally sequential. (b) Totally combinational. (c) Combined implementation. ("Recurrence step" in (b) and (c) does not include quotient conversion part.)

delay of the quotient-digit selection is a significant portion of the iteration delay, makes the combinational implementation of division less attractive.

In addition to this core, an initialization step and a termination step are required, which can be implemented by additional cycles or incorporated in the first and last iterations.

5.3.1   Examples of Algorithms and Implementations

We now illustrate several typical division algorithms and their implementations. These algorithms show a progression of radices, namely, radix 2, 4, 8, 16, and 512. The radix 2, 4, and 8 cases correspond directly to instances of the radix-r algorithm described in the previous section. For higher radices, the direct algorithm results in an impractical implementation, mainly because of the complexity of the quotient-digit selection function; consequently, we illustrate a radix-16 implementation that consists of two overlapped radix-4 iterations per cycle, and a radix-512 implementation, which uses prescaling of the divisor (and dividend) and selection by rounding.

In all cases, carry-save adders are used for the residual recurrence since this results in a faster iteration. Included for the radix-2 and the radix-4 cases are the quotient-digit selection functions, which are developed in the next section. Because of the redundancy provided by the quotient-digit set, these selection functions have as input the truncated residual and the truncated divisor.

Specifically, we illustrate and compare the following algorithms:

- **r2 scheme.** Radix 2 with carry-save residual (quotient digit set $\{-1, 0, 1\}$).
- **r4 scheme.** Radix 4 with carry-save residual (quotient digit set $\{-2, 1, 0, 1, 2\}$).
- **r8 scheme.** Radix 8 with carry-save residual ($-7 \leq q_j \leq 7$).
- **r16over scheme.** Radix 16 with two overlapped radix-4 stages.
- **r512 scheme.** Radix 512 with carry-save residual, scaling, and quotient-digit selection by rounding ($-511 \leq q_j \leq 511$).

The estimates of the execution time and area reflect what is typical for CMOS standard-cell libraries. More accurate evaluations need the use of specific data for the particular library. Since the actual execution time and area are technology dependent, we give here relative values for the schemes compared. These should be more technology independent and give an indication of the merits of each scheme. Specifically:

- Cells are modeled by a delay (as a function of the load) and an area. Delays and areas are given in units of 2-input NAND gates. The unit of delay assumes a fanout of three NAND gates. We include the delay and area of registers for the operands, the result, and the residual. This assumes that the implementation uses just one stage for the iterations. If, on the other hand, the stages are unfolded to produce a higher-radix divider, the residual register has to be counted only once.
- Interconnections are not included: we have not considered the delay, area, nor load of interconnections.
- Degree of optimization: The same modules have been used in all designs. Consequently, additional optimizations might be applied to most of them. However, the cycle time and area ratios should not change significantly.
- The execution time and the area are calculated for 53-bit operands and 54-bit result (which is typical for a double-precision floating-point implementation; the additional bit of the result is used for rounding to produce a final 53-bit quotient).
- All implementations are composed of the basic modules whose characteristics are given in Figure 5.4.

The detection of the negative-remainder and zero-remainder conditions could be performed by first converting the carry-save representation to conventional. However, that would require a carry-propagate adder, which is bulky and slow. Consequently, we consider an implementation directly from the carry-save representation.

Because the carry-save representation is redundant, the zero-remainder detection is difficult. However, the representation of -2^{-b}, where b is the number of fractional bits of the last residual, is unique. Moreover, in this representation the sum of the sum and the carry bits is 1 for all positions. Consequently, we first obtain

$$P = W_L - 2^b \qquad\qquad 5.34$$

where W_L is the last residual, represented in carry-save by WS and WC.

This is produced by a [3:2] addition with a third bit-vector of all 1s. The implementation simplifies to the following switching expressions:

$$PS_i = (WS_i \oplus WC_i)', \qquad PC_{i-1} = WS_i + WC_i \qquad\qquad 5.35$$

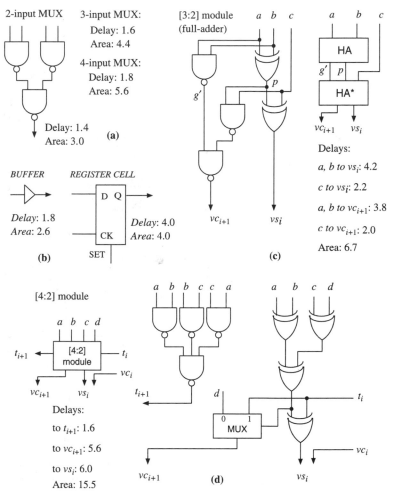

FIGURE 5.4 Basic modules: (a) Multiplexers. (b) Buffer and register cell. (c) Full-adder. (d) [4:2] module.

Then the zero-remainder condition is obtained by

$$p_i = PS_i \oplus PC_i, \qquad zero = \prod_{i=0}^{b} p_i \qquad\qquad \textbf{5.36}$$

The sign can also be detected using the PS and PC forms instead of WS and WC: if $ps + pc \geq 0$, then $ws + wc > 0$, and if $ps + pc > 0$, then $ws + wc \leq 0$. Therefore,

$$sign = (p_0 \oplus c_0)zero' \qquad \textbf{5.37}$$

where c_0 is the carry into the most-significant position (sign position).

The carry is obtained as $c_0 = g_{(1,b)} = G_{out}$ by a tree of P, G cells, which also produces $P_{out} = p_{(1,b)}$ resulting in the following:[9]

$$zero = p_0 P_{out} \qquad \textbf{5.38}$$

$$sign = (p_0 \oplus G_{out})zero' \qquad \textbf{5.39}$$

The implementation is shown in Figure 5.5(a).

Radix-2 Division with Residual in Carry-Save Form

The algorithm is summarized in Figure 5.6, and an illustration of the first four steps is given in Figure 5.7. The quotient-digit selection function for this scheme is discussed in Section 5.5 (Example 5.2).[10]

The corresponding implementation using the modules of Figure 5.4 is shown in Figure 5.8. The characteristics of this implementation are given in Table 5.3. Moreover, the critical path is shown in Figure 5.8 and summarized together with the area in Table 5.3.

Radix-4 Division with Residual in Carry-Save Form

The radix-4 algorithm with the residuals in carry-save form is similar to the radix-2 algorithm in Figure 5.6 with the following differences:

1. We consider the case where the quotient-digit set is $\{-2, -1, 0, 1, 2\}$. This is a redundant digit set and allows a simple implementation of $q_{j+1}d$.

2. For this case ($\rho < 1$) we initialize $WS[0] \leftarrow x/4$.

3. The next residual is

$$(WC[j + 1], \ WS[j + 1]) \leftarrow CS\,ADD(4WC[j], \ 4WS[j], \ -q_{j+1}d)$$

---

9. See Chapter 2 for a definition of these signals.
10. For this radix the input to the selection function does not include the truncated divisor.

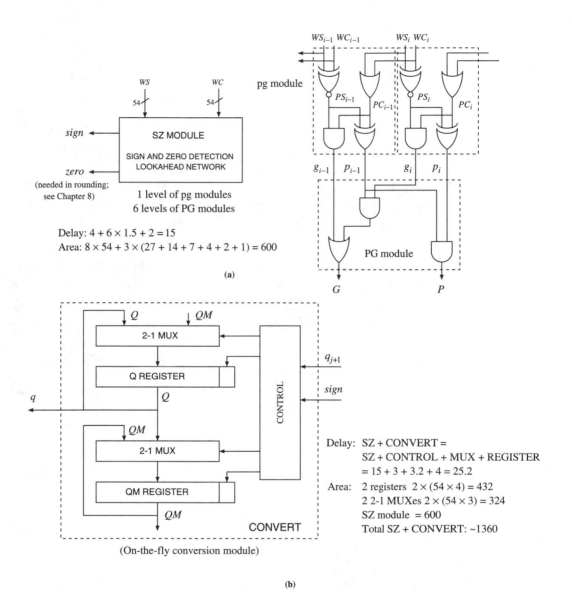

FIGURE 5.5 Implementation of (a) sign and zero-remainder detection network and (b) quotient conversion network.

(a) [*Initialize*]
    $WS[0] \leftarrow x/2; \ WC[0] \leftarrow 0; \ Q[-1] = 0, q_0 = 0$ (for the conversion)

(b) [*Recurrence*]
    for $j = 0 \ldots n + 1$ ($n + 2$ iterations because of initialization and guard bit)
    $q_{j+1} \leftarrow SEL(\widehat{y});$
    $(WC[j+1], \ WS[j+1]) \leftarrow CSADD(2WC[j], \ 2WS[j], \ -q_{j+1}d);$
    $Q[j] \leftarrow CONVERT(Q[j-1], \ q_j)$
    end for

(c) [*Terminate*]
    If $w[n+2] < 0$ then $q = 2(CONVERT(Q[n+1], \ q_{n+2} - 1))$
    else $q = 2(CONVERT(Q[n+1], \ q_{n+2}))$

where

- the residual is in redundant form, represented by the sum WS and stored-carry WC bit-vectors, i.e., $w[j] = (WC[j], \ WS[j])$,

- n is the precision in bits,

- $q_j \in \{-1, 0, 1\}$ is the jth quotient digit,

- SEL is the quotient-digit selection function (discussed in Section 5.5, Example 5.2):

$$q_{j+1} = SEL(\widehat{y}) = \begin{cases} 1 & \text{if } 0 \leq \widehat{y} \leq 3/2 \\ 0 & \text{if } \widehat{y} = -1/2 \\ -1 & \text{if } -5/2 \leq \widehat{y} \leq -1 \end{cases}$$

  with $\widehat{y}$ the value of the truncated carry-save shifted residual ($2w[j]$) with four bits (three integer bits and one fractional bit).

  Because of the range of $\widehat{y}$, $2w[j]$ requires also three integer bits and, therefore, $w[j]$ has two integer bits.

- $CSADD$ is carry-save addition,

- $-q_{j+1}d$ is in two's complement form, and

- $CONVERT$ is on-the-fly conversion function producing the accumulated quotient in conventional representation (discussed in Section 5.2.3).

FIGURE 5.6 Radix-2 algorithm with residual in carry-save form.

Dividend $x = (0.10011111)$, divisor $d = (0.11000101)$,
scaled residual $2w[0] = 2(x/2) = x$, $q_{computed} = q/2$

$2WS[0]^+ =$	000.10011111		
$2WC[0]^+ =$	000.00000001 *	$\widehat{y}[0] = 0.5$	$q_1 = 1$
$-q_1 d =$	11.00111010		

$WS[1] =$	11.10100100		
$WC[1] =$	00.00110110		

$2WS[1]^+ =$	111.01001000		
$2WC[1]^+ =$	000.01101100	$\widehat{y}[1] = -1$	$q_2 = -1$
$-q_2 d =$	00.11000101		

$WS[2] =$	11.11100001		
$WC[2] =$	00.10011000		

$2WS[2]^+ =$	111.11000010		
$2WC[2]^+ =$	001.00110001 *	$\widehat{y}[2] = 0.5$	$q_3 = 1$
$-q_3 d =$	11.00111010		

$WS[3] =$	01.11001001		
$WC[3] =$	10.01100100		

$2WS[3]^+ =$	011.10010010		
$2WC[3]^+ =$	100.11001001 *	$\widehat{y}[3] = 0$	$q_4 = 1$
$-q_4 d =$	11.00111010		

$WS[4] =$	00.01100001		
$WC[4] =$	11.00110100		

$2WS[4]^+ =$	000.11000010		
$2WC[4]^+ =$	110.01101000	$\widehat{y}[4] = -1.5$	$q_5 = -1$
$-q_5 d =$	00.11000101		

$w[5] =$	11.11101111

* least-significant 1 for two's complement of $q_{j+1} d$
\+ only two integer bits in the recurrence, because of the range of $w[j+1]$.

$$q[5] = .1\bar{1}11\bar{1} = .01101$$

FIGURE 5.7 Example of radix-2 division with residual in carry-save form. (On-the-fly conversion and termination not shown.)

Element	Delay	Area
Quotient-digit selection	6.8	50
Buffers	1.8	5
MUX	1.4	160
CSA	2.2	360
Registers (3)	4.0	650
Convert	(NC)	1360
Cycle time	16.2	
Total area		2585

Note: NC denotes a delay not in the critical path.

TABLE 5.3 Radix-2 stage.

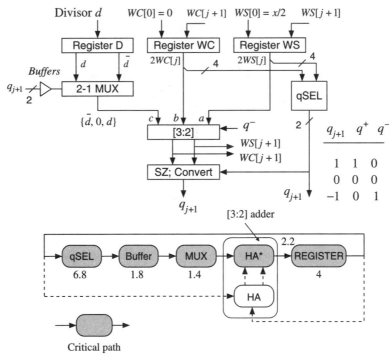

FIGURE 5.8 Implementation of radix-2 scheme and its critical path. The q^- at the input of the [3:2] adder is a carry used to produce the two's complement of d. Modules defined in Figure 5.4.

4. The quotient-digit selection has as arguments the truncated carry-save shifted residual $\widehat{y}$ and the truncated divisor $\widehat{d}$ (this is in contrast to the radix-2 case, in which the selection is independent of the divisor). As presented in detail in Section 5.5, the selection is described in terms of selection constants $m_k(i)$ so that

$$q_{j+1} = k \quad \text{if } m_k(i) \leq \widehat{y} < m_{k+1}(i) \quad k \in \{-2, -1, 0, 1, 2\}$$

where

- $i = 16\widehat{d}$ and $\widehat{d}$ is the divisor truncated to the fourth fractional bit. Since $1/2 \leq d < 1$, we get $8 \leq i \leq 15$

- $\widehat{y}$ is $4w[j]$ in carry-save form and truncated to the fourth fractional bit. Its range is $-44/16 \leq \widehat{y} \leq 42/16$, with three integer bits for a total of seven bits

- $m_3(i) = \max(\widehat{y}) + ulp$ and $m_{-2}(i) = \min(\widehat{y})$

As developed in Section 5.5, Example 5.3, the corresponding selection constants are given by the following table:

i	8	9	10	11	12	13	14	15
$m_2(i)^+$	12	14	15	16	18	20	20	24
$m_1(i)^+$	4	4	4	4	6	6	8	8
$m_0(i)^+$	−4	−6	−6	−6	−8	−8	−8	−8
$m_{-1}(i)^+$	−13	−15	−16	−18	−20	−20	−22	−24

+: real value = shown value/16

5. Because of the initialization, the final quotient is produced by multiplying the obtained quotient by four.

An example of execution is shown in Figure 5.9.

An implementation of the radix-4 scheme is shown in Figure 5.10. The critical path is shown in Figure 5.10 and is summarized together with the area in Table 5.4.

Dividend $x = (0.10101111)$, divisor $d = (0.11000101)$ ($i = 16(0.1100)_2 = 12$)
scaled residual $4w[0] = 4(x/4) = x$, $q_{computed} = q/4$

$$
\begin{array}{rll}
4WS[0]^+ = & 000.10101111 & \\
4WC[0]^+ = & 000.00000001 \ ^* \ \widehat{y}[0] = 10/16 & q_1 = 1 \\
-q_1d^+ = & 11.00111010 & \\
\hline
WS[1] = & 1.10010100 & \\
WC[1] = & 0.01010110 & \\
\hline
4WS[1]^+ = & 110.01010000 & \\
4WC[1]^+ = & 001.01011000 \quad \widehat{y}[1] = -6/16 & q_2 = 0 \\
-q_2d^+ = & 00.00000000 & \\
\hline
WS[2] = & 1.00001000 & \\
WC[2] = & 0.10100000 & \\
\hline
4WS[2]^+ = & 100.00100000 & \\
4WC[2]^+ = & 010.10000001 \ ^* \ \widehat{y}[2] = -22/16 & q_3 = -2 \\
-q_3d^+ = & 01.10001010 & \\
\hline
w[3] = & 0.00101011 & \\
\end{array}
$$

* least-significant 1 for two's complement of $q_{j+1}d$
+ only one integer bit used in the recurrence, because of the range of $w[j+1]$.

$$q[3] = .10\bar{2}_4 = .032_4$$

FIGURE 5.9 Example of radix-4 division with residual in carry-save form. (On-the-fly conversion and termination not shown.)

Radix-8 Division with Residual in Carry-Save Form

For the radix-8 implementation, we describe the case with quotient-digit set $\{-7, \ldots, 7\}$. To simplify the generation of dq_{j+1}, the quotient digit is decomposed into two components so that $q_{j+1} = q_{j+1}^H + q_{j+1}^L$ with $q_{j+1}^H = \{-8, -4, 0, 4, 8\}$ and $q_{j+1}^L = \{-2, -1, 0, 1, 2\}$. As a consequence of this, the recurrence is implemented with two carry-save adders, as shown in Figure 5.11.

The quotient-digit selection depends on the truncated shifted residual (eight bits) and the truncated divisor (four bits, of which three are used in the implementation since $d \geq 1/2$). Since the two components of q_{j+1} do not

Element	Delay	Area
Quotient-digit selection	10.8	160
Buffers	1.8	10
MUX	1.8	300
CSA	2.2	360
Registers (3)	4.0	650
Convert	(NC)	1360
Cycle time	20.6	
Total area		2840

Note: NC denotes a delay not in the critical path.

TABLE 5.4 Radix-4 stage.

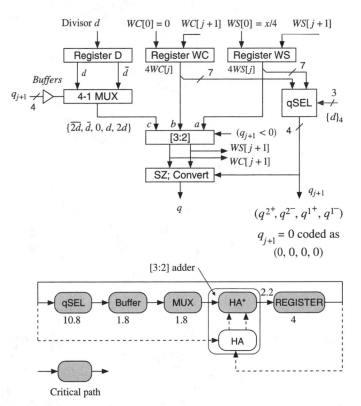

FIGURE 5.10 Implementation of radix-4 scheme and its critical path. Modules defined in Figure 5.4.

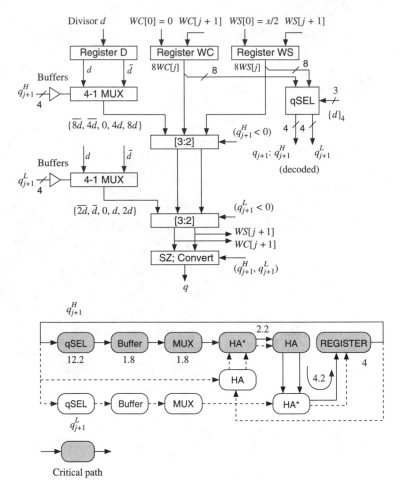

FIGURE 5.11 Implementation of radix-8 scheme and its critical path. Modules defined in Figure 5.4.

affect in the same way the critical path,[11] the design of the selection function[12] is done so as to minimize the critical path.

___

11. The path that includes one of the components traverses two carry-save adders, whereas the path of the other traverses only one carry-save adder.

12. The corresponding selection function is not shown in this text, but follows the method described in the next section. A specific set of selection constants is given in Nannarelli (1999).

Element	Delay	Area
Quotient-digit selection	(q_h) 12.2	610
Buffers	1.8	20
MUXes	1.8	600
CSAh	2.2	360
CSAl	4.2	360
Registers (3)	4.0	650
Convert	(NC)	1360
Cycle time	26.2	
Total area		3960

Note: NC denotes a delay not in the critical path.

TABLE 5.5 Radix-8 stage.

An implementation of the radix-8 scheme is shown in Figure 5.11 together with the critical path delay. A summary of delay and area is given Table 5.5.

Radix 16 with Two Radix-4 Overlapped Stages

Since the digit selection function for radix 16 is too complex (large delay) to implement directly, the unit for this radix is implemented with two radix-4 stages. Figure 5.12(a) shows the updating of the residual. In a straightforward implementation the delay would correspond to two times the delay of a radix-4 implementation, except for the delay of the register, which would be counted only once. To reduce the delay, the second radix-4 digit is computed conditionally so that the stages are overlapped. Specifically, the second digit is computed for all the possible values of the first digit and then the final value is selected when the first digit is known. To do this, it is necessary to first compute the conditional truncated residuals as

$$cond(w[j+1], k)_{trunc} = (4w[j])_{trunc} - kd_{trunc} \qquad \textbf{5.40}$$

for $-2 \leq k \leq 2$.

These conditional residual values are then input to the quotient-digit selection networks. The implementation of this conditional quotient-digit selection is shown in Figure 5.12(b). Note that, because of the carry-save addition, eight bits of the operands are needed to produce seven bits of the result.

The critical path is shown in Figure 5.13, and Table 5.6 summarizes the cycle time and area.

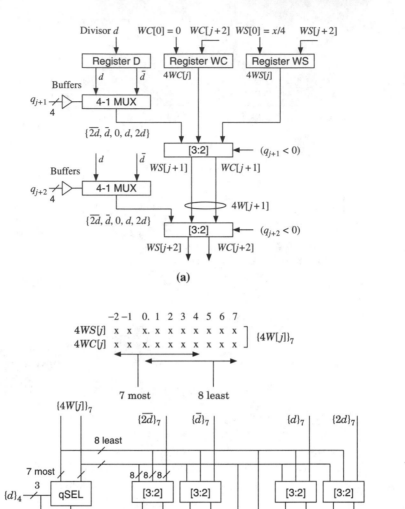

FIGURE 5.12 Implementation of radix-16 with radix-4 stages: (a) Generation of residuals. (b) Quotient-digit selection. (SZ and quotient conversion modules not shown.)

Element	Delay	Area
CSA	4.2	220
Quotient-digit selection	11.2	820
MUX	1.4	
Buffers	1.8	20
MUXes	1.8	600
CSA1	(NC)	360
CSA2	2.2	360
Registers (3)	4.0	650
Convert	(NC)	1360
Cycle time	26.6	
Total area		4390

Note: NC denotes a delay not in the critical path.

TABLE 5.6 Radix-16 stage (two overlapped radix-4 stages).

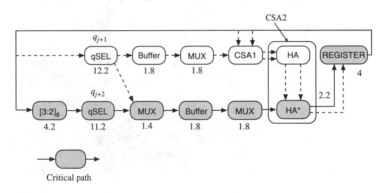

FIGURE 5.13 Critical path in radix-16 scheme.

Radix-512 with Scaling and Selection by Rounding

As indicated, the direct implementation of the quotient-digit selection is practical only for small radices, such as 2, 4, and 8. So, for very high radices, such as 512, it is necessary to modify the algorithm. It can be shown that if the divisor is sufficiently close to one, the quotient digit corresponds to the rounded shifted residual. So, one possibility is to prescale the divisor (and dividend) so that the scaled divisor is close to one and then do quotient-digit selection by rounding the shifted residual. We now summarize the algorithm and the implementation; for details see the references given at the end of the chapter.

We use a quotient-digit set $|q_{j+1}| \leq 511$ ($\rho = 1$). For a quotient of 54 bits, including the guard bit for rounding, the algorithm consists of the following cycles:

Cycle 1: Compute the scaling constant $M \approx 1/d$; since $Md \approx 1$, this constant is used to scale the divisor and dividend. Compare[13] x and d and set $g = 1$ if $x \geq d$ and $g = 0$ otherwise.

Cycle 2: Compute the scaled divisor $z = Md$ (in carry-save form); compute $v = 2^{-g}x$.

Cycle 3: Compute Mv and initialize $w[0] = Mv$ (in carry-save form); assimilate z.

Cycles 4–9: Perform iterations:
quotient-digit selection $q_{j+1} = round(\widehat{y})$
residual updating $w[j + 1] = 512w[j] - q_{j+1}z$

Cycle 10: Quotient correction (if residual is negative) and normalization.

The implementation is shown in Figure 5.14. In this figure:

- The constant M is computed in module M, in carry-save form.
- The multiplier-accumulator is used to scale the divisor (cycle 2) and the dividend (cycle 3) and then computes $q_{j+1}z$ (cycles 4–9).
- The recoder converts the multiplier from carry-save form to radix 4 with digit set $\{-2, -1, 0, 1, 2\}$. This multiplier can be either M or q_{j+1}, which is obtained by rounding $\widehat{y}$. The addition of 0.5 for rounding is also done in the recoder.

Table 5.7 gives the cycle time and area. The recoder delay corresponds to one AND-OR network plus one multiplexer. The cycle time is similar to the radix-8 case. However, because of the scaling, there are three cycles of overhead.[14]

---

13. This is done so that the initialization does not increase the number of iterations.
14. For details, see Ercegovac et al. (1994).

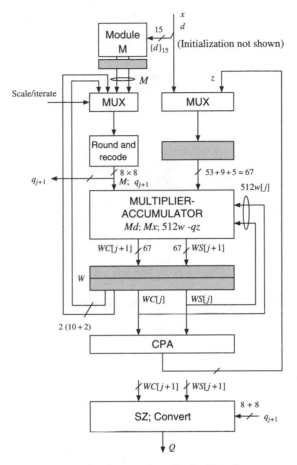

FIGURE 5.14 Implementation of radix-512 scheme.

Overall Comparisons

As an indication of the merits of the schemes presented, Table 5.8 summarizes the speedups and area factors, relative to the radix-2 case. The values have been rounded to give rough estimations. As can be seen from the table, increasing the radix produces a speedup; however, for a radix 2^k this speedup is significantly

Element	Delay	Area
M-module	(NC)	1800
MUX	1.4	
Recoder	6.0	70
Buffer	1.8	
Multiplier-accumulator	13.8	6100
Registers (3)	4.0	650
Convert	(NC)	1360
Cycle time	27	
Total area		9980

Note: NC denotes a delay not in the critical path.

TABLE 5.7 Delay and area for radix-512 scheme.

Scheme	r2	r4	r8	r16	r512
Cycle-time factor	1.0	1.3	1.6	1.6	1.7
Number of cycles*	57	29	20	15	10
Speedup	1.0	1.5	1.8	2.4	3.4
Area factor	1.0	1.1	1.5	1.7	3.9

*Correction: Two cycles for radix-2, one cycle for other cases.

TABLE 5.8 Comparison of schemes.

smaller than the ideal k because of the increase in the cycle time and of additional overheads. Moreover, there is an important increase in area.

5.4   Integer Division

Integer division (for unsigned operands) has integer operands $0 \leq x \leq r^n - 1$ and $0 \leq d \leq r^n - 1$ and produces an integer quotient q such that

$$q = \lfloor x/d \rfloor \qquad\qquad 5.41$$

It also produces the integer remainder

$$rem = (x) \bmod d \qquad\qquad 5.42$$

In Chapter 1 basic integer division algorithms are described. However, these algorithms require full-precision comparisons for the quotient-digit selection. In order to use the selection functions discussed in this chapter, the divisor is first normalized (shifted so that the most-significant bit is 1).[15] Consequently, for a shifting of m bits we get

$$d^* = 2^m d \qquad \textbf{5.43}$$

and the integer quotient

$$q = \lfloor x/d \rfloor = 2^m \lfloor x/d^* \rfloor \qquad \textbf{5.44}$$

The number of bits of the integer quotient is not larger than $m + 1$. Consequently, the number of iterations required to obtain these bits is

$$N = \lceil (m + 1)/k \rceil \qquad \textbf{5.45}$$

where $r = 2^k$ is the radix of the quotient digit.

We want to perform the integer division using the fractional division units discussed in the previous section. For this we define the fractional operands x_f and d_f so that

$$x_f = x \times r^{-n} \text{ (not normalized)} \qquad \textbf{5.46}$$
$$d_f = d^* \times r^{-n} \text{ (normalized)} \qquad \textbf{5.47}$$

Moreover, it is necessary to satisfy the following requirements:

1. To satisfy the residual bound, the initial residual is equal to $x_f/2$ (for $\rho = 1$) or $x_f/4$ (for $\rho < 1$). This requires that $m + 1 + v$ bits of the quotient be computed, where $v = 1$ (for $\rho = 1$) or $v = 2$ (for $\rho < 1$). The resulting number of iterations is

$$N = \lceil (m + 1 + v) \rceil \qquad \textbf{5.48}$$

2. To obtain a correct remainder the last bit of the quotient has to be aligned with a radix-r boundary. Since the quotient is in the range $1/2 < q < 2$ (one integer bit and m fractional bits), this is achieved by shifting x_f right by $v + s$ bits, so that $(m + v + s) \bmod k = 0$.

The quotient has to be aligned to the integer position. This can be done by placing the digits in the correct final position or by placing the digits aligned to

---

15. This is directly applicable when the radix is a power of two.

the left (to combine with fractional division) and then performing a right shift of $n - N$ digits.

As in the fractional division discussed before, the use of signed quotient digits requires the conversion to conventional representation and the correction to obtain a positive remainder.

Moreover, since the remainder should be less than the divisor and the divisor has $n \log_2 r - m$ bits, we obtain

$$
rem = \begin{cases} w[N]2^{n \log_2 r - m} & \text{if } w[N] \geq 0 \\ (w[N] + d_f)2^{n \log_2 r - m} & \text{if } w[N] < 0 \end{cases} \qquad \textbf{5.49}
$$

EXAMPLE 5.1   We now show an example of integer division for 8-bit operands using a radix-4 algorithm with $\rho = 2/3$. Consider the case $x = 125$ and $d = 6$ with binary representations

$$
x = 01111101, \quad d = 00000110
$$

We normalize d to produce $d^* = 11000000$ with $m = 5$. Since $\rho < 1$, we have $v = 2$ and $s = 1$. Consequently, we shift x_f by three positions and require $N = (m + 1 + v)/2 = 4$ iterations. The initial condition is

$$
w[0] = x_f/8 = .00001111101
$$

The iterations are shown in Figure 5.15.   ■

The details of an implementation are left as an exercise.

5.5   Quotient-Digit Selection Function

In previous sections we have described the recurrence step consisting of arithmetic shift, quotient-digit selection, multiple generation, subtraction, and quotient conversion. We now present the basic theoretical background required to design the quotient-digit selection function and give examples for radix 2 and radix 4.

The quotient-digit selection function determines the value of the quotient digit q_{j+1} as a function of the residual $w[j]$ and the divisor d. As indicated before, we use a symmetric signed-digit set for the values of the quotient digit; that is,

$$
q_{j+1} \in \mathcal{D}_a = \{-a, -a + 1, \ldots, -1, 0, 1, \ldots, a - 1, a\} \qquad \textbf{5.50}
$$

Initial residual $w[0] = x_f/8 = 0.00001111101$, $\widehat{d} = 0.1100 = 12/16$

$4WS[0] = 000.001111101$

$4WC[0] = 000.000000000$ $\widehat{y}[0] = 000.0011$ $q_1 = 00$

$WS[1] = 000.001111101$

$WC[1] = 000.000000000$

$4WS[1] = 000.111110100$

$4WC[1] = 000.000000001^*$ $\widehat{y}[1] = 000.1111$ $q_2 = 01$

$-d_f = 111.001111111$

$WS[2] = 111.110001010$

$WC[2] = 000.011101010$

$4WS[2] = 111.000101000$

$4WC[2] = 001.110101001^*$ $\widehat{y}[2] = 000.1110$ $q_3 = 01$

$-d_f = 111.001111111$

$WS[3] = 001.111111110$

$WC[3] = 110.001010010$

$4WS[3] = 111.111111000$

$4WC[3] = 000.101001001^*$ $\widehat{y}[3] = 000.1001$ $q_4 = 01$

$-d_f = 111.001111111$

$WS[4] = 000.011001110$

$WC[4] = 111.011110010$

Residual negative—correct the quotient and the residual

$+d_f = 000.110000000$

$w[4] = 000.101000000$ $\qquad\qquad q_4 = 00$

The quotient and the remainder are

$$q = 00010100 = (20)_{10}$$
$$rem = w[4] \times 2^3 = 101 = 5$$

FIGURE 5.15 Example of radix-4 integer division with residual in carry-save form and $n = 4$ (radix-4 digits).

with the redundancy factor ρ

$$\rho = \frac{a}{r-1}, \quad \frac{1}{2} < \rho \leq 1 \qquad \qquad \textbf{5.51}$$

The specific selection function depends on the way the residual is represented. Of particular practical interest is the case of redundant representation, either carry-save or signed-digit, because the addition in the recurrence is faster. Consequently, our goal is to present the quotient-digit selection for those cases. However, the case for nonredundant residual representation is simpler, so we discuss it first and then present the modifications required to use the redundant representations.

There are two fundamental conditions that must be satisfied by a selection function: *containment* (all residuals must be bounded) and *continuity* (for any value of the shifted residual there must exist a valid choice of the quotient digit). The containment condition determines a *selection interval* for each value of q_{j+1}. The continuity condition is used for choosing the specific selection function. We now discuss these concepts and then present several alternative selection functions.

5.5.1   Containment Condition and Selection Intervals

One basic requirement for the quotient-digit selection is to guarantee a bounded (contained) next residual. This containment condition determines the selection intervals, which are then used to design the selection function.

As developed in Section 5.2 the division recurrence is

$$w[j+1] = rw[j] - dq_{j+1} \qquad \qquad \textbf{5.52}$$

Moreover, for convergence, the residual has to be bounded so that

$$|w[j]| \leq \rho d \qquad \qquad \textbf{5.53}$$

where $\rho = a/(r-1)$ is the redundancy factor and $-a \leq q_j \leq a$.

Selection Intervals

Define the *selection interval* of $rw[j]$ for $q_{j+1} = k$ to be $[L_k, U_k]$. That is, L_k (U_k) is the smallest (largest) value of $rw[j]$ for which it is *possible* to choose $q_{j+1} = k$ and keep the next residual ($w[j+1]$) bounded. Therefore,

$$L_k \leq rw[j] \leq U_k \implies -\rho d \leq w[j+1] = rw[j] - k \cdot d \leq \rho d \qquad \textbf{5.54}$$

Consequently,

$$U_k - k \cdot d = \rho d, \qquad L_k - k \cdot d = -\rho d \qquad\qquad \textbf{5.55}$$

and

$$U_k = (k + \rho)d \qquad L_k = (k - \rho)d \qquad\qquad \textbf{5.56}$$

The division recurrence, the residual bounds, and the selection-interval bounds can be represented in *Robertson's diagram* (Figure 5.16(a)). This diagram has as axes the shifted residual $r w[j]$ and the next residual $w[j+1]$. It represents the recurrence by the lines with parameter $q_{j+1} = k$ for $k = -a, \ldots, a$ and the residual bounds by the rectangle $w[j+1] = \rho d$, $w[j+1] = -\rho d$, $r w[j] = r \rho d$, and $r w[j] = -r \rho d$. The selection interval for $q_{j+1} = k$ is obtained from the projection of the corresponding line on the $r w[j]$ axis. The diagram of Figure 5.16(a) illustrates the computation of $w[6] = r w[5] - kd$.

Another diagram, which is useful in the design of the quotient-digit selection function, is the $r w[j]$ versus d diagram, called the *P-D diagram* (Figure 5.16(b)). The bounds of the selection intervals U_k and L_k are plotted as lines originating from $(0, 0)$ with slope $k + \rho$ and $k - \rho$, respectively. The regions delineated by these lines are helpful in analyzing the quotient-digit selection function, as described later.

5.5.2 Continuity Condition, Overlap, and Quotient-Digit Selection

We now relate the quotient-digit selection function to the selection intervals. As stated, the function is of the form

$$q_{j+1} = SEL(w[j], d) \qquad\qquad \textbf{5.57}$$

We can represent this function by the set $\{s_k\}$, $-a \le k \le a$, such that

$$q_{j+1} = k \qquad \text{if } s_k \le r w[j] \le s_{k+1} - ulp \qquad\qquad \textbf{5.58}$$

That is, s_k is defined as the *minimum* value of $r w[j]$ for which $q_{j+1} = k$ is chosen. As indicated by the function *SEL* above, the s_k's are functions of the divisor d.

To satisfy the containment condition, s_k must be inside the selection interval; that is,

$$L_k \le s_k \le U_k \qquad\qquad \textbf{5.59}$$

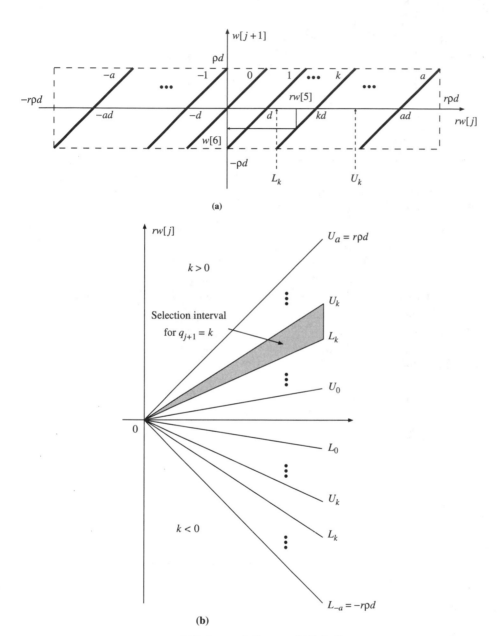

FIGURE 5.16 (a) Robertson's diagram. (b) P-D diagram.

Moreover, to satisfy the continuity condition, it is necessary to select $q_{j+1} = k - 1$ for $rw[j] = s_k - ulp$. Consequently,

$$s_k - ulp \leq U_{k-1} \qquad \textbf{5.60}$$

Since $U_k \geq U_{k-1} + ulp$, the combined restriction on s_k is

$$L_k \leq s_k \leq U_{k-1} + ulp \qquad \textbf{5.61}$$

For simplicity, in some cases, we use the more conservative bound

$$L_k \leq s_k \leq U_{k-1} \qquad \textbf{5.62}$$

Consequently, the values s_k have to be inside the overlap between consecutive selection intervals, as shown in Figure 5.17. This is the basic condition required by the quotient-digit selection functions we describe later.

The subscript $k - 1$ for U, in contrast with the subscript k for L in (5.61), results from the choice of s_k as the minimum of the interval for $q_{j+1} = k$. If, on the other hand, we selected s_k to be the maximum of the interval, then the subscripts would be reversed. We will comment further on this asymmetry later.

The amount of overlap is given by

$$U_{k-1} - L_k = (k - 1 + \rho)d - (k - \rho)d = (2\rho - 1)d \qquad \textbf{5.63}$$

This overlap depends on ρ and on d. Note that the overlap is zero for nonredundant quotient-digit set ($\rho = 1/2$). The main reason for using a redundant quotient-digit set is to provide a suitable overlap to simplify the quotient-digit selection. Moreover, for this same reason, it is common to restrict the range of the divisor so that $d \geq 1/2$ *(normalized divisor)*.[16] This restriction is shown in Figure 5.17. As indicated before, for floating-point representation, the divisor is usually in normalized form, and in cases in which the original value is not normalized, it is possible to normalize it by shifting both the divisor and the dividend. Unless noted otherwise, we assume that the divisor is normalized.

---

16. We normalize the divisor to $d \geq 1/2$ even for higher radices, since the higher radix is used only to reduce the number of steps in the algorithm, but the representation of the divisor remains in radix 2. This restricts the radix to be a power of 2.

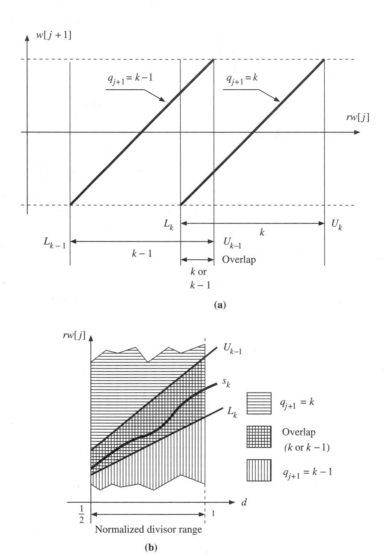

FIGURE 5.17 Overlap between selection intervals and selection function.

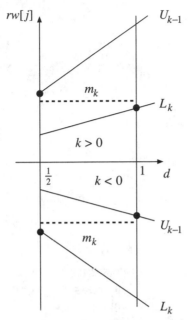

FIGURE 5.18 Bounds on m_k.

5.5.3 Quotient-Digit Selection Using Selection Constants

The simplest selection function is to make the s_k's constants, independent of the divisor. We call these constants m_k. From (5.61), the constants have to satisfy

$$\max(L_k) \leq m_k \leq \min(U_{k-1}) + ulp \qquad \textbf{5.64}$$

where the max and min have to be obtained for the range $2^{-1} \leq d < 1$.

As shown in Figure 5.18 and using the expressions for L and U:

- For $k > 0$

$$(k - \rho) \leq m_k \leq (k - 1 + \rho)2^{-1} + ulp \qquad \textbf{5.65}$$

which requires

$$\rho \geq \frac{k + 1}{3} \qquad \textbf{5.66}$$

- For $k \leq 0$

$$(k - \rho)2^{-1} \leq k - 1 + \rho \qquad\qquad \textbf{5.67}$$

which requires

$$\rho \geq \frac{(-k) + 2}{3} \qquad\qquad \textbf{5.68}$$

For a quotient-digit set $-a \leq q_j \leq a$ ($\rho = a/(r-1)$), the worst case is $k = -a$, resulting in

$$\rho \geq \frac{2}{4 - r} \qquad\qquad \textbf{5.69}$$

Since $\rho \leq 1$, we get that $r = 2$ is the only radix for which the selection function can be independent of the divisor.

Using Truncated Residual

The selection function corresponds to comparisons between the selection constants m_k and $rw[j]$. If the selection constants are of the form $A_k 2^{-c}$, with A_k integer, then for $rw[j]$ in two's complement representation the comparisons are done using $\{r[w]\}_c$, the truncated $rw[j]$ with c fractional bits. This is because $rw[j] \geq \{rw[j]\}_c$, independently of the sign of the residual since in two's complement representation the portion discarded by truncation is always positive.[17] We use two's complement representation for the residual, unless noted otherwise.

As a consequence of this, to simplify the implementation of the selection function the selection constants chosen should correspond to the smallest c possible.

Radix-2 Division with Nonredundant Residual

This algorithm is an extension of nonrestoring division (in which the quotient-digit set is $\{-1, 1\}$). Now the quotient-digit set is signed digit with the inclusion of 0. This algorithm is called SRT.[18]

Since the nonrestoring algorithm already uses selection constants ($m_1 = 0$), it seems unnecessary to include the value $q_j = 0$. The purpose of introducing the

---

17. On the other hand, if the representation is in sign-and-magnitude, the portion discarded by the truncation has the sign of the residual.
18. After D. Sweeney (Cocke and Sweeney 1957), J. E. Robertson (Robertson 1957), and K. D. Tocher (Tocher 1958).

quotient-digit value 0 is to eliminate the need for subtraction/addition when this value 0 is selected (skipping over zeros).

From expression (5.56) the selection intervals are[19]

$$
\begin{array}{ll}
U_1 = 2d & L_1 = 0 \\
U_0 = d \geq \dfrac{1}{2} & L_0 = -d \leq -\dfrac{1}{2} \\
U_{-1} = 0 & L_{-1} = -2d
\end{array}
\qquad \textbf{5.70}
$$

Consequently, from (5.64), the selection constants have to satisfy

$$
0 \leq m_1 \leq \frac{1}{2} \qquad -\frac{1}{2} \leq m_0 \leq 0 \qquad \textbf{5.71}
$$

A possible quotient-digit selection function would be to choose $m_1 = m_0 = 0$; however, this is the same as the nonrestoring case and does not use the quotient-digit value 0. To maximize the region for 0 (and therefore its frequency), we choose

$$
m_1 = \frac{1}{2}, \qquad m_0 = -\frac{1}{2}. \qquad \textbf{5.72}
$$

The corresponding quotient-digit selection function is

$$
q_{j+1} = \left\{
\begin{array}{ll}
1 & \text{if } \frac{1}{2} \leq 2w[j] \\
0 & \text{if } -\frac{1}{2} \leq 2w[j] < \frac{1}{2} \\
-1 & \text{if } 2w[j] < -\frac{1}{2}
\end{array}
\right. \qquad \textbf{5.73}
$$

This selection function is illustrated in Robertson's diagram (Figure 5.19(a)) and in the corresponding P-D diagram (Figure 5.19(b)). It requires only the comparison with the constants $\frac{1}{2}$ and $-\frac{1}{2}$. The selection rules effectively correspond to checking if the shifted residual is normalized; that is, if $|2w[j]| \geq \frac{1}{2}$.

Staircase Selection Function

For radix larger than 2 it is not possible to find one constant m_k for the whole range of the divisor. In this case, the range of the divisor is divided into intervals

---

19. Since $\rho = 1$, the bound used for these selection intervals would allow the value $w[n] = d$, which would require a restoration step. This is avoided if $w[0] < d$ and $m_0 \leq 0$ (to avoid $q_{j+1} = -1$ for $2w[j] = 0$).

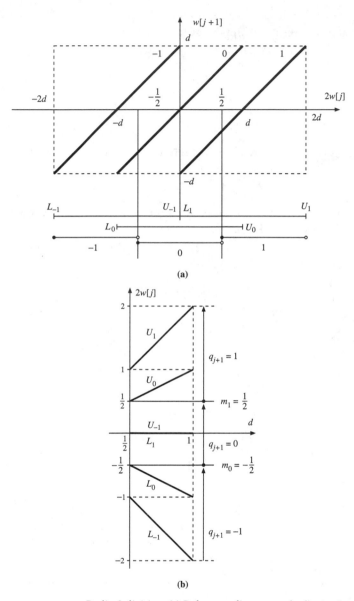

FIGURE 5.19 Radix-2 division: (a) Robertson diagram and selection intervals. (b) P-D diagram and selection function.

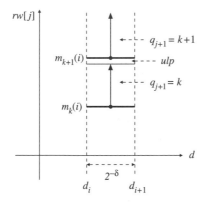

FIGURE 5.20 Definition of $m_k(i)$.

$[d_i, d_{i+1})$ with

$$d_0 = \frac{1}{2}, \qquad d_{i+1} = d_i + 2^{-\delta} \qquad \textbf{5.74}$$

so that the δ most-significant fractional bits of the divisor represent the interval. Moreover, in the interval $[d_i, d_{i+1})$ the quotient-digit selection is described by the set of *selection constants* $m_k(i)$. That is,

$$\text{for } d \in [d_i, d_{i+1}), \qquad q_{j+1} = k \quad \text{if } m_k(i) \leq rw[j] \leq m_{k+1}(i) - ulp \quad \textbf{5.75}$$

as illustrated in Figure 5.20.

Since a single selection constant is used for the whole interval $[d_i, d_{i+1})$, from (5.64) we get (as shown in Figure 5.21) that

$$\max(L_k(d_i), L_k(d_{i+1})) \leq m_k(i) \leq \min(U_{k-1}(d_i), U_{k-1}(d_{i+1})) + ulp \quad \textbf{5.76}$$

The quotient-digit selection is a function of the δ most-significant fractional bits of d. Actually, only $\delta - 1$ bits are needed because $d \geq \frac{1}{2}$. In addition, for

$$m_k(i) = A_k(i)2^{-c} \qquad \textbf{5.77}$$

with $A_k(i)$ integer, the selection function uses $\{rw[j]\}_c$, which corresponds to the truncated $rw[j]$ with c fractional bits. The use of these selection constants is illustrated in Figure 5.22.

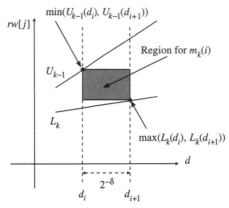

FIGURE 5.21 Selection constant region.

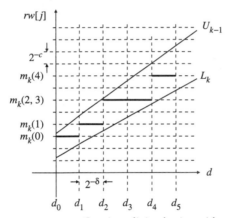

FIGURE 5.22 Quotient-digit selection with selection constants.

In terms of the low-precision selection constants, from (5.74), (5.76), and (5.77), the quotient-digit selection must satisfy

$$\text{for } k > 0 \quad L_k(d_i + 2^{-\delta}) \le A_k(i)2^{-c} \le U_{k-1}(d_i) + ulp$$
$$\text{for } k \le 0 \qquad L_k(d_i) \le A_k(i)2^{-c} \le U_{k-1}(d_i + 2^{-\delta}) + ulp$$

5.78

for all i and all k.

The design problem consists, therefore, in finding selection constants and divisor intervals so that c and δ are minima. Unfortunately, there is no single

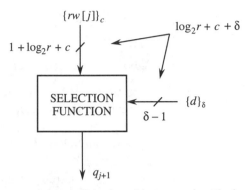

FIGURE 5.23 Selection with truncated residual and divisor.

solution since as δ is reduced c increases. Consequently, a possible optimization criterion is to minimize $c + \delta$ since this relates to the number of bits of the input of the quotient-digit selection function, as shown in Figure 5.23. Note that the integer part of $\{r\,w[j]\}_c$ requires $1 + \log_2(r\,\rho d)$ bits ($< 1 + \log_2 r$ for $\rho \leq 1$).

A lower bound on δ, the number of divisor bits required, is obtained by requiring a nonnegative overlap. Consider the case $k > 0$ (a similar argument can be made for $k \leq 0$); then

$$U_{k-1}(d_i) - L_k(d_i + 2^{-\delta}) \geq 0 \qquad \textbf{5.79}$$

From (5.56) we obtain the expressions for U_{k-1} and L_k so that

$$(\rho + k - 1)d_i - (-\rho + k)(d_i + 2^{-\delta}) \geq 0 \qquad \textbf{5.80}$$

which results in

$$(2\rho - 1)d_i \geq (k - \rho)2^{-\delta} \qquad \textbf{5.81}$$

This must be true for all values of d_i and k, the worst case being the smallest value of d_i and the largest value of k. Since $d \geq \frac{1}{2}$ and $k \leq a$, expression (5.81) becomes

$$2^{-\delta} \leq \frac{2\rho - 1}{2(a - \rho)} = \frac{2\rho - 1}{2\rho(r - 2)} \qquad \textbf{5.82}$$

However, the use of this minimum value of δ can result in a large value of c; that is, many bits of the shifted residual. Consequently, the values of δ and c have to be selected so that the implementation is simplified; the actual values for this to occur depend on the particular technology used. The bound is helpful in reducing the number of alternatives to consider.

Radix-4 Division with a = 2 and Nonredundant Residual

In the radix-4 case,[20] two possibilities exist for the redundant digit set: $a = 2$ and $a = 3$. The case $a = 2$ has the advantage that the multiples of the divisor that have to be generated are d, $2d$, $-d$, and $-2d$, which are simple to generate, whereas the alternative $a = 3$ also requires the multiples $3d$ and $-3d$, which are more complex to obtain. On the other hand, because of the greater redundancy, the case $a = 3$ results in a simpler quotient-digit selection function. The choice of the digit set depends on specific implementation constraints. Here we describe the case $a = 2$.

Since $\rho = \frac{2}{3}$, from (5.56) the selection intervals are

$$U_k = \left(\frac{2}{3} + k \right) d \qquad L_k = \left(-\frac{2}{3} + k \right) d \qquad \text{5.83}$$

Figure 5.24 shows the corresponding Robertson's diagram and P-D diagram.

From (5.82) we get the bound on δ to be

$$2^{-\delta} \le \frac{2\rho - 1}{2(a - \rho)} = \frac{1}{8} \qquad \text{5.84}$$

Consequently, a truncated divisor of at least three bits is required. However, in this case the use of three bits results in a large value of c. For example, for the selection constant $m_2(1)$ for the region of the divisor between $\frac{4}{8}$ and $\frac{5}{8}$, we get $L_2(\frac{5}{8}) = U_1(\frac{4}{8}) = \frac{5}{6}$. Therefore, the only possible value of $m_2(1)$ is $\frac{5}{6}$, which requires a selection constant of full precision. Because of this, we use a truncated divisor of *four* bits instead.

We now use this value of $\delta = 4$ to determine the minimum value of c and the resulting selection function. We do this by considering all cases of d_i and k to satisfy expression (5.78). The results of the analysis are shown in Table 5.9.

As expected, there is symmetry between positive and negative constants. Since the selection constant with highest precision is of the form $A \cdot 2^{-3}$, three fractional bits of the shifted residual are needed. The selection constants $m_2(i)$ are shown in the P-D diagram of Figure 5.25(a). Since the shifted residual is

---

20. This algorithm is called Robertson's Division Algorithm (Robertson, 1958).

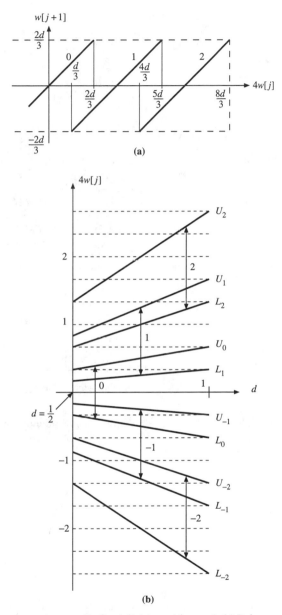

FIGURE 5.24 Radix-4 division with $a = 2$: (a) Robertson's diagram. (b) P-D diagram.

$[d_i, d_{i+1})^+$	[8, 9)	[9, 10)	[10, 11)	[11, 12)
$L_2(d_{i+1}), U_1(d_i)^\#$	36, 40	40, 45	44, 50	48, 55
$m_2(i)^*$	6	7	8	8
$L_1(d_{i+1}), U_0(d_i)^\#$	9,16	10, 18	11, 20	12, 22
$m_1(i)$	2	2	2	2
$L_0(d_i), U_{-1}(d_{i+1})^\#$	−16, −9	−18, −10	−20, −11	−22, −12
$m_0(i)$	−2	−2	−2	−2
$L_{-1}(d_i), U_{-2}(d_{i+1})^\#$	−40, −36	−45, −40	−50, −44	−55, −48
$m_{-1}(i)$	−6	−7	−8	−8
$[d_i, d_{i+1})^+$	[12, 13)	[13, 14)	[14, 15)	[15, 16)
$L_2(d_{i+1}), U_1(d_i)^\#$	52, 60	56, 65	60, 70	64, 75
$m_2(i)$	10	10	10	12
$L_1(d_{i+1}), U_0(d_i)^\#$	13, 24	14, 26	15, 28	16, 30
$m_1(i)$	4	4	4	4
$L_0(d_i), U_{-1}(d_{i+1})^\#$	−24, −13	−26, −14	−28, −15	−30, −16
$m_0(i)$	−4	−4	−4	−4
$L_{-1}(d_i), U_{-2}(d_{i+1})^\#$	−60, −52	−65, −56	−70, −60	−75, −64
$m_{-1}(i)$	−10	−10	−10	−12

Note: +: real value = shown value/16; #: real value = shown value/48; *: real value = shown value/8.

TABLE 5.9 Selection intervals and m_k constants (radix 4, nonredundant residual).

bounded by

$$|4w[j]| \leq 4\rho d < \frac{8}{3} \qquad \textbf{5.85}$$

three integer bits are needed for the two's complement representation. Therefore, the selection function is implemented using three bits of the divisor and six bits of the shifted residual, as illustrated in Figure 5.25(b).

5.5.4   Use of Redundant Adder

The quotient-digit selection discussed previously requires that the residual be computed to full precision, although a truncated version is used in the selection

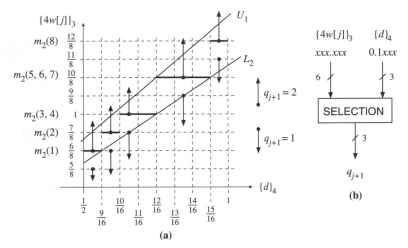

FIGURE 5.25 Radix-4 with nonredundant residual. Quotient-digit selection: (a) a fragment of the P-D diagram and (b) implementation.

function due to the limited-precision selection constants. In this implementation a substantial fraction of the step time is due to the addition required for the computation of the residual.

The overlap between selection intervals can be further used to reduce the step time by basing the selection on an *estimate* of the residual. We derive the requirements for such use and apply the results to estimates involved when using a redundant (carry-free) adder.

If we call y the actual value of the shifted residual and $\widehat{y}$ its *estimate*, we can write

$$\epsilon_{min} \le y - \widehat{y} \le \epsilon_{max} \qquad\qquad \textbf{5.86}$$

where ϵ_{min} and ϵ_{max} are the minimum error and maximum error, respectively. Note that usually ϵ_{min} is nonpositive.

We now develop expressions that have to be satisfied to design a quotient-digit selection function for a general estimate. The basic constraint that must be satisfied is that if we choose $q_{j+1} = k$ for an estimate $\widehat{y}$, then this choice must be correct for the interval

$$y \in [\widehat{y} + \epsilon_{min},\ \widehat{y} + \epsilon_{max}] \qquad\qquad \textbf{5.87}$$

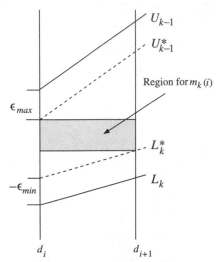

FIGURE 5.26 Constraints for selection based on estimates.

Consequently, the restricted selection interval is $[L_k^*, U_k^*]$ such that

$$L_k^* = L_k - \epsilon_{min}$$
$$U_k^* = U_k - \epsilon_{max}$$

5.88

The range of values of $\widehat{y}$ for which m_k can be chosen is determined as before by replacing U and L by U^* and L^*, respectively, namely,

$$\max(L_k^*(d_i), L_k^*(d_{i+1})) \leq m_k(i) \leq \min(U_{k-1}^*(d_i), U_{k-1}^*(d_{i+1})) + ulp \quad 5.89$$

This expression is illustrated in Figure 5.26.

From (5.89) we get the minimum overlap required

$$\min(U_{k-1}^*(d_i), U_{k-1}^*(d_{i+1})) - \max(L_k^*(d_i), L_k^*(d_{i+1})) \geq 0 \qquad \textbf{5.90}$$

The range of the estimate $\widehat{y}$ determines the number of bits of the representation. Since the range of $rw[j]$ is

$$|rw[j]| \leq r\rho d < r\rho \quad (\text{for } d < 1) \qquad \textbf{5.91}$$

we get

$$-r\rho - \epsilon_{max} < \widehat{y} < r\rho - \epsilon_{min} \qquad \textbf{5.92}$$

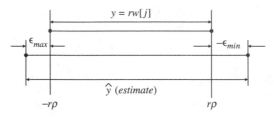

FIGURE 5.27 Range of estimate.

as shown in Figure 5.27. Note, however, that in specific cases the maximum errors might not occur for the maximum values of $r\,w[j]$. In such cases, a more detailed analysis is required to obtain a better bound on $\widehat{y}$.

One way of having an estimate of the residual is to use a redundant representation (carry-save or signed-digit), produced by a redundant (carry-free) addition, as shown in Figure 5.28(a). The quotient-digit selection function uses an estimate of this shifted residual obtained by truncating the redundant representation to t bits. The error introduced by this truncation depends on the type of redundant adder (carry-save or signed-digit), as discussed now.

Carry-Save Adder

For the carry-save case, the representation is in a two's complement form (as that of the nonredundant case). Consequently, the error due to the truncation is always positive, as illustrated in Figure 5.28(b). That is,

$$\epsilon_{min} = 0 \qquad\qquad \textbf{5.93}$$

Moreover, the maximum positive error corresponds to the maximum value of the discarded portion. Consequently,

$$\epsilon_{max} = 2^{-t+1} - ulp \qquad\qquad \textbf{5.94}$$

Using these values for the error, we get the restricted selection interval

$$\begin{aligned} U_k^* &= U_k - 2^{-t+1} + ulp \\ L_k^* &= L_k \end{aligned} \qquad\qquad \textbf{5.95}$$

Moreover, since the estimate is the truncated shifted residual with t fractional bits, the selection constants cannot have more than t fractional bits. They should be located on the grid of granularity 2^{-t}, as shown in Figure 5.29. Consequently,

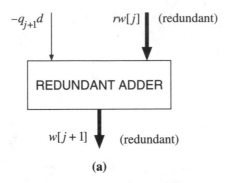

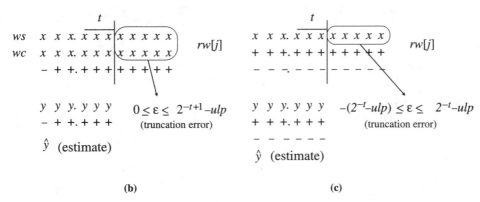

FIGURE 5.28 Use of redundant adder: (a) Redundant adder. (b) Carry-save case. (c) Signed-digit case.

for the region in which m_k can be located we use $\widehat{U}$ and $\widehat{L}$ located on the grid. That is,

$$\max(\widehat{L}_k(d_i), \widehat{L}_k(d_{i+1})) \leq m_k(i) \leq \min(\widehat{U}_{k-1}(d_i), \widehat{U}_{k-1}(d_{i+1})) \quad \textbf{5.96}$$

We now relate $\widehat{U}, \widehat{L}$ with U^*, L^*. Since U^*_{k-1} is the largest value for which it is still possible to select $q_{j+1} = k - 1$, the *upper bound* of the region for m_k is the next grid value that is *larger*[21] than U^*_{k-1}. That is,

$$\widehat{U}_{k-1} = \lfloor U^*_{k-1} + 2^{-t} \rfloor_t = \lfloor U_{k-1} - 2^{-t} \rfloor_t \quad \textbf{5.97}$$

---

21. This is equivalent to saying that in expression (5.89) the *ulp* for $\widehat{y}$ is 2^{-t}.

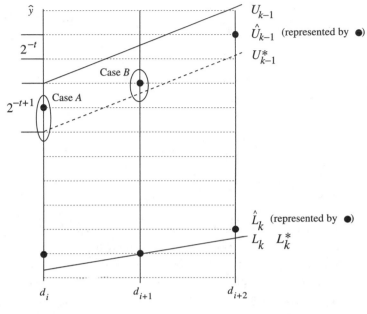

Case A: U_{k-1}^* is on the grid;

$\hat{U}_{k-1} = U_{k-1}^* + 2^{-t}$ on the grid

Case B: U_{k-1}^* is off the grid;

$\hat{U}_{k-1} > U_{k-1}^*$ on the grid

FIGURE 5.29 $\widehat{U}$ and $\widehat{L}$ for residual in carry-save form. Note that when L^* passes through a grid point this is also $\widehat{L}$, whereas this is not the case for $\widehat{U}$.

where $\lfloor x \rfloor_t$ corresponds to the carry-save representation of x truncated at fractional bit t.

On the other hand, since L_k^* is the smallest value for which it is possible to select $q_{j+1} = k$, the *lower bound* of the region for m_k is the grid value that is *equal or larger* than L_k^*, so that

$$\widehat{L}_k = \lceil L_k^* \rceil_t = \lceil L_k \rceil_t \qquad\qquad \textbf{5.98}$$

These bounds are shown by dots in Figure 5.29.

We now obtain a lower bound for t and δ by requiring a nonnegative overlap. That is (for positive k),

$$\widehat{U}_{k-1}(d_i) - \widehat{L}_k(d_{i+1}) \geq 0 \qquad \textbf{5.99}$$

A lower bound is now obtained by replacing $\widehat{U}$ by its upper bound and $\widehat{L}$ by its lower bound. This results in

$$U_{k-1}(d_i) - 2^{-t} - L_k(d_{i+1}) \geq 0 \qquad \textbf{5.100}$$

Introducing the corresponding expressions and the worst-case condition $d_i = \frac{1}{2}$ and $k = a$, we get

$$\frac{2\rho - 1}{2} - (a - \rho)2^{-\delta} \geq 2^{-t} \qquad \textbf{5.101}$$

Finally, we determine the range of the estimate. Since the estimate involves truncation of the carry-save representation to t fractional bits, the difference between the estimate and the truncated two's complement representation corresponds to a possible carry into fractional bit t. This carry affects the negative range of $\widehat{y}$, so that the range becomes

$$\lfloor -r\rho - 2^{-t} \rfloor_t \leq \widehat{y} \leq \lfloor r\rho - ulp \rfloor_t \qquad \textbf{5.102}$$

where $\lfloor z \rfloor_t = 2^{-t} \lfloor 2^t z \rfloor$. The term ulp is required to use $\leq$ instead of $<$. Note that because of the asymmetry of the error, this range is also asymmetric. As noted before, the actual range might be smaller because the neglected carry might not occur for the maximum value of $rw[j]$.

EXAMPLE 5.2   Radix-2 Division with Carry-Save Adder.
For this case, from (5.101)

$$\frac{1}{2} - 0 \times 2^{-\delta} \geq 2^{-t} \qquad \textbf{5.103}$$

This bound indicates that it is possible to have a single set of selection constants for the whole range of the divisor. That is, the quotient-digit selection function is independent of the value of the divisor. The bound also indicates that $t \geq 1$.

Now we see whether $t = 1$ results in valid selection constants. These constants have to satisfy (5.96):

$$\max(\widehat{L}_k(d_i), \widehat{L}_k(d_{i+1})) \leq m_k(i) \leq \min(\widehat{U}_{k-1}(d_i), \widehat{U}_{k-1}(d_{i+1})) \qquad \textbf{5.104}$$

for the whole range of the divisor. From (5.70), (5.97), and (5.98), we get[22]

$$\widehat{L}_1(1) \quad = 0$$

$$\widehat{U}_0\left(\frac{1}{2}\right) = 0$$

$$\widehat{L}_0\left(\frac{1}{2}\right) = -\frac{1}{2} \qquad \qquad \textbf{5.105}$$

$$\widehat{U}_{-1}(1) \quad = -\frac{1}{2}$$

Consequently,

$$(\widehat{L}_1(1) = 0) \le m_1 \le (\widehat{U}_0(1/2) = 0)$$

$$\left(\widehat{L}_0(1/2) = -\frac{1}{2}\right) \le m_0 \le \left(\widehat{U}_{-1}(1) = -\frac{1}{2}\right) \qquad \textbf{5.106}$$

This results in the selection constants $m_1 = 0$ and $m_0 = -\frac{1}{2}$, as shown in the P-D diagram of Figure 5.30(a). Therefore, $t = 1$ results in valid selection constants.

The range of the estimate is obtained from (5.102) as

$$\lfloor -2 - 2^{-1} \rfloor_1 \le \widehat{y} \le \lfloor 2 - ulp \rfloor_1 \qquad \textbf{5.107}$$

which results in

$$-\frac{5}{2} \le \widehat{y} \le \frac{3}{2} \qquad \textbf{5.108}$$

The corresponding quotient-digit selection function is[23]

$$q_{j+1} = \begin{cases} 1 & \text{if } 0 \le \widehat{y} \le \frac{3}{2} \\ 0 & \text{if } \widehat{y} = -\frac{1}{2} \\ -1 & \text{if } -\frac{5}{2} \le \widehat{y} \le -1 \end{cases} \qquad \textbf{5.109}$$

The estimate has four bits (three integer bits and one fractional bit), as shown in Figure 5.30(b). The corresponding algorithm was summarized in Figure 5.6 and an example of execution was given in Figure 5.7. The quotient-digit selection function can be implemented in two ways:

---

22. Since $\rho = 1$ the bound of the residual could produce $w[n] = d$, requiring a restoration step; as indicated for the case with nonredundant residual, this is avoided by making $w[0] < d$ and $m_0 \le 0$.

23. Note the difference from (5.73).

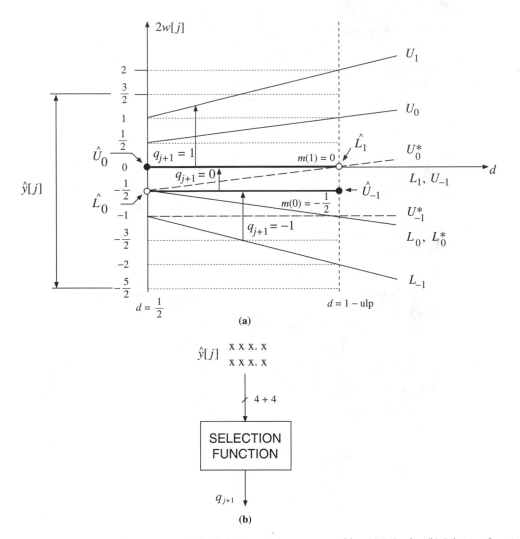

FIGURE 5.30 Radix-2 division with carry-save adder: (a) P-D plot. (b) Selection function.

1. By first converting into non-redundant representation the four most-significant binary positions of $2w[j]$ to produce $\widehat{y}$ and then using the four resulting bits as inputs to a combinational network for the selection.

2. By using the eight bits of the four most-significant positions of the carry-save representation of $2w[j]$ as inputs to the combinational network, as shown in Figure 5.30(b).

Since this combinational network is relatively simple, we use this second scheme to obtain a faster implementation. A possible implementation in which the digit q_{j+1} is represented in sign-and-magnitude by (q_s, q_m) is as follows:

$$q_m = (p_{-1}p_0p_1)' \qquad \textbf{5.110}$$

$$q_s = p_{-2} \oplus (g_{-1} + p_{-1}g_0 + p_{-1}p_0g_1)$$

where

$$p_i = c_i \oplus s_i \qquad g_i = c_i \cdot s_i$$

and $(c_{-2}, c_{-1}, c_0, c_1), (s_{-2}, s_{-1}, s_0, s_1)$ are the carry and sum components of the carry-save representation of $\widehat{y}[j]$. Note that in this case the representation of $q_{j+1} = 0$ is $(q_s, q_m) = (1, 0)$. ∎

EXAMPLE 5.3 Radix-4 Division with Carry-Save Adder.

Another algorithm that has received significant attention is the radix-4 case with digit set $\{-2, \ldots, 2\}$ and carry-save adder. This digit set is advantageous because the multiples dq_{j+1} are easy to implement. We obtain a lower-bound relation between δ and t from (5.101):

$$\frac{1}{6} - \frac{4}{3}2^{-\delta} \geq 2^{-t} \qquad \textbf{5.111}$$

If we use $\delta = 4$ we get

$$2^{-t} \leq \frac{1}{6} - \frac{1}{12} = \frac{1}{12} \qquad \textbf{5.112}$$

so we can try $t = 4$. The selection intervals are obtained from

$$L_k = \left(k - \frac{2}{3}\right)d \qquad U_k = \left(k + \frac{2}{3}\right)d \qquad \textbf{5.113}$$

To determine the quotient-digit selection function, we use expressions (5.97) and (5.98) and obtain Table 5.10. A fragment of this selection function is shown in the P-D diagram of Figure 5.31(a).

$[d_i, d_{i+1})^+$	$[8, 9)$	$[9, 10)$	$[10, 11)$	$[11, 12)$
$\widehat{L}_2(d_{i+1}), \widehat{U}_1(d_i)^+$	12, 12	14, 14	15, 15	16, 17
$m_2(i)^+$	**12**	**14**	**15**	**16**
$\widehat{L}_1(d_{i+1}), \widehat{U}_0(d_i)^+$	3, 4	4, 5	4, 5	4, 6
$m_1(i)$	**4**	**4**	**4**	**4**
$\widehat{L}_0(d_i), \widehat{U}_{-1}(d_{i+1})^+$	$-5, -4$	$-6, -5$	$-6, -5$	$-7, -5$
$m_0(i)$	**-4**	**-6**	**-6**	**-6**
$\widehat{L}_{-1}(d_i), \widehat{U}_{-2}(d_{i+1})^+$	$-13, -13$	$-15, -15$	$-16, -16$	$-18, -17$
$m_{-1}(i)$	**-13**	**-15**	**-16**	**-18**
$[d_i, d_{i+1})^+$	$[12, 13)$	$[13, 14)$	$[14, 15)$	$[15, 16)$
$\widehat{L}_2(d_{i+1}), \widehat{U}_1(d_i)^+$	18, 19	19, 20	20, 22	22, 24
$m_2(i)$	**18**	**20**	**20**	**24**
$\widehat{L}_1(d_{i+1}), \widehat{U}_0(d_i)^+$	5, 7	5, 7	5, 8	6, 9
$m_1(i)$	**6**	**6**	**8**	**8**
$\widehat{L}_0(d_i), \widehat{U}_{-1}(d_{i+1})^+$	$-8, -6$	$-8, -6$	$-9, -6$	$-10, -7$
$m_0(i)$	**-8**	**-8**	**-8**	**-8**
$\widehat{L}_{-1}(d_i), \widehat{U}_{-2}(d_{i+1})^+$	$-20, -19$	$-21, -20$	$-23, -21$	$-25, -23$
$m_{-1}(i)$	**-20**	**-20**	**-22**	**-24**

+: real value = shown value/16; $\widehat{L}_k = \lceil L_k \rceil_4$, $\widehat{U}_k = \lfloor U_k - \frac{1}{16} \rfloor_4$.

TABLE 5.10 Selection intervals and constants m_k (radix 4, carry-save residual).

Because of the selection constants $\frac{15}{16}$, $-\frac{13}{16}$, and $-\frac{15}{16}$, the input of $\widehat{y}$ to the selection function has four fractional bits ($c = 4$), which in this case is equal to t. The range of the estimate is obtained from (5.91) to be

$$\left\lfloor -\frac{8}{3} - \frac{1}{16} \right\rfloor_4 \le \widehat{y} \le \left\lfloor \frac{8}{3} - ulp \right\rfloor_4 \qquad \textbf{5.114}$$

which results in

$$-\frac{44}{16} \le \widehat{y} \le \frac{42}{16} \qquad \textbf{5.115}$$

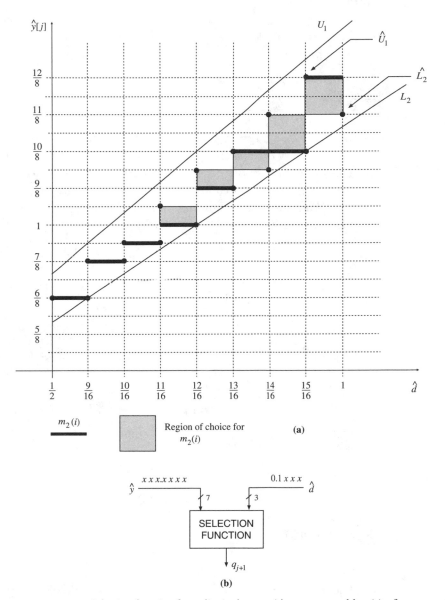

FIGURE 5.31 Selection function for radix-4 scheme with carry-save adder: (a) a fragment of P-D diagram and (b) implementation.

requiring three integer bits. Consequently, the selection function has the inputs shown in Figure 5.31(b). An example of the execution of this algorithm was given in Figure 5.9.   ■

Signed-Digit Adder

Instead of using a carry-save adder, it is possible to use a signed-digit adder, resulting in a residual in signed-digit representation. For the case of radix-2 representation, we now determine the error produced by using as estimate the shifted residual truncated to t fractional digits. In this case, the discarded part when truncating can be positive or negative (see Figure 5.28(c)). Consequently, the errors are

$$\epsilon_{max} = 2^{-t} - ulp, \qquad \epsilon_{min} = -(2^{-t} - ulp) \qquad \text{5.116}$$

The restricted selection interval is

$$U_k^* = U_k - (2^{-t} - ulp)$$
$$L_k^* = L_k + 2^{-t} - ulp \qquad \text{5.117}$$

By a similar reasoning as for the carry-save case, we obtain

$$\widehat{U}_{k-1} = \lfloor U_{k-1} \rfloor_t$$
$$\widehat{L}_k \quad = \lceil L_k + 2^{-t} \rceil_t \qquad \text{5.118}$$

The relation between δ and t is, from (5.90),

$$\frac{2\rho - 1}{2} - (a - \rho)2^{-\delta} \geq 2^{-t} \qquad \text{5.119}$$

Note that the relation between δ and t is the same as for the carry-save case, but the actual value of the selection constants might be different.

Finally, from expression (5.92) we have

$$\lfloor -r\rho \rfloor_t \leq \widehat{y} \leq \lfloor r\rho + 2^{-t} - ulp \rfloor_t \qquad \text{5.120}$$

The design of the selection function and an example of execution is left as an exercise.

5.6   Concluding Remarks

As for other digital modules, the design of a digit recurrence division unit is a trade-off among several characteristics, such as execution time, area, and energy. We have shown that the main parameters to consider in this trade-off are the radix of the quotient digit obtained each iteration and the quotient-digit set. We have given some example implementations that show the speedup achieved for higher radices, as well as the increase in area.

A higher radix reduces the number of iterations, but complicates the selection function and the multiplication of the divisor by a quotient digit. The use of a redundant quotient-digit set simplifies the selection function because it allows the utilization of a truncated residual and a truncated divisor. Moreover, to have a fast addition, a redundant residual is used.

The direct implementation of the selection function is practical for radices up to 8. For radix 16 an implementation of interest uses two overlapped radix-4 stages. For higher radices, prescaling of the divisor and selection by rounding provides a good speedup with a reasonable area.

We have presented a method to design quotient-digit selection functions and given some examples. The method can be applied to other radices, quotient-digit sets, and residual representations. It is also the basis to the implementation of other digit recurrence algorithms, such as square root and reciprocal square root.

Other methods to perform the division operation are presented in Chapter 7, together with some comparison comments.

5.7   Exercises

Examples of execution

5.1   **[Radix 2 and radix 4]** Divide 126×2^{-8} by $\frac{6}{8}$ and produce an 8-bit result, using the following algorithms:

- radix 2, $q_j \in \{0, 1\}$, conventional (nonredundant) residual
- radix 2, $q_j \in \{-1, 0, 1\}$, carry-save residual
- radix 4, $q_j \in \{-2, -1, 0, 1, 2\}$, carry-save residual

5.2   **[Radix 16]** For $x = 0.1001001110100101$ and $d = 0.110$ perform two iterations of the division recurrence using a radix-16 implementation with overlapped radix-4 stages.

5.3   **[Scaling and selection by rounding]** For $x = 126 \times 2^{-8}$ and $d = 6 \times 2^{-3}$ produce an 8-bit quotient using a radix-16 algorithm with scaling and selection by rounding. For the scaling factor M use 1.3125.

On-the-Fly Conversion

5.4   **[Example of conversion]** Perform the on-the-fly conversion of the signed-digit result $1\bar{2}\bar{1}1\bar{1}00\bar{1}2$, where the digit set is $\{-2, -1, 0, 1, 2\}$.

5.5   **[Conversion for digit set {0, 1, 2}]** Develop an algorithm to convert on-the fly a radix-2 positive redundant representation with digit set $\{0, 1, 2\}$ into conventional representation.

Characteristics of the Implementation

5.6   **[Delay and area]** For a 24-bit division unit give expressions for the delay and area in terms of delay and area of the component modules, for the following two cases:

(a)   Radix 2, carry-save residual
(b)   Radix 4, carry-save residual and digit set $\{-2, -1, 0, 1, 2\}$

5.7   **[Retiming the recurrence]** An alternative implementation of digit recurrence division (for instance, radix 4 with carry-save adder) is to retime the recurrence so that the quotient-digit selection is performed at the end of one cycle and the digit used in the next. This retiming creates two slices in the implementation: a most-significant slice, which includes the quotient-digit selection, and the rest. This might reduce the critical path by eliminating the need of a buffer to distribute the quotient digit to the most-significant slice. It also allows this slice to be optimized for delay, and the other part optimized for area and/or for energy dissipation.

To illustrate the characteristics of this retiming, design a radix-4 implementation for 54 bits using, in addition to the components described in this chapter, a faster variety with a delay that is 20% smaller. Also assume that the buffer required for the most-significant slice has a delay of 40% of the one used without retiming. After you optimize for speed, use the standard modules in the noncritical components to reduce area and energy (since these components have a smaller area and consume less energy than the faster variety). Determine the delay of a cycle and compare with the non-retimed version.

5.8 **[Overlapped radix-2 stages]** Consider a radix-2 division algorithm with the quotient-digit set $\{-1, 0, 1\}$ and redundant residuals in carry-save form. The quotient-digit selection is performed using the selection constants as described in the text. The cycle time is

$$t_{cycle} = t_{qsel} + t_{buff} + t_{mux} + t_{HA} + t_{reg} = 4 + 1 + 1 + 1 + 2 = 9t_g$$

To reduce the total time, we propose to obtain all three possible values of q_{j+2}, corresponding to $q_{j+1} = -1, 0, 1$, and select the correct one once q_{j+1} is known, all of this in the same cycle. The next residual is obtained as

$$w[j+2] = 2(2w[j] - q_{j+1}d) - q_{j+2}d$$

In other words, two quotient bits are generated per cycle.

(a) Design the network for the selection of q_{j+1} and q_{j+2}. Assume that the selection function is already implemented; that is, you can use the module that produces a quotient-digit based on the $(4 + 4)$ MS bits of the redundant residual. Show all details; in particular, show the details of the conditional selection.

(b) Design the network to produce the next residual (assume 8 bits in the fractional part). Show all details.

(c) Determine the cycle time of the new scheme and compare the total time to obtain a quotient of 8 bits with the scheme described in the text. Discuss your findings.

5.9 **[Gate-level design]** Design a radix-2 12-bit division unit at the gate level. You may use full-adder, multiplexer, register modules, and individual gates. Provide necessary design details to establish delays of critical paths. Use a carry-save adder to form residuals in redundant form. Assume that the dividend and divisor are positive fractions. Give an estimate of the overall delay in gate delay units (t_g) and cost in number of different modules. Assume that a full-adder has a delay of $4t_g$; a 3-1 multiplexer, $2t_g$; and a register load, $3t_g$.

Integer Division

5.10   Perform the integer division of $x = 120$ by $d = 9$ using a radix-4 algorithm with residual in carry-save form.

5.11   Give an algorithm that combines fractional division and integer division. Show the combined implementation, highlighting the modules that are required by the inclusion of integer division.

Quotient-Digit Selection

5.12   [Using signed-digit adder] Determine a quotient-digit selection function for a radix-2 algorithm using a signed-digit adder. Show the execution for $x = 128 \times 2^{-8}$ and $d = 6 \times 2^{-3}$ (obtain a quotient of 8 bits).

5.13   [Radix 4 with $\rho = 1$] Determine a good quotient-digit selection function for a radix-4 division algorithm with $\rho = 1$ and carry-save residual. Give only the portion for selection of $q_{j+1} = 2$, but use a value of δ and t suitable for the whole selection function.

5.14   [Restricted divisor range] Determine a good quotient-digit selection function for a radix-4 division algorithm with $\rho = \frac{2}{3}$ and carry-save residual, if the divisor is restricted to the range $[\frac{63}{64}, 1)$. Use a value of δ and t suitable for this restricted range.

5.15   [Radix-8 selection] Using selection constants, determine the minimum number of bits of the residual estimate to perform radix-8 division ($\rho = 1$) with the divisor in the range (a) $[\frac{1}{2}, 1)$ and (b) $[\frac{3}{4}, \frac{5}{4})$, for nonredundant residuals. Draw the corresponding P-D diagrams (first quadrant only).

5.16   [Divisor range $[1, 2)$] Consider a radix-4 division algorithm with the quotient-digit set $\{-2, \ldots, 2\}$, the divisor in the range $[1, 2)$ (instead of the usual $[\frac{1}{2}, 1)$), and nonredundant residuals. The quotient-digit selection is performed using selection constants.

(a)   Determine the size (the number) of the divisor intervals.

(b)   Determine the best selection constants $m_2(i)$. Show details of your derivations. What is the total number of bits needed to select $q_{j+1} = 2$?

How does it compare with the case $\frac{1}{2} \leq d < 1$ in which three bits of d and six bits of the shifted residual are required?

(c)   Summarize the effect of using the divisor range $[1, 2)$. Is it a good idea?

5.17   **[Scaling and selection by rounding]** Consider a high radix r digit recurrence division method. Assume that residuals are in nonredundant form. The quotient digit $q_{j+1} \in \{-a, \ldots, a\}, a \leq r - 1$, is selected as the integer part of the rounded shifted partial residual. That is,

$$q_{j+1} = integer(r\,w[j] + 0.5)$$

For convergence, such an algorithm requires that the divisor be in the range

$$1 \leq d \leq 1 + \beta, \qquad \beta \geq 0$$

(a)   Determine the range of the divisor necessary for the convergence of the method. That is, determine β. Make sure that $|q_{j+1}| \leq a$.

(b)   Discuss a possible implementation and its cost/performance advantages and disadvantages relative to a fast radix-2 division algorithm.

(c)   Illustrate the method for $r = 100$, dividend $x = 0.83703960$, and divisor $d = 1.00827040$ by finding the first three radix r quotient digits.

5.18   Consider a radix-4 digit recurrence division algorithm with the residual recurrence

$$w[j + 1] = 4w[j] - q_{j+1}d$$

and divisor range $[\frac{1}{2}, 1)$.

(a)   Show that it is possible to perform division using the quotient-digit set $\{-3, -1, 1, 3\}$.

(b)   Under what conditions is (a) true?

(c)   How does the division algorithm in (a) compare with an algorithm using the digit set $\{-2, -1, 0, 1, 2\}$? What are the trade-offs?

5.8   Further Readings

Division methods for hardware implementation were considered in the early literature as direct mappings of the paper-and-pencil method for long division in radix 2. The two main algorithms used, restoring and nonrestoring division, are reviewed in Chapter 1. These algorithms are slow, and subsequent research led to a variety of methods for the design of fast dividers. One of the earliest discussions of digit recurrence division methods and implementation aspects appeared in Robertson (1957, 1964). Early literature on division includes survey articles by Reitwiesner (1960), Garner (1965), and Tung (1972). Parts of this chapter are based on the monograph by Ercegovac and Lang (1994), which presents an in-depth study of digit recurrence methods for division and square root.

SRT Division

The pioneering radix-2 division algorithm with the redundant quotient-digit set $\{-1, 0, 1\}$ and comparison constants $\pm\frac{1}{2}$ was proposed independently by Cocke and Sweeney (1957), Robertson (1957), and Tocher (1958) and named SRT division after Sweeney, Robertson, and Tocher. A similar algorithm is also discussed in Nadler (1956). Since the main objective in the early days was to reduce the number of costly additions by maximizing the frequency of zero quotient digits, analysis of the average number of consecutive zeros in the quotient was studied in great detail (Freiman 1961; Shively 1963; Robertson 1970). Some improvements to the SRT method leading to an increased number of zeros in the quotient are presented in MacSorley (1961), Wilson and Ledley (1961), and Metze (1962). In Robertson (1960), a variant of the SRT method with constants $\pm\frac{1}{4}$ allows the use of a truncated redundant shifted residual avoiding full-precision carry-propagation in the adder. Although developed for radix 2, the term SRT is often used for division with radices greater than 2.

Redundant Quotient-Digit Set and Residual

A method with a redundant quotient-digit set for higher-radix nonrestoring division is discussed in Robertson (1958). The method uses redundancy to simplify the quotient-digit selection function and allow the use of estimates of the partial remainders and the divisor. The problem of selecting quotient digits was studied in Robertson (1965) and later extended in Atkins (1968, 1970a) to a general formulation of the quotient-digit selection based on short-precision estimates of the scaled residual (shifted partial remainder) and of the divisor. The original treatment of graphical representation of the division process and the selection problem was introduced by Robertson (1958, 1964, 1965) and further developed by Atkins (1968, 1970a). There are a number of studies of the derivation and complexity of the quotient-digit tables (Atkins 1968, 1970a; Paal 1973; Tan 1978; Bushard 1983; Ercegovac and Lang 1994; Burgess and Williams 1995; Oberman and Flynn 1998). History of research in higher-radix nonrestoring division until 1975 is summarized in Atkins (1975).

Quotient Conversion

On-the-fly conversion of redundant into conventional representations was introduced in Ercegovac and Lang (1987a) and its combinational alternatives described in Ercegovac and Lang (1986, 1990a) and Ciminiera and Montuschi (1993b). A reduced-area on-the-fly conversion scheme is reported in Takagi and Horiyama (1999).

Divider with Stages

The idea of implementing higher-radix division as a composition of lower-radix stages has been used in the design of the Illiac III arithmetic unit (Atkins 1970b), where a radix-256 quotient digit is obtained from four radix-4 division steps. Radix-16 implementation using overlapping of stages is developed in Taylor (1985) and for radix 8 in Fandrianto (1989) and Prabhu and Zyner (1995). A self-timed division scheme with overlapped stages is described in Williams and Horowitz (1991a) and Williams (1991).

Radix-8 Divider

A direct radix-8 implementation is presented in Ware et al. (1982). To reduce the step time, conditional residual generation is done for all eight possible values of the quotient-digit. Nannarelli (1999) provides a set of selection constants for radix-8 division. In Carter and Robertson (1990) a direct radix-16 implementation is presented using a signed-digit adder array.

Design and Performance

Design and performance issues in dividers are discussed in Zurawski (1980), Oberman (1996), and Oberman and Flynn (1997). The area and performance of various types of division algorithms, including the digit recurrence type, are discussed in a survey paper (Soderquist and Leeser 1996). Implementations of radix-2 and radix-4 SRT dividers with carry-save adders are described frequently in the literature. Many sequential and combinational (array) alternatives have been investigated and implemented (for example, Taylor and Patterson 1981; Zurawski and Gosling 1981; Ercegovac et al. 1987; Bose et al. 1987; Zurawski and Gosling 1987; Peng et al. 1987; Montuschi and Ciminiera 1992; Harris et al. 1997). Dividers with signed-digit adders are reported in Avižienis (1961), Tung (1968, 1970), Takagi (1987), Kuninobu et al. (1987), and Carter and Robertson (1990). A systematic approach to design array dividers implementing the SRT method is presented in McQuillan (1992) and McQuillan and McCanny (1994).

Overredundant Quotient-Digit Set

The use of overredundant quotient-digit sets in the design of digit recurrence dividers is reported in Montuschi and Ciminiera (1994a) and applied in a radix-8 divider in Montuschi and Ciminiera (1994b). A radix-4 division scheme with overredundant digit set and prescaling is discussed in Montuschi and Ciminiera (1991a).

Combinational Divider

Combinational linear arrays for radix-2 division have been frequently considered. Early schemes, often called cellular or iterative dividers, correspond to unfolded nonrestoring division (Deegan 1971; Gardiner and Hont 1972; Agrawal 1979). Cappa and Hamacher (1973), Gaviland and Hamacher (1973), and Williams and Hamacher (1981) describe linear arrays of carry-save adders with short CLAs producing estimates of the residual for the quotient-digit selection. A VLSI implementation of such an array divider is reported in Tsunekava et al. (1998). Two's complement multiplication-division arrays are analyzed in Kutsuwa et al. (1987). Pipelined division arrays are discussed in Deverell (1975). Combinational VLSI implementations of radix-2 division with residuals in carry-save form are described in Zuras and McAllister (1986) and Vanmeulebroecke et al. (1990). Linear array dividers using signed-digit adders are discussed in Tung (1970) and Soceneantu and Toma (1972).

Division with Scaling of Operands

The idea of scaling operands to make the quotient-digit selection independent of the divisor was proposed in Svoboda (1963) and Klir (1963) for a decimal computer. The quotient digit is obtained as the most-significant digit of the partial remainder and the maximum number of scaling multiplications is 3, with an average of about 1.8. A standard nonrestoring division recurrence with carry-propagate adder is used. This scheme was extended in Tung (1968) to an arbitrary radix with signed-digit adder and applied in signed-digit division algorithm. Svoboda's scaling is one-sided; that is, the (positive) divisor is transformed into the range $1 + \epsilon$. A generalization of division via transformation of the operands' range is provided in Krishnamurthy (1970). An alternative two-sided scaling approach is discussed in Ercegovac (1977, 1983). A radix-2 SRT with simplified selection based on Svoboda's approach is discussed in Burgess (1991), McQuillan and McCanny (1992), and Montalvo et al. (1998). The scaling technique and the radix-4 division are considered in Ercegovac et al. (1988), Ercegovac and Lang (1990b), Burgess (1994), and Srinivas and Parhi (1995). A VLSI implementation of floating-point radix-16 divider with prescaling appears in Inui et al. (1999).

Division with Prediction

Quotient-digit prediction with scaling suitable for division with redundant residual was presented in Ercegovac and Lang (1985), and implementations for radix-2 and radix-4 are described in Ercegovac and Lang (1987b, 1989) and Modiri and Lang (1988). Quotient prediction without prescaling is discussed in Montuschi and Ciminiera (1995).

Very High Radix Division

Very high radix methods with prescaling and selection by rounding are reported in Ercegovac et al. (1993, 1994) and Montuschi and Lang (2001). Organizations of higher-radix division are discussed in Montuschi (1992). In Matula (1991) a radix-2^{17} division unit is described, based on scaling the residual (multiplying it by a short reciprocal of the divisor) so that digit selection can be done by truncation. The unit uses an 18×69 rectangular multiplier.

Miscellaneous Division Schemes

Algorithms with skipping over zero quotient bits are discussed in Ligomenides (1977), Montuschi and Ciminiera (1991b, 1993), and Mandelbaum (1990). Division methods and implementations with speculation of quotient digits are described in Cortadella and Lang (1993, 1994), Cornetta and Cortadella (1999), Wey and Wang (1999), and Wey (2000). Integer algorithms and implementations can be found in Purdy and Purdy (1987), Magenheimer et al. (1988), and Wang et al. (2000).

Implementation

Examples of implementation of dividers in VLSI are Ware et al. (1982), Bose et al. (1987), Peng et al. (1987), Moes et al. (1993), Eisig et al. (1993), and Prabhu and Zyner (1952). Design and implementation of division suitable for FPGA technologies is developed in Louie and Ercegovac (1993, 1994). VLSI layout of dividers is discussed in Guyot et al. (1995).

Variable-Time Divider

Self-timed and asynchronous dividers are discussed in Williams (1991), Williams and Horowitz (1991a, 1991b), Renaudin et al. (1996), and Cornetta and Cortadella (2001). Low-power self-timed dividers are reported in Lee and Choi (1996) and Won and Choi (2000).

Online Division

Digit-serial division schemes (online division) and related literature are discussed in Chapter 9.

Low-Power Dividers

Design and implementation of low-power dividers for radix 4 are given in Nannarelli and Lang (1996, 1999a) and for radix 8 in Nannarelli and Lang (1998a). Comparison of radix-4, radix-8, and radix-16 low-power dividers is presented in Nannarelli and Lang (1999b). Kuhlmann and Parhi (1998) discuss a low-power design of an SRT divider. Nannarelli and Lang (1998b) present power-delay trade-offs in the design of digit recurrence dividers. The low-power design of division and square root is the subject of a doctoral dissertation (Nannarelli 1999).

Verification

Formal verification of implementations of the SRT division is considered in Bryant (1996), Clarke et al. (1999), and Ruess et al. (1999).

Delay and Area Bounds

Theoretical aspects of time and size complexity of the division operation is considered in Beame et al. (1986), where the bounds on the depth of circuits are developed. Optimal size of integer division circuits is discussed in Reif and Tate (1989). The area-time optimality of division networks is studied in Mehlhorn and Preparata (1987). No practical implementations based on this work exist.

5.9 Bibliography

Agrawal, D. P. (1979). High-speed arithmetic arrays. *IEEE Transactions on Computers*, C-28(3):215–24.

Atkins, D. E. (1968). Higher-radix division using estimates of the divisor and partial remainders. *IEEE Transactions on Computers*, C-17(10):925–34.

Atkins, D. E. (1970a). *A Study of Methods for Selection of Quotient Digits during Digital Division*. PhD thesis, Department of Computer Science, University of Illinois at Urbana-Champaign. Technical report UIUCDCS-R-397.

Atkins, D. E. (1970b). Design of the arithmetic units of ILLIAC III: Use of redundancy and higher radix methods. *IEEE Transactions on Computers*, C-19(8):720–33.

Atkins, D. E. (1975). Higher radix, non-restoring division: History and recent developments. In *Proceedings of the 3rd IEEE Symposium on Computer Arithmetic*, pages 158–60.

Avižienis, A. (1961). Signed digit number representations for fast parallel arithmetic. *IRE Transactions on Electronic Computers*, EC-10(9):389–400.

Beame, P., S. Cook, and H. Hoover (1986). Log depth circuits for division and related problems. *SIAM Journal on Computing*, 15:994–1003.

Bose, B. K., L. Pei, G. S. Taylor, and D. A. Paterson (1987). Fast multiply and divide for a VLSI floating-point unit. In *Proceedings of the 8th IEEE Symposium on Computer Arithmetic*, pages 87–94.

Bryant, R. (1996). Bit-level analysis of an SRT divider circuit. In *Proceedings of the 33rd Design Automation Conference*, pages 661–65.

Burgess, N. (1991). Radix-2 SRT division with simple quotient digit selection. *Electronics Letters*, 27(21):1910–11.

Burgess, N. (1994). Prescaled maximally-redundant radix-4 SRT divider. *Electronics Letters*, 30(23):1926–28.

Burgess, N., and T. Williams (1995). Choices of operand truncation in the SRT division algorithm. *IEEE Transactions on Computers*, 44(7):933–38.

Bushard, L. B. (1983). A minimum table size result for higher radix nonrestoring division. *IEEE Transactions on Computers*, C-32(6):521–26.

Cappa, M., and V. C. Hamacher (1973). An augmented iterative array for high-speed binary division. *IEEE Transactions on Computers*, C-22(2):172–75.

Carter, T. M., and J. E. Robertson (1990). Radix-16 signed-digit division. *IEEE Transactions on Computers*, C-39(12):1424–33.

Ciminiera, L., and P. Montuschi (1996). Carry-save multiplication schemes without final addition. *IEEE Transactions on Computers*, 45(9):1050–55.

Clarke, E. M., S. M. German, and X. Zhao (1999). Verifying the SRT division algorithm using theorem proving techniques. *Formal Methods in System Design*, 14(1):7–44.

Cocke, J., and D. W. Sweeney (1957). High speed arithmetic in a parallel device. Technical report, IBM.

Cornetta, G., and J. Cortadella (1999). A radix-16 SRT division unit with speculation of the quotient digits. In *Proceedings of the 9th Great Lakes Symposium on VLSI*, pages 74–77.

Cornetta, G., and J. Cortadella (2001). A multi-radix approach to asynchronous division. In *ASYNC 2001, Proceedings of the 7th International Symposium on Asynchronous Circuits and Systems*, pages 25–34.

Cortadella, J., and T. Lang (1993). Division with speculation of quotient digits. In *Proceedings of the 11th IEEE Symposium on Computer Arithmetic*, pages 87–94.

Cortadella, J., and T. Lang (1994). High-radix division and square-root with speculation. *IEEE Transactions on Computers*, 43(8):919–31.

Deegan, I. (1971). Concise cellular array for multiplication and division. *Electronics Letters*, 7(23):702–4.

Deverell, J. (1975). Pipeline iterative arithmetic arrays. *IEEE Transactions on Computers*, C-24(3):317–22.

Eisig, D., J. Rotstain, and I. Koren (1993). The design of a 64-bit integer multiplier/divider unit. In *Proceedings of the 11th Symposium on Computer Arithmetic*, pages 171–78.

Ercegovac, M. D. (1977). A general hardware-oriented method for evaluation of functions and computations in a digital computer. *IEEE Transactions on Computers*, C-26(7):667–80.

Ercegovac, M. D. (1983). A higher radix division with simple selection of quotient digits. In *Proceedings of the 6th IEEE Symposium on Computer Arithmetic*, pages 94–98.

Ercegovac, M. D., and T. Lang (1985). A division algorithm with prediction of quotient digits. In *Proceedings of the 7th IEEE Symposium on Computer Arithmetic*, pages 51–56.

Ercegovac, M. D., and T. Lang (1986). Alternative on-the-fly conversion of redundant into conventional representations. Technical report CSD-860027, Computer Science Department, University of California, Los Angeles.

Ercegovac, M. D., and T. Lang (1987a). On-the-fly conversion of redundant into conventional representations. *IEEE Transactions on Computers*, C-36(7):895–97.

Ercegovac, M. D., and T. Lang (1987b). Simple radix-4 division with divisor scaling. Technical report CSD-870015, Computer Science Department, University of California, Los Angeles.

Ercegovac, M. D., and T. Lang (1989). Fast radix-2 division with quotient-digit prediction. *Journal of VLSI Signal Processing*, 2(1):169–80.

Ercegovac, M. D., and T. Lang (1990a). Fast multiplication without carry-propagate addition. *IEEE Transactions on Computers*, C-39(11):1385–90.

Ercegovac, M. D., and T. Lang (1990b). Simple radix-4 division with operands scaling. *IEEE Transactions on Computers*, C-39(9):1204–7.

Ercegovac, M. D., and T. Lang (1994). *Division and Square Root: Digit-Recurrence Algorithms and Implementations*. Kluwer Academic Publishers.

Ercegovac, M. D., T. Lang, and R. Modiri (1988). Implementation of fast radix-4 division with operands scaling. In *Proceedings of the ICCD '88 Conference*, pages 486–89, New York.

Ercegovac, M. D., T. Lang, and P. Montuschi (1993). Very high radix division with selection by rounding and prescaling. In *Proceedings of the 11th IEEE Symposium on Computer Arithmetic*, pages 112–19.

Ercegovac, M. D., T. Lang, and P. Montuschi (1994). Very-high radix division with prescaling and rounding. *IEEE Transactions on Computers*, 43(8):909–18.

Ercegovac, M. D., T. Lang, J. G. Nash, and L. P. Chow (1987). An area-time efficient binary divider. In *Proceedings of the ICCD '87 Conference*, pages 645–48, New York.

Fandrianto, J. (1989). Algorithm for high-speed shared radix-8 division and radix-8 square root. In *Proceedings of the 9th IEEE Symposium on Computer Arithmetic*, pages 68–75.

Freiman, C. V. (1961). Statistical analysis of certain binary division algorithms. *Proceedings of IRE*, 49:91–103.

Gardiner, A. B., and J. Hont (1972). Cellular-array arithmetic unit with multiplication and division. *Proceedings of the IEE*, 119(6):559–60.

Garner, H. L. (1965). Number systems and arithmetic. In *Advances in Computers*, volume 6, pages 131–94. Academic Press, New York.

Gaviland, J., and V. C. Hamacher (1973). High-speed multiplier/divider iterative arrays. In *1973 Sagamore Computer Conference on Parallel Processing*, pages 91–100.

Guyot, A., L. Montalvo, A. Houelle, H. Mehrez, and N. Vaucher (1995). Comparison of the layout synthesis of radix-2 and pseudo-radix-4 dividers. In *Proceedings of the 8th International Conference on VLSI Design*, pages 386–91.

Harris, D., S. F. Oberman, and M. H. Horowitz (1997). SRT division architectures and implementations. In *Proceeding of the 13th IEEE Symposium on Computer Arithmetic*, pages 18–25.

Inui, S., T. Uesugi, H. Saito, Y. Hagihara, A. Yoshikawa, M. Nishida, and M. Yamashina (1999). A 250 MHz CMOS floating-point divider with operand pre-scaling. In *Symposium on VLSI Circuits*, pages 17–18.

Klir, J. (1963). A note on Svoboda's algorithm for division. *Information Processing Machines (Stroje na Zpracovani Informaci)*, (9):35–39.

Krishnamurthy, E. V. (1970). On range-transformation techniques for division. *IEEE Transactions on Computers*, C-19(2):157–60.

Kuhlmann, M., and K. K. Parhi (1998). Power comparison of SRT and GST dividers. In *Proceedings of the SPIE—Advanced Signal Processing Algorithms, Architectures, and Implementations VIII*, volume 3461, pages 584–94.

Kuninobu, S., T. Nishiyama, H. Edamatsu, T. Taniguchi, and N. Takagi (1987). Design of high speed MOS multiplier and divider using redundant binary representation. In *Proceedings of the 8th IEEE Symposium on Computer Arithmetic*, pages 80–86.

Kutsuwa, T., M. Mun, and K. Ebata (1987). Configuration and evaluation of two's complement multiplication-division arrays. *IEEE Transactions Circuits and Systems.*, CAS-34:304–8.

Lee, K., and K. Choi (1996). Self-timed divider based on RSD number system. *IEEE Transactions on Very Large Scale Integration (VLSI) Systems*, 4(2):292–95.

Ligomenides, P. A. (1977). The skip-and-set fast division algorithm. *IEEE Transactions on Computers*, C-26:1030–32.

Louie, M. E., and M. D. Ercegovac (1993). On digit-recurrence division implementation for field programmable gate arrays. In *Proceedings of the 11th IEEE Symposium on Computer Arithmetic*, pages 202–9.

Louie, M. E., and M. D. Ercegovac (1994). Implementing division with field programmable gate arrays. *Journal of VLSI Signal Processing*, 7(3):271–85.

MacSorley, O. L. (1961). High-speed arithmetic in binary computers. *IRE Proceedings*, 49:67–91.

Magenheimer, D. J., L. Peters, K. W. Pettis, and D. Zuras (1988). Integer multiplication and division on the HP precision architecture. *IEEE Transactions on Computers*, C-37:980–90.

Mandelbaum, D. M. (1990). A systematic method for division with high average bit skipping. *IEEE Transactions on Computers*, C-39(1):127–30.

Matula, D. W. (1991). Design of a highly parallel IEEE Standard floating point arithmetic unit. In *Proceedings of the Symposium on Combinatorial Optimization in Science and Technology at RUTCOR/DIMACS*.

McQuillan, S., and J. V. McCanny (1994). Fast algorithms for division and square root. *Journal of VLSI Signal Processing*, 8(2):151–68.

McQuillan, S. E. (1992). *Algorithms and Architectures for High Performance Arithmetic Processors*. PhD thesis, The Queen's University of Belfast.

McQuillan, S. E., and J. V. McCanny (1992). VLSI module for high-performance multiply, square root and divide. *IEE Proceedings E: Computers and Digital Techniques*, 139(6):505–10.

Mehlhorn, K., and F. P. Preparata (1987). Area-time optimal division for $t = \omega(\log n)^{1+\epsilon}$. *Information and Computation*, 72(3):270–82.

Metze, G. (1962). A class of binary divisions yielding minimally represented quotients. *IRE Transactions Electronic Computers*, EC-11(6):761–64.

Modiri, R., and T. Lang (1988). Alternative implementations of a radix-4 divider with scaling. Technical report CSD-880069, Computer Science Department, University of California, Los Angeles.

Moes, E. A. J., R. Nouta, and G. J. Hekstra (1993). Divider architectures for VLSI implementation. *International Journal of High Speed Electronics and Systems*, 4(1):1–33.

Montalvo, L. A., K. K. Parhi, and A. Guyot (1998). New Svoboda-Tung division. *IEEE Transactions on Computers*, 47(9):1014–20.

Montuschi, P. (1992). Parallel architectures for higher-radix division. *IEE Proceedings E: Computers and Digital Techniques*, 139(2):101–10.

Montuschi, P., and L. Ciminiera (1991a). Algorithm and architectures for radix-4 division with over-redundant digit set and simple digit selection hardware. In *Conference Record of the 25th Asilomar Conference on Signals, Systems and Computers*, pages 418–22.

Montuschi, P., and L. Ciminiera (1991b). Simple radix 2 division and square root with skipping of some addition steps. In *Proceedings of the 10th IEEE Symposium on Computer Arithmetic*, pages 202–9.

Montuschi, P., and L. Ciminiera (1992). Design of a radix 4 division unit with simple selection table. *IEEE Transactions on Computers*, 41(12):1606–11.

Montuschi, P., and L. Ciminiera (1993). Reducing iteration time when result digit is zero for radix-2 SRT division and square root with redundant remainders. *IEEE Transactions on Computers*, 42(2):239–46.

Montuschi, P., and L. Ciminiera (1994a). Over-redundant digit sets and the design of digit-by-digit division units. *IEEE Transactions on Computers*, 43(3):269–77.

Montuschi, P., and L. Ciminiera (1994b). Radix-8 division with over-redundant digit set. *Journal of VLSI Signal Processing*, 7(3):259–70.

Montuschi, P., and L. Ciminiera (1995). Quotient prediction without prescaling. *IEE Proceedings: Computers and Digital Techniques*, 142(1):15–22.

Montuschi, P., and T. Lang (2001). Boosting very-high radix division with prescaling and selection by rounding. *IEEE Transactions on Computers*, 50(1):13–27.

Nadler, M. (1956). A high speed electronic arithmetic unit for automatic computing machines. *Acta Technica* (6):464–78.

Nannarelli, A. (1999). *Low Power Division and Square Root*. PhD thesis, University of California, Irvine.

Nannarelli, A., and T. Lang (1996). Low-power radix-4 divider. In *International Symposium on Low Power Electronics and Design*, pages 205–8.

Nannarelli, A., and T. Lang (1998a). Low-power radix-8 divider. In *Proceedings International Conference on Computer Design. VLSI in Computers and Processors*, pages 420–26.

Nannarelli, A., and T. Lang (1998b). Power-delay tradeoffs for radix-4 and radix-8 dividers. In *1998 International Symposium on Low Power Electronics and Design*, pages 109–11.

Nannarelli, A., and T. Lang (1999a). Low-power divider. *IEEE Transactions on Computers*, 48(1):2–14.

Nannarelli, A., and T. Lang (1999b). Low-power division: Comparison among implementations of radix 4, 8 and 16. In *Proceedings of the 14th IEEE Symposium on Computer Arithmetic*, pages 60–67.

Oberman, S. F. (1996). *Design Issues in High Performance Floating Point Arithmetic units*. PhD thesis, Department of Electrical Engineering, Stanford University.

Oberman, S. F., and M. J. Flynn (1997). Design issues in division and other floating-point operations. *IEEE Transactions on Computers*, 46(2):154–61.

Oberman, S. F., and M. J. Flynn (1998). Minimizing the complexity of SRT tables. *IEEE Transactions on Very Large Scale Integration (VLSI) Systems*, 6(1):141–49.

Paal, F. (1973). Implementation of truncated comparison and quotient prediction in the Q-P (quotient predictor) division algorithms. In *Proceedings of the 7th Asilomar Conference on Circuits, Systems and Computers*, pages 734–36.

Peng, V., S. Samudrala, and M. Gavrielov (1987). On the implementation of shifters, multipliers, and dividers in VLSI floating-point units. In *Proceedings of the 8th IEEE Symposium on Computer Arithmetic*, pages 95–102.

Prabhu, J. A., and G. B. Zyner (1995). 167 MHz radix-8 divide and square root using overlapped radix-2 stages. In *Proceedings of the 12th IEEE Symposium on Computer Arithmetic*, pages 155–62.

Purdy, C. N., and G. B. Purdy (1987). Integer division in linear time with bounded fan-in. *IEEE Transactions on Computers*, C-36:640–44.

Reif, J. H., and S. R. Tate (1989). Optimal size integer division circuits. In *Proceedings of the 21st Annual ACM Symposium on Theory of Computing*, pages 264–73.

Reitwiesner, G. W. (1960). Binary arithmetic. In *Advances in Computers*, volume 1, pages 232–308. Academic Press, New York.

Renaudin, M., B. E. Hassan, and A. Guyot (1996). A new asynchronous pipeline scheme: application to the design of a self-timed ring divider. *IEEE Journal of Solid-State Circuits*, 31(7):1001–13.

Robertson, J. E. (1957). Arithmetic unit (chapter 8). In *On the Design of Very High-Speed Computers*. Technical report no. 80, Computer Science Department, University of Illinois at Urbana-Champaign.

Robertson, J. E. (1958). A new class of digital division methods. *IRE Transactions Electronic Computers*, EC-7(3):88–92.

Robertson, J. E. (1960). *Theory of Computer Arithmetic Employed in the Design of the New Computer at the University of Illinois*. File no. 319, Computer Science Department, University of Illinois at Urbana-Champaign.

Robertson, J. E. (1964). *Introduction to Digital Computer Arithmetic*. File no. 599, Department of Computer Science, University of Illinois at Urbana-Champaign.

Robertson, J. E. (1965). *Methods of Selection of Quotient Digits during Digital Division*. File no. 663, Computer Science Department, University of Illinois at Urbana-Champaign.

Robertson, J. E. (1970). The correspondence between methods of digital division and multiplier recoding procedures. *IEEE Transactions on Computers*, C-19(8):692–701.

Ruess, H., N. Shankar, and M. K. Srivas (1999). Modular verification of SRT division. *Formal Methods in System Design*, 14(1):45–73.

Shively, R. R. (1963). *Stationary Distributions of Partial Remainders in SRT Digital Division*. PhD thesis, University of Illinois.

Soceneantu, A., and C. I. Toma (1972). Cellular logic array for redundant binary division. In *Proceedings of the IEE*, volume 119, pages 1452–56.

Soderquist, P., and M. Leeser (1996). Area and performance tradeoffs in floating-point division and square root implementations. *ACM Computing Surveys*, 28(3):518–64.

Srinivas, H. R., and K. K. Parhi (1995). A fast radix-4 division algorithm and its architecture. *IEEE Transactions on Computers*, 44(6):826–31.

Svoboda, A. (1963). An algorithm for division. *Information Processing Machines (Stroje na Zpracovani Informaci)*, 9:25–34.

Takagi, N. (1987). *Studies on Hardware Algorithms for Arithmetic Operations with a Redundant Binary Representation*. PhD thesis, Department of Information Science, Kyoto University.

Takagi, N., and T. Horiyama (1999). A high-speed reduced-size adder under left-to-right input arrival. *IEEE Transactions on Computers*, 48(1):76–80.

Tan, K. G. (1978). The theory and implementation of high-radix division. In *Proceedings of the 4th IEEE Symposium on Computer Arithmetic*, pages 154–63.

Taylor, G. S. (1985). Radix-16 SRT dividers with overlapped quotient-selection stages. In *Proceedings of the 7th IEEE Symposium on Computer Arithmetic*, pages 64–71.

Taylor, G. S., and D. A. Patterson (1981). VAX hardware for the proposed IEEE Floating-Point Standard. In *Proceedings of the 5th IEEE Symposium on Computer Arithmetic*, pages 190–96.

Tocher, K. D. (1958). Techniques of multiplication and division for automatic binary computers. *Quart. J. Mech. Appl. Math.*, XI(Pt. 3):364–84.

Tsunekava, Y., M. Hinosugi, and M. Miura (1998). Design and VLSI evaluation of a high-speed cellular array divider with a selection function. *Electrical Engineering in Japan*, 124(4):760–97.

Tung, C. (1968). A division algorithm for signed-digit arithmetic. *IEEE Transactions on Computers*, C-17(9):887–89.

Tung, C. (1970). Signed-digit division using combinational arithmetic nets. *IEEE Transactions on Computers*, C-19(8):746–48.

Tung, C. (1972). Arithmetic (Chapter 3). In *Computer Science*. Wiley-Interscience, New York.

Vanmeulebroecke, A., E. Vanzieleghem, T. Denyer, and P. G. A. Jespers (1990). A new carry-free division algorithm and its application to a single-chip 1024-b RSA processor. *IEEE Journal of Solid-State Circuits*, SC-25(3):748–65.

Wang, C.-C., C.-J. Huang, and G.-C. Lin (2000). Cell-based implementation of radix-4/2 64b dividend 32b divisor signed integer divider using the COMPASS cell library. *IEE Proceedings—Computers and Digital Techniques*, 147(2):109–15.

Ware, F. A., W. McAllister, J. R. Carlson, D. K. Sun, and R. J. Vlach (1982). 64 bit monolithic floating-point processors. *IEEE Journal of Solid-State Circuits*, SC-17(5):898–907.

Wey, C. L. (2000). Design of fast high-radix SRT dividers and their VLSI implementation. *IEE Proceedings—Computers and Digital Techniques*, 147(4):275–81.

Wey, C. L., and C.-P. Wang (1999). Design of a fast radix-4 SRT divider and its VLSI implementation. *IEE Proceedings—Computers and Digital Techniques*, 146(4):205–10.

Williams, J., and V. C. Hamacher (1981). A linear-time divider array. *Canadian Electr. Engineering Journal*, 6:14–20.

Williams, T. E. (1991). *Self-Timed Rings and Their Application to Division*. PhD thesis, Stanford University. Computer Systems Laboratory technical report no. CSL-TR-91-482.

Williams, T. E., and M. Horowitz (1991a). A 160ns 54-bit CMOS division implementation using self-timed and symmetrically overlapped SRT stages. In *Proceedings of the 10th IEEE Symposium on Computer Arithmetic*, pages 210–17.

Williams, T. E., and M. A. Horowitz (1991b). A zero-overhead self-timed 160-ns 54-b CMOS divider. *IEEE Journal of Solid-State Circuits*, 26(11):1651–61.

Wilson, J. B., and R. S. Ledley (1961). An algorithm for rapid binary division. *IRE Transactions Electronic Computers*, EC-10(4):662–70.

Won, J.-H., and K. Choi (2000). Low power self-timed radix-2 division. In *ISLPED'00: Proceedings of the 2000 International Symposium on Low Power Electronics and Design*, pages 210–12.

Zuras, D., and W. H. McAllister (1986). Balanced delays trees and combinatorial division in VLSI. *IEEE Journal of Solid-State Circuits*, SC-21:814–19.

Zurawski, J. H. P. (1980). *High Performance Evaluation of Division and Other Elementary Functions*. PhD thesis, University of Manchester, England.

Zurawski, J. H. P., and J. B. Gosling (1981). Design of high-speed digital divider units. *IEEE Transactions on Computers*, C-30(9):691–99.

Zurawski, J. H. P., and J. B. Gosling (1987). Design of a high-speed square root, multiply and divide unit. *IEEE Transactions on Computers*, C-36(1):13–23.

**IN THIS CHAPTER WE PRESENT AND DISCUSS
THE FOLLOWING MAIN TOPICS:**

- Fractional square root: recurrence and step
- General digit recurrence algorithm, implementation and timing
- Examples of algorithms and implementations of the fractional square root algorithm for radices 2 and 4
- Combined radix-2 division and square root
- Integer square root
- Result-digit selection function: theory and radix-2 and radix-4 implementations

| # Square Root by Digit Recurrence

This method of performing the square root operation is conceptually very similar to the method for division discussed in the previous chapter. Consequently, we develop the algorithm in a similar fashion, providing less detail since we assume familiarity with the developments for division. Table 6.1 gives a summary of the main definitions. Moreover, we concentrate on algorithms that use estimates for the result-digit selection and redundant addition in the recurrence.

As in division, the algorithm is presented for *fractional* operand x and result s. For floating-point representation and normalized operand, it is necessary to scale the operand to have an even exponent to allow the computation of the result exponent. Consequently,

$$s = \sqrt{x}, \qquad \frac{1}{4} \leq x < 1, \qquad \frac{1}{2} \leq s < 1 \qquad \textbf{6.1}$$

6.1 Recurrence and Step

Each iteration of the recurrence produces one digit of the result, most-significant digit first. Let us call $S[j]$ the value of the result after j iterations, that is,

$$S[j] = \sum_{i=0}^{j} s_i r^{-i} \qquad \textbf{6.2}$$

The digit s_0 should be 1 for $\rho < 1$ to represent a result value greater than ρ; it can be either 1 or 0 for $\rho = 1$.

The final result is then

$$s = S[n] = \sum_{i=0}^{n} s_i r^{-i} \qquad \textbf{6.3}$$

Operand	$\frac{1}{4} \leq x < 1$
Result	$\frac{1}{2} \leq s < 1$
Result after j iterations	$S[j] = \sum_{i=0}^{j} s_i r^{-i}$
Result-digit set	$s_i \in \{-a, \ldots, -1, 0, 1, \ldots, a\}$
Redundancy factor	$\rho = a/(r-1)\ \frac{1}{2} < \rho \leq 1$
Selection interval for $s_{j+1} = k$	$L_k[j] \leq rw[j] \leq U_k[j]$
Estimate of redundant shifted residual	$\widehat{y}$ with t fractional bits
Selection constants	$m_k(i)$ with c fractional bits
Result estimate	$\widehat{S}[j]$ with δ fractional bits

TABLE 6.1 Summary of definitions.

and the result has to be correct for n-digit precision; that is,

$$|x^{1/2} - s| < r^{-n} \qquad \text{6.4}$$

The use of absolute value allows positive and negative remainders necessary for efficient implementation. We define an error function ϵ so that its value after j steps (iterations) is

$$\epsilon[j] = x^{1/2} - S[j] \qquad \text{6.5}$$

As in division, since the minimum (maximum) digit value is $-a\,(a)$, we get[1]

$$-\rho r^{-j} < \epsilon[j] < \rho r^{-j} \qquad \text{6.6}$$

Introducing (6.5) in (6.6) and transforming to eliminate the square root operation (add $S[j]$ and obtain the square), we get

$$\rho^2 r^{-2j} - 2\rho r^{-j} S[j] + S[j]^2 < x < \rho^2 r^{-2j} + 2\rho r^{-j} S[j] + S[j]^2 \qquad \text{6.7}$$

Subtracting $S[j]^2$ we obtain

$$\rho^2 r^{-2j} - 2\rho r^{-j} S[j] < x - S[j]^2 < \rho^2 r^{-2j} + 2\rho r^{-j} S[j] \qquad \text{6.8}$$

That is, $S[j]$ is computed such that $x - S[j]^2$ is bounded according to (6.8).
We now define a residual (or scaled partial remainder) w so that

$$w[j] = r^j (x - S[j]^2) \qquad \text{6.9}$$

---

1. As in division, can make $\leq$ for $\rho < 1$.

From (6.8) the bounds on the residual are

$$-2\rho S[j] + \rho^2 r^{-j} < w[j] < 2\rho S[j] + \rho^2 r^{-j} \qquad \textbf{6.10}$$

and the initial condition is

$$w[0] = x - S[0]^2 = x - s_0 \quad \text{for } s_0 = 0 \text{ or } 1 \qquad \textbf{6.11}$$

In terms of the residual we obtain the recurrence

$$w[j+1] = r w[j] - 2S[j]s_{j+1} - s_{j+1}^2 r^{-(j+1)} \qquad \textbf{6.12}$$

Expression (6.12) is the basic recurrence on which the square root algorithms are based. The result digit is chosen, in a way that satisfies the bounds (6.10) for $w[j+1]$, by the function

$$s_{j+1} = SEL_{SQR}(\widehat{y}[j], \widehat{S}[j]) \qquad \textbf{6.13}$$

where $\widehat{y}[j]$ and $\widehat{S}[j]$ are estimates of $rw[j]$ and $S[j]$, respectively.

Each square root iteration consists of four subcomputations (Figure 6.1(a)):

1. An arithmetic left shift of $w[j]$ by one position to produce $rw[j]$.

2. Determination of the result digit s_{j+1} using the result-digit selection function SEL_{SQR}.

3. Formation of the adder input

$$F[j] = -\left(2S[j]s_{j+1} + s_{j+1}^2 r^{-(j+1)}\right) \qquad \textbf{6.14}$$

4. Addition of $F[j]$ to $rw[j]$ to produce $w[j+1]$. As in division, to have a fast iteration, a redundant adder is used for this addition. This adder can be of the signed-digit or of the carry-save type. Since the digits of $S[j]$ are produced in signed-digit form, if a carry-save adder is used in the recurrence, it is necessary to convert the signed-digit form to two's complement form by means of a variant of the on-the-fly conversion.

The four subcomputations are executed in sequence as indicated in the timing diagram of Figure 6.1(b). Note that no time has been allocated for the arithmetic shift since it is performed by suitable wiring. Moreover, the relative magnitudes of the delay of each of the components depend on the specific implementation.

As in division, different specific versions are possible, depending on the radix, the redundancy factor, the type of representation of the residual, and the result-digit selection function.

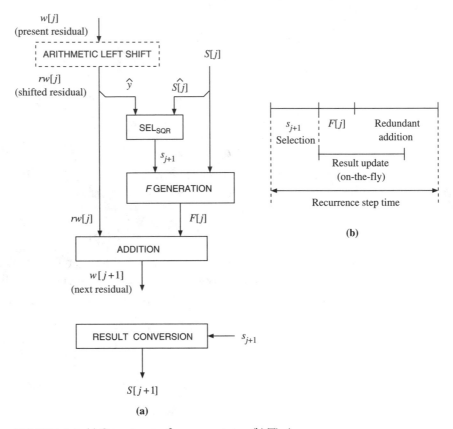

FIGURE 6.1 (a) Components of square root step. (b) Timing.

6.2   Generation of Adder Input $F[j]$

As part of the implementation of the recurrence (6.12), it is necessary to form the adder input F with value

$$F[j] = -2S[j]s_{j+1} - s_{j+1}^2 r^{-(j+1)} \qquad \textbf{6.15}$$

so that

$$w[j + 1] = rw[j] + F[j]$$

Since the digit of the result is produced in a signed-digit form, the partial result $S[j]$ is also in this form. Depending on the type of adder, $S[j]$ has to be converted to adapt to the adder. In particular, for the case of a carry-save adder the input F has to be in two's complement representation. The conversion is done on-the-fly using a variation of the scheme presented in Chapter 5. It requires that two conditional forms $A[j]$ and $B[j]$ are kept, such that

$$A[j] = S[j] \qquad\qquad\qquad\qquad\qquad\qquad \textbf{6.16}$$

$$B[j] = S[j] - r^{-j} \qquad\qquad\qquad\qquad\qquad \textbf{6.17}$$

These forms are updated with each result digit as follows:

$$A[j+1] = \begin{cases} A[j] + s_{j+1} r^{-(j+1)} & \text{if } s_{j+1} \geq 0 \\ B[j] + (r - |s_{j+1}|) r^{-(j+1)} & \text{otherwise} \end{cases} \qquad \textbf{6.18}$$

$$B[j+1] = \begin{cases} A[j] + (s_{j+1} - 1) r^{-(j+1)} & \text{if } s_{j+1} > 0 \\ B[j] + (r - 1 - |s_{j+1}|) r^{-(j+1)} & \text{otherwise} \end{cases} \qquad \textbf{6.19}$$

In a sequential implementation this conversion requires two registers for A and B, appending of one digit, and loading. For controlling this appending and loading, a shift register K is used, containing a moving 1. This implementation is shown in Figure 6.2.

In terms of these forms, the value of F is given by the following expressions. For $s_{j+1} > 0$:

$$F[j] = -2S[j]s_{j+1} - s_{j+1}^2 r^{-(j+1)} = -(2A[j] + s_{j+1} r^{-(j+1)})s_{j+1} \quad \textbf{6.20}$$

For $s_{j+1} < 0$:

$$F[j] = 2S[j]|s_{j+1}| - s_{j+1}^2 r^{-(j+1)} = 2(B[j] + r^{-j})|s_{j+1}| - s_{j+1}^2 r^{-(j+1)}$$
$$= \left(2B[j] + (2r - |s_{j+1}|)r^{-(j+1)}\right)|s_{j+1}|$$

Note that these expressions are implemented by concatenation and multiplication by one radix-r digit. The implementation is especially simple for radix 2 and radix 4 (with digit set $\{-2, \ldots, 2\}$), as shown in the examples of Section 6.3.1.

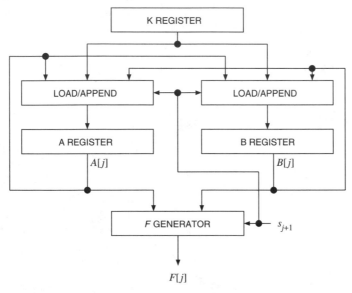

FIGURE 6.2 Network for generating F. (Adapted from (Ercegovac and Lang, 1990).)

6.3   Overall Algorithm, Implementation, and Timing

The overall algorithm is shown in Figure 6.3 and its implementation at the block-diagram level in Figure 6.4. The cycle time is

$$T_{cycle} = t_{SEL} + t_{F\text{-}GEN} + t_{ADD} + t_{load} \qquad \textbf{6.21}$$

6.3.1   Examples of Implementations

We now describe two example implementations, one radix-2 and one radix-4. The corresponding selection functions are derived in Section 6.6.

Radix-2 Square Root with Carry-Save Adder

In this case the quotient-digit set is $\{-1, 0, 1\}$ with $\rho = 1$. We choose to make $s_0 = 0$, resulting in the initial condition

$$w[0] = x$$

1. **[Initialize]** (*all assignments in parallel*)

   $w[0] \leftarrow x - s_0$; $s_0 = 0$ for $\rho = 1$ and $s_0 = 1$ for $\rho < 1$

   $A[0] \leftarrow s_0.000...000$;

   $B[0] \leftarrow (1 - s_0).000...000$;   since B[0] = A[0] − 1

   $K[0] \leftarrow 0.100...000$;

2. **[Recurrence]** (*all assignments in parallel*)

   for $j = 0 \ldots n$

   $\qquad s_{j+1} = SELECT(\widehat{y}[j], \widehat{S}[j])$

   $\qquad F[j] = f(A[j], B[j], s_{j+1})$ (expressions (6.20)

   $\qquad w[j + 1] \leftarrow rw[j] + F[j]$;

   $\qquad A[j + 1] \leftarrow g_a(A[j], B[j], K[j], s_{j+1})$;   (expressions (6.18))

   $\qquad B[j + 1] \leftarrow g_b(A[j], B[j], K[j], s_{j+1})$;   (expressions (6.19))

   $\qquad K[j + 1] \leftarrow \text{shift-right}(K[j])$

   end for

3. **[Termination]** *Correct result* (*same as for division*)

FIGURE 6.3 Square root algorithm.

Since $s_j \in \{-1, 0, 1\}$ we have $s_j^2 = |s_j|$ and hence the recurrence is

$$w[j + 1] = 2w[j] - 2S[j]s_{j+1} - 2^{-(j+1)}|s_{j+1}| \qquad \textbf{6.22}$$

resulting in

$$F[j] = -(2S[j]s_{j+1} + 2^{-(j+1)})|s_{j+1}| \qquad \textbf{6.23}$$

As discussed before, we use the conditional forms A and B for the conversion of $S[j]$ to two's complement representation and for the formation of $F[j]$. However, since in this case the only nonzero values of s_{j+1} are 1 and −1, we define

$F[j]$	Value	Bit-String
$F_1[j]$	$-2S[j] - 2^{-(j+1)}$	$\{\overline{2S[j]}\}, 1, 1_{j+1}, 0, \ldots, 0$
$F_{-1}[j]$	$2S[j] - 2^{-(j+1)}$	
	$= 2S[j] - 2^{-(j-1)} + 3 \times 2^{-(j+1)}$	$\{2S[j] - 1\}, 1, 1_{j+1}, 0, \ldots, 0$

TABLE 6.2 F_1 and F_{-1} forms for radix 2.

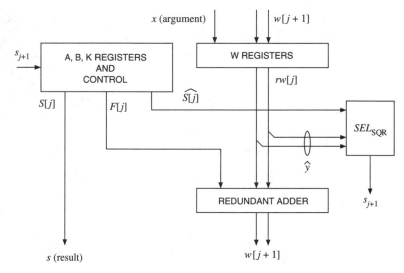

FIGURE 6.4 Block diagram of digit recurrence square root scheme. (Adapted from Ercegovac and Lang (1990).)

F_1 and F_{-1} so that[2]

$$F[j] = \begin{cases} F_1[j] & \text{if } s_{j+1} = 1 \\ 0 & \text{if } s_{j+1} = 0 \\ F_{-1}[j] & \text{if } s_{j+1} = -1 \end{cases} \qquad \textbf{6.24}$$

Table 6.2 describes the values and the corresponding bit-strings of the F_1 and F_{-1} forms. In this table, $\{\overline{2S[j]}\}$ is the bit-string produced by the bit-complement of

---

2. In this case, A and B are used only for the conversion of $S[j]$.

s_{j+1}	$F_1[j+1]$	$F_{-1}[j+1]$
1	$(X, 0, 1, 1_{j+2}, 0, \ldots, 0)$	$(\overline{X}, 0, 1, 1_{j+2}, 0, \ldots, 0)$
0	$(X, 1, 1, 1_{j+2}, 0, \ldots, 0)$	$(Y, 1, 1, 1_{j+2}, 0, \ldots, 0)$
-1	$(\overline{Y}, 0, 1, 1_{j+2}, 0, \ldots, 0)$	$(Y, 0, 1, 1_{j+2}, 0, \ldots, 0)$

TABLE 6.3 Updating of F_1 and F_{-1} forms for radix 2.

$2S[j]$ in radix-2 conventional representation, and $\{2S[j] - 1\}$ is the bit-string of $2(S[j] - 2^{-j})$. The subscript of 1_{j+1} indicates that the corresponding 1 is in the $(j+1)$th position.

The bit-strings $F_1[j] = (X, 1, 1_{j+1}, 0, \ldots, 0)$ and $F_{-1}[j] = (Y, 1, 1_{j+1}, 0, \ldots, 0)$ are updated according to Table 6.3. The initial conditions are $F_1[0] = -2S[0] - 2^{-1} = -0.5 = 111.1$ and $F_{-1}[0] = 2S[0] - 2^{-1} = -0.5 = 111.1$.[3]

The updating of the registers is controlled by the control register K with contents

$$K[j] = (1, 1, 1, \ldots, 1, 1_j, 0, \ldots, 0) \qquad \textbf{6.25}$$

with the initial condition $K[0] = 1111.0000\ldots0$.

The result-digit selection developed in Section 6.6 is

$$s_{j+1} = \begin{cases} 1 & \text{if } 0 \le \widehat{y} \le 3 \\ 0 & \text{if } \widehat{y} = -1 \\ -1 & \text{if } -5 \le \widehat{y} \le -2 \end{cases} \qquad \textbf{6.26}$$

where $\widehat{y}$ is an estimate of $2w[j]$ with $t = 0$ fractional bits.

EXAMPLE 6.1 We show an example of execution of the radix-2 algorithm for $x = 0.10110111$ in Figure 6.5. ∎

Radix-4 with Carry-Save Adder

We develop the radix-4 case with carry-save adder and result-digit set $\{-2, -1, 0, 1, 2\}$.

---

3. The three integer bits are needed since $\widehat{y}$ requires four integer bits, so $w[j+1]$ requires three integer bits.

$2w[0]^+$	0001.01101110		$S[0] = 0$
	0000.00000000	$\widehat{y} = 1 \; s_1 = 1$	$S[1] = 0.1$
$F_1[0]$	111.10000000		
$w[1]$	110.11101110		
	010.00000000		
$2w[1]^+$	1101.11011100		
	0100.00000000	$\widehat{y} = 1 \; s_2 = 1$	$S[2] = 0.11$
$F_1[1]$	110.11000000		
$w[2]$	111.00011100		
	001.10000000		
$2w[2]^+$	1110.00111000		
	0011.00000000	$\widehat{y} = 1 \; s_3 = 1$	$S[3] = 0.111$
$F_1[2]$	110.01100000		
$w[3]$	011.01011000		
	100.01000000		
$2w[3]^+$	0110.10110000		
	1000.10000000	$\widehat{y} = -2 \; s_4 = -1$	$S[4] = 0.1101$
$F_{-1}[3]$	001.1011000		
$w[4]$	111.10000000		
	001.01100000		
$2w[4]^+$	1111.00000000		
	0010.11000000	$\widehat{y} = 1 \; s_5 = 1$	$S[5] = 0.11011$
$F_1[4]$	110.01011000		
$w[5]$	011.10011000		
	100.10000000		

$^+$ only three integer bits in the recurrence because of the range of $w[j]$.

FIGURE 6.5 Example of Radix-2 algorithm execution.

s_{j+1}	Value (in terms of $S[j]$)	$F[j]$ Value (in terms of $A[j]$ and $B[j]$)	Bit-string
0	0	0	$0\ldots00000$
1	$-2S[j]-4^{-(j+1)}$	$-2A[j]-4^{-(j+1)}$	$\overline{a}\ldots\overline{a}a\,111$
2	$-4S[j]-4\times 4^{-(j+1)}$	$-4A[j]-4\times 4^{-(j+1)}$	$\overline{a}\ldots\overline{a}\,1100$
-1	$2S[j]-4^{-(j+1)}$	$2B[j]+7\times 4^{-(j+1)}$	$b\ldots bb\,111$
-2	$4S[j]-4\times 4^{-(j+1)}$	$4B[j]+12\times 4^{-(j+1)}$	$b\ldots b\,1100$

TABLE 6.4 Generation of $F[j]$ for radix 4.

The recurrence for this case is

$$w[j+1]=4w[j]-\left(2S[j]s_{j+1}+4^{-(j+1)}s_{j+1}^{2}\right)=4w[j]+F[j] \quad \textbf{6.27}$$

The adder input $F[j]$ is formed as discussed in Section 6.2. The resulting bit-strings are given in Table 6.4, where $a\ldots aa$ and $b\ldots bb$ are the bit-strings representing $A[j]$ and $B[j]$, respectively (shifted one position). As in the radix-2 case, the trailing location of the string is controlled by the moving 1 of register K.

The result digit is a function of $\widehat{y}$, the carry-save $rw[j]$ truncated to four fractional bits, and $\widehat{S}[j]$, the partial result also truncated to four fractional bits. As in division, the selection function is defined in terms of selection constants $m_k(i)$ such that

$$s_{j+1}=k \quad \text{if } m_k(i)\leq \widehat{y}<m_{k+1}(i) \text{ and } \widehat{S}[j]=2^{-1}+i\times 2^{-4} \quad \textbf{6.28}$$

A selection function is given in Table 6.5. Since the selection constants are all multiples of 2^{-3}, only three fractional bits of $\widehat{y}$ are used for selection, as shown in Figure 6.6. To use the same selection function for all values of j, the following transformation is performed:

$$(\widehat{S_1},\widehat{S_2},\widehat{S_3},\widehat{S_4})=\begin{cases}(1,1,0,-) & \text{if}\,(j=0)\\(1,1,1,1) & \text{if}\,(A_0=1)\text{ and }(j\neq 0)\\(1,A_2,A_3,A_4) & \text{if}\,(A_0=0)\text{ and }(j\neq 0)\end{cases} \quad \textbf{6.29}$$

where (A_0,A_1,A_2,A_3,A_4) are the most-significant bits of A, the conventional representation of $S[j]$. Since for $j=0$, $A_0=1$ and $A_2=A_3=A_4=0$, we obtain the implementation of Figure 6.6.

i	0	1	2	3	4	5	6	7
$\widehat{S}j$	$\frac{8}{16}$	$\frac{9}{16}$	$\frac{10}{16}$	$\frac{11}{16}$	$\frac{12}{16}$	$\frac{13}{16}$	$\frac{14}{16}$	$\frac{15}{16}$
$m_2(i)^+$	12	14	16	16	18	20	20	22
$m_1(i)^+$	4	4	4	4	6	6	8	8
$m_0(i)^+$	-4	-5	-6	-6	-6	-8	-8	-8
$m_{-1}(i)^+$	-13	-14	-16	-17	-18	-20	-22	-23

+: real value is given value divided by eight.

TABLE 6.5 Selection function for radix-4 square root.

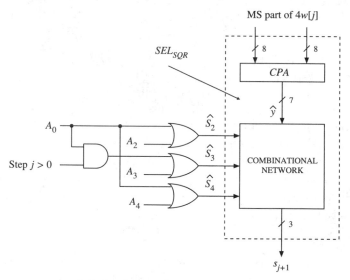

FIGURE 6.6 Selection function for $r = 4$. (Adapted from Ercegovac and Lang (1990).)

The overall algorithm follows directly the algorithm given earlier in this chapter with $r = 4$, and it is not repeated here. The cycle time is

$$
\begin{aligned}
T_{cycle} = t_{SEL}+ & \quad \text{(8-bit CPA + 10-input q-sel)} \\
t_{F\text{-}GEN}+ & \quad \text{(4-to-1 multiplexer)} \\
t_{HA}+ & \quad \text{(HA part of [3:2] carry-save adder)} \\
t_{load} & \quad \text{(register loading)}
\end{aligned}
$$

This is comparable to the cycle time of a radix-4 division with carry-save adder.

$$w[0] = x - 1 = 1.10110111$$

$$S[0] = 1$$

$$
\begin{array}{llll}
4w[0]^+ & 1110.11011100 & \widehat{S} = .1101 \\
& 0000.00000000 & \widehat{y} = 1110.110 = -10/8 \; s_1 = -1 & S[1] = 0.11 \\
F_{-1}[0]^+ & 001.11000000 \\
\hline
w[1] & 11.00011100 \\
& 01.10000000 \\
\\
4w[1]^+ & 1100.01110000 & \widehat{S} = .1100 \\
& 0110.00000000 & \widehat{y} = 0010.011 = 19/8 \; s_2 = 2 & S[2] = 0.1110 \\
F_2[1]^+ & 100.11000000 \\
\hline
w[2] & 10.10110000 \\
& 00.10000000 \\
\\
4w[2]^+ & 1010.11000000 & \widehat{S} = .1110 \\
& 0010.00000000 & \widehat{y} = 1100.110 = -26/8 \; s_3 = -2 & S[3] = 0.110110 \\
F_{-2}[2]^+ & 011.01110000 \\
\hline
w[3] & 11.10110000 \\
& 00.10000000 \\
\end{array}
$$

+ only two integer bits used in recurrence, because the range of $|w[j]| < 2$.

FIGURE 6.7 Example of radix-4 algorithm execution.

EXAMPLE 6.2   We give an example of execution of the radix-4 algorithm for $x = 0.10110111$ in Figure 6.7. ∎

6.4   Combination of Division and Square Root

Since the recurrences of the square root and division operations have many similarities, it is possible to implement a combined unit that performs both operations. We now describe such a unit for radix 2; a generalization to higher radices is possible. Table 6.6 shows the operand and result ranges.

Operation	Operand 1	Operand 2	Result
Division	Dividend $\frac{1}{4} \leq x < \frac{1}{2}$*	Divisor $\frac{1}{2} \leq d < 1$	Quotient $\frac{1}{4} < q < 1$
Square root	$\frac{1}{4} \leq x < 1$†		$\frac{1}{2} \leq s < 1$

* Because of initial condition.
† To accommodate odd exponents.

TABLE 6.6 Operands and results ranges for combined unit.

	Division	Square Root
Recurrence $w[j+1] =$	$2w[j] - dq_{j+1}$ $\|w[j]\| < 1$ $w[0] = x/2$	$2w[j] - S[j]s_{j+1} - s_{j+1}^2 2^{-(j+2)}$ $\|w[j]\| < 1$ $w[0] = x/2$
Estimate fraction bits t	1	1
Selection m_1 m_0 m_{-1}		0 $-\frac{1}{2}$ $-\frac{5}{2}$

TABLE 6.7 Algorithms.

Since the bound of the residual for square root is about twice that of division, to combine both recurrences it is convenient to modify the residual for square root so that

$$w[j](new) = 2^{-1}w[j](old) \qquad \textbf{6.30}$$

After making this modification (and calling $w[j](new)$ just $w[j]$), we get the algorithms described in Table 6.7. From this table we see that we can implement a generic recurrence of the form

$$w[j+1] = 2w[j] + F[j] \qquad \textbf{6.31}$$

	Division	Square Root
u_{j+1}	q_{j+1}	s_{j+1}
$F_1[j]$	$-d$	$-S[j] - 2^{-(j+2)}$
$F_{-1}[j]$	d	$S[j] - 2^{-(j+2)}$

TABLE 6.8 Correspondence.

where

$$F[j] = \begin{cases} F_1[j] & \text{if } u_{j+1} = 1 \\ 0 & \text{if } u_{j+1} = 0 \\ F_{-1}[j] & \text{if } u_{j+1} = -1 \end{cases} \qquad \textbf{6.32}$$

and $F_1[j]$, $F_{-1}[j]$, and u_{j+1} are related to the operations by the correspondence of Table 6.8. Moreover, the result-digit selection function is $u_{j+1} = Sel(\widehat{y})$, where $\widehat{y}$ is an estimate of $2w[j]$ obtained by assimilating the carry-save representation up to one fractional bit. From Table 6.7 we see that the selection function can be made to be the same for both operations (and corresponds to that described for division in Chapter 5).

The generation of the inputs to the adder is performed as described in this chapter for square root,[4] whereas for division the registers have to be loaded as indicated by the correspondence of Table 6.8. Conversion of the result is performed as described in Chapter 5. Figure 6.8 shows a block diagram and the cycle time of the combined implementation.

6.5   Integer Square Root

Integer square root (for unsigned operands) has an integer operand $0 \le x \le 2^n - 1$ and produces an integer result s such that

$$s = \lfloor x^{1/2} \rfloor \qquad \textbf{6.33}$$

To use the staircase selection functions discussed in this chapter, it is necessary that the result be in the range $[\frac{1}{2}, 1)$. This is achieved by shifting the operand m

---

4. Because of the modification of the residual for square root, now the values of $F[j]$ are one half of those reported in Table 6.2.

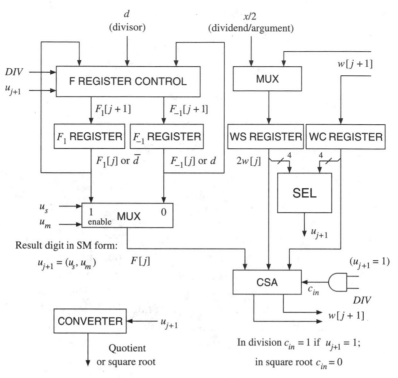

FIGURE 6.8 Overall implementation of the radix-2 combined unit. (Adapted from Ercegovac and Lang (1991).)

bits (and placing the binary point on the left) so that

$$x^* = 2^m x(2^{-n}) = 2^{-(n-m)}x \qquad\qquad 6.34$$

producing

$$s^* = 2^{-(n-m)/2}s \qquad\qquad 6.35$$

with $\frac{1}{2} \le s^* < 1$. Then

$$s = 2^{(n-m)/2}s^* \qquad\qquad 6.36$$

To obtain s from s^* by shifting, it is necessary that $n - m$ be even. Consequently, $\frac{1}{4} \le x^* < 1$.

The number of bits of the integer square root is $(n - m)/2$. Consequently, the number of iterations required to obtain these bits is

$$NI = \lceil (n - m)/2k \rceil \qquad \qquad \textbf{6.37}$$

where $r = 2^k$ is the radix of the result digit.

The result has to be aligned to the integer position. This can be done by placing the digits in the correct final position or placing the digits aligned to the left (to combine with fractional square root) and then performing a right shift of $(n - m)/2$ bits.

EXAMPLE 6.3 We now show an example of integer square root, radix 4 with $\rho = \frac{2}{3}$ and 8-bit operand. Consider the case $x = 27$ with binary representation $x = 00011011$.

Since $n = 8$ is even, we shift $m = 2$ bits and produce $x^* = .01101100$. The number of bits of the integer result is $(8 - 2)/2 = 3$. Consequently, two radix-4 iterations are necessary.

The radix-4 square root algorithm uses $S[0] = 1$ and $w[0] = x^* - 1 = 1.01101100$ (two's complement). As selection function we use Table 6.5 for carry-save representation of the residual.

The iterations are as follows (for simplicity we show the residual in conventional representation):

$$
\begin{aligned}
w[0] &= \quad 11.01101100 & & & & S[0] = 1 \\
4w[0] &= 1101.10110000 & \hat{y} = 1101.1011 & s_1 = -2 & & S[1] = 0.10(0.5) \\
+F &= \quad 011.00000000 \\
w[1] &= \quad 00.10110000 \\
4w[1] &= 0010.11000000 & \hat{y} = 0010.1100 & s_2 = 2 & & S[2] = 0.1010 \\
(w[2] & \text{ not needed})
\end{aligned}
$$

So, $s = 2^3(0.101) = 101 = (5)_{10}$. ∎

6.6 Result-Digit Selection

We now develop the details of the selection function design. We give examples for radix 2 and radix 4.

6.6.1   Selection Intervals

As in division, two fundamental conditions must be satisfied by a result-digit selection: *containment* and *continuity*. These conditions determine the selection intervals and the selection constants. We now develop expressions for these selection intervals. The bounds of the residual $w[j]$ (called $\min(w[j])$ and $\max(w[j])$ below) are defined by (6.10). Note that these bounds depend on j, whereas in division they are constants. From recurrence (6.12), the interval of $rw[j]$ where $s_{j+1} = k$ can be selected is

$$
\begin{aligned}
U_k[j] &= \max(w[j+1]) + 2S[j]k + k^2 r^{-(j+1)} \\
&= 2\rho S[j+1] + \rho^2 r^{-(j+1)} + 2S[j]k + k^2 r^{-(j+1)}
\end{aligned}
$$

$$
\begin{aligned}
L_k[j] &= \min(w[j+1]) + 2S[j]k + k^2 r^{-(j+1)} \\
&= -2\rho S[j+1] + \rho^2 r^{-(j+1)} + 2S[j]k + k^2 r^{-(j+1)}
\end{aligned}
$$

6.38

Since $S[j+1] = S[j] + kr^{-(j+1)}$, we get

$$
U_k[j] = 2S[j](k+\rho) + (k+\rho)^2 r^{-(j+1)}
$$

$$
L_k[j] = 2S[j](k-\rho) + (k-\rho)^2 r^{-(j+1)}
$$

6.39

As in division, variations of Robertson's diagram and P-D plot can be used to represent the selection interval bounds.

The continuity condition

$$
U_{k-1} \geq L_k
$$

6.40

results in

$$
(2\rho - 1)(2S[j] + (2k - 1)r^{-(j+1)}) \geq 0
$$

6.41

The overlap between consecutive selection intervals is given by

$$
U_{k-1} - L_k = (2\rho - 1)(2S[j] + (2k - 1)r^{-(j+1)})
$$

6.42

As in division, this overlap is used to simplify the selection function.

Note that the bounds, selection intervals, and the overlap depend on j. This is in contrast to division, in which they are independent of j. Since now the selection intervals depend on three parameters, namely, $S[j]$, j, and k, the notation becomes more complicated. In general, we use parentheses for $S[j]$, square brackets for j, and subscripts for k; however, we skip any one of these if it is unnecessary in a particular context. The dependence on j makes the implementation more complicated than in division, as discussed later.

6.6.2 Staircase Selection Using Redundant Adder

The basic relations that allow the use of estimates of the residual in the result-digit selection are identical to those we developed for division in the previous chapter. Since in square root the result-digit selection depends on the partial result $S[j]$ instead of on the divisor, the corresponding expressions are obtained by replacing d by $S[j]$. Instead of repeating the development of these relations, we ask the reader to refer to Chapter 5. On the other hand, some relations are different since they depend on the specific form of the recurrence; we develop these relations here.

Since the use of a redundant adder increases the speed of the implementation with a small increase in complexity, we concentrate on this type of implementation.

We now determine a staircase result-digit selection function using an estimate of the partial result and an estimate of the shifted residual obtained by truncating the redundant form.

Estimate of $S[j]$

The estimate of the result used in the result-digit selection divides the range of the result $S[j]$ into intervals, as illustrated in Figure 6.9. Specifically, if δ is the number of fractional bits of estimate $\widehat{S}[j]$, then the value

$$\widehat{S}[j] = 2^{-1} + i \times 2^{-\delta} \qquad 0 \le i \le 2^{\delta-1} - 1 \qquad \textbf{6.43}$$

defines the interval I_i. Note that the value of $\widehat{S}[j]$ for $i = 0$ is 2^{-1}, since the result is normalized.

Since the result is being produced one digit per iteration in signed-digit form, several alternatives can be used to form the estimate. This is in contrast to

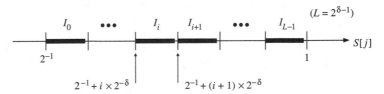

FIGURE 6.9 Generic intervals.

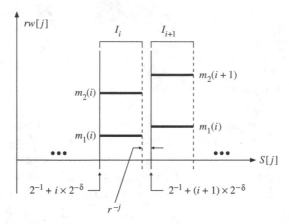

FIGURE 6.10 Selection intervals obtained by truncating conventional form.

division, where the estimate is always obtained by truncating the conventional representation of the divisor. The alternative considered here uses the truncated conventional representation of $S[j]$, which is obtained by the on-the-fly conversion. This is similar to the case of division, in which the divisor is in conventional representation. Therefore, if $\widehat{S}[j]$ is obtained by truncating $S[j]$ to δ fractional bits, then as shown in Figure 6.10, the ith interval I_i is defined by

$$\widehat{S}[j] = 2^{-1} + i \times 2^{-\delta} \leq S[j] < 2^{-1} + (i+1) \times 2^{-\delta} \qquad \textbf{6.44}$$

However, since $S[j]$ has j fractional radix-r digits, the upper bound of the interval is restricted so that I_i corresponds to

$$\widehat{S}[j] = 2^{-1} + i \times 2^{-\delta} \leq S[j] \leq 2^{-1} + (i+1) \times 2^{-\delta} - r^{-j} \qquad \textbf{6.45}$$

This restriction of the upper limit of the interval is significant for small j.

Determination of the Selection Constants $m_k(i)$

We now summarize the condition required for the result-digit selection, which is the same as for division. We then apply it to the square root case.

Using the estimate of the result, the result-digit selection is described by the set of selection constants

$$\{m_k(i) \mid 0 \leq i \leq 2^{\delta-1} - 1, \quad k \in \{-a, \ldots, -1, 0, 1, \ldots, a\}\} \qquad \textbf{6.46}$$

That is, there is one selection constant per interval and per value of the result digit. Using these selection constants, the result-digit selection is defined by

$$s_{j+1} = k \quad \text{if } m_k(i) \le \widehat{y} < m_{k+1}(i) \text{ and } \widehat{S}[j] = 2^{-1} + i \times 2^{-\delta} \qquad \textbf{6.47}$$

where $\widehat{y}$ is an estimate of the shifted residual $rw[j]$ obtained by truncating the redundant form to t fractional bits.

As discussed in Chapter 5, the use of the estimate by truncation of the redundant form of $y = rw[j]$ results in the following condition for the selection constants:

$$\max(\widehat{L}_k(I_i)) \le m_k(i) \le \min(\widehat{U}_{k-1}(I_i)) \qquad \textbf{6.48}$$

Moreover, for the carry-save representation case[5]

$$\begin{aligned} \widehat{U}_{k-1} &= \lfloor U_{k-1} - 2^{-t} \rfloor_t \\ \widehat{L}_k &== \lceil L_k \rceil_t \end{aligned} \qquad \textbf{6.49}$$

From these expressions, a feasible result-digit selection requires that

$$\min(\lfloor U_{k-1}(I_i) \rfloor_t) - \max(\lceil L_k(I_i) \rceil_t) \ge 2^{-t} \qquad \textbf{6.50}$$

Since L_k and U_{k-1} depend on j, the iteration index, the selection constants can, in general, be different for different j.

We now apply expressions (6.48) to the square root case. For this, we use the expressions (6.39) for the selection interval and expression (6.45) for the definition of the interval I_i. Consequently, for $k > 0$, the minimum U is produced at the lower end of the interval I_i and the maximum L at the upper end of the interval. Therefore,

$$\begin{aligned} \min(U_{k-1}(I_i)) &= 2(2^{-1} + i \times 2^{-\delta})(k - 1 + \rho) \\ &\quad + (k - 1 + \rho)^2 r^{-(j+1)} \\[8pt] \max(L_k(I_i)) &= 2(2^{-1} + (i + 1) \times 2^{-\delta})(k - \rho) \\ &\quad + (k - \rho)(k - \rho - 2r)r^{-(j+1)} \end{aligned} \qquad \textbf{6.51}$$

---

5. For the expressions that hold for signed-digit representation, see Chapter 5 and Exercise 6.12.

For $k \le 0$, the minimum U is produced at the upper end of the interval I_i and the maximum L at the lower end of the interval. Substituting we get

$$\min(U_{k-1}(I_i)) = 2(2^{-1} + (i+1) \times 2^{-\delta})(k-1+\rho)$$
$$+ (1 - \rho - k)(2r + 1 - \rho - k)r^{-(j+1)}$$

$$\max(L_k(I_i)) = 2(2^{-1} + i \times 2^{-\delta})(k-\rho) + (k-\rho)^2 r^{-(j+1)} \qquad \textbf{6.52}$$

These expressions are used in expression (6.48) to determine the result-digit selection. However, since they depend on j, a different selection function might result for different j. To have a single selection function for all j we develop bounds that are independent of j.

For $k > 0$, in the expression for $\min(U_{k-1}(I_i))$ the term depending on j is always positive and approaching zero for large j. Therefore, in a bound this term can be neglected. Similarly, for $\max(L_k(I_i))$ the term depending on j is negative $(k - \rho - 2r < 0)$ so it can also be neglected.

For $k \le 0$, the same situation occurs for $\min(U_{k-1}(I_i))$, but now for $\max(L_k(I_i))) j$ is positive, so it cannot be neglected, and for a bound we have to use its maximum value (for $j = 0$).

The corresponding expressions independent of j are as follows. For $k > 0$:

$$\min(U_{k-1}(I_i)) = 2(2^{-1} + i \times 2^{-\delta})(k - 1 + \rho)$$
$$\max(L_k(I_i)) = 2(2^{-1} + (i+1) \times 2^{-\delta})(k - \rho) \qquad \textbf{6.53}$$

For $k \le 0$:

$$\min(U_{k-1}(I_i)) = 2(2^{-1} + (i+1) \times 2^{-\delta})(k - 1 + \rho)$$
$$\max(L_k(I_i)) = 2(2^{-1} + i \times 2^{-\delta})(k - \rho) + (k - \rho)^2 r^{-1} \qquad \textbf{6.54}$$

To determine if a single selection function is possible, we use (6.50).[6] The worst case is $i = 0$ and $k = -a + 1$, resulting in

$$2(2^{-1} + 2^{-\delta})(-a + \rho) - 2(2^{-1})(-a + 1 - \rho) - (-a + 1 - \rho)^2 r^{-1} \ge 2^{-t}$$

$$\textbf{6.55}$$

---

6. To simplify the analysis we consider the best case, which occurs when both $\min(U_{k-1})$ and $\max(L_k)$ are multiples of 2^{-t}.

This simplifies to

$$(2\rho - 1) - (\rho r - 1)^2 r^{-1} \geq 2^{-t} + 2 \times 2^{-\delta}(a - \rho) \qquad \textbf{6.56}$$

For $r = 2$ ($\rho = 1$), this results in $2^{-1} \geq 2^{-t}$, so that a single selection function is possible. On the other hand, there is no solution for $r \geq 4$ (because the term in the left-hand side becomes negative), so that no single selection function for all j exists. A possible alternative is to find a value J so that a single selection can be used for $j \geq J$ and then consider separately the cases for $j < J$.

For the case $j \geq J$, the same considerations given before produce the following. For $k > 0$ (same as (6.53)):

$$\min(U_{k-1}(I_i)) = 2(2^{-1} + i \times 2^{-\delta})(k - 1 + \rho)$$
$$\max(L_k(I_i)) = 2(2^{-1} + (i + 1) \times 2^{-\delta})(k - \rho) \qquad \textbf{6.57}$$

For $k \leq 0$ (expression (6.54) replacing -1 by $-(J + 1)$ in the last term):

$$\min(U_{k-1}(I_i)) = 2(2^{-1} + (i + 1) \times 2^{-\delta})(k - 1 + \rho)$$
$$\max(L_k(I_i)) = 2(2^{-1} + i \times 2^{-\delta})(k - \rho) + (k - \rho)^2 r^{-(J+1)} \qquad \textbf{6.58}$$

Introducing these expressions in (6.50) (for the worst case $i = 0, k = -a + 1$), we get (similar to (6.56))

$$(2\rho - 1) - (\rho r - 1)^2 r^{-(J+1)} \geq 2^{-t} + 2 \times 2^{-\delta}(a - \rho) \qquad \textbf{6.59}$$

For specific values of r and ρ we can determine the values of J, t, and δ. These are then used to determine a selection function for $j \geq J$. For example, for $r = 4$ possible solutions are ($a = 2, \delta = 4, t = 3, J = 3$) and ($a = 3, \delta = 4, t = 3$, and $J = 1$).

The case $j < J$ has to be treated separately. A possible solution is to have a table lookup to produce $S[J]$ directly from x; that is, obtain the first J digits of the result from the table.[7] As we see later in the example for $r = 4$, a detailed analysis of the cases for $j < J$ eliminates the need for this initial table lookup.

Finally, the range of $\widehat{y}$ for carry-save residual is given by

$$\lfloor r \min(w[j]) - 2^{-t} \rfloor_t \leq \widehat{y} \leq \lfloor r \max(w[j]) \rfloor_t \qquad \textbf{6.60}$$

where $\min(w[j])$ and $\max(w[j])$ are the bounds of $w[j]$.

---

7. In this case, it is possible to obtain an algorithm in which J is not an integer.

6.6.3   Selection Function for Radix 2 with Carry-Save Adder

To obtain the result-digit selection, we begin by obtaining the values of L_k and U_k from expressions (6.39). That is,

$$
\begin{aligned}
U_1[j] &= 4S[j] + 2^{-j+1} & L_1[j] &= 0 \\
U_0[j] &= 2S[j] + 2^{-j-1} & L_0[j] &= -2S[j] + 2^{-j-1} \\
U_{-1}[j] &= 0 & L_{-1}[j] &= -4S[j] + 2^{-j+1}
\end{aligned}
\tag{6.61}
$$

We first determine the intervals of $S[j]$ for the staircase function and the value of t. As shown in (6.56), a single interval over the whole range $\frac{1}{2} \le S[j] < 1$ exists so that the result-digit selection is independent of the result. Since $\rho = 1$, it is possible to make $s_0 = 0$, resulting in $S[0] = 0$. To obtain the selection constants according to (6.48), the extreme values are

$$
\min\left(U_0\left(\frac{1}{2}\right), U_0(1)\right) \ge
\begin{cases}
2^{-1} & \text{for } j = 0 \quad (\text{since } S[0] = 0) \\
1 & \text{for } j > 0 \quad \left(\text{since } S[j] \ge \frac{1}{2}\right)
\end{cases}
$$

$$
\max\left(L_1\left(\frac{1}{2}\right), L_1(1)\right) = 0
$$

$$
\tag{6.62}
$$

$$
\min\left(U_{-1}\left(\frac{1}{2}\right), U_{-1}(1)\right) = 0
$$

$$
\max\left(L_0\left(\frac{1}{2}\right), L_0(1)\right) = -1
$$

This last value is obtained as follows. We have

$$
\max(L_0) = -\min(2S[j]) + 2^{-j-1}
\tag{6.63}
$$

This value is only important when we select $s_{j+1} = -1$. Since $s \ge \frac{1}{2}$, the selection $s_{j+1} = -1$ implies that $S[j] \ge \frac{1}{2} + 2^{-j}$ (so that $S[j + 1] = S[j] - 2^{-j-1} \ge \frac{1}{2}$). Consequently,

$$
\max(L_0) = -2\left(\frac{1}{2} + 2^{-j}\right) + 2^{-j-1} \le -1
\tag{6.64}
$$

The selection constants and the value of t are now obtained so that they satisfy (6.48) and (6.49). A possible choice is

$$t = 0 \quad m_1 = 0 \quad m_0 = -1 \qquad \textbf{6.65}$$

The only case in which (6.48) is not satisfied for this choice is for $j = 0$ because for this value of j, $\min(\widehat{U}_0) < \max(\widehat{L}_1)$. However, for $j = 0$ the residual is not in carry-save form, but in conventional form, so that it is sufficient to have $U_0 \geq L_0$. Moreover, since in the result $s \geq \frac{1}{2}$, the first bit of s is always 1 (which is obtained by the selection function, since $x > 0$).

The range of the shifted residual is given by (6.60). Since $|2w[j]| < 4S < 4$, the range of the estimate is $-5 \leq \widehat{y} \leq 3$. That is, 4 bits of the carry-save residual are needed for selection.

Consequently, the result-digit selection is

$$s_{j+1} = \begin{cases} 1 & \text{if } 0 \leq \widehat{y} \leq 3 \\ 0 & \text{if } \widehat{y} = -1 \\ -1 & \text{if } -5 \leq \widehat{y} \leq -2 \end{cases} \qquad \textbf{6.66}$$

6.6.4   Selection Function for Radix 4 with Carry-Save Adder

We develop the radix-4 case with carry-save adder and result-digit set $\{-2, -1, 0, 1, 2\}$.

As indicated by expression (6.56) it is not possible to have a single selection function for all values of j. Consequently, we obtain a value of J so that a single selection function is possible for $j \geq J$ (and uses small t and δ). From (6.59) we get

$$\frac{1}{3} - \frac{25}{36}4^{-J} \geq 2^{-t} + \frac{8}{3}2^{-\delta} \qquad \textbf{6.67}$$

A possible solution is $J = 3, t = 3, \delta = 4$. However, the corresponding selection function is not feasible (Exercise 6.13); this can occur because expression (6.59) has been obtained for the best case in which the limits of the interval are multiples of 2^{-t} (Exercise 6.13). Therefore, we now develop a result-digit selection with $t = 4$. It has to satisfy expression (6.48), that is,

$$\max(\lceil L_k(I_i) \rceil_4) \leq m_k(i) \leq \min(\lfloor U_{k-1}(I_i) - 2^{-4} \rfloor_4) \qquad \textbf{6.68}$$

Moreover, the maximum and minimum in the interval are described by expressions (6.57) and (6.58) with $J = 3$ and $\delta = 4$. That is, for $k > 0$:

$$\min(U_{k-1}(I_i)) = 2(2^{-1} + i \times 2^{-4})\left(k - \frac{1}{3}\right)$$

$$\max(L_k(I_i)) = 2(2^{-1} + (i+1) \times 2^{-4})\left(k - \frac{2}{3}\right)$$

6.69

For $k \leq 0$:

$$\min(U_{k-1}(I_i)) = 2(2^{-1} + (i+1) \times 2^{-4})\left(k - \frac{1}{3}\right)$$

$$\max(L_k(I_i)) = 2(2^{-1} + i \times 2^{-4})\left(k - \frac{2}{3}\right) + \left(k - \frac{2}{3}\right)^2 4^{-4}$$

6.70

Table 6.9 shows the limits of the intervals and possible selection constants. The following notation is used:

$$m\widehat{U}_{k-1}(i) = \min(\lfloor U_{k-1}(I_i) - 2^{-4} \rfloor_4)$$

$$ML_k(i) = \max(\lceil L_k(I_i) \rceil_4)$$

6.71

Now we have to consider the case $j < 3$. One possible approach is to have a module to determine from a truncated x directly the value of $S[3]$ and to use it to perform the first iterations to determine $w[3]$. Another possibility is to analyze the case $j < 3$ and try to match the corresponding selection functions with that for $j \geq 3$. The particular choice of selection constants in Table 6.9 was made so that the same constants hold for all j.[8] The resulting selection function and its implementation are shown in Section 6.3.1.

The range of $\widehat{y}$ is obtained from (6.60). Since in this case we initialize to $S[0] = 1$, from (6.10) we obtain $\max(w[j]) = \max(w[1]) < \frac{4}{3} + \frac{1}{9} = \frac{13}{9}$ and $\min(w[j] > -\frac{4}{3}$. Conequently,

$$-\frac{88}{16} \leq \widehat{y} \leq \frac{92}{16}$$

6.72

Consequently, $\widehat{y}$ has four integer bits.

---

8. This is developed in Ercegovac and Lang (1990).

i	0	1	2	3
$\widehat{S}[j]^*$	8	9	10	11
$ML_2(i)\, m\widehat{U}_1(i)^*$	24 25	27 29	30 32	32 35
$m_2(i)^*$	**24**	**28**	**32**	**32**
$ML_1(i)\, m\widehat{U}_0(i)$	6 9	7 11	8 12	8 13
$m_1(i)$	**8**	**8**	**8**	**8**
$ML_0(i)\, m\widehat{U}_{-1}(i)$	−10 −7	−12 −8	−13 −9	−14 −9
$m_0(i)$	**−8**	**−10**	**−12**	**−12**
$ML_{-1}(i)\, m\widehat{U}_{-2}(i)$	−26 −25	−30 −28	−33 −31	−36 −33
$m_{-1}(i)$	**−26**	**−28**	**−32**	**−34**
i	4	5	6	7
$\widehat{S}[j]$	12	13	14	15, 16
$ML_2(i)\, m\widehat{U}_1(i)$	35 39	38 42	40 45	43 49
$m_2(i)$	**36**	**40**	**40**	**44**
$ML_1(i)\, m\widehat{U}_0(i)$	9 15	10 16	10 17	11 19
$m_1(i)$	**12**	**12**	**16**	**16**
$ML_0(i)\, m\widehat{U}_{-1}(i)$	−16 −10	−17 −11	−18 −11	−20 −12
$m_0(i)$	**−12**	**−16**	**−16**	**−16**
$ML_{-1}(i)\, m\widehat{U}_{-2}(i)$	−40 −36	−43 −39	−46 −41	−50 −44
$m_{-1}(i)$	**−36**	**−40**	**−44**	**−46**

* Real value is indicated value divided by 16.

TABLE 6.9 Selection function for $r = 4$ with carry-save adder ($j \geq 3$).

6.7 Exercises

Examples of Execution

6.1 Obtain the 8-bit square root of 144×2^{-8} using the following algorithms:

(a) Radix 2, $s_j \in \{-1, 0, 1\}$, conventional (nonredundant) residual
(b) Radix 2, $s_j \in \{-1, 0, 1\}$, carry-save residual
(c) Radix 4, $s_j \in \{-2, -1, 0, 1, 2\}$, carry-save residual

Characteristics of the Implementation

6.2   Design a 12-bit radix-2 square root unit at the gate level. You may use full-adders, multiplexers, registers, and gates. Provide the necessary design details to establish delays of critical paths. Use a carry-save adder to form residuals in redundant form. Give an estimate of the overall delay in gate delay units (t_g) and cost. Assume that a full-adder has a delay of $4t_g$, a 3-1 multiplexer $2t_g$, and a register load $3t_g$.

6.3   Develop the following two ways of generating the adder input F when a signed-digit adder is used:

(a)   Use $S[j]$ in its original signed-digit form.
(b)   Convert $S[j]$ to two's complement representation.

Discuss the design trade-offs in these alternatives.

6.4   [**Retiming of the recurrence**]. As Exercise 5.7, but for square root.

6.5   [**Overlapped radix-2 stages**]. Consider a radix-2 square root algorithm with the result-digit set $\{-1, 0, 1\}$ and redundant residuals in carry-save form. The result-digit selection is performed using the selection constants as described in the text. The cycle time is

$$t_{cycle} = t_{SEL_{SQRT}} + t_{buff} + t_{mux} + t_{HA} + t_{reg} = 4 + 1 + 1 + 1 + 2 = 10t_g$$

To reduce the total time, we propose to obtain all three possible values of s_{j+2}, corresponding to $s_{j+1} = -1, 0, 1$, and select the correct one once s_{j+1} is known, all of this in the same cycle. In other words, two result bits are generated per cycle.

(a)   Design the network for the selection of s_{j+1} and s_{j+2}. Assume that the selection function is already implemented. Show all details; in particular, show the details of the conditional selection.
(b)   Design the network to produce the next residual (assume 8 bits in the fractional part). Show all details.
(c)   Determine the cycle time of the new scheme and compare the total time to obtain a result of 8 bits with the scheme described in the text. Discuss your findings.

Combination of Division and Square Root

6.6   Using the combined radix-2 algorithm described in the chapter, perform the following operations:

(a)   Division of $x = 0.10110011$ by $d = 0.10001111$

(b)   Square root of $x = 0.01110111$

6.7   Compare the cycle time and the cost of the combined radix-2 implementation with an implementation only for square root.

Integer Square Root

6.8   Perform the integer division algorithm for radix 4 with residual in carry-save representation for $x = 53$ and $d = 9$. Consider that operands and result are represented by 8-bit vectors.

6.9   Give an algorithm that combines fractional square root with integer square root. Show the combined implementation, highlighting the modules required to include integer square root.

Result-Digit Selection

6.10   Determine a result-digit selection function for a radix-2 algorithm in which a signed-digit adder is used in the recurrence.

6.11   Develop a bit-level implementation for the radix-2 selection function for a digit s_j represented in sign-and-magnitude by (s_s, s_m) and the residual is in carry-save representation.

6.12   For the selection function, consider using the truncated signed-digit representation or (equivalently) the fixed value of $S[\lceil \delta / \log(r) \rceil]$ (that is, the value of $S[j]$ immediately after at least δ fractional bits are produced). Show that in this case (for simplicity we consider the case in which exactly δ fractional bits are produced) the ith interval corresponds to the following range of $S[j]$:

$$2^{-1} + i \times 2^{-\delta} - \rho \times 2^{-\delta} < S[j] < 2^{-1} + i \times 2^{-\delta} + \rho \times 2^{-\delta}$$

and give a figure illustrating this method.

6.13   Following the derivation in Section 6.6.4, develop a radix-4 selection function for $J = 3, t = 3$, and $\delta = 4$. Determine the selection constants. How many fractional bits are required?

6.14   Determine a result-digit selection function for a radix-4 square root algorithm with $\rho = 1$ and carry-save residual. Give only the portion for selection of $s_{j+1} = 2$.

6.8   Further Readings

A survey of square-root algorithms is presented in Montuschi and Mezzalama (1990). Parts of this chapter are based on the monograph Ercegovac and Lang (1994) which presents an in-depth study of digit-recurrence methods for division and square root.

Early work in square rooting algorithms with redundancy in the result-digit set to maximize the number of zero digits is developed in Metze (1967).

Radix 2

Radix-2 algorithms and implementations are discussed in Taylor (1981) and Majerski (1985).

Radix 4 and Radix 8

Specific radix-4 implementations are presented in Fandrianto (1987) and Ercegovac and Lang (1989, 1991) and a radix-8 alternative in Fandrianto (1989). The radix-4 algorithm described in this chapter which uses a single selection function for all iterations, is developed in Ercegovac and Lang (1990).

Higher Radix

Higher radix algorithms and implementations are considered in Ciminiera and Montuschi (1990). A very-high-radix implementation, with prescaling and selection by rounding, is described in Lang and Montuschi (1992, 1999).

Combined Division and Square Root

Combined division/square root implementations are presented in a number of places (Taylor 1981; Zurawski and Gosling 1987; Fandrianto 1987, 1989; Ercegovac and Lang 1991; McQuillan and McCanny 1994; Prabhu and Zyner 1995). In Srinivas and Parhi (1999) the residuals are kept in signed-digit

representation and the result digit is overredundant, which simplifies the selection function. Self-timed designs are presented in Matsubara et al. (1995) and .
Guyot et al. (1996).

Reciprocal Square Root

Digit recurrence algorithms and implementations for reciprocal square root have recently been presented in Lang and Antelo (2001) and Takagi (2001). A very-high-radix version is presented in Antelo et al. (1998).

Low-Power Design

Radix-4 combined division/square root low-power units are described in Kuhlmann and Parhi (1998) and Nannarelli and Lang (1999). A self-timed low-power design for combined division/square-root is presented in Matsubara and Ide (1997).

Combinational Implementation

Combinational implementations of square root as linear arrays are reported in Majithia and Kitai (1971), Majithia (1972), Agrawal (1979), Cappuccino et al. (1998, 1999), and Corsonello et al. (2000). Combinational implementations of combined division/square root schemes are described in McQuillan et al. (1991, 1993).

Miscellaneous

A square root scheme for integers is presented in Hashemian (1990). A radix-2 square root implementation for field-programmable gate arrays is developed in Louie and Ercegovac (1993). Skipping of zero result digits is considered in Montuschi and Ciminiera (1993).

Area/Delay Analysis

Area/performance of square root units is discussed in Soderquist and Leeser (1996).

Verification

Verification of square root implementations is presented in Leeser and O'Leary (1995) and of combined multiplication, division, and square root implementations in Walter (1995).

6.9   Bibliography

Agrawal, D. P. (1979). High-speed arithmetic arrays. *IEEE Transactions on Computers*, C-28(3):215–24.

Antelo, E., T. Lang, and J. D. Bruguera (1998). Computation of $\sqrt{x/d}$ in a very high radix combined division/square-root unit with scaling. *IEEE Transactions on Computers*, 47(2):152–61.

Cappuccino, G., G. Cocorullo, P. Corsonello, and S. Perri (1999). High speed self-timed pipelined datapath for square rooting. *IEE Proceedings—Circuits, Devices and Systems*, 146(1):16–22.

Cappuccino, G., P. Corsonello, and G. Cocorullo (1998). High performance VLSI modules for division and square root. *Microprocessors and Microsystems*, 22(5):239–46.

Ciminiera, L., and P. Montuschi (1990). Higher radix square rooting. *IEEE Transactions on Computers*, 39(10):1220–31.

Corsonello, P., S. Perri, and G. Cocorullo (2000). Performance comparison between static and dynamic CMOS logic implementations of a pipelined square-rooting circuit. *IEE Proceedings—Circuits, Devices and Systems*, 147(6):347–55.

Ercegovac, M. D., and T. Lang (1989). Radix-4 square root without initial PLA. In *Proceedings of the 9th IEEE Symposium on Computer Arithmetic*, pages 162–68.

Ercegovac, M. D., and T. Lang (1990). Radix-4 square root without initial PLA. *IEEE Transactions on Computers*, C-39(8):1016–24.

Ercegovac, M. D., and T. Lang (1991). Module to perform multiplication, division and square root in systolic arrays for matrix computations. *Journal of Parallel and Distributed Computing*, 11(3):212–21.

Ercegovac, M. D., and T. Lang (1994). *Division and Square Root: Digit-Recurrence Algorithms and Implementations*. Kluwer Academic Publishers.

Fandrianto, J. (1987). Algorithm for high-speed shared radix-4 division and radix-4 square-root. In *Proceedings of the 8th IEEE Symposium on Computer Arithmetic*, pages 73–79.

Fandrianto, J. (1989). Algorithm for high-speed shared radix-8 division and radix-8 square root. In *Proceedings of the 9th IEEE Symposium on Computer Arithmetic*, pages 68–75.

Guyot, A., M. Renaudin, B. El Hassan, and V. Levering (1996). Self timed division and square-root extraction. In *Ninth International Conference on VLSI Design*, pages 376–81.

Hashemian, R. (1990). Square rooting algorithms for integer and floating-point numbers. *IEEE Transactions on Computers*, C-39(8):1025–29.

Kuhlmann, M., and K. Parhi (1998). Power comparison of SRT and GST dividers. In *Proceedings of SPIE—Advanced Signal Processing Algorithms, Architectures, and Implementations VIII*, volume 3461, pages 584–94.

Lang, T., and E. Antelo (2001). Correctly rounded reciprocal square root by digit recurrence and radix-4 implementation. In *Proceedings of the 15th IEEE Symposium on Computer Arithmetic*, pages 94–100.

Lang, T., and P. Montuschi (1992). Higher radix square root with prescaling. *IEEE Transactions on Computers*, 41(8):996–1009.

Lang, T., and P. Montuschi (1999). Very high radix square root with prescaling and rounding and a combined division/square root unit. *IEEE Transactions on Computers*, 48(8):827–41.

Leeser, M., and J. O'Leary (1995). Verification of a subtractive radix-2 square root algorithm and implementation. In *International Conference on Computer Design: VLSI in Computers and Processors*, pages 526–31.

Louie, M. E., and M. D. Ercegovac (1993). A digit-recurrence square root implementation for field programmable gate arrays. In *IEEE Workshop on FPGAs for Custom Computing Machines*.

Majerski, S. (1985). Square-root algorithms for high-speed digital circuits. *IEEE Transactions on Computers*, C-34(8):724–33.

Majithia, J. C. (1972). Cellular array for extraction of squares and square roots of binary numbers. *IEEE Transactions on Computers*, C-21(9):1023–24.

Majithia, J. C., and R. Kitai (1971). A cellular array for the nonrestoring extraction of square roots. *IEEE Transactions on Computers*, C-20(12):1617–18.

Matsubara, G., and N. Ide (1997). A low power zero-overhead self-timed division and square root unit combining a single-rail static circuit with a dual-rail dynamic circuit. In *Third International Symposium on Advanced Research in Asynchronous Circuits and Systems*, pages 198–209.

Matsubara, G., N. Ide, H. Tago, S. Suzuki, and N. Goto (1995). 30-ns 55-b shared radix-2 division and square root using a self-timed circuit. In *Proceedings of the 12th IEEE Symposium on Computer Arithmetic*, pages 98–105.

McQuillan, S., and J. V. McCanny (1994). Fast algorithms for division and square root. *Journal of VLSI Signal Processing*, 8(2):151–68.

McQuillan, S. E., J. V. McCanny, and R. Hamill (1993). New algorithms and VLSI architectures for SRT division and square root. In *Proceedings of the 11th IEEE Symposium on Computer Arithmetic*, pages 80–86.

McQuillan, S. E., J. V. McCanny, and R. F. Woods (1991). High performance VLSI architecture for division and square root. *Electronics Letters*, V27(1):19–21.

Metze, G. (1967). Minimal square rooting. *IEEE Transactions on Computers*, EC-14(2):181–85.

Montuschi, P., and L. Ciminiera (1993). Reducing iteration time when result digit is zero for radix-2 SRT division and square root with redundant remainders. *IEEE Transactions on Computers*, 42(2):239–46.

Montuschi, P., and M. Mezzalama (1990). Survey of square rooting algorithms. *IEE Proceedings E: Computers and Digital Techniques*, 137(1):31–40.

Nannarelli, A., and T. Lang (1999). Low-power radix-4 combined division and square root. In *Proceedings of the IEEE International Conference on Computer Design: VLSI in Computers and Processors (ICCD'99)*, pages 236–42.

Prabhu, J. A., and G. B. Zyner (1995). 167 MHz radix-8 divide and square root using overlapped radix-2 stages. In *Proceedings of the 12th IEEE Symposium on Computer Arithmetic*, pages 155–62.

Soderquist, P., and M. Leeser (1996). Area and performance tradeoffs in floating-point division and square root implementations. *ACM Computing Surveys*, 28(3):518–64.

Srinivas, H. R., and K. K. Parhi (1999). A radix 2 shared division/square root algorithm and its VLSI architecture. *Journal of VLSI Signal Processing Systems for Signal, Image, and Video Technology*, 21(1):37–60.

Takagi, N. (2001). A hardware algorithm for computing reciprocal square root. In *Proceedings of the 15th IEEE Symposium on Computer Arithmetic*, pages 94–100.

Taylor, G. S. (1981). Compatible hardware for division and square root. In *Proceedings of the 5th IEEE Symposium on Computer Arithmetic*, pages 127–34.

Walter, C. D. (1995). Verification of hardware combining multiplication, division and square root. *Microprocessors and Microsystems*, 19(5):243–45.

Zurawski, J. H. P., and J. B. Gosling (1987). Design of a high-speed square root, multiply, and divide unit. *IEEE Transactions on Computers*, C-36(9):13–23.

CHAPTER 7

Reciprocal, Division, Reciprocal Square Root, and Square Root by Iterative Approximation

The methods of this chapter compute a function by iteratively improving an initial approximation. For the operations considered here, the most complex operation involved in the iteration is multiplication; because of this the methods are also called *multiplicative methods*. This contrasts with the digit recurrence method of Chapter 5, in which the recurrence involves a digit selection, a digit multiplication, and an addition. Other methods for these functions are presented in Chapters 10 and 11.

The methods presented here have a *quadratic convergence rate*, which loosely means that the number of bits of accuracy of the approximation doubles after each iteration. In contrast, the digit recurrence method has a linear convergence. As a consequence of this quadratic convergence, the number of iterations for a desired accuracy is smaller than for linear convergence. However, since full-precision multiplications are involved, the time of an iteration is larger and a detailed analysis has to be performed to compare the total execution time.

One application in which these methods might be attractive is in the floating-point unit of a processor (see Chapter 8) because they essentially use the already existing floating-point multiplier and do not require additional hardware. However, to perform the operations efficiently, it is necessary to do some modifications to the multiplier, and these modifications might increase the area and affect the performance. Moreover, the rounding of the floating-point results is simpler to perform with the digit recurrence method, as discussed in Chapter 8.

We consider first the computation of the reciprocal function. This function is important in itself and is the basis for the methods for division. A similar situation occurs for reciprocal square root and square root.

We consider the case of operands and result in sign-and-magnitude representation. Moreover, since the determination of the sign is straightforward and independent of the operation on magnitudes, we concentrate on the latter.

7.1　Reciprocal

For the reciprocal function, we describe two related methods: the application of the general Newton-Raphson method to obtain the zero of a function and the multiplicative normalization method.

7.1.1　Newton-Raphson Method for Reciprocal Approximation

This is based on a general method to obtain the zero of a function, that is, the value of x for which $f(x) = 0$. If $x[j]$ is an approximation of the zero, then a better approximation is

$$x[j+1] = x[j] - \frac{f(x[j])}{f'(x[j])} \qquad \text{7.1}$$

where $f'(x[j])$ is the derivative of $f(x)$ with respect to x, evaluated at $x[j]$. A graphical interpretation of this method for the reciprocal function is shown in Figure 7.1.

This general method is applied to the reciprocal function as follows.[1] Calling $R[j]$ the approximation of reciprocal, we apply (7.1) to the function $f(R) = 1/R - d$ (whose zero is $1/d$). Since $f'(R) = -1/R^2$ we obtain the recurrence

$$R[j+1] = R[j](2 - R[j]d) \qquad \text{7.2}$$

The recurrence is initiated with an initial approximation $R[0]$. Each iteration requires two multiplications and one subtraction from the value 2.

---

1. For an alternative development of the recurrence, see Exercise 7.1.

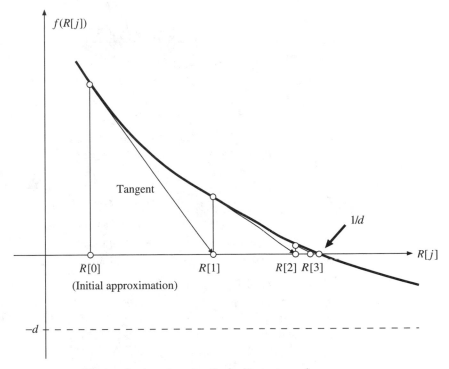

FIGURE 7.1 Newton-Raphson iteration for finding reciprocal.

The convergence of this method is quadratic; that is, if the error at step j is $\epsilon[j]$, then the error at step $j + 1$ is $\epsilon[j]^2$. This can be shown as follows. Since $R[j]$ is an approximation of $1/d$, the relative error is

$$\epsilon[j] = 1 - d\,R[j] \qquad\qquad 7.3$$

Then from (7.2),

$$R[j + 1] = \left(\frac{1 - \epsilon[j]}{d}\right)(2 - (1 - \epsilon[j]))$$

$$= \frac{1 - \epsilon[j]^2}{d} \qquad\qquad 7.4$$

j	$R[j]$	$dR[j]$	$2 - dR[j]$	$R[j+1]$	$\epsilon[j+1] =$ $1 - dR[j+1]$
0	1	5×2^{-3}	11×2^{-3}	11×2^{-3}	0.14
1	11×2^{-3}	55×2^{-6}	73×2^{-6}	$803 \times 2^{-9} = 1.5683594$	0.020
2	803×2^{-9}	4015×2^{-12}	4177×2^{-12}	$3354131 \times 2^{-21} = 1.5993743...$	0.00039

TABLE 7.1 Steps in Newton-Raphson approximation of reciprocal.

so that[2]

$$\epsilon[j+1] = 1 - dR[j+1] = \epsilon[j]^2 \qquad 7.5$$

The algorithm converges for $|\epsilon[0]| < 1$. Moreover, the convergence is from below, since the error is always positive. The number of iterations required to achieve a desired precision depends on the initial approximation. Specifically, if the relative error of the initial approximation is

$$\epsilon[0] \leq 2^{-k} \qquad 7.6$$

then to get a relative error

$$\epsilon[m] \leq 2^{-n} \qquad 7.7$$

the number of iterations is

$$m = \left\lceil \log_2 \left(\frac{n}{k} \right) \right\rceil \qquad 7.8$$

Consequently, the choice of the initial approximation is critical for the speed of the algorithm.

EXAMPLE 7.1   Consider the calculation of an approximation of the reciprocal of $d = \frac{5}{8}$.

We need an appropriate initial approximation. We consider the issue of obtaining this approximation later; for now let us take $R[0] = 1$ (which assures convergence for $R[m] < 2$). Then the procedure is illustrated by Table 7.1. The exact result is $1/d = \frac{8}{5} = 1.6$. ∎

---

2. This analysis implies that all operations are performed with full precision, that is, there is no roundoff error; these additional errors are considered in Section 7.1.4.

7.1.2   Multiplicative Normalization Method

This technique (presented more generally in Chapter 10), consists of two multiplicative recurrences, one of which converges to one while the other converges to the desired function. Specifically, for reciprocal we can write

$$R = \frac{1}{d} = \frac{1}{d}\frac{P[0]}{P[0]}\frac{P[1]}{P[1]} \cdots \frac{P[m]}{P[m]} = \frac{R[m]}{d[m]} \qquad \text{7.9}$$

so that $R = R[m]$ if $d[m] = 1$. Consequently, we define an approximation

$$R[j] = \prod_{i=0}^{j} P[i] \qquad \text{7.10}$$

and the variable that tends to one

$$d[j] = dP[j] \qquad \text{7.11}$$

The approximation is refined by the two recurrences

$$R[j + 1] = R[j]P[j + 1] \qquad \text{7.12}$$

$$d[j + 1] = d[j]P[j + 1] \qquad \text{7.13}$$

and the sequence $P[j]$ is selected so that $d[j]$ tends to the value 1. The initial conditions are $R[0] = P[0], d[0] = dP[0]$, and $P[0]$ an (initial) approximation of $1/d$.

The method is illustrated in Figure 7.2.

Determination of $P[j]$

We now determine the factors $P[j]$ for quadratic convergence. Since $R[j]$ is the approximation of $1/d$, the relative error is

$$\epsilon[j] = 1 - R[j]d = 1 - d[j] \qquad \text{7.14}$$

Therefore, for quadratic convergence ($\epsilon[j + 1] = \epsilon[j]^2$)

$$(1 - d[j + 1]) = (1 - d[j])^2 \qquad \text{7.15}$$

Consequently, since $d[j + 1] = d[j]P[j + 1]$ we get

$$P[j + 1] = 2 - d[j] \qquad \text{7.16}$$

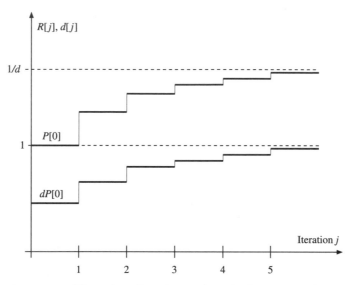

FIGURE 7.2 Illustration of iterations in the multiplicative normalization method.

The algorithm is the following[3]:

1. Obtain approximation $P[0]$ to $1/d$
2. $d[0] = dP[0];$ $R[0] = P[0]$
3. For $j = 0, 1, 2, 3, \ldots, m - 2$ do

$$P[j + 1] = 2 - d[j]$$
$$d[j + 1] = d[j]P[j + 1]; \quad R[j + 1] = R[j]P[j + 1]$$

4. $P[m] = 2 - d[m - 1]; \quad R[m] = R[m - 1]P[m]$

Implementation

As in the Newton-Raphson method, an iteration of the recurrence requires two multiplications and a two's complement operation. However, in this case the two multiplications are independent; consequently, it is possible to use more efficiently a pipelined multiplier.

---

3. This algorithm for reciprocal has also been called "by series expansion" since it also can be obtained by using the MacLaurin series for $1/(1 + x)$.

An implementation using a two-stage pipelined multiplier is shown in Figure 7.3.

7.1.3 Initial Approximation

Both methods described require an initial approximation of $1/d$. As stated, the accuracy of this approximation determines the number of iterations. However, it is necessary to consider also the delay of the module to produce this approximation, as well as its area.

A variety of methods have been used to obtain the initial approximation. The selection of the most appropriate method depends on the accuracy, delay, and area requirements. Some of the alternatives are the following (many variations have been developed)[4]:

- Use a constant value, independent of the operand d. The minimum relative error is obtained by using the middle point in the range of $1/d$.
- A linear interpolation in the whole range. That is, the initial approximation is

$$R[0] = a - bd \qquad\qquad 7.17$$

  where a and b are constants. The implementation is specially simple if b is a power of two. A good solution for $1/2 \leq d \leq 1$ is $a = 2.928$ and $b = 2$. Since the maximum relative error is about 0.1, the error is not increased much if $R[0]$ is truncated, say, to 5 bits. This truncation reduces the size of the initial multiplications.
- A set of constants, one per interval of d. This is called a *table lookup* because the constants are usually stored in a table. The input to the module that produces the initial value is the number of the interval, that is, the truncated d. The constants are selected to produce the smallest maximum error in each interval. It can be shown that using k bits of d produces an approximation of $k + 1$ bits and a maximum error of 2^{-k} (Exercises 7.5 and 7.6).
- If the size of the table is too large for the required initial error, a piecewise linear approximation can be used. Let

$$d = d_t 2^{-k} + d_p 2^{-p} + d_r 2^{-n} \quad k < p < n \qquad 7.18$$

---

4. These alternatives are discussed further in Chapter 10.

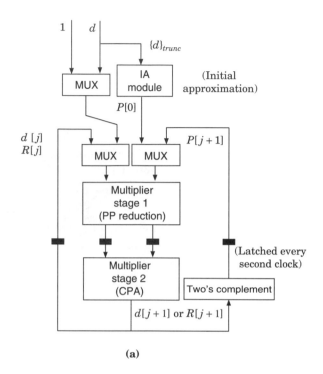

(a)

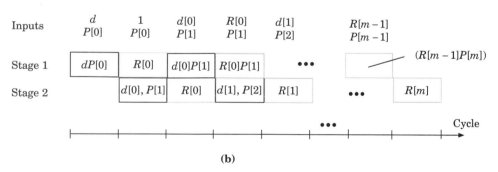

(b)

FIGURE 7.3 Multiplicative normalization for reciprocal: (a) Implementation with a two-stage multiplier. (b) Timing diagram.

The most-significant k bits of d are used to access the table and produce coefficients a and b. Then

$$R[0] = a + bd_p 2^{-p} \qquad \textbf{7.19}$$

Again, the constants are selected for smallest maximum error. This method requires a table lookup, a small multiplication, and an addition. The approximation can be truncated without significantly increasing the error. Typically, for an error of 2^{-g} the number of input bits of the table is about $g/2$.

- Instead of performing a multiplication, the bipartite method obtains two values from tables and performs an addition. On the other hand, the number of input bits of the tables is about $2g/3$, for an error of 2^{-g} (see Exercise 7.7 and Chapter 10).

7.1.4 Implementation and Additional Errors

We now consider the implementation of the reciprocal approximation algorithms and the effect of the additional error introduced by a practical implementation.

In both methods the implementation consists of the following components (see Figure 7.3):

- A module to compute the initial approximation. The area requirements and delay depend on the precision of this approximation and on the method used.
- A multiplier. To achieve the error expressions derived before, the multiplications have to be performed with full precision. This means that the number of bits of the products increases in each iteration. For instance for the Newton-Raphson method, if a and n are the number of bits of $R[0]$ and d, respectively, then the width of the products is as follows:

	$R[j]$	$R[j]d$	$R[j+1] = R[j](2 - R[j]d)$
$j = 0$	a	$a + n$	$2a + n$
$j = 1$	$2a + n$	$2a + 2n$	$4a + 3n$
$j = 2$	$4a + 3n$	$4a + 4n$	$8a + 7n$
$\ldots$			

As can be seen, the widths of the resulting products are very large and would make the implementation impractical. Consequently, the products are truncated or rounded in a way that has a small effect on the final error. We consider two alternatives after discussing the additional errors.

- A complementer (two's complement representation). A ones' complement (bit complement) can be used instead, but the error introduced has to be taken into account to determine the error of the approximation.

We now consider the total relative error $\epsilon_T[j]$, which includes the effects mentioned above. We saw that both algorithms considered converge quadratically; that is, the algorithmic (relative) error $\epsilon_A[j + 1]$ is

$$\epsilon_A[j + 1] = \epsilon_T[j]^2 \qquad \text{7.20}$$

We call the additional relative errors, introduced by the implementation of the iteration, *generated errors* and denote them by $\epsilon_G[j]$, for iteration j. The total relative error is then

$$\epsilon_T[j] = \epsilon_A[j] + \epsilon_G[j] \qquad \text{7.21}$$

We now consider separately each method.

Newton-Raphson Method

Including the generated error, from (7.20) and (7.21) we obtain the total error in iteration j

$$\epsilon_T[j] = \epsilon_T[j - 1]^2 + \epsilon_G[j] \qquad \text{7.22}$$

The generated error includes the following components[5]:

- Roundoff of $y[j - 1] = dR[j - 1]$
- Error in $z[j - 1] = 2 - y[j - 1]$
- Roundoff of $R[j - 1]z[j - 1]$

From (7.22) we conclude that the final error is positive (approximation from below[6]) if the last ϵ_G is not negative. Specifically, a final error that is positive and

---

5. Usually these errors are considered as absolute errors; the corresponding relative errors are obtained by dividing the absolute errors by $(1/d)$.
6. This is important in some methods to produce the correctly rounded result (see Chapter 8).

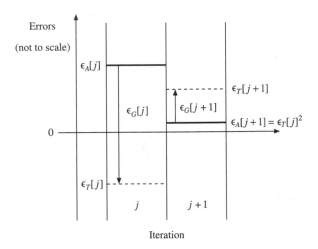

FIGURE 7.4 Error reduction in NR iterations.

less than 2^{-s} is obtained if

$$\epsilon_G[m] \geq 0 \qquad\qquad \textbf{7.23}$$

$$\epsilon_T[m] = \epsilon_T[m-1]^2 + \epsilon_G[m] < 2^{-s} \qquad\qquad \textbf{7.24}$$

Moreover, since

$$\epsilon_T[j] = \epsilon_T[j-1]^2 + \epsilon_G[j] = (\epsilon_T[j-2]^2 + \epsilon_G[j-1])^2 + \epsilon_G[j] \qquad \textbf{7.25}$$

an error in an iteration is reduced quadratically by the iterations that follow.[7] Figure 7.4 illustrates this error behavior.

Multiplicative Normalization Method

In this method we need to distinguish between the error in $d[j]$ and the error in $R[j]$.

---

7. Because of this, this algorithm is said to have the self-correcting property. Moreover, the total error is larger than the generated error in the last iteration; consequently, the multiplier needs to be of higher precision than the precision required for the result.

1.   Error in $d[j]$:

$$ed_T[j] = ed_T[j-1]^2 + ed_G[j] \qquad\qquad \textbf{7.26}$$

where $ed_G[j]$ includes the error in the two's complement operation and the roundoff error in the multiplication. As in the NR method, the final error $ed_T[m]$ is positive as long as the generated error in the last iteration is not negative. Also generated errors are reduced quadratically.

2.   Error in $R[j]$: If the multiplications $R[i]d$ are performed in full precision, then the error is the same as that of $d[j]$ (including there the errors in roundoff and two's complement). To that error it is necessary to add the generated errors for all $R[i]$ with $i \leq j$. This error corresponds to the roundoff of the multiplication $R[i]P[i]$. That is,

$$e R_T[j] = ed_T[j] + \sum_{i=0}^{j} e R_G[i] \qquad\qquad \textbf{7.27}$$

Note that here $e R_G[i]$ is not reduced by the iterations that follow. Consequently, the error in $R[j]$ is two sided (from below or from above), as illustrated in Figure 7.5.

Using Reduced Multipliers

As mentioned, full-precision multiplications are very wide. In practice, the following two alternatives have been found attractive:

*   Using a floating-point multiplier that produces a rounded product. In this case, all multiplications produce products of the same number of bits. This alternative is used when the algorithm is implemented in a floating-point unit that has this type of multiplier.
    The width of the rounded product is selected so as to achieve the required final error. In the Newton-Raphson method, because of the quadratic reduction of errors, this width is determined by the multiplications in the last iteration. On the other hand, in the multiplicative normalization method, the errors in the $R[j]$ recurrence accumulate so that the width has to take into account the rounding errors in all iterations.
*   Using an $n \times k$ rectangular multiplier. In this case, as the precision increases, the multiplications are performed by a sequence of several rectangular multiplications. This might have the advantage that the

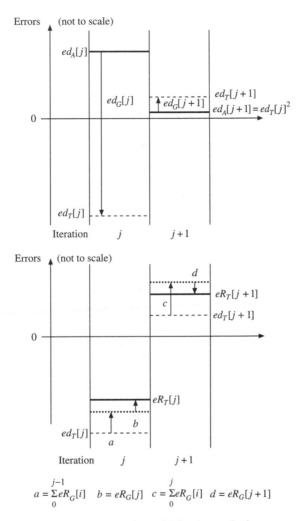

FIGURE 7.5 Errors in the multiplicative method.

rectangular multiplier is smaller and faster than the square multiplier.

However, a rounding error might occur for each rectangular multiplication.

We now compare the number of cycles in both approaches for the Newton-Raphson method.

EXAMPLE 7.2   For the Newton-Raphson method, we compare the number of cycles used in Scheme A (full multiplier) and Scheme B (rectangular multiplier) to obtain a reciprocal of d (53 bits) with an error of no more than 2^{-54}. The initial approximation $R[0]$ has an error of no more than 2^{-8} and has a width of 9 bits.

The multiplier in Scheme A is a standard floating-point multiplier, while the multiplier in Scheme B is a dedicated multiplier. To achieve the final error, the operation requires at least three iterations. In order not to increase the delay, the additional errors should not increase the number of iterations (see Exercise 7.9). We neglect the delay of obtaining $2 - R[i]d$.

Scheme A: Full multiplier $55 \times 55 \to 55$ (rounded); assuming 1 cycle for obtaining $R[0]$ and 3 cycles per multiplication results in a total of $1 + 3 \times 2 \times 3 = 19$ cycles.

Scheme B: Rectangular multiplier $55 \times 16 \to 55$; assuming 1 cycle per multiplication, the algorithm is performed as follows:

- $R[1] = R[0](2 - dR[0])$. Since $R[0]$ has 9 bits, the multiplications are performed by one 55×16 multiplication each (2 cycles). The result $R[1]$ is rounded to 16 bits (15 fractional bits).
- $R[2] = R[1](2 - dR[1])$. Since $R[1]$ has 16 bits, each multiplication is performed as one 55×16 multiplication (2 cycles). The result $R[2]$ is rounded to 32 bits.
- $R[3] = R[2](2 - dR[2])$. Since $R[2]$ has 32 bits, the multiplications are done by two 55×16 multiplications each (4 cycles).

This results in a total of $1 + 2 + 2 + 4 = 9$ cycles.   ■

A similar analysis is done for the multiplicative normalization method (see Exercise 7.10).

7.2   Division

The reciprocal of the divisor can be used to obtain the quotient by performing a multiplication by the dividend x. That is,

$$Q = R[m]x \qquad\qquad 7.28$$

In the multiplicative normalization method, instead of performing this multiplication at the end, it is possible to initialize $q[0] = x$ and obtain Q instead of R.

In floating-point units, the quotient has to be correctly rounded. The most prevalent method to do this rounding (see Chapter 8) computes the remainder produced by the approximation and performs a correction step. However, unlike in digit recurrence methods, in these iterative methods, the remainder is not obtained directly. Instead, at the end of operation, the product of the computed quotient and the divisor is formed and subtracted from the dividend.

7.3   Square Root

We now present two iterative methods for square root, which follow the same strategies as those discussed for division.

7.3.1   Newton-Raphson Method

According to expression (7.1) the Newton-Raphson method for square root is obtained by making $f(S) = S^2 - x$, which has a root at $S = x^{1/2}$. Since $f'(S) = 2S$, we get the following iteration

$$S[j + 1] = 2^{-1}\left(S[j] + \frac{x}{S[j]}\right) \qquad\qquad 7.29$$

Each iteration requires one division, one addition, and one shift.

Because of the division involved, it might be better to use an alternative with only multiplication and addition. This alternative is to compute instead the reciprocal square root of x and then multiply by x. The function to use is $f(R) = 1/R^2 - x$, resulting in $f'(R) = -2/R^3$ and

$$R[j + 1] = 2^{-1}R[j](3 - xR[j]^2) \qquad\qquad 7.30$$

An iteration requires three multiplications and one addition. Similar considerations as for reciprocal apply, with respect to the effect of limited-precision multiplications. Also, the subtraction from 3 can be approximated by bit inversion since $3 - xR[j]^2 = 1 + (2 - xR[j]^2)$ and the term $2 - xR[j]^2$ corresponds to a two's complement, which can be approximated by a bit inversion.

7.3.2   Multiplicative Normalization Method

Similar to the multiplicative method for reciprocal, this method consists in the determination of a sequence $P[j]$ such that

$$x \prod_{j=0}^{m} P[j]^2 \approx 1 \qquad\qquad 7.31$$

then

$$\frac{1}{\sqrt{x}} \approx \prod_{j=0}^{m} P[j] \qquad\qquad 7.32$$

and

$$\sqrt{x} \approx x \prod_{j=0}^{m} P[j] \qquad\qquad 7.33$$

The algorithm consists of m iterations to perform the recurrences

$$x[j+1] = x[j]P[j]^2 \qquad\qquad 7.34$$

$$S[j+1] = S[j]P[j] \qquad\qquad 7.35$$

The variable $x[j]$ tends to 1 and $R[j]$ tends to $\sqrt{x}$. The initial conditions are $x[0] = S[0] = x$. Note that reciprocal square root can be computed by making $S[0] = 1$. $P[0]$ is an (initial) approximation of $1/\sqrt{x}$.

Now for the determination of $P[j]$. Since $x[j]$ is close to 1, define $\epsilon[j]$ such that

$$x[j] = 1 - \epsilon[j] \qquad\qquad 7.36$$

Then for quadratic convergence we want

$$x[j+1] = 1 - O(\epsilon[j]^2) \qquad\qquad 7.37$$

To achieve this make

$$P[j]^2 = (1 + 2^{-1}\epsilon[j])^2 \qquad\qquad 7.38$$

resulting in

$$x[j+1] = (1 - \epsilon[j])(1 + 2^{-1}\epsilon[j])^2 = 1 - \frac{3}{4}\epsilon[j]^2 - \frac{1}{4}\epsilon[j]^3 \qquad 7.39$$

From (7.38) we get

$$P[j] = 1 + 2^{-1}\epsilon[j] = 1 + 2^{-1}(1 - x[j]) \qquad \textbf{7.40}$$

That is, $P[j]$ is obtained by complementing $x[j]$ and shifting the fractional part one bit to the right.

The iteration consists then of

1. $P[j]^2 = P[j]P[j]$
2. $x[j+1] = x[j]P[j]^2 \quad S[j+1] = S[j]P[j]$
3. $P[j+1] = 1 + 2^{-1}(1 - x[j+1])$

Each iteration has three multiplications, but two of them can be performed concurrently, or in pipelined fashion.

7.3.3 Implementation and Error Issues

The implementation of the square root algorithms and the error analysis are similar to that for the reciprocal operation.

7.4 Example of Implementation of Division and Square Root

We describe an implementation of a multiplicative method for division and square root used in a floating-point unit.[8] The division and square root algorithms shown are for operands/results of 53 bits (double precision). The internal precision of the implementation is 76 bits to support extended precision of 64 bits and rounding requirements. The details of rounding are covered in Chapter 8 and will not be explained here.

The main part of the implementation, shown in Figure 7.6, is a pipelined multiplier with a latency of four cycles. The operands have 76 bits, producing a 152-bit internal product rounded to 76 bits. A radix-8 multiplier recoding with the digit set $\{-4, -3, \ldots, 3, 4\}$ produces 26 multiples, which are reduced to the sum and carry vectors by a tree of [4:2] adders. The $3\times$ multiple is produced by a

---

8. This example is based on the AMD-K7 Floating-Point Unit implementation (Oberman 1999).

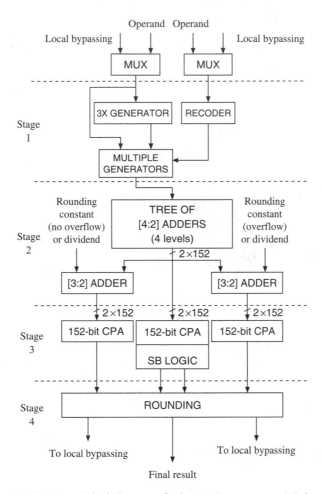

FIGURE 7.6 Block diagram of a division/square root unit (adapted from Oberman 1999).

separate 78-bit adder in parallel with the multiplier recoding. The multiplication unit consists of four pipelined stages:

- **Stage 1:** Performs recoding, generation of 3× multiple, and generation of 26 radix-8 multiples.
- **Stage 2:** Produces the product of 152 bits in a carry-save form. Rounding constants are also added using additional [3:2] adders for the cases of

overflow and no overflow in the product (see Chapter 8, Section 8.5). Moreover, these [3:2] adders are used to subtract the dividend in producing the remainder that is needed in rounding.

- **Stages 3 and 4:** Preparation for rounding and rounding operations are performed.

The initial approximations for the divisor reciprocal and the square root reciprocal are obtained by a bipartite method, using several tables with a total size of 69K bits and one adder.

The initial approximation of the divisor reciprocal is obtained from a pair of tables T1 and T2, each of 1K entries and of width 16 bits and 7 bits, respectively. The values obtained from the tables are added to produce the reciprocal approximation, accurate to at least 14.94 bits. A separate pair of tables T3 and T4, each having 2K entries, is used to obtain an approximation of the reciprocal square root. The width of T3 is 16 bits and of T4 is 7 bits, and the approximation, accurate to at least 15.84 bits, is obtained by adding the two words.

The division operation is shown in Figure 7.7. It is based on the multiplicative method described in Section 7.2. The latency of the operation is 20 cycles, and a new instance of the operation can begin after 17 cycles.

The square root operation is shown in Figure 7.8. It is based on the multiplicative method described in Section 7.3.2. The latency of the operation is 27 cycles, and a new instance of the operation can begin after 24 cycles.

7.5   Concluding Remarks

The methods described in this chapter provide an alternative to the digit recurrence methods presented in Chapters 5 and 6. Other methods are discussed in Chapters 10 and 11. The choice of method and of specific parameters depends on many considerations, such as latency, throughput, area, and energy requirements, as well as the sharing of components with other operations. In Chapter 8 (floating-point operations), we perform a comparison between methods; this is appropriate to do there since the methods of this chapter are mostly used in floating-point units and therefore the comparison should include the effect of using a floating-point multiplier and of the implementation of the corresponding roundoff modes. Let us comment here only that, as illustrated in the previous section, although the

---

1. [*Initialize*]

    $P[0] \leftarrow RECIP(\hat{d})$

    $d[0] \leftarrow d; \; q[0] \leftarrow x$

2. [*Iterate*]

    for $j = 0, 1$

    $d[j+1] \leftarrow d[j] \times p[j]; \; q[j+1] \leftarrow q[j] \times p[j]$

    $p[j+1] = CMPL(d[j+1])$

    end for

3. [*Terminate*]

    $q[3] \leftarrow q[2] \times p[2]$

    $REM \leftarrow d \times q[3] - x$

    $q \leftarrow ROUND(q[3], REM, mode)$

where

- *RECIP* produces the initial approximation of $1/d$ in three cycles.
- *CMPL(a)* performs bit complementation of a.
- *REM* is a negated remainder (see Exercise 7.18).
- *ROUND* produces a quotient rounded according to the specified *mode* (rounding modes are discussed in Chapter 8). The sign and zero conditions of the remainder are also used.

FIGURE 7.7 Multiplicative division algorithm (double precision).

multiplicative method has quadratic convergence and therefore results in a small number of iterations, the total number of cycles can be large.

We have discussed two related methods: Newton-Raphson and multiplicative normalization. Both have a quadratic convergence rate and have multiplication and addition as the primitive operations. In most instances the multiplicative normalization method would be preferred because the two multiplications per iteration are independent and can use more effectively a pipelined multiplier. On the other hand, the Newton-Raphson method has the self-correcting property, which limits the effect of generated errors and provides the possibility of having a one-sided approximation.

1. [*Initialize*]

$$P[0] \leftarrow RECSQR(\hat{x})$$
$$T[0] \leftarrow P[0]^2$$
$$x[0] \leftarrow x; \quad R[0] \leftarrow x$$

2. [*Iterate*]

   for $j = 0, 1$

   $$x[j+1] \leftarrow x[j] \times T[j]; \quad R[j+1] \leftarrow R[j] \times P[j]$$
   $$P[j+1] = CMPL3(x[j+1])$$
   $$T[j+1] = P[j+1]^2$$

   end for

3. [*Terminate*]

   $$R[3] \leftarrow R[2] \times P[2]$$
   $$REM \leftarrow R[3] \times R[3] - x$$
   $$s \leftarrow ROUND(R[3], REM, mode)$$

where

- *RECSQR* produces the initial approximation of the reciprocal square root of x in three cycles.
- *CMPL3*(a) produces $2^{-1}(3 - a)$.
- *REM* is a negated remainder.
- *ROUND* produces a result rounded according to the specified *mode* (rounding modes are discussed in Chapter 8). The sign and zero conditions of the remainder are also used.

FIGURE 7.8 Multiplicative square root algorithm (double precision).

7.6 Exercises

Newton-Raphson Method for Reciprocal

7.1 Obtain the recurrence for the Newton-Raphson method of reciprocal approximation directly from the definition of the relative error ($\epsilon[j] = 1 - dR[j]$) and the requirement of quadratic convergence.

7.2   Perform the steps in the calculation of the reciprocal of $d = 29/256$ by the Newton-Raphson method. Use $R[0] = 2 - d$ truncated to four fractional bits. Perform sufficient iterations so that the maximum error in the range $\frac{1}{2} \leq d < 1$ is less than 2^{-12}.

7.3   Show that the rate of convergence of the reciprocal approximation can be improved by including additional power terms in the recurrence (Ferrari 1967). Specifically, show that the approximation using the recurrence

$$R[j + 1] = R[j](1 + (1 - dR[j]) + (1 - dR[j])^2 + \cdots + (1 - dR[j])^k)$$

has a convergence rate such that

$$\epsilon[j + 1] = \epsilon[j]^{k+1}$$

Describe the implementation for $k = 2$ and determine the latency in multiplication times. Compare with the case $k = 1$.

Multiplicative Normalization for Reciprocal

7.4   Obtain a reciprocal approximation of $d = \frac{29}{256}$ by the multiplicative normalization method. Use $P[0] = 2 - d$ truncated to four fractional bits. Do sufficient iterations so that the maximum error in the range $\frac{1}{2} \leq d < 1$ is less than 2^{-12}.

Initial Approximations

7.5   In a *faithful* reciprocal table $2^a \times b$ (i.e., with a-bit input and b-bit output), the table outputs differ from $1/x$ by less than 1 ulp, $1 \leq x < 2$ (Das Sarma and Matula 1995).

(a)   Show that a table with $a = b, b \geq 3$ has a maximum error greater than 1 ulp. That is, the table is not faithful. *Hint*: Analyze the second smallest input interval.

(b)   Show that if $a \geq b + 1, b \geq 1$, a maximum error for any output is strictly less than 1 ulp.

7.6   Generate a reciprocal approximation table using the midpoint reciprocal method (Ferrari 1967; Das Sarma and Matula 1995) for $a = 5$ and $b = 4$:

$$\textbf{for } i = 2^a \textbf{ to } 2^{a+1} - 1 \textbf{ step } 1:$$
$$T(i) = \lfloor \tfrac{2^{a+b+1}}{i+0.5} + 0.5 \rfloor$$

The table entries are the reciprocals of midpoints of the input intervals rounded to the nearest value. The values should be divided by 2^a to correspond to the input range $1 \leq x < 2$.

7.7   A bipartite table function approximation (Das Sarma and Matula 1995; Schulte and Stine 1997) can be formulated as follows. Partition the bit-vector of the argument $1 \leq x < 2$ into three parts: $x = (X1, X2, X3)$ where $X1 = (1.x_1, \ldots, x_k)$, $X2 = (x_{k+1}, \ldots, x_{2k})$, and $X3 = (x_{2k+1}, \ldots, x_{3k})$. $(n = 3k)$. Define two functions $f_A(X1, X2)$ and $f_B(X1, X3)$ such that $f_A + f_B$ approximates the reciprocal $1/x$ to the desired accuracy. The functions f_A and f_B can be derived using Taylor series expansion and stored in separate tables. The result is obtained by adding the outputs of the two tables.

(a)   Obtain a bipartite approximation for $1/x$ assuming the truncated input $1.x_1 \ldots x_5$ and output $(0.1z_1 \ldots z_4)$.

(b)   Compare the size of the tables for f_A and f_B with that of a single table reciprocal approximation.

7.8   This exercise deals with a reciprocal approximation method based on MacLauren's series expansion. This approximation can be used as an initial approximation for the iterative methods.

Let $R = 1/d$ where $1 \leq d < 2$ and $\frac{1}{2} < R \leq 1$. The corresponding bit-vectors are

$$D = (1, d_1, d_2, \ldots, d_k, d_{k+1}, \ldots, d_n)$$

$$R = (0, 1, r_2, \ldots, r_n)$$

Decompose d into $d = d_A + 2^{-k}d_B$, where $1 \leq d_A < 2 - 2^{-k}$ and $0 \leq d_B < 1 - 2^{-(n-k)}$. Then

$$D_A = (1, d_1, d_2, \ldots, d_k)$$

$$D_B = (0, d_{k+1}, \ldots, d_n)$$

(a)   Let $r_A = 1/d_A$. Show that

$$R = r_A[1 - 2^{-k}d_B r_A + 2^{-2k}(d_B r_A)^2 - 2^{-3k}(d_B r_A)^3 + \cdots]$$

(b)   Consider computing the approximation $\hat{R}$ by

$$\hat{R} = \{\hat{r}_A - 2^{-k}\hat{r}_A^2 \hat{d}_B\}_t$$

where $\{x\}_t$ denotes x truncated to t fractional bits and $\hat{r}_A = \{r_A\}_u, \hat{r}_A^2 = \{r_A^2\}_s$, and $\hat{d}_B = \{d_B\}_v$. Use a table, a multiplier, and an adder. Determine relations between $k, u, v, s,$ and t to obtain an economical implementation in terms of table and multiplier size.

Use of Truncated Multipliers and Error Calculation

7.9   Determine a bound on the maximum final error for the implementations of Example 7.2. Consider rounding errors and the error introduced by performing ones' complement instead of two's complement.

7.10   Repeat Example 7.2 for implementations using the multiplicative normalization approach. Consider that the 55×55 multiplier is implemented by a three-stage pipeline.

7.11   Consider computing

$$R[1] = R[0](2 - d \cdot R[0])$$

using (i) d truncated to t bits, (ii) the product $d \cdot R[0]$ rounded to t bits, and (iii) the final product rounded to t bits. Consider rounding to nearest mode in both cases.

   Let the error of the initial approximation be in the range $-2^{-16} < e_{R[0]} < 2^{-16}$. Determine the range of the error in $R[1]$ for $t = 32$.

7.12   Repeat the error analysis of Exercise 7.11 if ones' complement is used instead of two's complement.

Division

7.13   Perform the division operation for the following operands:

$$x = (0.010100011110)_2$$

$$d = (0.101101000011)_2$$

Consider two cases:

(a)   Use the Newton-Raphson approximation of the reciprocal.

(b)   Use the multiplicative normalization with initial condition $R[0] = x$.

   Use as initial approximation $2.98 - 2d$. Perform sufficient iterations to get a maximum error less than 2^{-12} for the entire range $\frac{1}{2} \le x, d < 1$.

7.14   Design a double-precision division unit (53-bit quotient) based on the Newton-Raphson reciprocal approximation method using 56×12 rectangular multipliers (Wong and Goto 1994). The initial approximation has an error of 2^{-10} and 10 fractional bits. Determine the number of cycles and show a timing diagram.

Square Root

7.15   Obtain the square root of $x = 0.125$ using the Newton-Raphson method for reciprocal square root. Perform sufficient iterations so that the maximum error is less than 2^{-12} in the range $\frac{1}{4} \le x < 1$.

7.16   Repeat Exercise 7.15 using the direct multiplicative method.

7.17   Repeat Example 7.2 for square root using the direct multiplicative method. Consider a three-stage pipeline 55×55 multiplier.

7.18   Explain why a negated remainder is computed in the algorithms shown in Figures 7.7 and 7.8. Consider the organization of the corresponding implementation in Figure 7.6.

7.7   Further Readings

General treatments of iterative methods for division, reciprocal, square root, and reciprocal square root are presented in Flynn (1970), Krishnamurthy (1970), Ramamoorthy et al. (1972), and Markstein (2000).

Multiplicative Method

A multiplicative division method is reported in Goldschmidt (1964). Anderson et al. (1967) describe an implementation of multiplicative methods for division and square root, also known as the Goldschmidt method.

Initial Approximation

The problem of initial approximations has been frequently considered in the literature (Shaham and Riesel 1972; Parhami 1987; Das Sarma 1995; Schwarz and Flynn 1996; Ito et al. 1997). Table methods for initial approximations have been the subject of many studies (Parker and Hamblen 1992; Das Sarma and Matula 1994, 1995; Schulte and Swartzlander 1994; Schulte and Stine 1997; Matula 2001).

Area/Delay Analysis

Area/performance comparisons of digit recurrence and multiplicative division and square root implementations are provided in Soderquist and Leeser (1996).

Implementation

An overview of implementation issues is presented in Oberman and Flynn (1997). Specific implementations have been presented in many papers (Anderson et al. 1967; Markstein 1990; Kabuo et al. 1994; Oberman 1999; Naini et al. 2001). Acceleration of the multiplicative method is proposed in Ercegovac et al. (2000).

Miscellaneous

Alternative convergence methods for division based on Taylor-series approximations of the reciprocal are presented in Wong and Flynn (1992) and Agarwal et al. (1999). A hybrid scheme with Newton-Raphson and digit recurrence methods is described in Montuschi et al. (1994). Use of rectangular multipliers is discussed in Briggs and Matula (1993) and Wong and Goto (1994). Error analysis of the Newton-Raphson method for reciprocals is considered in Fowler and Smith (1989).

Verification

Correctness proofs are considered in Rusinoff (1998) and Cornea-Hasegan et al. (1999). Iterative methods are often used in software routines to implement high-precision floating-point division and square root using limited-precision hardware (Karp and Markstein 1997).

7.8   Bibliography

Agarwal, R. C., F. G. Gustavson, and M. S. Schmookler (1999). Series approximation methods for divide and square root in the power3 microprocessor. In *Proceedings of the 14th IEEE Symposium on Computer Arithmetic*, pages 116–23.

Anderson, S. F., J. G. Earle, R. E. Goldschmidt, and D. M. Powers (1967). The IBM 360/370 model 91: floating-point execution unit. *IBM Journal of Research and Development*, pages 34–53.

Briggs, W. S., and D. W. Matula (1993). A 17×69 bit multiply and add unit with redundant binary feedback and single cycle latency. In *Proceedings of the 11th IEEE Symposium on Computer Arithmetic*, pages 163–71.

Cornea-Hasegan, M. A., R. A. Golliver, and P. Markstein (1999). Correctness proofs outline for Newton-Raphson based floating-point divide and square root algorithms. In *Proceedings of the 14th IEEE Symposium on Computer Arithmetic*, pages 96–105.

Das Sarma, D. (1995). *Highly Accurate Initial Reciprocal Approximations for High Performance Division Algorithms*. PhD thesis, Southern Methodist University.

Das Sarma, D., and D. W. Matula (1994). Measuring the accuracy of ROM reciprocal tables. *IEEE Transactions on Computers*, 43(8):932–40.

Das Sarma, D., and D. W. Matula (1995). Faithful bipartite ROM reciprocal tables. In *Proceedings of the 12th IEEE Symposium on Computer Arithmetic*, pages 17–28.

Ercegovac, M. D., L. Imbert, D. W. Matula, J.-M. Muller, and G. Wei (2000). Improving Goldschmidt division, square root and square root reciprocal. *IEEE Transactions on Computers*, 49(7):759–63.

Ferrari, D. (1967). A division method using a parallel multiplier. *IEEE Transactions Electronic Computers*, EC-16(4):224–26.

Flynn, M. J. (1970). On division by functional iteration. *IEEE Transactions on Computers*, C-19(8):702–6.

Fowler, D. L., and J. E. Smith (1989). An accurate, high speed implementation of division by reciprocal approximation. In *Proceedings of the 9th IEEE Symposium on Computer Arithmetic*, pages 60–67.

Goldschmidt, R. E. (1964). *Applications of Division by Convergence*. Master's thesis, Massachusetts Institute of Technology.

Ito, M., N. Takagi, and S. Yajima (1997). Efficient initial approximation for multiplicative division and square root by a multiplication with operand modification. *IEEE Transactions on Computers*, 46(4):495–98.

Kabuo, H., T. Taniguchi, A. Miyoshi, H. Yamashita, M. Urano, H. Edamatsu, and S. Kuninobu (1994). Accurate rounding scheme for the Newton-Raphson method using redundant binary representation. *IEEE Transactions on Computers*, 43(1):43–51.

Karp, A. H., and P. Markstein (1997). High-precision division and square root. *ACM Transactions on Mathematical Software*, 23(4):561–89.

Krishnamurthy, E. V. (1970). On optimal iterative schemes for high-speed division. *IEEE Transactions on Computers*, C-19(3):227–31.

Markstein, P. (2000). *IA-64 and Elementary Functions: Speed and Precision*. Hewlett-Packard Professional Books. Prentice Hall.

Markstein, P. W. (1990). Computation of elementary functions on IBM RISC System/6000 processor. *IBM Journal Research and Development*, pages 111–19.

Matula, D. W. (2001). Improved table lookup algorithms for postscaled division. In *Proceedings of the 15th IEEE Symposium on Computer Arithmetic*, pages 101–8.

Montuschi, P., L. Ciminiera, and A. Giustina (1994). Division unit with Newton-Raphson approximation and digit-by-digit refinement of the quotient. *IEE Proceedings—Computers and Digital Techniques*, 141(6):317–24.

Naini, A., A. Dhablania, W. James, and D. Das Sarma (2001). 1 GHz HAL SPARC64^r dual floating point unit with RAS features. In *Proceedings of the 15th IEEE Symposium on Computer Arithmetic*, pages 173–83.

Oberman, S. F. (1999). Floating-point division and square root algorithms and implementation in the AMD-K7 microprocessor. In *Proceedings of the 14th IEEE Symposium on Computer Arithmetic*, pages 106–15.

Oberman, S. F., and M. J. Flynn (1997). Design issues in division and other floating-point operations. *IEEE Transactions on Computers*, 46(2):154–61.

Parhami, B. (1987). On the complexity of table-lookup for iterative division. *IEEE Transactions on Computers*, C-36:1233–36.

Parker, A., and J. O. Hamblen (1992). Optimal value for the Newton-Raphson division algorithm. *Information Processing Letters*, 42(3):141–44.

Ramamoorthy, C. V., J. R. Goodman, and K. H. Kim (1972). Some properties of iterative square-rooting methods using high-speed multiplication. *IEEE Transactions on Computers*, C-21(8):837–47.

Rusinoff, D. (1998). A mechanically checked proof of IEEE compliance of a register-transfer-level specification of the AMD-k7 floating-point multiplication, division, and square root instructions. *LMS Journal of Computation and Mathematics*, 1:148–200.

Schulte, M. J., J. Omar, and E. Swartzlander, Jr. (1994). Optimal initial approximations for the Newton-Raphson division algorithm. *Computing*, 53(3–4): 233–42.

Schulte, M. J., and J. E. Stine (1997). Symmetric bipartite tables for accurate function approximation. In *Proceedings of the 13th IEEE Symposium on Computer Arithmetic*, pages 175–83.

Schwarz, E. M., and M. J. Flynn (1996). Hardware starting approximation method and its application to the square root operation. *IEEE Transactions on Computers*, 45(12):1356–69.

Shaham, Z., and Z. Riesel (1972). A note on division algorithms based on multiplication. *IEEE Transcations on Computers*, C-21(5):513–14.

Soderquist, P., and M. Leeser (1996). Area and performance tradeoffs in floating-point division and square root implementations. *ACM Computing Surveys*, 28(3):518–64.

Wong, D., and M. J. Flynn (1992). Fast division using accurate quotient approximations to reduce the number of iterations. *IEEE Transactions on Computers*, 41(8):981–95.

Wong, W. F., and E. Goto (1994). Fast hardware-based algorithms for elementary function computations using rectangular multipliers. *IEEE Transactions on Computers*, 43(3):278–94.

IN THIS CHAPTER WE DISCUSS THE FOLLOWING TOPICS:

| Floating-Point
Representation,
Algorithms, and
Implementations

Many scientific and engineering applications require computations with real numbers. To represent these numbers, fixed-point representations can be used. However, in many cases the range of this representation does not correspond to the range required by the applications, producing frequent overflows and underflows. A standard solution is the use of floating-point representations. In this chapter we present this representation, consider its properties, and discuss the algorithms and implementations for the basic arithmetic operations.

8.1 Floating-Point Representation

As indicated, a floating-point representation is used to represent real numbers. Since, as in a fixed-point representation, the floating-point representation is encoded in a finite number of bits, it is possible to represent only a finite subset of the infinite set of real numbers. For a specific floating-point system, a real number that is (exactly) represented in the system is called a *floating-point number*. The rest of the real numbers either fall outside the range of the representation (overflow and underflow) or are represented by floating-point numbers that have a value that approximates the real number. The process of approximation is called *roundoff* and produces a roundoff error.

8.1.1   Significand, Exponent, and Base

The representation of the floating-point number x consists of two components, the *significand* M_x^* (also called the *mantissa*)[1] and the *exponent* E_x, such that

$$x = M_x^* \times b^{E_x} \qquad\qquad 8.1$$

where b is a constant called the *base*. The sign of the number is determined by the sign of the significand. The exponent is a signed integer.

The signed significand can be represented using any representation system, such as sign-and-magnitude or two's complement. Today the most used representation is sign-and-magnitude. In such a case, a floating-point number x is represented by a triple (S_x, M_x, E_x), that is,

$$x = (-1)^{S_x} \times M_x \times b^{E_x} \qquad\qquad 8.2$$

where $S_x \in \{0, 1\}$ is the sign and M_x denotes the magnitude of the significand. We assume this representation in the rest of the chapter. Moreover, we refer to the magnitude of the significand as "the significand" and use the term "signed significand" when we include the sign.[2] The representation of the exponent is discussed later.

8.1.2   Advantage: Dynamic Range

The objective of using floating-point representation is to increase the dynamic range, with respect to a fixed-point representation. This *dynamic range* is defined as the ratio between the largest and the smallest (nonzero and positive) number that can be represented.

For a fixed-point representation using n radix-r digits for the magnitude, the dynamic range is

$$DR_{fxpt} = r^n - 1 \qquad\qquad 8.3$$

---

1. We include the superscript * to indicate a signed significand and distinguish it from the magnitude, which we denote M_x.
2. Although our presentation is limited to the sign-and-magnitude representation, the modifications for other representations are straightforward.

In contrast, for the floating-point representation,

$$DR_{flpt} = \frac{M_{max} b^{E_{max}}}{M_{min} b^{E_{min}}} \qquad \textbf{8.4}$$

For instance, if the n digits are partitioned so that m digits are used for the significand and $n - m$ digits for the exponent, and $b = r$ we get

$$DR_{flpt} = (r^m - 1) r^{(r^{n-m} - 1)} \qquad \textbf{8.5}$$

As an example, if $n = 32, m = 24, r = 2$

$$DR_{fxpt} = 2^{32} - 1 \approx 4.3 \times 10^9$$

$$DR_{flpt} = (2^{24} - 1) 2^{2^8 - 1} \approx 9.7 \times 10^{83}$$

As mentioned before, a large dynamic range is required in many applications to avoid overflows and underflows. If the dynamic range of the fixed-point representation is not sufficient, complicated scaling operations have to be included in the program. Thus, a floating-point system is preferable in such applications.

8.1.3 Disadvantages: Less Precision, Roundoff Error, and Complex Implementation

The precision of a representation corresponds to the number of digits of the significand. Since in the floating-point representation the total number of digits is partitioned between the significand and the exponent, for the same total number of digits, the floating-point representation has a smaller precision than the fixed-point representation. In the example given for the illustration of the dynamic range, the precision of the fixed-point representation is 32 bits, whereas that of the floating-point representation is 24 bits.

Moreover, floating-point arithmetic introduces roundoff errors and makes analysis of accuracy of the results more difficult.

Another disadvantage of using a floating-point representation is the more complex implementation of floating-point operations, which leads to a larger area and a slower execution.

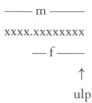

FIGURE 8.1 Representation of significand.

8.1.4   Range of Significand and Unit in the Last Position (ulp)

For a fixed-radix representation of the significand, its range is mainly determined by the position of the radix point. In general, the m digits are divided into integer digits and fractional digits (see Figure 8.1). We denote by f the number of fractional digits, so that the number of integer digits is $m - f$. Then, for magnitude representation

$$M = \sum_{i=-f}^{m-f-1} d_i r^i \qquad\qquad 8.6$$

The corresponding range of the (nonzero) significand is

$$r^{-f} \le M \le r^{m-f} - r^{-f} \qquad\qquad 8.7$$

One of the representations used is to have only fractional digits ($f = m$). That is,

$$M = \sum_{i=1}^{m} d_i r^{-i} \qquad\qquad 8.8$$

Notice the change in the indexing convention. In this case, the range of the (nonzero) significand is

$$r^{-m} \le M \le 1 - r^{-m} \qquad\qquad 8.9$$

Another common choice is to have one integer digit ($f = m - 1$) so that

$$M = \sum_{i=0}^{m-1} d_i r^{-i} \qquad\qquad 8.10$$

resulting in a range

$$r^{-(m-1)} \le M \le r - r^{-(m-1)} \qquad\qquad 8.11$$

The difference between two consecutive values of the significand is called an *ulp* (unit in the last place). In terms of the representation described (see Figure 8.1)

$$ulp = r^{-f} \qquad \textbf{8.12}$$

The error introduced by representing a real number with a floating-point number is typically given in ulps. For example, if the floating-point result is 1.374×10^{-4} (with an ulp of 10^{-3}) and the exact result obtained using infinite precision is 0.00013755, the error is 1.5 ulps.

8.1.5   Normalized, Unnormalized, and Denormalized Representation

The floating-point representation is redundant; that is, a floating-point number can have several representations. For example the number 1 can be represented as $M = 1, E = 0$ or $M = 0.5, E = 1$, with $b = 2$. This redundancy is not convenient, for instance, for the comparison of values. Consequently, a unique representation is used. Moreover, to improve the potential "accuracy" of the computations it is convenient to eliminate nonsignificant leading zeros. For this, the *normalized* representation is defined so that the most-significant digit of the significand is always different from zero (except for the zero value).

On the other hand, using a normalized representation reduces the range of floating-point numbers since now the smallest significand is

$$\overbrace{10\ldots0.\underbrace{0\ldots0}_{f}}^{m} = r^{m-f-1} \qquad \textbf{8.13}$$

so that the smallest floating-point number is

$$r^{m-f-1} \times b^{E_{min}} \qquad \textbf{8.14}$$

To avoid this reduction in range, unnormalized representation is also included for the values that cannot be represented in normalized form. That is, in this case unnormalized significands are only allowed with the minimum exponent. These unnormalized numbers are called *denormalized numbers (denormals)*. With the use of denormals, as numbers decrease in magnitude they gradually include more most-significant zeros in their significand; this is called *gradual underflow*.

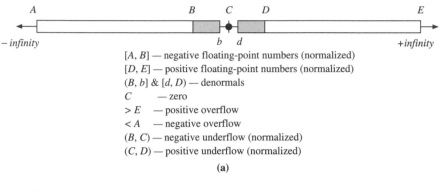

[A, B] — negative floating-point numbers (normalized)
[D, E] — positive floating-point numbers (normalized)
(B, b] & [d, D) — denormals
C   — zero
> E   — positive overflow
< A   — negative overflow
(B, C) — negative underflow (normalized)
(C, D) — positive underflow (normalized)

(a)

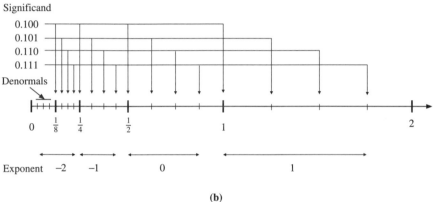

(b)

FIGURE 8.2 (a) Regions in floating-point representation. (b) Example for $m = f = 3, r = 2$, and $-2 \leq E \leq 1$ (only positive region).

For the value zero also a unique representation is used. The usual convention is to represent zero by $M = 0$ and the minimum exponent.

8.1.6   Values Represented and Their Distribution

The set of floating-point numbers (values represented by a floating-point system) depends on the range of the significand and of the exponent. Figure 8.2(a) shows the different regions in which a floating-point system divides the real numbers. The points A, B, and so on in the figure are defined in the following table:

Floating-Point System

	Normalized	Unnormalized
A	$-(r^{m-f} - r^{-f}) \times b^{E_{max}}$	
B	$-r^{m-f-1} \times b^{E_{min}}$	$-r^{-f} \times b^{E_{min}}$
C	± 0	
D	$r^{m-f-1} \times b^{E_{min}}$	$r^{-f} \times b^{E_{min}}$
E	$(r^{m-f} - r^{-f}) \times b^{E_{max}}$	

The overflow regions correspond to values that have a larger magnitude than what can be represented. Similarly, the underflow regions correspond to small values that cannot be represented.

As an example, Figure 8.2(b) shows the values represented for a floating-point system with a normalized fractional significand of $f = 3$ radix-2 digits, and an exponent in the range $-2 \le E \le 1$. For simplicity, only positive values are shown.

As indicated in Figure 8.2, the floating-point numbers are not uniformly distributed along the real number line. They are more dense close to 0. Density depends on the exponent base and the partitioning of bits among significand and exponent. The difference between two consecutive values is (for same exponents E and $r = b$)

$$\Delta = r^{-f} r^{E} = r^{E-f} \qquad\qquad \textbf{8.15}$$

Tables 8.1, 8.2, and 8.3 and Figure 8.3 illustrate the distributions of floating-point numbers for three representations with $n = 6$ bits, a normalized fractional significand of $m = f$ bits, and an integer exponent of e bits (for positive significand and exponent).

8.1.7   Choice of b

As illustrated by Tables 8.1 and 8.3, the choice of the base b affects the range and number of values represented as well as the distribution of these values. Moreover, it has an impact in the implementation of floating-point addition, where variable shifters are required (see Section 8.4).

In summary, larger b results in a larger range and more values but in less density. Moreover, larger b simplifies the shifter required for floating-point addition.

Significand	2^E			
	1	2	4	8
0.1000	$\frac{1}{2}$	1	2	4
0.1001	$\frac{9}{16}$	$\frac{9}{8}$	$\frac{9}{4}$	$\frac{9}{2}$
0.1010	$\frac{10}{16}$	$\frac{10}{8}$	$\frac{10}{4}$	5
0.1011	$\frac{11}{16}$	$\frac{11}{8}$	$\frac{11}{4}$	$\frac{11}{2}$
0.1100	$\frac{12}{16}$	$\frac{12}{8}$	3	6
0.1101	$\frac{13}{16}$	$\frac{13}{8}$	$\frac{13}{4}$	$\frac{13}{2}$
0.1110	$\frac{14}{16}$	$\frac{14}{8}$	$\frac{14}{4}$	7
0.1111	$\frac{15}{16}$	$\frac{15}{8}$	$\frac{15}{4}$	$\frac{15}{2}$

TABLE 8.1 Distribution for $b = 2, m = f = 4,$ and $e = 2$.

Significand	2^E							
	1	2	4	8	16	32	64	128
0.100	$\frac{1}{2}$	1	2	4	8	16	32	64
0.101	$\frac{5}{8}$	$\frac{5}{4}$	$\frac{5}{2}$	5	10	20	40	80
0.110	$\frac{6}{8}$	$\frac{3}{2}$	3	6	12	24	48	96
0.111	$\frac{7}{8}$	$\frac{7}{4}$	$\frac{7}{2}$	7	14	28	56	112

TABLE 8.2 Distribution for $b = 2, m = f = 3,$ and $e = 3$.

Many studies have been made to quantify these trade-offs, and the present conclusion is that best is $b = 2$.

8.1.8   Representation of Significand

As mentioned in the introduction, the significand is a signed number, so that a representation system for signed numbers is required. The most used are sign-and-magnitude and two's complement. Today sign-and-magnitude is preferred because it is considered a more natural representation, and it simplifies somewhat the implementation of multiplication and most aspects of floating-point addition. Although two's complement representation simplifies the addition of significands, this is a relatively small portion of overall floating-point addition. In the rest of the chapter we assume a sign-and-magnitude representation.

Significand	4^E			
	1	4	16	64
0.0100	$\frac{1}{4}$	1	4	16
0.0101	$\frac{5}{16}$	$\frac{5}{4}$	5	20
0.0110	$\frac{6}{16}$	$\frac{6}{4}$	6	24
0.0111	$\frac{7}{16}$	$\frac{7}{4}$	7	28
0.1000	$\frac{1}{2}$	2	8	32
0.1001	$\frac{9}{16}$	$\frac{9}{4}$	9	36
0.1010	$\frac{10}{16}$	$\frac{10}{4}$	10	40
0.1011	$\frac{11}{16}$	$\frac{11}{4}$	11	44
0.1100	$\frac{12}{16}$	3	12	48
0.1101	$\frac{13}{16}$	$\frac{13}{4}$	13	52
0.1110	$\frac{14}{16}$	$\frac{14}{4}$	14	56
0.1111	$\frac{15}{16}$	$\frac{15}{4}$	15	60

TABLE 8.3 Distribution for $b = 4, m = f = 4$ ($r = 2$), and $e = 2$.

For $r = 2$ and normalized format, the most-significant bit of the significand is always 1. Consequently, it does not have to be included in the representation. This is called the *hidden bit*.

8.1.9 Representation of Exponent

The exponent E is a signed integer, which can be represented by any of several systems, such as sign-and-magnitude, true-and-complement, and biased. The biased representation is preferred in this case because

- it simplifies the comparison of floating-point numbers, making it the same as fixed-point comparison
- the minimum exponent is represented by 0, so that the representation of the floating-point value 0 is all zeros (0 sign, 0 exponent, 0 significand)

In a biased representation with bias B, the signed integer E is represented by the positive integer denoted by E_B such that

$$E_B = E + B \qquad \textbf{8.16}$$

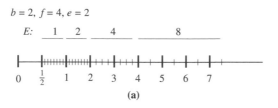

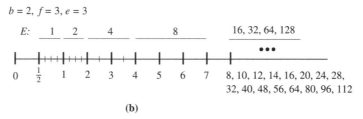

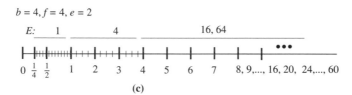

FIGURE 8.3 Examples of distributions of floating-point numbers.

To represent the minimum exponent by $E_B = 0$, we obtain

$$B = -E_{min}$$ **8.17**

Moreover, for a symmetric exponent range

$$-B \leq E \leq B$$ **8.18**

resulting in

$$0 \leq E_B \leq 2B$$ **8.19**

If e is the number of bits of the binary representation of E_B, then

$$2B \leq 2^e - 1$$ **8.20**

Consequently, for B integer we obtain

$$B \leq \frac{1}{2}(2^e - 2) = 2^{e-1} - 1 \qquad \textbf{8.21}$$

For instance, for $e = 8$ we can make $B = 127$ and

$$E_B = E + 127 \qquad \textbf{8.22}$$

for a symmetric exponent range of $-127 \leq E \leq 127$. Note that the maximum value of E_B is 255, so that this value can be used to represent $E = 128$ (nonsymmetric range) or as a singularity condition.

8.1.10   Special Values

These are values that are not representable in the floating-point system, but are useful. Two examples, which are included in the IEEE Standard presented later, are NAN (not a number) and infinity (positive and negative). For instance, the result of the square root of a negative number is set to NAN. Moreover, an operation that has a NAN as an operand produces a NAN result. These features allow computations to continue in the presence of NANs, without special checks. The availability of infinities allows the use of arithmetic on infinities.

8.1.11   Exceptions

There are cases in which a floating-point operation produces a value that is not representable in the floating-point number system. In such cases, a flag is set. The computation can continue or a trap to an exception handler performed (depending whether the exception handler is enabled). The most important exceptions are the following:

- Overflow (exponent). Occurs when the magnitude of the result is larger than the largest floating-point number.
- Underflow. Occurs when the (nonzero) magnitude of the result is smaller than the magnitude of the smallest floating-point number.

8.2   Roundoff Modes and Error Analysis

The result of a floating-point operation is a real number that, to be represented exactly, might require a significand with an infinite number of digits. Since the representation of the significand has only f fractional digits, it is necessary to

FIGURE 8.4 Relation between x, $Rmode(x)$, and floating-point numbers $F1$ and $F2$.

obtain a representation that is close to the exact result. This is achieved by performing a *roundoff operation* (also called *rounding*). We use the following notation:

- The exact (infinite precision) results are denoted by x, y, and so on.
- The floating-point number that represents x by applying the roundoff mode *Rmode* is denoted by $Rmode(x)$.

For a satisfactory roundoff scheme the following relations have to be satisfied:

1.   Ordering. If $x \leq y$, then $Rmode(x) \leq Rmode(y)$.

2.   Representability. If x is representable in the floating-point system (x is a floating-point number), then $Rmode(x) = x$.

3.   Containment. If $F1$ and $F2$ are two consecutive floating-point numbers such that $F1 \leq x \leq F2$, then $Rmode(x)$ should be either $F1$ or $F2$, as illustrated in Figure 8.4. Observe that $F1$ and $F2$ have the sign of x.

Several roundoff modes are used. We now give the definition of the modes used in the IEEE Standard and then discuss them in more detail. Other modes are described in the references.

Consider the real number x and the consecutive floating-point numbers $F1$ and $F2$ such that $F1 \leq x \leq F2$, as shown in Figure 8.4. Then:

- Round to nearest (tie to even). $Rnear(x)$ is the floating-point number that is closest to x. If there is a tie, the significand of $Rnear(x)$ should be even (least-significant bit equal to 0).[3] That is,

$$Rnear(x) = \begin{cases} F1 & \text{if } |x - F1| < |x - F2| \\ F2 & \text{if } |x - F1| > |x - F2| \qquad \textbf{8.23} \\ even(F1, F2) & \text{if } |x - F1| = |x - F2| \end{cases}$$

---

3. It is also possible to define a round to nearest mode with tie to odd; we use the convention of the IEEE Standard.

- Round toward zero. For this mode, $Rzero(x)$ is the closest to 0 among $F1$ and $F2$. That is,

$$Rzero(x) = \begin{cases} F1 & \text{if } x \geq 0 \\ F2 & \text{if } x < 0 \end{cases} \qquad \textbf{8.24}$$

- Round toward plus infinity. For this mode, $Rpinf(x)$ is the largest among $F1$ and $F2$, so

$$Rpinf(x) = F2 \qquad \textbf{8.25}$$

- Round toward minus infinity. For this mode, $Rninf(x)$ is the smallest among $F1$ and $F2$, so

$$Rninf(x) = F1 \qquad \textbf{8.26}$$

The roundoff modes are characterized by numerical and implementation characteristics. The numerical characteristics can be described by the following set of errors:

1. The (maximum) absolute representation error $ABRE$ ($MABRE$). The absolute representation error is defined as the difference between the represented value and the exact value. That is,

$$ABRE = Rmode(x) - x \qquad \textbf{8.27}$$

so that

$$MABRE = max_x(|ABRE|) \qquad \textbf{8.28}$$

2. The bias (RB). This is defined as the average absolute error considering an unsigned significand[4] and measures the tendency toward errors of a particular sign. To compute this average, it is necessary to consider a frequency distribution of the values of the unsigned significand. The usual assumption is a uniform frequency distribution,[5] in which case

$$RB = \lim_{t \to \infty} \frac{\sum_{M \in \{M_{m+t}\}} (Rmode(M) - M)}{\#M} \qquad \textbf{8.29}$$

---

4. If the signed significand is used, the bias would be zero for most rounding modes.
5. Although this uniform distribution might not occur in typical applications.

where $\{M_{m+t}\}$ is the set of all unsigned significands with $m + t$ bits, and $\#M$ is the number of significands in the set.

3. The relative representation error (RRE), defined as

$$RRE = \frac{Rmode(x) - x}{x} \qquad \textbf{8.30}$$

We now discuss further the errors and implementation characteristics of the previously defined modes. We consider the case of sign-and-magnitude representation of the significands.[6] For this, we describe x *exactly* by the triple $(S_x,\ E_x,\ M_x)$, with M_x normalized but having infinite precision.[7] Moreover, we decompose M_x into two components M_f and M_d such that

$$M_x = M_f + M_d \times r^{-f} \qquad \textbf{8.31}$$

with $0 \leq M_d < 1$. Namely, M_f has the precision of the significand in the floating-point system and M_d represents the rest.

8.2.1   Round to Nearest (Unbiased, Tie to Even)

Since in this mode the value represented is the closest possible to the exact value, it produces the smallest absolute error. Because of this, it is the default mode of the IEEE Standard.

In terms of the operation on the infinite precision significand,[8] round to nearest can be described as follows:

$$Rnear(x) = \begin{cases} M_f + r^{-f} & \text{if } M_d \geq \frac{1}{2} \\ M_f & \text{if } M_d < \frac{1}{2} \end{cases} \qquad \textbf{8.32}$$

---

6. From this discussion, the determination for other representations, such as two's complement, is straightforward.

7. Because of this infinite precision x might not be a floating-point number. On the other hand, note that x is inside the range of the floating-point numbers.

8. The definitions of the roundoff modes refer to the "infinite precision significand" (exact value). However, in the implementation of the floating-point operations, this exact value might not be obtained; consequently, it is necessary to implement the operations so that the approximation to the exact value obtained is suitable for the rounding. We discuss this further for each operation.

Equivalently, the round to nearest consists of adding $(r^{-f})/2$ to the infinite precision significand and keeping the resulting f fractional digits. That is,

$$Rnear(x) = \left(\left\lfloor \left(M_x + \frac{r^{-f}}{2} \right) r^f \right\rfloor \right) r^{-f} \qquad \text{8.33}$$

For $M_d \geq \frac{1}{2}$, the addition of r^{-f} can produce a significand that cannot be represented (significand overflow). In such a case, the resulting significand is multiplied by b^{-1} and the exponent incremented by 1.

EXAMPLE 8.1 The exact value 1.100100011101 is rounded to nearest with 8-bit precision as follows:

$$
\begin{array}{r}
1.100100011101 \\
+ \qquad\qquad 1 \\
\hline
1.10010010
\end{array}
$$ ∎

The absolute error is

$$ABRE[Rnear] = \begin{cases} -M_d r^{-f} \times b^E & \text{if } M_d < \frac{1}{2} \\ (1 - M_d) r^{-f} \times b^E & \text{if } M_d \geq \frac{1}{2} \end{cases} \qquad \text{8.34}$$

The maximum absolute error occurs when $M_d = \frac{1}{2}$, resulting in

$$MABRE[Rnear] = \frac{r^{-f}}{2} \times b^{E_{max}} \qquad \text{8.35}$$

We now consider the bias. As indicated above, the absolute errors for $M_d = a$ and for $M_d = 1 - a$ (for $a < \frac{1}{2}$) have the same magnitude but different sign. Consequently, with respect to the bias, these errors cancel each other. The only remaining case is for $a = \frac{1}{2}$, which produces a positive error. To have a bias equal to 0, this case is treated in a special manner. As indicated before, the IEEE Standard specifies that in this case the rounding is done to even.[9] That is, the

---

9. Rounding to odd in the tie case also has a bias of zero. However, round to even is preferable because it leads to less error when the result is divided by 2—a common computation.

unbiased round to nearest is

$$Rnear(x) = \begin{cases} M_f & \text{if } M_d < \frac{1}{2} \\ M_f + r^{-f} & \text{if } M_d > \frac{1}{2} \\ M_f & \text{if } M_d = \frac{1}{2} \text{ and } M_f = \text{even} \\ M_f + r^{-f} & \text{if } M_d = \frac{1}{2} \text{ and } M_f = \text{odd} \end{cases} \qquad 8.36$$

Consequently, for this mode

$$RB[Rnear] = 0 \qquad 8.37$$

This roundoff mode is illustrated in Figure 8.5(a) for $f = 2$.

In summary, round to nearest (unbiased) produces the smallest possible absolute error[10] and has a zero bias. However, the implementation of this mode requires an addition, so it is slow. We will see ways of reducing the delay for specific operations.

8.2.2   Round Toward Zero (Truncation)

In terms of the operation on the infinite precision significand, the rounded significand is obtained by discarding M_d. That is,

$$Rzero(x) = (\lfloor M \times r^f \rfloor) r^{-f} = M_f \qquad 8.38$$

The absolute error is

$$ABRE[Rzero] = -M_d r^{-f} \times b^E \qquad 8.39$$

Since $M_d < 1$ the maximum absolute error is

$$MABRE[Rzero] \approx r^{-f} \times b^{E_{max}} \qquad 8.40$$

This absolute error is larger than for round to nearest. Moreover, since for unsigned significand the absolute error is always negative, the bias is significant. Its value is

$$RB[Rzero] \approx -\frac{1}{2} r^{-f} \qquad 8.41$$

This roundoff mode is illustrated in Figure 8.5(b) for $f = 2$. The implementation of this mode is simple.

---

10. Since it selects the nearest floating-point number.

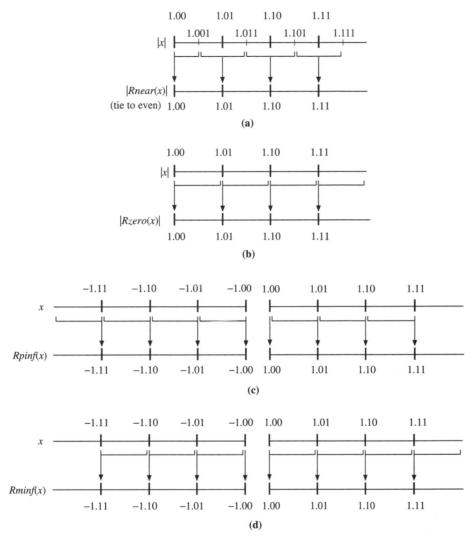

FIGURE 8.5 Rounding to (a) nearest, tie to even, (b) zero, (c) plus infinity, and (d) minus infinity.

8.2.3   Round Toward Plus and Minus Infinity

These two directed modes are useful for interval arithmetic, in which the operands and the result of an operation are intervals. This permits the monitoring of the accuracy of the result.

In terms of the infinite precision significand and the sign,

$$Rpinf(x) = \begin{cases} M_f + r^{-f} & \text{if } M_d > 0 \text{ and } S = 0 \\ M_f & \text{if } M_d = 0 \text{ or } S = 1 \end{cases} \qquad 8.42$$

$$Rninf(x) = \begin{cases} M_f + r^{-f} & \text{if } M_d > 0 \text{ and } S = 1 \\ M_f & \text{if } M_d = 0 \text{ or } S = 0 \end{cases} \qquad 8.43$$

As in the round to nearest, the addition of r^{-f} can produce a significand overflow.

These roundoff modes are illustrated in Figure 8.5(c) and (d) for $f = 2$. Note that in an implementation using sign-and-magnitude, these modes require the use of the sign. The determination of errors and bias are left as an exercise.

8.3   IEEE Standard 754

As we have seen, there are many parameters that define a floating-point representation system. This resulted in a variety of floating-point processors with different representations, producing different results to the execution of the same program. In some cases, because of anomalies, the results might be very different. To avoid this, the IEEE Floating-point Standard 754 was developed. It is claimed that this standard

- minimizes anomalies
- enhances portability
- enhances numerical quality
- allows different implementations

We now describe the main components of the IEEE Standard 754, which is used today by most floating-point processors.[11]

---

11. The reasons for the choices, as well as additional details, are presented in several of the references at the end of this chapter.

8.3.1 Representation and Formats

The two parts of the representation are as follows:

First, the *significand* is in sign-and-magnitude representation. Consequently, it is represented by two components:

- *Sign S*. One bit. $S = 1$ if negative.
- *Magnitude* (also called the significand). Represented in radix 2 with one integer bit. That is, the normalized significand is represented by

$$1.F \qquad \qquad \textbf{8.44}$$

where F of f bits (depending on the format) is called the *fraction* and the most-significant 1 is the *hidden bit*. The range of the (normalized) significand is

$$1 \le 1.F \le 2 - 2^{-f} \qquad \qquad \textbf{8.45}$$

Second, the *exponent* is base 2 and in biased representation. The number of bits of the exponent field is e, depending on the format. The representation is biased with bias $B = 2^{e-1} - 1$.

The three components are packed into one word, in which the order of the fields is S, E, F.[12] This order makes comparisons simpler.

The value zero, denormals, and the special values NAN and infinities are represented as follows:

- The representation of floating-point zero is $E = 0$ and $F = 0$. The sign S differentiates between positive and negative zero. Because of this representation and the hidden bit, the value 1.0×2^{-B} is not represented.
- The representation $E = 0$ and $F \ne 0$ is used for denormals; in this case[13] the floating-point value represented is $v = (-1)^S 2^{-(B-1)}(0.F)$.
- The maximum exponent representation ($E = 2^e - 1 = 2B + 1$) is used to represent not-a-number (NAN) for $F \ne 0$ and plus and minus infinity for $F = 0$.

The system has two formats: basic and extended. Moreover, the basic format allows representation in single and double precision. We now describe these formats.

---

12. To simplify the notation we use here E (instead of E_B) to denote the biased representation of the exponent.
13. Note that in this case the hidden bit is not used.

In each case we give the three components with the number of bits in parentheses. We call v the value represented.

1. Basic: single (32 bits) and double (64 bits)

    - Single: $S(1)$, $E(8)$, $F(23)$

        (a) If $1 \leq E \leq 254$, then $v = (-1)^S 2^{E-127}(1.F)$ (normalized fp number).
        (b) If $E = 255$ and $F \neq 0$, then $v = NAN$ (not a number).
        (c) If $E = 255$ and $F = 0$, then $v = (-1)^S \infty$ (plus and minus infinity).
        (d) If $E = 0$ and $F \neq 0$, then $v = (-1)^S 2^{-126}(0.F)$ (denormal, gradual underflow).
        (e) If $E = 0$ and $F = 0$, then $v = (-1)^S 0$ (positive and negative zero).

    - Double: $S(1)$, $E(11)$, $F(52)$

        -   Similar representation to single, replacing 255 by 2047, and so on.

2. Extended: single (at least 43 bits $= S(1)$, $E(11)$, $F(31)$) and double (at least 79 bits $= S(1)$, $E(15)$, $F(63)$).

8.3.2   Rounding

Rounding modes are:

- Default: Round to nearest, to even when tie
- Directed: Round toward plus infinity; Round toward minus infinity; and Round toward 0 (truncate)

8.3.3   Operations

Operations include:

- Numerical: Add, Sub, Mult, Div, Square root, Rem
- Conversions: Floating to integer; Binary to decimal (integer); Binary to decimal (floating)
- Miscellaneous: Change formats; Compare and set condition code

8.3.4 Exceptions

The IEEE standard defines the following five exceptions. By default these exceptions set a flag and the computation continues. The implementation can include a trap handler for each exception that, when enabled, is called when an exception occurs.

- Overflow (when rounded value is too large to be represented). Result is set to ±infinity.
- Underflow (when rounded value is too small to be represented).
- Division by zero.
- Inexact result (result is not an exact floating-point number). Infinite precision result different from floating-point number.
- Invalid. This flag is set when a NAN result is produced.

8.4 Floating-Point Addition

We now consider the algorithm and implementations for floating-point addition. The algorithm is given in generic terms, whereas the implementations are tuned to the IEEE standard.

Let x and y be the operands represented by (S_x, M_x, E_x) and (S_y, M_y, E_y), respectively. The significands are normalized. We consider addition or subtraction, so that the result

$$z = x \pm y$$

is represented by (S_z, M_z, E_z), where M_z is also normalized. Let $M_x^* = (-1)^{S_x} M_x$ and define similarly M_y^* and M_z^*. The high-level description of this operation is composed of the following four steps:

1. Add/subtract significand and set exponent:

$$M_z^* = \begin{cases} (M_x^* \pm (M_y^* \times b^{(E_y - E_x)})) \times b^{E_x} & \text{if } E_x \geq E_y \\ ((M_x^* \times b^{(E_x - E_y)}) \pm M_y^*) \times b^{E_y} & \text{if } E_x < E_y \end{cases} \qquad \textbf{8.46}$$

$$E_z = \max(E_x, E_y) \qquad \textbf{8.47}$$

That is, the significand of the number with the smallest exponent has to be multiplied by b to the power of the difference between the exponents (this operation is called *alignment*) and then added/subtracted to the other significand. This is illustrated as follows:

$$E_x - E_y = 4$$

M_x	1.xxxxxxxxxxx
$M_y\left(2^{(E_y-E_x)}\right)$	0.0001yyyyyyyyyy

z.zzzzzzzzzzzzzzz

The exponent of the result is equal to the largest of the exponents of the operands.

2. Normalize significand and update exponent. The result of Step 1 might be unnormalized (as described later). Consequently, it has to be normalized and the exponent has to be updated accordingly.

3. Round, normalize, and adjust exponent.

4. Set flags for special cases.

We now give the basic algorithm corresponding to this description.

8.4.1   Basic Algorithm

The above high-level description results in the following basic algorithm:

1. **Subtract exponents** $(d = E_x - E_y)$.

2. **Align significands**. This consists of the following:
   - Shift right d positions the significand of the operand with the smallest exponent.
   - Select as the exponent of the result the largest exponent.

3. **Add (subtract) significands and produce sign of result**. This is a signed addition. The effective operation (add or subtract) is determined by the floating-point operation (ADD or SUBTRACT) and the signs of the operands, as follows:

Floating-Point Operation	Signs of Operands	Effective Operation (EOP)
ADD	equal	add
ADD	different	subtract
SUBTRACT	equal	subtract
SUBTRACT	different	add

From now on we refer to the effective operation.

The sign of the result depends on the signs of the operands, the operation, and the relative magnitude of the operands (see exercise 8.15).

4. **Normalization of result.** Three situations can occur:

(a)   The result is already normalized: no action is needed.

$$
\begin{array}{r}
1.10011111 \\
0.00101011 \\
\text{ADD} \quad \overline{} \\
1.11001010
\end{array}
$$

(b)   When the effective operation is an addition, there might be an overflow of the significand. The normalization consists of the following:

   •   Shift right the significand one position.
   •   Increment by one the exponent.

$$
\begin{array}{r}
1.1001111 \\
0.0110110 \\
\text{ADD} \quad \overline{} \\
10.0000101 \\
\text{NORM} \quad 1.00000101
\end{array}
$$

(c)   When the effective operation is subtraction, the result might have leading zeros. The normalization consists of the following:
   •   Shift left the significand by a number of positions corresponding to the number of leading zeros.
   •   Decrement the exponent by the number of leading zeros.

$$
\begin{array}{r}
1.1001111 \\
1.1001010 \\
\text{SUB} \quad \overline{} \\
0.0000101 \\
\text{NORM } 1.0100000
\end{array}
$$

5. **Round.** Perform the rounding according to the specified mode. This might require an addition. If an overflow occurs because of this addition, it is necessary to normalize by a right shift and increment the exponent.

6. **Determine exception flags and special values.** Exponent overflow (special value $\pm$ infinity), exponent underflow (special value gradual underflow), inexact, and the special value zero.

8.4.2   Basic Implementation

The previous algorithm is implemented by the block diagram of Figure 8.6. To make the description more specific, we consider a representation of the type of the IEEE standard. Namely:

-   The significand is normalized and represented in sign-and-magnitude. The magnitude is $M = 1.F$, where the 1 corresponds to the hidden bit and is appended to the fraction at the beginning of the operation.
-   The base of the exponent is 2. This results in the use of radix-2 shifters.[14]

Note the following:

1. To have only one alignment shifter, it is convenient to swap the significands of the operands, according to the sign of the exponent difference.

2. The adder is a sign-and-magnitude adder. Since direct implementation of sign-and-magnitude addition is complicated, several options using two's complement addition are presented in Section 8.4.7.

---

14. Although the IEEE standard uses a bised representation for the exponents, in all operations we give the descriptions using the unbiased value. When using biased representations we include the subscript B.

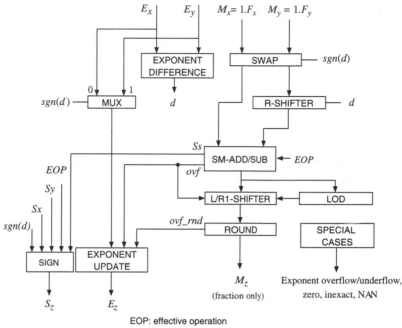

FIGURE 8.6 Basic implementation of floating-point addition.

3. The normalization step requires

   - the detection of the position of the leading 1, done with the block labeled LOD (Leading-One-Detector)
   - a shift performed by the shifter (no shift, right shift of one position, or left shift of up to m positions)
   - the appropriate updating of the exponent

4. The rounding step uses several guard bits as discussed in the next section. The overflow due to rounding results in the significand of the form 10.000...0. A right shift to produce the correct significand 1.000..0 leaves the fraction part unchanged and requires no implementation since the integer bit (hidden bit) is implicit. Of course, the exponent is incremented in the case of rounding overflow.

8.4.3   Guard Bits and Rounding

Because of the right shift of one of the operands during the alignment step, the result of the addition/subtraction may have more fractional bits than the operands. Moreover, during the normalization a left shift of the result might be performed. Finally, during the rounding step these additional bits are disposed of and the result has a significand of f fractional bits.

To get the correct final result after the normalization and rounding, a possibility is to obtain all the fractional bits of the addition. However, we now show that this is not necessary, and a few additional fractional bits are sufficient. These additional bits are called *guard bits*.

To determine the number of guard bits, we first review the requirements for the normalized result of the addition for the different rounding modes:

- For rounding toward zero (truncation), only the f fractional bits are required.
- For rounding to nearest, one additional bit is required ($f + 1$ fractional bits). Moreover, for unbiased rounding to even, it is necessary to know when the rest of the bits are all zero.
- For rounding toward infinity, it is necessary to know when all the bits to be discarded are zero.

In summary, to be able to perform any of the rounding modes, $f + 1$ fractional bits of the normalized result are required, plus the indication of whether the rest of the fractional bits of the normalized result are all zero.

So the question we consider now is how many bits have to be produced by the effective addition/subtraction before normalization. We consider two cases: effective addition and effective subtraction.

In effective addition, the result of the addition is either normalized or produces one additional integer bit. Consequently, the normalization might require a 1-bit right shift and no left shift is required.

Therefore, $f + 1$ fractional bits of the result are required. Moreover, it is necessary to determine whether all the discarded bits are zero. Since these discarded bits are produced by the alignment and are obtained by adding 0 to the bits shifted out by the alignment, it is sufficient to determine whether all the bits shifted out by the alignment are 0. This situation is represented by the *sticky bit T*, which corresponds to the OR of the discarded bits.

EXAMPLE 8.2

$$1.0101110$$
$$0.00010101010$$

ADD ——————————

$$1.01110001 \qquad T = OR(010) = 1$$ ∎

The second case is effective subtraction. Here we consider two subcases:

1. The difference of exponents d is larger than 1. Then, as shown in the following example, the smallest operand is aligned so that there are more than one leading zeros. As a consequence, the result of the subtraction is either normalized or, if not normalized, has only one leading zero. Since, for this last case, the normalization is performed by a left shift of one position, in addition to the bit for rounding to nearest, another bit is required in the result of the addition. Consequently, $f + 2$ fractional bits of result are required.

   Moreover, during the subtraction, a borrow into position $f + 2$ is produced if any of the shifted-out bits is different from zero. This borrow is determined by the sticky bit, defined in the previous situation, which also serves for the unbiased rounding to nearest and for rounding toward infinity.

   Therefore, the width of the subtraction has to be of $f + 3$ fractional bits, the last bit being the sticky bit.

EXAMPLE 8.3 After alignment:

$$1.0000011$$
$$0.000011011001$$

SUB ——————————

During alignment compute $T = OR(001) = 1$ resulting in

$$1.0000011$$
$$0.0000110111$$

SUB ——————————

$$0.1111100001$$

NORM $\quad 1.1111000010$ ∎

2. The difference of exponents is either 0 or 1. Only in this case, the result might have more than one leading zeros (up to m). Consequently, a left shift of up to m positions is required. However, as shown in the example, since the alignment shift was only of zero or one position, at most one nonzero bit is shifted in during the normalization. Consequently, only one additional bit is required in the result of the subtraction.

EXAMPLE 8.4

$$
\begin{array}{r}
1.0000011 \\
0.11111001 \\
\text{SUB} \quad \overline{} \\
0.00001101 \\
\text{NORM} \quad 1.10100000
\end{array}
$$
■

In summary, in all cases it is sufficient to perform the addition with three additional bits. These are called guard (G), round (R), and sticky (T).

We now consider the use of these bits to perform the different rounding modes. First the normalization is performed and the bits after normalization are labeled as follows:[15]

$$
\begin{array}{c}
\text{LGRT} \\
1.\text{XXXXXXXXXXXX} \\
\underline{f}
\end{array}
$$

Note that during the shift right of normalization the sticky bit has to be recomputed as the OR of the previous value of T and the previous value of R.

Round to Nearest (Tie to Even)

As indicated by (8.36) in rounding to nearest mode we round up (add 1 to position L) if $G = 1$ and R and T are not both 0, and round to even if $G = 1$ and

---

15. Actually, after normalization only two bits are needed if the sticky bit is recomputed as $R + T$.

$R = T = 0$. Consequently, calling *rnd* the value to be added to bit L, we get

- If $G = 1$ and R and T are not both zero, $rnd = G(R + T)$.
- If $G = 1$ and $R = T = 0$ then $rnd = G(R + T)'L$.

Combining both cases,

$$rnd = G(R + T) + G(R + T)'L = G(L + R + T) \qquad \textbf{8.48}$$

Note that in this implementation, we need to determine first the bits L, G, R, T to compute *rnd* and perform the increment. In some implementations it might be preferable to do as follows:

1. Always add 1 in position G (which produces the rounding to nearest).
2. Correct the bit in position L if there is a tie. Namely, make $L = 0$ if $G(R + T)' = 1$.

Round toward Zero

In this case the result after normalization is truncated at bit L.

Round toward Infinity

For round toward positive infinity, add one to L when the sign is positive and G, R, and T are not all zero. That is,

$$rnd = sgn'(G + R + T) \qquad \textbf{8.49}$$

where *sgn* is the sign of the result.

Similarly for round toward negative infinity:

$$rnd = sgn(G + R + T) \qquad \textbf{8.50}$$

8.4.4 Exceptions and Special Values

We now discuss the exceptions and special values that may occur in floating-point addition and subtraction.

- **Overflow:** This situation can occur when the exponent is incremented during normalization (because of overflow of addition requiring a right shift of significand) and because of overflow of significand during the rounding step. It is detected by an exponent $E \geq 255$. The overflow flag is set, and the result is set to $\pm$infinity.

- **Underflow:** This situation can occur when the exponent is decremented during normalization (left shift of significand). The underflow flag is set, and the result exponent is set to $E = 0$. The fraction is left unnormalized (denormal, gradual underflow).
- **Zero:** This situation occurs when the significand of the result of addition is 0. The result is $E = 0$ and $F = 0$.
- **Inexact:** This situation is detected before the rounding; the result is inexact if $G + R + T = 1$. The inexact flag is set.
- **NAN:** If one operand (or both) is a NAN, then the result is set to NAN.

8.4.5   Denormal and Zero Operands

When an operand is a denormal number ($E = 0$ and $F \neq 0$), then there is no hidden 1. Consequently, the operand of addition should be set to $E = 1$ and $0.F$. The rest of the algorithm remains unchanged.

The zero operand ($E = 0$ and $F = 0$) is treated in the same way as a denormal number.

8.4.6   Delay and Pipelining

The delay, or latency, of the floating-point addition corresponds to the critical path, obtained from the delay graph shown in Figure 8.7. As in any combinational network, the critical path might not be the sum of the critical paths of the individual modules. However, if networks with logarithmic delay are used for the adders and the LOD, a reasonable first approximation is to add the critical path delays. Since the delay is large and floating-point additions are frequent, the unit is usually pipelined. The number of stages depends on the clock cycle.

8.4.7   Alternative Implementations

There are several modifications that have been developed for the implementation of floating-point addition. The main objective of these modifications is to reduce the latency, and the approach is to combine mutually exclusive steps and/or to perform in parallel independent steps. Since a variety of possibilities exist, we illustrate the approaches by two designs, one single-path implementation and

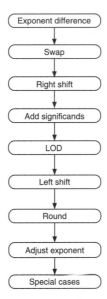

FIGURE 8.7 Dependence graph for basic implementation of floating-point addition.

one double-path implementation. Other variations are described in the references listed at the end of the chapter.

Single-Path Implementation[16]

This implementation is shown in Figure 8.8. It results from the following changes to the basic implementation considered before:

1. The sign-and-magnitude addition is performed using a two's complement adder. When an effective subtraction is performed, one of the operands is complemented (bit-inversion put carry-in to the adder) and the result is complemented if negative. To avoid the complementation of the result, which would require a bit-inversion and the addition of one ulp, the

---

16. Although this implementation has two partial paths, it is called a "single-path" implementation to distinguish it from the next one, which has two completely distinct significand datapaths.

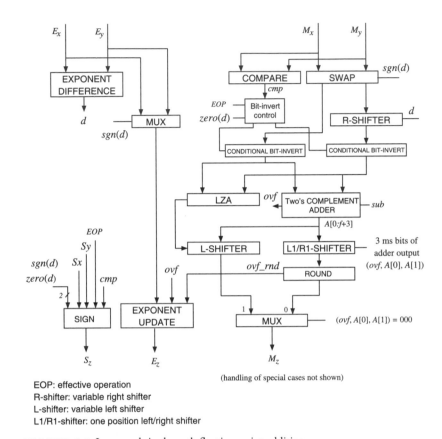

FIGURE 8.8 Improved single-path floating-point addition.

smallest operand is complemented so that the result is always positive. To determine the smallest operand, two cases are considered:

- The exponents of the two operands are different. In this case, during alignment, the significand corresponding to the operand with smallest exponent is shifted right. Consequently, this corresponds to the smallest operand to the adder and is complemented.
- The exponents of both operands are the same. In this case, it is necessary to compare the significands. This comparison is performed in parallel with the alignment module, so it does not increase the delay.

2. Leading-zeros anticipation (LZA).[17] This module determines the position of the leading one in the result, concurrently with the actual addition. In this way, it eliminates the delay of the leading-one detector from the critical path.

3. Performing the rounding in parallel with the massive left shift. The massive left shift (more than one position) is required only when the output of the adder has at least two leading zeros (actually, three including the carry-out). As discussed before, this can only occur when there is an (effective) subtraction and the difference in exponents is 0 or 1. Therefore, the shift left of at least two positions introduces a 0 into bits G, R, and T, and no roundup is required.

EXAMPLE 8.5

$$
\begin{array}{r}
1.0000101 \\
0.11111101 \\
\mathrm{SUB} \quad \rule{2cm}{0.4pt} \\
0.00001101
\end{array}
$$

NORM   1.11010000 = Round down in all modes   ∎

Consequently, two concurrent paths can be designed, as follows:

- Path 1, including a 1-bit left/right shift (to incorporate also normalization in effective addition) and the rounding
- Path 2, massive left shift (2 or more positions)

After these paths a selection is performed so that the left path in the figure is chosen when the three most-significant bits of the output of the adder (including the carry-out) are 0.

Note that in this implementation we have included the comparator and the bit inverters, required by the scheme that uses a two's complement adder with an output that is always positive.

---

17. This has also been called *leading-one prediction* (LOP). For algorithms and implementations, see the references at the end of the chapter.

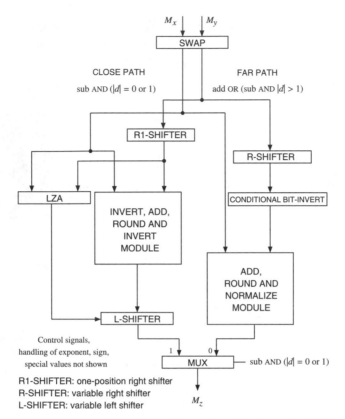

FIGURE 8.9 Double-path implementation of floating-point addition.

Double-Datapath Implementation

In the single-datapath implementation the critical path includes two variable shifters: one for alignment of the operands and the other for normalization of the result. However, as indicated before, the normalization of the result requires a shift of more than one position only when the operation is subtraction and the exponent difference is zero or one; moreover, for this case the alignment is at most of one position. Consequently, as shown in Figure 8.9, it is possible to define two disjoint paths:

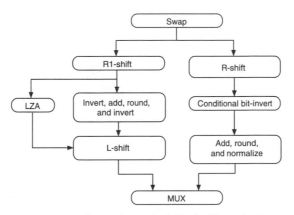

FIGURE 8.10 Dependence graph for double-path scheme.

- CLOSE, for subtraction and exponent difference of zero or one
- FAR, for addition and for subtraction with exponent difference larger than one

In the CLOSE path, there is a simple shifter for alignment of at most one position, the adder, the variable left shifter for normalization, and the module for rounding.[18] In contrast, the FAR path has a variable right shifter for alignment, the adder, a one-position left-right shifter for normalization, and the module for rounding.

The dependence graph of the double-path scheme (significand part only) is shown in Figure 8.10.

To balance the delay of both paths, the following has to be considered:

1. To achieve a higher throughput, the floating-point adder is pipelined, as shown in Figure 8.11. As can be seen, to pipeline the double-datapath implementation it is necessary to have two adders, one per path, because the addition occurs in different stages of the pipeline: in the CLOSE path simultaneously with the variable right shifter of the FAR path, and in the FAR path simultaneously with the variable left shifter of the CLOSE path.

---

18. Note that in Figure 8.9 (for both paths) the rounding is included in the module together with the adder and is performed before the normalization; this is explained later.

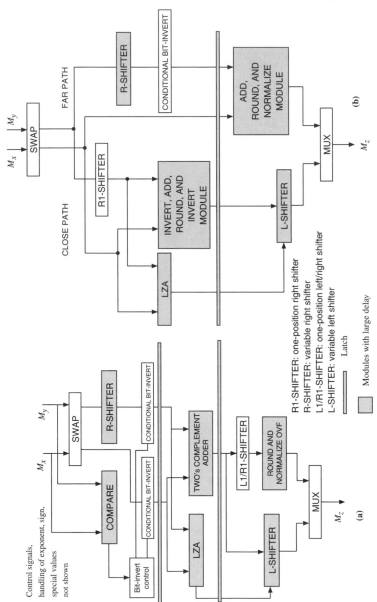

FIGURE 8.11 Pipelined implementations: (a) Single-path scheme. (b) Double-path scheme.

2. To reduce the latency, the rounding is combined with the adder and performed before the normalization. This combined addition + rounding is performed by having a compound adder (which produces the sum and the sum plus 1) and the correct rounded result is selected from the two possible outputs. Specifically:

- For the CLOSE path, roundup might be required when the exponent difference is one and the output of the adder is normalized.

EXAMPLE 8.6

$$
\begin{array}{r}
1.1100100 \\
0.10000001 \\
\text{SUB} \quad \rule{3cm}{0.4pt} \\
1.01000111 \\
\end{array}
$$

ROUND   1.0100100   ∎

- For the FAR path, roundup might be required both when the result of addition is normalized or unnormalized (with one additional integer bit for addition and with one leading zero for subtraction). Since the rounding is done before normalization, the selection of the correct output has to take into account the various positions of the rounding bit.

To illustrate the operation of the ADD, ROUND, and NORMALIZE module we show the case of effective addition and rounding to nearest.[19] In this case, the ADD part produces two outputs: *Sum* (of inputs) and *Sum + one* (sum-plus-one), both up to bit position $m - 1$ (L). Moreover, we have also bits m (G), $m + 1$ (R), and the sticky bit (T), which correspond to the operand that has been

---

19. The complete description can be obtained from references at the end of this chapter. The operation has additional complications for subtraction because of the 1 to be added for two's complement, and for round toward infinity when overflow occurs because an additional output of sum plus two is needed.

shifted right. Then, for the selection among the two outputs of the adder, two situations have to be considered:

– The result of addition (*Sum*) is normalized. In this situation, the rounded result is (*L* is bit position $m - 1$ of *Sum*)

$$rounded = \begin{cases} Sum & \text{if } (G' + L'R'T') = 1 \\ Sum + one & \text{if } G(R + T + L) = 1 \end{cases} \qquad \textbf{8.51}$$

– The result of addition (*Sum*) has one additional integer bit. In this situation, the rounded result is (L^* is position $m - 2$ of *Sum*)

$$rounded = \begin{cases} R1SHIFT(Sum) & \text{if } (L' + (L^*)'G'R'T') = 1 \\ R1SHIFT(Sum + one) & \text{if } L(L^* + G + R + T) = 1 \end{cases}$$
$$\textbf{8.52}$$

3. The magnitude subtraction is performed with a two's complement adder, as discussed for the single-datapath case. However, for the case in which the exponents are the same (CLOSE path) it is not possible to perform a comparison of the significants before the adder, since this adder is in the first stage of the pipeline. Consequently, to avoid the two's complement of the result when it is negative (which would require an incrementer), we do as follows:

    • Bit-invert one operand.
    • Select the sum plus one output if the result is positive and the bit invert of the sum if the result is negative.

    Note that because of the swap at the input, the difference can be negative only when the exponents are equal, and therefore this situation does not conflict with the roundup.

    In the FAR datapath, the swap assures that the output is always positive.

4. Leading-zeros anticipation (LZA) is included in the CLOSE path (as discussed for the single-datapath case).

As seen in Figure 8.11, the use of a double-path implementation might reduce the latency by one pipeline stage. However, it increases significantly the area.

8.5   Floating-Point Multiplication

We now consider the algorithm and implementations for floating-point multiplication. The algorithm is given in generic terms, whereas the implementations are geared to the IEEE standard.

Let x and y be the operands represented by (M_x^*, E_x) and (M_y^*, E_y), respectively. The significands are signed and normalized, and the result

$$z = x \times y \qquad\qquad 8.53$$

is represented by (M_z^*, E_z), where M_z^* is also signed and normalized. The high-level description of this operation is composed of the following four steps:

1. Multiply significands and add exponents.

$$M_z^* = M_x^* \times M_y^* \qquad\qquad 8.54$$
$$E_z = E_x + E_y \qquad\qquad 8.55$$

2. Normalize M_z^* and update exponent.

3. Round.

4. Determine exception flags and special values.

8.5.1   Basic Implementation

As in addition/subtraction we now consider an implementation in which operands and result significands are in sign-and-magnitude representation. Moreover, to be specific we consider the representation of the significand as in the IEEE Standard 754, that is, normalized and in the range $[1, 2)$. The number of bits of the significand is $m = f + 1$ bits, of which the most-significant bit is hidden.

The first three steps are implemented as follows:

1. Multiplication of magnitudes, addition of exponents, and generation of sign.

    *   Multiplication of magnitudes produces a magnitude P of $2m$ bits. Since only m bits are required in the result, from the second half we require only one guard bit and the sticky bit, for rounding. No additional guard bit is necessary, as discussed below.

- The implementation of the addition of the exponents depends on the representation. In a biased representation, the addition is performed by adding the representation of the exponents and subtracting the bias. That is,

$$E_{B,z} = E_{B,x} + E_{B,y} - B \qquad \textbf{8.56}$$

- The sign of the result is

$$S_z = S_x \oplus S_y \qquad \textbf{8.57}$$

2. Normalization. Since $1 \le M_x, M_y < 2$, the result of the multiplication is in the range $[1, 4)$. Consequently, it might be necessary to normalize by shifting one position to the right and incrementing the exponent.

   Since no normalization left shift is required, the result of multiplication requires only one guard bit (and the sticky bit) for rounding. Note that the sticky bit has to be updated during the normalization shift, so that the new sticky bit is equal to the OR of the previous guard bit and the previous sticky bit. The output of multiplier module P is in positions:

$$(-1)0.123\ldots(m-2)(m-1)m(m+1)\ldots(2m-2) \qquad \textbf{8.58}$$

If $P[-1] = 0$, P is normalized:

$$L = P[m-1], \quad G = P[m], \quad T = OR(P[m+1], \ldots, P[2m-2])$$
$$\textbf{8.59}$$

If $P[-1] = 1$, normalize P by shifting right one position:

$$L = P[m-2], \quad G = P[m-1], \quad T = OR(P[m], \ldots, P[2m-2])$$
$$\textbf{8.60}$$

3. Rounding. The four rounding modes are implemented as in floating-point addition, but now with only one guard bit (G) and the sticky bit (T).

Round to Nearest (to Even if Tie)

The rounding is done by adding *rnd* to the L position (least-significant position of the result), where

$$rnd = GT + GT'L = G(T + L) \qquad \textbf{8.61}$$

with G and T being the two bits following L *after* the normalization.

As discussed for floating-point addition, it is also possible to do the rounding by adding 1 to the G position (after normalization) and updating the L bit for a tie. This method will be used in the modified implementations discussed later.

Round toward Zero

In this case the result after normalization is truncated at bit L.

Round toward Infinity

For round toward positive infinity, add *rnd* to L for

$$rnd = sgn'(G + T) \qquad\qquad \textbf{8.62}$$

where *sgn* is the sign of the result.

Similarly for round toward negative infinity:

$$rnd = sgn(G + T) \qquad\qquad \textbf{8.63}$$

Figure 8.12 illustrates the basic implementation discussed above.

8.5.2 Exceptions and Special Values

The exceptions and special values that happen in floating-point multiplication are:

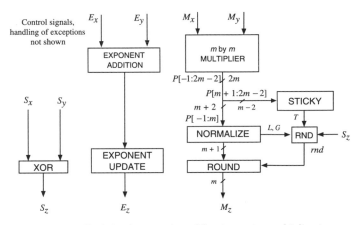

FIGURE 8.12 Basic implementation of floating-point multiplication.

- **Overflow:** This situation can occur because the resulting exponent is too large. This is detected after the exponent update. The overflow flag is set and the result value is $\pm$infinity.
- **Underflow:** The resulting exponent can be too small to be represented. In such a case, the underflow flag is set and the exponent is set to $E = 0$. Moreover, the significand is shifted right to represent a denormal.
- **Zero:** The result of multiplication is zero when one of the operands has value 0 and the other is not $\pm$infinity. The zero result is set.
- **Inexact:** The result is inexact if, after normalization, $G + T = 1$.
- **NAN:** The result is a NAN if one (or both) of the operands is a NAN or if one of the operands is a 0 and the other $\pm$infinity.

8.5.3   Denormals

As in addition, denormal operands do not have a hidden 1. When one (or both) operands are denormal, then the output of the multiplier will have leading zeros. Consequently, a variable left shift is necessary for normalization, as in floating-point addition (and a subtraction in the exponent).

When there is an exponent underflow, the significand is shifted right to allow for gradual underflow (denormal result) and the exponent is set to 0.

8.5.4   Delay and Pipelining

The delay, or latency, of floating-point multiplication corresponds to the sum of the delays of the modules in the significand path of Figure 8.12. To increase the throughput, the unit is usually pipelined. The number of stages depends on the clock rate. Since the multiplier module has a larger delay than the other modules, it might be decomposed into several components, such as recoder, adder tree, and final adder, and these components included in different pipeline stages.

8.5.5   Alternative Implementation

To reduce the latency of floating-point multiplication, the following items can be included, as shown in Figure 8.13:

- Computing only the most-significant half (plus the guard bit) of the result of multiplication in conventional representation and simplifying the computation of the effect of the second half on this first half.

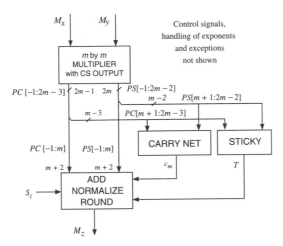

FIGURE 8.13 Alternative implementation.

One way to do this is to use the standard approach to implement multipliers in which the bit array is first reduced to two rows (carry-save representation of product) and then this carry-save representation is converted to a conventional representation. As shown in Figure 8.13, for the floating-point multiplier only the first part is converted, including the carry c_m produced by the second part. In this case, the delay of generating c_m is in the critical path.

- Overlapping the computation of the sticky bit with the multiplication. The basic way to compute the sticky bit requires that the second half of the product is produced in conventional representation. This would require a carry-propagate adder for the last part. The following two methods eliminate the need for this adder:

    1. The sticky bit can be determined directly from the operands for the multiplication. This results from the fact that for a prime radix, such as radix 2, the number of trailing zeros of the product is equal to the sum of the trailing zeros of the operands. Consequently, the value of the sticky bit is obtained from this sum. The implementation of this method requires detectors of the number of trailing zeros, an adder, and a comparator.

    2. The sticky bit can be determined from the carry-save representation of the second half of the product. This method is based on the general

method to determine whether the sum of two operands is zero, without actually performing the addition. This method can be described by adding -1 to the sum and detecting the value -1. Since the representation of -1 is $1111..1$ and this value can only occur when for all bit positions the sum bit plus the carry bit add to 1, the algorithm is as follows:

$$
\begin{array}{ll}
S & s\ s\ s\ s\ s\ s\ s\ s \\
C & c\ c\ c\ c\ c\ c\ c\ c \\
-1 & 1\ 1\ 1\ 1\ 1\ 1\ 1\ 1 \\
\hline
& z\ z\ z\ z\ z\ z\ z\ z \\
& t\ t\ t\ t\ t\ t\ t
\end{array}
$$

Consequently,

$$
\begin{aligned}
z_i &= (s_i \oplus c_i)' \\
t_i &= s_{i+1} + c_{i+1}
\end{aligned}
\qquad \textbf{8.64}
$$

Now we compute

$$
w_i = z_i \oplus t_i \qquad \textbf{8.65}
$$

and the sticky bit is

$$
T = NAND(w_i) \qquad \textbf{8.66}
$$

- Combining in one module the carry-propagate addition (with inclusion of the carry from the second part), the normalization, and the rounding.

    A method to reduce the delay by removing the carry from the second half from the critical path is described at the end of this section.

    As shown in Figure 8.14, the carry from the second part (c_m) is added at the guard bit position.

Now consider the rounding. We consider rounding to nearest (up if a tie) and then include the effect of sticky. The other rounding modes are left as an exercise (Exercise 8.31). Since we want to perform the rounding before the normalization, we need to consider two situations:

Product P[− 1:2m − 2]

——(m + 2)——   ——(m − 2)——

xxxxxxxxxxxxxxxx xxxxxxxxxxxxxxx

xxxxxxxxxxxxxxxx xxxxxxxxxxxxxxx

c_m

c_m is the carry produced by
the least-significant m−2 bits of product P
and added in position m.

FIGURE 8.14 Adding carry from the least-significant half.

Bit position: (−1) (0) .123. . . (m−2) (m−1) m

  0   1 .xxx. . .   x   x   x

c_m

1

(a)

Bit position: (−1) (0) .123. . . (m−2) (m−1) m

  1   x .xxx. . .   x   x   x

1   c_m

(b)

FIGURE 8.15 Rounding position: (a) Normalized product. (b) Unnormalized product.

1. The product is normalized. Then the rounding is performed by adding 1 to bit position *m* (see Figure 8.15(a)).

2. The product is not normalized (that is, it has to be shifted 1 position right for normalization). Then the rounding is performed *before normalization* by adding 1 to bit position *m* − 1, as shown in Figure 8.15(b).

Since the product is in carry-save form, it is not known whether it is normalized. Therefore, to combine the rounding with the addition, both additions should be performed and then the correct one selected when it is determined whether the result is normalized.

	(-1)	$0.$	1	2	3	$\dots$	$(m-2)$	$(m-1)$	(m)
PS	x	x	x	x	x		x	x	x
PC	x	x	x	x	x		x	x	x
								c_m	c'_m $<=>(c_m+1)2^{-m}$

	(-1)	$0.$	1	2	3	$\dots$	$(m-2)$	$(m-1)$	(m)
PS*	x	x	x	x	x		x	x	x
PC*	x	x	x	x	x		x	x	

Get P0 and $P1 = P0 + 2^{-m}$:

	(-1)	$0.$	1	2	3	$\dots$	$(m-2)$	$(m-1)$	(m)
PS*	x	x	x	x	x		x	x	x
PC*	x	x	x	x	x		x	x	0

	(-1)	$0.$	1	2	3	$\dots$	$(m-2)$	$(m-1)$	(m)
P0	ovf	x	x	x	x		x	x	x
P1	x	x	x	x	x		x	x	x

After selection:

P	1. x x x $\dots$ x	L

FIGURE 8.16 Adding carry c_m and rounding.

Consequently, calling *PM* the most-significant part of the product, obtained by adding *PS* plus *PC* up to position m, we need to compute

$$P0 = PM + (c_m + 1) \times 2^{-m} \qquad\qquad \textbf{8.67}$$

and

$$P1 = PM + (c_m + 2) \times 2^{-m} \qquad\qquad \textbf{8.68}$$

and then select

$$P = \begin{cases} P0 & \text{if } P0[-1] = 0 \\ 2^{-1}P1 & \text{if } P0[-1] = 1 \end{cases} \qquad\qquad \textbf{8.69}$$

This is illustrated in Figure 8.16.

The complete process can be implemented as shown in Figure 8.17, which consists of the following parts:

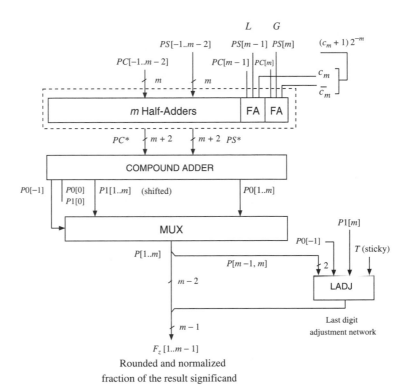

FIGURE 8.17 Adding carry, normalization, and rounding implementation.

1. A row of HAs and FAs to add $(c_m + 1)2^{-m}$ to $PS[-1, m]$ and $PC[-1, m]$.

2. A compound adder that produces the sum $P0$ and the sum plus 1 ($P1$).

3. A multiplexer that selects $P0$ or the normalized (shifted) $P1$ depending whether $P0$ does not overflow or overflows, respectively.

4. A module LADJ that determines the least-significant bit of the significand. The tie situation (round to even) needs the use of the sticky bit. This sticky bit includes the guard bit when there was an overflow in the addition. Consequently, the sticky bit is updated by

$$T^* = T + P1[m] \cdot P0[-1] \quad \text{update sticky bit} \qquad \textbf{8.70}$$

The expression for the adjustment of the least-significant bit in the tie case is based on the fact that the 1 added for rounding complements the bit in position m. So, if tie corresponds to $(bit_m, T) = 10$ *before* rounding, it corresponds to $(bit_m, T) = 00$ *after* rounding. Consequently,

$$L = P[m-1](P[m] + T^*) \qquad\qquad 8.71$$

Removing c_m from Critical Path

The carry of the second half can be taken out of the critical path as follows. As indicated, it is necessary to add to the position m of the carry-save representation of the product either $(c_m + 1)$ or $(c_m + 2)$, depending on the overflow condition. Then a compound adder produces the sum and the sum plus one, and the correct result is selected. Consequently, taking into account the sum and carry bits in position m, it is necessary to produce a value (Σ) from 1 to 5 in that position. This addition produces a carry (c_{m-1}) to position $m - 1$, which can have a value from 0 to 2. Therefore, a direct implementation of this would require the computation of the sum, sum plus one, and sum plus two, and the selection after Σ is known. This implementation is not convenient because of the three values required.

The implementation can be simplified to select only among sum and sum plus one because three of the five bits that compose Σ are known in advance, specifically the sum and carry bits of the carry-save representation and the one added for rounding to nearest. In order to reduce the values of the carry c_{m-1} to $(0, 1)$, a 1 is preadded to the carry-save representation in position $m - 1$ (reducing the range of Σ by 2). The different cases are shown in Table 8.4.

The preaddition requires a row of half-adders as shown in Figure 8.18. The control of selection is done using the four bits composing Σ (see Exercise 8.33). Note that, in contrast to the previous implementation, in this case a right shifter is included after the selection. This is because, when there is overflow, P1 is not always selected.

Carry + Sum in Position m	Range of Σ before Preadd	Range of c_{m-1}	Preadd 1?	Range of Σ after Preadd	Range of c_{m-1}
0	$[1, 3]$	$[0, 1]$	no	$[1, 3]$	$[0, 1]$
1	$[2, 4]$	$[1, 2]$	yes	$[0, 2]$	$[0, 1]$
2	$[3, 5]$	$[1, 2]$	yes	$[1, 3]$	$[0, 1]$

TABLE 8.4 Preaddition cases.

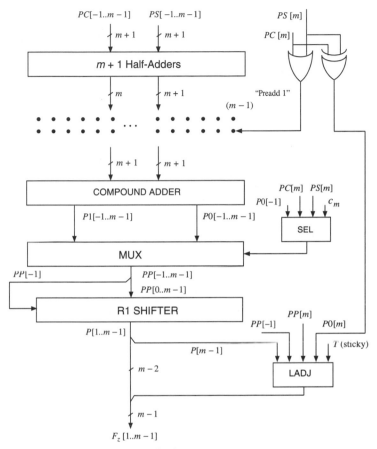

FIGURE 8.18 Adding carry, normalization, and rounding implementation with carry out of critical path.

8.5.6 Floating-Point Multiply-Add Fused (MAF)

We describe the algorithm and implementation for fused floating-point multiplication and addition $z = xy + w$. This operation reduces the number of interconnections between floating-point units, the number of adders and normalizers, and provides additional accuracy compared to separate multiply and add units. The increased accuracy is a consequence of having to perform a single, instead of two, round/normalize steps. It also helps compilers produce more efficient code. On the other hand, it increases the precision and delay of the adder and requires a more complex normalizer. The MAF unit can also be used to perform floating-point addition and floating-point multiplication by setting $y = 1.0$ or $w = 0.0$, respectively.

The algorithm is given in generic terms, whereas the implementations are geared to the IEEE standard. Let x, y, and w be the operands represented by (M_x^*, E_x), (M_y^*, E_y), and (M_w^x, E_w), respectively. The significands are signed and normalized, and the result

$$z = (x \times y) + w \qquad\qquad 8.72$$

is represented by (M_z^*, E_z), where M_z^* is also signed and normalized. The high-level description of this operation is composed of the following five steps:

1. Multiply significands M_x^* and M_y^*, add exponents E_x and E_y, and determine the alignment shift and shift M_w^*. Produce the intermediate result exponent $E_z = \max(E_x + E_y, E_w)$.
2. Add the product and the aligned M_w^*.
3. Normalize the adder output and update the result exponent.
4. Round.
5. Determine exception flags and special values.

Implementation

We now consider an implementation in which operands and result significands are in sign-and-magnitude representation. Moreover, to be specific in the normalization step, we consider the representation of the significand as in the IEEE Standard 754, that is, normalized and in the range $[1, 2)$. Again, the number of bits of the significand is $m = f + 1$, with a hidden most-significant bit.

The organization of a MAF unit is shown in Figure 8.19. For biased exponent representation, we have

$$\max(E_x + E_y, \ E_w) = \max\left(E_{B,x} + E_{B,y} - B, \ E_{B,w}\right) \qquad\qquad 8.73$$

The alignment of the addend M_w with respect to the double-precision product is performed concurrently with the multiplication step. Since the product is not

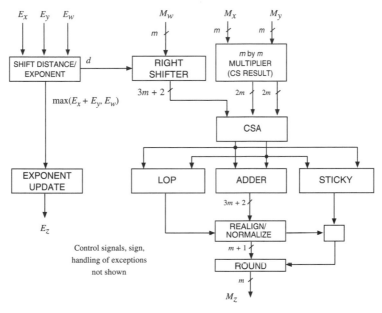

FIGURE 8.19 Basic implementation of MAF operation.

shifted for alignment, the alignment requires a left shift of at most $m + 3$ positions (Figure 8.20(a)) and a right shift of at most $2m - 1$ positions (Figure 8.20(b)). The maximum left shift is obtained by observing that the guard (position m) and the round (position $m + 1$) bits are 0 when the result significand corresponds to M_w. Consequently, two additional positions are included, resulting in the shift of $m + 3$ positions. The maximum right shift assures that M_w is shifted to the right of the least-significant bit of the product $M_x \times M_y$.

To avoid the bidirectional shifter, the addend M_w is positioned $m + 3$ bits to the left of the product, as indicated in Figure 8.21(a), and shifted right by the distance

$$d = E_x + E_y - E_w + m + 3 \qquad \text{8.74}$$

which for biased exponent representation is performed as

$$d = E_{B,x} + E_{B,y} - E_{B,w} - B + m + 3 \qquad \text{8.75}$$

No shift is performed for $d \leq 0$ and the maximum shift is $3m + 1$.

$$\overline{\hspace{2em} m \hspace{2em}} \quad \overline{\hspace{2em} 2m \hspace{2em}}$$

Product x*y: 00xx.xxxxx...xxxxxxxxxx

Addend: 1.xxxxxxxxxxxxx

|— m − 1 + 4 ——|

(a)

$$\overline{\hspace{4em} 2m \hspace{4em}}$$

Product x*y: xx.xxxxx...xxxxxxxxxx

Addend: 01xxxxxxxxxxxxx

Shift distance: |— 2m − 2 + 1 —|

(b)

FIGURE 8.20 Position of addends using bidirectional shift: (a) Maximum left shift. (b) Maximum right shift.

The two zero bits shown in the figure are used as the guard and round bits when M_w is not shifted, that is, the result significand is M_w. Figures 8.21(b), (c), and (d) show the different alignment situations.

The multiplier produces $2m$-bit carry and sum vectors that are reduced, together with the aligned M_w, in a $(3m + 2)$-bit adder to produce a result (possibly unnormalized). Since the leftmost $m + 2$ bits of the adder input produced by the multiplier are always 0, the corresponding adder positions can be implemented as an incrementer with the carry-out of the lower part of the adder as the increment input. This implementation is shown in Figure 8.22. The sticky bit is adjusted after normalization.

The output of the adder may require a realignment/normalization left shift to place the leading 1 in the leftmost position, as shown in Figure 8.23. Since the product is not shifted for alignment, the left shift can be up to about $2m$ positions, which is twice as much as in the floating-point addition. The additional m positions are due to the initial position of the adder operands, as shown in Figure 8.21(a). As in floating-point addition, for fast implementation, the leading-one position of the adder output is computed by a LZA module, concurrently with the addition. Moreover, the sticky bit is updated by the normalization.

The rounding is performed after realignment/normalization. It involves obtaining the guard, round, and sticky bits and performing the actual rounding.

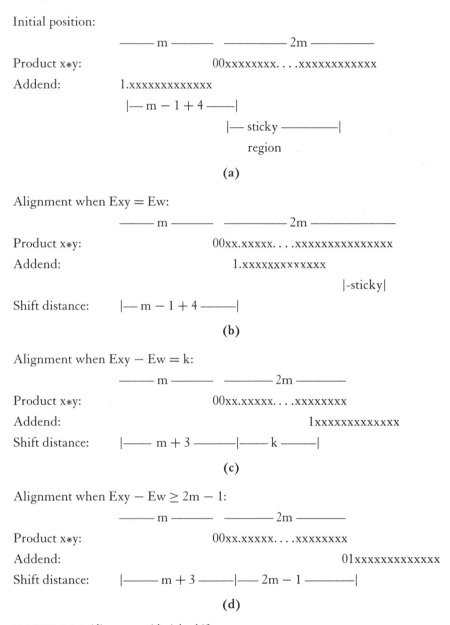

FIGURE 8.21 Alignment with right shifter.

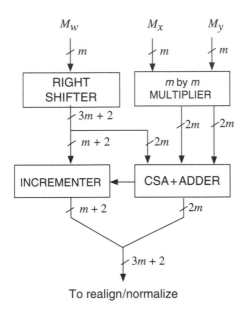

FIGURE 8.22 Implementation of MAF adder.

Adder output $\quad$ |——— m+2 ——|——— 2m ———|

Before shift $\qquad$ 00000000000000001.xxxxxxxxxxxxxxxxxx

After shift $\qquad$ 1.xxxxxxxxxxxxLGRT

FIGURE 8.23 Left shifting of the adder output.

In floating-point addition the delay of the rounding is reduced by performing it together with the addition and before the normalization. In the MAF case, this delay reduction is difficult because, at the adder output, the radix point can be anywhere in the leftmost $m + 3$ bits, so too many cases have to be considered.

To increase the throughput, the MAF unit is usually pipelined. For example, in a three-stage pipeline, Stage 1 implements the multiplication, alignment, and 3-2 carry-save addition; Stage 2 performs 2-1 addition and predicts the leading one in the sum; and, finally, Stage 3 performs normalization and rounding.

8.6   Floating-Point Division and Square Root

We now consider algorithms and implementations for floating-point division and square root. The algorithms are given in generic terms, while implementations are determined by the IEEE standard.

8.6.1   Division: Algorithm and Basic Implementation

The operands are x and d, represented by (M_x^*, E_x) and (M_d^*, E_d), with M_x^* and M_d^* signed and normalized. The result

$$q = x/d \qquad\qquad 8.76$$

is represented by (M_q^*, E_q), with M_q also signed and normalized. The high-level description of the floating-point division algorithm is composed of the following steps:

1.   Divide significands and subtract exponents.

$$\begin{aligned} M_q^* &= M_x^*/M_d^* \\ E_q &= E_x - E_d \end{aligned} \qquad 8.77$$

2.   Normalize M_q^* and update exponent.

3.   Round.

4.   Determine exception flags and special values.

Figure 8.24 shows the basic implementation. For biased representation of the exponents, we produce the intermediate result exponent as

$$E_{B,q} = E_{B,x} - E_{B,d} + B \qquad 8.78$$

For the division of the significands the methods discussed in Chapters 5 and 7 are used.

The normalization step depends on the range in the representation of the significands. For instance, for the IEEE standard the range is $[1, 2)$ so that Step 1 results in a range $(\frac{1}{2}, 2)$. Consequently, normalization is required when the value is less than 1 (a left shift of one position and a decrement of the exponent). To perform this normalization, a guard bit (G) is needed.

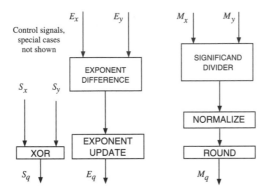

FIGURE 8.24 Basic implementation of floating-point division.

To perform the four rounding modes (after normalization), it is necessary to compute the significand of the quotient as a truncation of $f + 1$ fractional bits of the infinite precision quotient (that is, including another bit, the round bit R), as well as the sticky bit T. This sticky bit is needed for rounding to plus and to minus infinity and to determine whether the result is exact. However, it is not needed for rounding to nearest, since the tie condition cannot occur. This can be seen as follows:

The tie condition corresponds to an $(f+1)$-bit exact quotient when dividing operands with f-bit fractions. To show that this cannot happen, consider dividing $M_x = 1.x_1 \ldots x_f$ by $M_d = 1.d_1 \ldots d_f$. The "exact" quotient with $f+1$ fractional bits would be $M_q = 1.q_1 \ldots q_f 1 \times 2^{-e}$ where $e = \{0, 1\}$, because this quotient can be normalized or unnormalized. Then, $(M_d \times 2^f) \times (M_q \times 2^{f+1+e}) = M_x \times 2^f 2^{f+1+e}$, or

$$1d_1 \ldots d_f \times 1q_1 \ldots q_f 1 \times 2^{1-e} = 1x_1 \ldots x_f \times 2^{f+2}$$

Since the second factor of the left side is odd, the product cannot have more than $f + 1$ least-significant zeros. However, the right-hand side has at least $f + 2$ least-significant zeros. Consequently, the identity cannot be true, and the tie case cannot occur.

The actual rounding step depends on the method used to obtain the significand of the result (digit recurrence or iterative approximation). We consider these two situations now.

8.6.2 Division: Rounding

As indicated, to perform the rounding it is necessary to produce the normalized infinite precision quotient (significand) truncated to $f + 1$ fractional bits. Moreover, it is necessary to obtain the sticky bit.

Rounding for Digit Recurrence

In the digit recurrence method the truncated quotient is directly obtained by the iterations of the recurrence, after the correction step to make the remainder positive and the normalization. The sticky bit is provided by the condition "remainder not equal to zero." Consequently, the rounding requires the detection of the zero remainder and a conditional incrementation of the significand, depending on the roundoff mode and on bits G and T. The number of bits to be computed by the iterations is the number of bits of the normalized significand (m) plus two (the guard and the round bits) plus p, where, because of the initialization, $p = 1$ for $\rho = 1$ and $p = 2$ for $\rho < 1$ (see Section 5.2.2).

To reduce the overhead of correction, normalization, and rounding, it is possible to combine these steps in one cycle, together with the on-the-fly conversion of the last digit. We show now how to do this for the simplified case[20] in which the following applies:

- We consider only rounding to nearest (remember that the tie case cannot occur).
- The number of bits $m + 2 + p$ is a multiple of $K = \log_2 r$. These bits are computed in $(m + 2 + p)/K$ iterations.

Let us call q_L the digit obtained in the last iteration of the recurrence. For the correction step, it is necessary to determine the *sign* of the residual, defined by

$$sign = \begin{cases} 1 & \text{if residual is negative} \\ 0 & \text{otherwise} \end{cases} \qquad \textbf{8.79}$$

Since a negative residual makes it necessary to decrement the result, the correct value of the last digit becomes $(q_L - sign)$, as shown in Figure 8.25. If the value of q_L is in the range $[-a, a]$, the corrected digit is in the range $[-a - 1, a]$.

---

20. The more general case is described in references listed at the end of this chapter.

$$
\begin{array}{c}
\text{G R} \\
\underline{\hspace{1.5em} \text{K} \hspace{1.5em}}
\end{array}
$$

q_L: x x ... x x x

                   $-sign$

                   $+norm$

                   $+1$

$$\overline{\hspace{6em}}$$

q*_L: y y y ... y y y

FIGURE 8.25 Adjustment of the last quotient digit due to correction and rounding.

We now consider the rounding. If it is done after normalization, it is performed by adding one in position G. However, we need to consider also the case when the quotient is not normalized: in such a case a one should be added in position R. Consequently, both cases are included by adding in position R one plus *norm*, where *norm* $= 1$ if the quotient is normalized and 0 otherwise. Consequently, the corrected and rounded digit is

$$
q_L^* = q_L + norm + (1 - sign), \quad q_L^* \in [-a, a + 1] \qquad \textbf{8.80}
$$

Note that the value of q_L^* can be larger than $r - 1$, requiring one additional bit. Figure 8.25 shows this process.

To determine *norm*, it is necessary to determine whether the result would be normalized after performing the correction and the conversion, but before the rounding. Since this conversion is not performed, the situation has to be detected using the already converted part[21] ($Q[L - 1]$, $QM[L - 1]$), the digit q_L, and *sign*. Specifically,

$$
norm = \begin{cases}
1 & \text{if } q_L - sign \geq 0 \text{ and } Q[L - 1] \text{ is normalized} \\
1 & \text{if } q_L - sign < 0 \text{ and } QM[L - 1] \text{ is normalized} \\
0 & \text{otherwise}
\end{cases} \qquad \textbf{8.81}
$$

---

21. For the on-the-fly conversion, see Section 5.2.3.

or

$$norm = \begin{cases} Q[L-1]_{msb} & \text{if } q_L - sign \geq 0 \\ QM[L-1]_{msb} & \text{if } q_L - sign < 0 \end{cases} \qquad \textbf{8.82}$$

Since q_L^* can be larger than $r - 1$, it is necessary to incorporate to the on-the-fly conversion a third form

$$QP[L-1] = Q[L-1] + r^{-(L-1)}$$

so that the rounded significand before normalization and truncation to m bits is

$$MM_q = \begin{cases} (QP[L-1], u) & \text{if } q_L^* \geq r & \text{(condition } K1) \\ (Q[L-1], u) & \text{if } 0 \leq q_L^* \leq r - 1 & \text{(condition } K2) \\ (QM[L-1], u) & \text{if } q_L^* < 0 & \text{(condition } K3) \end{cases} \qquad \textbf{8.83}$$

with $u = q_L^* \bmod r$.

The final m-bit significand is

$$M_q[0:m-1] = \begin{cases} MM_q[0:m-1] & \text{if } MM_q[0] = 1 \quad \text{(normalized case)} \\ & \text{discard } MM_q[m, m+1] \text{ bits} \\ MM_q[1:m] & \text{if } MM_q[0] = 0 \quad \text{(unnormalized case)} \\ & \text{shift left and discard } MM_q[m+1] \text{ bit} \end{cases} \qquad \textbf{8.84}$$

where $V[a:b]$ denotes bits $V_a, \ldots, V_b$.

The updating of $Q[j]$ and $QM[j]$ are described in Chapter 5 for the on-the-fly conversion. For the rounding, the form $QP[j] = Q[j] + r^{-j}$ is also needed,[22] and the updating is done according to the following expression:

$$QP[j+1] = \begin{cases} (QP[j], 0) & \text{if } q_{j+1} = r - 1 \\ (Q[j], (q_{j+1} + 1)) & \text{if } -1 \leq q_{j+1} \leq r - 2 & \textbf{8.85} \\ (QM[j], (r - |q_{j+1}| + 1)) & \text{if } q_{j+1} < -1 \end{cases}$$

Table 8.5 describes the updating for the radix-4 case, with signed-digit set $q_j \in \{-3, \ldots, 3\}$.

---

22. Note that this form is only needed when $q_L^* \geq r$, which can occur when $a \geq r - 2$. For instance, this form is not needed in the radix-16 implementation with $a = 10$ (see Exercise 8.43).

q_{j+1}	$Q[j+1]$	$QM[j+1]$	$QP[j+1]$
3	$(Q[j], 3)$	$(Q[j], 2)$	$(QP[j], 0)$
2	$(Q[j], 2)$	$(Q[j], 1)$	$(Q[j], 3)$
1	$(Q[j], 1)$	$(Q[j], 0)$	$(Q[j], 2)$
0	$(Q[j], 0)$	$(QM[j], 3)$	$(Q[j], 1)$
-1	$(QM[j], 3)$	$(QM[j], 2)$	$(Q[j], 0)$
-2	$(QM[j], 2)$	$(QM[j], 1)$	$(QM[j], 3)$
-3	$(QM[j], 1)$	$(QM[j], 0)$	$(QM[j], 2)$

TABLE 8.5 Updating of forms for radix 4.

EXAMPLE 8.7   The following example illustrates the rounding-to-nearest process for a radix-4 division with $a = 2$. Since $a = r - 2$, it is necessary to include QP. We show the conversion of quotient digits q_{L-1} and q_L. Note that we have already shifted the quotient by two bit positions, which is required by the fact that the initial condition is $x/4$.

$Q[L-2]$	$1.xx\ldots x\,23$
$QP[L-2]$	$1.xx\ldots x\,30$
$QM[L-2]$	$1.xx\ldots x\,22$
q_{L-1}	-1
$Q[L-1]$	$1.xx\ldots x\,223$
$QP[L-1]$	$1.xx\ldots x\,230$
$QM[L-1]$	$1.xx\ldots x\,222$
q_L	-2
	residual
sum	01011001
carry	00101000

The residual is negative (*sign* $= 1$). Moreover, q_L is negative and $QM[L-1]$ is normalized, so that *norm* $= 1$. Consequently,

$$q_L^* = -2 + 1 + 0 = -1 \qquad\qquad \textbf{8.86}$$

Therefore, $MM_q = (QM[L-1], 3)$. Since MM_q is normalized, we get

$$M_q = QM[L-1] = 1.xx \ldots x\,222 \qquad \textbf{8.87}$$

To verify that this is correct let us compare with the result when the rounding is done after conversion, correction, and normalization. The conversion would produce

$$Q[L] = (QM[L-1], 2) = 1.xx..x\,2222 \qquad \textbf{8.88}$$

Since the residual is negative, correction produces

$$Q[L] = 1.xx..x\,2221 \qquad \textbf{8.89}$$

The result is already normalized, so for rounding, 2 has to be added to the last digit, namely,

$$Q[L] \quad 1.xx\ldots x\,222|01|$$
$$1$$

Then the result is truncated, producing $M_q = 1.xx\ldots x\,222$. ∎

Implementation. A possible implementation of the rounding-to-nearest scheme is shown in Figure 8.26. It consists of three left-shift registers to keep the forms $Q[j]$, $QM[j]$, and $QP[j]$, logic to generate the digit to concatenate, and the loading controls.

The rounding is performed by the selection described by expression (8.84). In addition, it is necessary to have a network to detect the sign of the residual from its redundant representation, as discussed in Section 5.3.1.

Rounding for Iterative Approximation

The algorithms presented in Chapter 7 result in approximations that do not produce directly the truncated quotient required for rounding, even if the error is small enough. This is illustrated by the following example.

EXAMPLE 8.8   In this example q is the infinite precision quotient (here 16 bits) and q_c are two approximations (also 16 bits) with error less than 2^{-10}. We see that for the first approximation the truncation to four bits does not produce the truncated q:

$$\begin{array}{l} \quad\quad\quad 1\ 2\ 3\ 4\ 5\ 6\ 7\ 8\ 9\ 0\ 1\ 2\ 3\ 4\ 5 \\ q \quad 1.1\ 0\ 1\ 1|0\ 0\ 0\ 0\ 0\ 0\ 0\ 0\ 1\ 0\ 1 \\ \\ qc \quad 1.1\ 0\ 1\ 0|1\ 1\ 1\ 1\ 1\ 1\ 1\ x\ x\ x\ x \\ \quad\quad 1.1\ 0\ 1\ 1|0\ 0\ 0\ 0\ 0\ 0\ x\ x\ x\ x\ x \end{array}$$

■

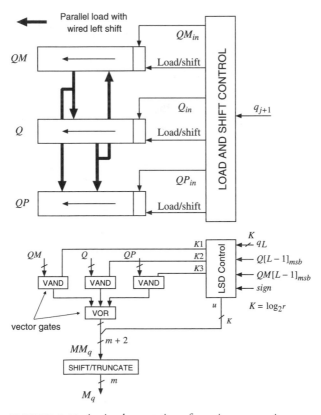

FIGURE 8.26 An implementation of quotient conversion, correction, normalization, and rounding.

Moreover, the algorithm does not produce the sticky bit (zero remainder).[23]

As illustrated in the previous example, the cases for which it is not correct to truncate the approximation correspond to approximations that have a string of all zeros or all ones after the truncation bit. It has been shown that for division the maximum length of this string is about $2f$ bits (for the exact number, see the references at the end of this chapter). Consequently, it is correct to truncate an approximation that has an error of about 2^{-2f}. This method has the disadvantage that producing such an accurate approximation might require additional iterations and a wider multiplier and adder.

An alternative is to produce the truncated approximation with an error of about 2^{-f} and then compute the corresponding remainder and correct the approximation for the cases in which the remainder is not correct (not positive or not bounded by $d \times 2^{-(f+1)}$). This also allows the detection of the zero remainder condition. We now consider two cases depending on whether the approximation is always from one side (for instance, from above) or from either side.

Case 1: Approximation from One Side. As discussed in Chapter 7, the one-sided approximation is achieved by the implementation of the Newton-Raphson algorithm with a suitable roundoff error in the last iteration. In this case, the approximation is from below. However, the rounding is simpler if the approximation is from above.[24] To achieve this approximation from above with an error bound of 2^{-w}, the approximation from below with the same error bound is incremented by 2^{-w}.

So, we consider that the approximation is from above. Moreover, we assume that the approximation has an error of less than $2^{-(f+1)}$. That is, if q is the infinite precision quotient and q_c is the approximation, then

$$0 \leq q_c - q < 2^{-(f+1)} \qquad \textbf{8.90}$$

This situation is described by Figure 8.27, which represents the discrete real-numbers line.

As can be seen, q_t, the truncated q_c to $f + 1$ fractional bits, has the two possible values A and B, and only A is the correct q_T. Consequently, to determine

---

23. As indicated in the previous section, this remainder can only be 0 if $q_{f+1} = 0$.
24. For the case from below, see Exercise 8.39.

q is infinite precision value

qT truncation of q (to granularity $2^{-(f+1)}$)

qc computed value (from above error $<2^{-(f+1)}$)

y region of qc

qt truncation of qc (two possible values A and B)

```
. . . . . . . . . qqqqqqqq. . . . . . . .    Discrete real-number line
    |         |        |         |     points of granularity  2^{-(f+1)}
          q T
qc                yyyyyyyyyyyyyyyy
qt                A         B
rt                +         −
```

FIGURE 8.27 Quotient approximation.

whether the computed q_t is correct, it is necessary to obtain

$$r_t = x - q_t \times d \qquad\qquad 8.91$$

Because the approximation is from above, $r_t \geq 0$ for $q_t = A$ and $r_t < 0$ for $q_t = B$. Consequently,

$$q_T = q_t \quad \text{if } r_t \geq 0$$
$$q_T = q_t - 2^{-(f+1)} \quad \text{if } r_t < 0$$

The sticky bit is zero if $r_t = 0$ or if $r_t = -d \times 2^{-(f+1)}$.

    All four rounding modes can be achieved by observing the sign and zero value of r_t and position $f + 1$ of q_t (called the guard bit g) and selecting q_f, $q_f + 2^{-f}$, or $q_f - 2^{-f}$, where q_f is q_t truncated to position f.

    For instance, for rounding to nearest the rounded value is $q_T + 2^{-(f+1)}$ truncated to f fractional bits. Consequently, we have the following cases:

- $r_t \geq 0$. Then $q_T = q_t$ and the rounded value is $q_f + g2^{-f}$.
- $r_t < 0$. Then $q_T = q_t - 2^{-(f+1)}$ and the rounded value is q_f.

Case 2: Two-Sided Approximation. As described in the previous chapter, the direct method produces a one-sided approximation if the precision of the multiplications is sufficient, but a two-sided one for less precision.

```
·········qqqqqqqq·········
   |       |       |      |
          q T
```

qc yyyyyyyyyyyyyyyyyyyyyyyy

qt A   B   C

rt +   +   −

FIGURE 8.28 Two-sided quotient approximation.

For a two-sided approximation,

$$|q - q_c| < 2^{-(f+1)} \qquad\qquad \textbf{8.92}$$

The situation is described by Figure 8.28.

As can be seen, q_t can take three possible values. So, $r_t = x - q_t d$ is computed and

$$q_T = q_t \quad \text{if } 0 \le r_t < 2^{-(f+1)}d$$
$$q_T = q_t + 2^{-(f+1)} \quad \text{if } r_t \ge 2^{-(f+1)}d$$
$$q_T = q_t - 2^{-(f+1)} \quad \text{if } r_t < 0$$

This requires a comparison of r_t with $2^{-(f+1)}d$. A variation does not require this comparison, but an approximation with an error of less than $2^{-(f+2)}$, that is,

$$|q - q_c| < 2^{-(f+2)} \qquad\qquad \textbf{8.93}$$

We then produce

$$q^* = q_c + 2^{-(f+2)} \qquad\qquad \textbf{8.94}$$

resulting in Figure 8.29.

Consequently, q_T can be obtained by detecting the sign of r_t^* as in Case 1. Moreover, the rounded value is obtained in a similar manner.

8.6.3   Square Root: Algorithm and Implementation

The operand x is represented by $(M_x^*, \ E_x)$, with M_x^* signed and normalized. The result

$$s = \sqrt{x} \qquad\qquad \textbf{8.95}$$

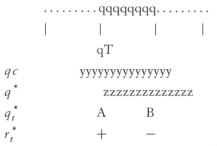

FIGURE 8.29 Case with error less than $2^{-(f+2)}$.

is represented by (M_s^*, E_s), with M_s also signed and normalized. The high-level description of the algorithm is composed of the following steps:

1. Obtain the square root of the significand and produce the exponent of the result.

$$M_s^* = \sqrt{M_x^*}$$
$$E_s = E_x/2$$

8.96

   To obtain an integer, exponent E_x should be even. Consequently, if E_x is odd, we utilize $M_x/2$ and $E_x + 1$.

2. Normalize M_s^* and update exponent.

3. Round.

4. Determine exception flags and special values.

The implementation of this operation is very similar to floating-point division. Consequently, here we give a short summary.

Figure 8.30 shows the basic implementation. For a biased exponent representation, the intermediate result exponent is computed as

$$E_{B,s} = \left\lceil (E_{B,x} + B)/2 \right\rceil$$

8.97

To obtain the square root of the significand, the methods discussed in Chapters 6 and 7 are used.

The normalization step depends on the range in the representation of the significands. For instance, for the IEEE standard the range is $[\frac{1}{2}, 2)^{25}$ so that

---

25. Because of the division by two for odd exponent. An alternative is to have an operand range of $[1, 4)$, producing a result in the range $[1, 2)$, so no normalization and exponent update is required; in this case, $E_z = \lfloor E_x/2 \rfloor$ and the significand input is either M_x or $2M_x$.

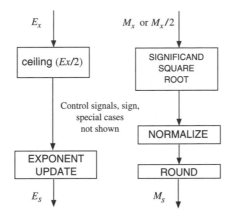

FIGURE 8.30 Basic implementation of floating-point square root.

Step 1 results in a range $(1/\sqrt{2}, \sqrt{2})$. Consequently, normalization is required when the value is less than 1 (a left shift of one position and a decrement of the exponent). To perform this normalization a guard bit (G) is needed.

To perform the four rounding modes (after normalization), it is necessary to compute the significand of the quotient as a truncation of $f + 1$ fractional bits of the infinite precision quotient (that is, including another bit, the round bit R), as well as the sticky bit T. This sticky bit is needed for rounding to plus and to minus infinity and to determine whether the result is exact. However, it is not needed for rounding to nearest, since the tie condition cannot occur.

The actual rounding step depends on the method used to obtain the significand of the result (digit recurrence or iterative approximation). The corresponding methods are variations of those discussed for division (for additional details see the references at the end of this chapter).

8.6.4 Comparison between Digit Recurrence and Multiplicative Methods

In Chapters 5, 6, and 7 we have presented two methods to perform division and square root. Moreover, in this chapter we have extended these methods to perform the corresponding floating-point operations. We now discuss the elements required for a comparison between these methods. This is not intended to determine which of the methods is preferable, since this depends on many characteristics of the implementation and on system requirements.

The main aspects that are considered in the evaluation of an implementation are timing aspects, such as execution time, throughput, and effect on other operations; cost issues, such as the additional hardware required to perform the operations; and energy consumed. We now compare the methods in terms of timing and cost. We consider the case in which for the digit recurrence method dedicated hardware is used (no sharing with other operations), whereas for the multiplicative method a (modified) floating-point multiplier is used (shared with floating-point multiplication).

Timing

The execution time depends on the number of cycles and the cycle time. In the digit recurrence method, the convergence is linear and each iteration is usually performed in one cycle. Consequently, the number of cycles corresponds to the ratio between the number of bits of the quotient and the number of bits per iteration. Moreover, one cycle is needed to terminate the operation. Since the rounding is simple, this cycle can also include the rounding process.

On the other hand, in the multiplicative method the convergence is quadratic so that the number of iterations is much smaller than in the digit recurrence method. However, each iteration includes floating-point multiplications,[26] so that it is performed in several cycles. Moreover, the rounding process is more complex and requires some additional cycles.

The cycle time depends on the desired clock rate and affects the radix to be used in the digit recurrence method and the number of stages in the pipelined floating-point multiplier.

To get a rough feeling of the corresponding execution times for double-precision floating-point operations, we consider the following situation, in which the cycle times are assumed to be the same so that the relative execution times correspond to the ratio of number of cycles:

- A radix-16 digit recurrence implementation (see Chapter 5), resulting in about $54/4 + 1 = 15$ cycles. The same number of cycles is required for division and for square root.

---

26. In this situation, each iteration performs complete floating-point multiplications so that the improvements resulting from using limited multiplications are not applicable.

- A multiplicative implementation with a four-stage pipelined floating-point multiplier. The number of iterations depends on the initial approximation. If the error of this approximation is about 2^{-15}, two iterations are required and the total number of cycles is about 20 for division[27] and 25 for square root.

Moreover, since digit recurrence implementation uses dedicated hardware, it does not affect the execution of the other operations. On the other hand, the multiplicative method utilizes the multiplier so it prevents the execution of multiplications while the division/square root is being executed. In addition, it requires a modification of the multiplier, which can make it slower.

Additional Hardware

The digit recurrence implementation uses dedicated hardware. The number of equivalent gates is estimated in Chapter 5.

The multiplicative method could be implemented using directly the floating-point multiplier or fused multiply-add. However, to implement it more efficiently, in most cases some modifications to the multiplier are included. Specifically, it might be necessary (1) to increase the size of the multiplier to achieve the required approximation accuracy, and (2) to incorporate some feedback paths to implement the recurrence and the rounding. Moreover, a module is required to obtain the initial approximation.

8.7   Concluding Remarks

Floating-point representation produces a high dynamic range that simplifies the design and programming of numerical computations. However, with respect to fixed-point representations, it reduces the available precision and makes the implementation of operations slower and more complex. Moreover, because it eliminates the need for specific scaling operations, it might lead unaware users to unsatisfactory results.

The definition of a specific floating-point representation consists of a variety of parameters, such as the number of bits for each component, the base of

---

27. For instance the algorithm described in Oberman (1999) has 3 cycles to obtain the initial approximation, 9 cycles for the two iterations, 4 cycles for a final multiplication, and 4 cycles to compute the residual and correct the result.

the exponent, the range and representation of the significand and of the exponent, as well as the definition and representation of special cases. Although these parameters can be optimized for a particular application, the tendency today is to use the IEEE Floating-Point Arithmetic Standard 754 (ANSI and IEEE 1985). This provides portability among systems and assures that experts have considered the possible anomalies and designed a good compromise among the different characteristics.

The algorithms and implementations of the basic operations for floating-point representation are based on the corresponding fixed-point ones, as presented in the previous chapters. On top of these, it is necessary to incorporate the effect of the exponents, the requirement for normalization and rounding, and the detection of special cases. Division and square root are implemented either using digit recurrence schemes, discussed in Chapters 5 and 6, or iterative (convergence) schemes, described in Chapter 7. In the latter case, additional modules such as tables are needed to provide initial approximations. The resulting algorithms/implementations are complex and have been the object of much research and development in order to produce the desired objectives, in terms of delay, throughput, area, and energy.

8.8   Exercises

Floating-Point Representation

8.1   **[Range and precision]** How many radix-10 digits are needed in a fixed-point format to represent the approximations of both Planck's constant (6.63×10^{-27}) and Avogadro's number (6.02×10^{23})? How many radix-10 digits are needed to represent these constants in a floating-point number format (consider a base 10 for the exponent, and radix-10 biased representation of the exponent, such that $E_{biased} = E + 50$) ?

8.2   **[Range and precision]** Consider computing x^p for $2^{-1} \leq x < 1$ and $p = 64$ for a 32-bit fixed-point representation with the radix point at the left.

(a)   Determine the range of x for which at least 16 significant bits in the result are representable.

(b)   What is the maximum value of x for which no significant bits of the result are representable?

8.3 [**Spacing of FLPT numbers**] Consider a floating-point representation with a 40-bit word, composed of a normalized significand in sign-and-magnitude with a fractional part of 8 hexadecimal digits, and a sign-and-magnitude exponent with 7 bits (exponent base 16).

(a) Determine the maximum and minimum difference between successive floating-point numbers.

(b) Determine the maximum relative spacing

$$\alpha = \max \left(\frac{x' - x}{x} \right)$$

where x' and x are two successive floating-point numbers.

8.4 [**FLPT representation with different bases**] Consider two floating-point representations, both with m bits for the normalized significand and e bits for the exponent. Determine the ratio between the number of floating-point numbers that are represented by systems A and B for

(a) System A has base 16 and system B base 2.

(b) System A has base 16 and system B has base 4.

8.5 [**ulp and relative error**] A value x is represented by a floating-point number with an error in the significand of $\frac{1}{2}$ ulp. Determine the relative error in ulps. For details on the relation between relative error and ulp, see Goldberg (1991).

8.6 [**Unnormalized form**] How many different floating-point numbers represent exactly the value $\frac{3}{4}$ if the format consists of 24-bit fractions and 8-bit exponent? The base is 2.

8.7 [**Normalized form**] How many normalized significands can be represented in a base-64 floating-point system with 48-bit significands?

8.8 [**Effect of base b**] Show that for a given machine word of $n = f + e$ bits, the choice of base $b = 2$ always provides as much accuracy and more exponent range than some $b = 2^k$. The accuracy (maximum relative spacing) is defined as

$$\alpha(f, k) = \max \left(\frac{x' - x}{x} \right)$$

where x' and x are two successive floating-point numbers. The exponent range is defined as $E(f, k) = k(2^{n-f} - 1)$. For details, see Brown and Richman (1969).

8.9 [Biased arithmetic] In a binary biased number representation a number x is represented by $x + B$ where the bias $B = 2^{n-1}$ or $B = 2^{n-1} - 1$ and n is the number of bits in the bit-vector. Develop bit-level algorithms for the following operations and the two choices for B: (1) conversion from/to two's complement; (2) change of sign; (3) addition; (4) subtraction; and (5) overflow detection.

(a) Compare (1) through (5) for the two choices of B.

(b) Compare (2) through (5) for the two choices of B with the same operations in the two's complement number representation system.

Roundoff Modes and Error Analysis

8.10 [Rounding] Consider the following rounding schemes: round-to-nearest-even (RNE), round-to-nearest-odd (RNO), round towards zero (truncate) (RZ), and round toward +inf (RP). Show the final rounded result in the following three cases (fill in the blanks in the table):

Sign	Exponent	Fraction	Guard	Round Mode
0	00011111	1111111111	1	
				RNE
				RNO
				RZ
				RP
0	11111110	1111111111	1	
				RNE
				RNO
				RZ
				RP
1	11111110	1111111111	1	
				RNE
				RNO
				RZ
				RP

8.11 [Bias for the four rounding modes] Derive the expressions for the total and average bias for each rounding mode.

IEEE Floating-Point Standard

8.12 **[Representation]** Complete the following table assuming the IEEE FLPT Standard single-precision format:

Hex-Vector	Value
	0.0
80000000	
A73FF801	
	-2^{48}
7F7FFFFF	
00800000	
	plus infinity
FF800000	
7FC00000	

8.13 **[Errors in the rounding modes]** Determine the absolute and relative error in representing the number 0.1 (decimal) using the IEEE Standard single-precision format with significands of 8 bits instead of 24 bits for each rounding mode.

8.14 **[Denorms]** How many denormalized numbers are there in the IEEE Standard single-precision format, and what is their range?

Floating-Point Addition

8.15 **[FLPT addition—sign of result]** Determine a switching expression for the sign of the result of an addition/subtraction in terms of the signs of the operands, their relative magnitudes, and the operation (addition/subtraction).

8.16 **[Examples of execution for basic implementation]** Perform the following operations using the basic implementation. Include the guard bits, perform all four rounding modes, and determine if there is an exponent overflow. The representation is IEEE Standard single precision, with significands of 10 bits instead of 24 bits. Indicate the outputs of each module in Figure 8.6.

Operation	X	Y
Add	000110001001111000	000110011100011101
Add	000110001001111000	100110011100011101
Sub	000110001001111000	000110001001110111
Sub	011111110111100011	111111100001010101

8.17 **[FLPT addition—exceptions]** For the basic implementation of Figure 8.6 determine expressions for the five special cases.

8.18 **[FLPT addition—special values]** Give sets of operands (in IEEE Standard single-precision representation with 10-bit significands) and operations (add/subtract) that produce each of the five special cases. Indicate the output of each module in Figure 8.6.

8.19 **[Denormals]**

(a) Modify the basic implementation of Figure 8.6 to allow denormal operands and produce a denormal result.

(b) Give one example of the execution of addition and one example of subtraction for the case in which one of the operands is denormal. Show the output of each module.

8.20 **[Calculation of delay for basic implementation]** For critical path delays of the modules in Table 8.6, determine the delay of the floating-point adder in Figure 8.6

Module	Delay (ns)
Exponent difference	$0.3 \lceil \log_2 e \rceil + 0.5$
Swap (includes buffer for control)	0.5
Right shift	$0.2 \lceil \log_2 m \rceil$
Add significands $(s + m)$	$0.3 \lceil \log_2 m \rceil + 1.0$
LOD	$0.3 \lceil \log_2 m \rceil$
Left shift (includes buffer)	$0.2 \lceil \log_2 m \rceil + 0.2$
Round	$0.2 \lceil \log_2 m \rceil$
Right shift (one position, including buffer)	0.5
Special cases	0.8

TABLE 8.6 Delay of modules.

(as stated in the text, as an approximation the delay can be obtained by the sum of the delays in the critical path) for single precision and for double precision.

Pipeline the floating-point adder (for single precision and for double) for a clock rate of 200 MHz. To account for clock skew and other delays, the stage delay should not be larger than 80% of the clock cycle.

8.21 **[Executing FLPT addition and subtraction on improved single-path implementation]** Perform the following operations using the implementation of Figure 8.8. Include the guard bits, perform all four rounding modes, and determine if there is an exponent overflow. The representation is IEEE Standard single precision, with significands of 10 bits instead of 24 bits. Indicate the outputs of each module in Figure 8.8.

Operation	X	Y
Add	000110001001111000	001001100100011101
Add	000110001001111000	101001100100011101
Sub	000110001001111000	000110001001110111
Sub	011111110111100011	111111100001010101

8.22 **[Design details]** Determine switching expressions for the shift control of the L1/R1 shifter of Figure 8.8.

8.23 **[Executing add/sub on double-path implementation]** Perform the following operations using the implementation of Figure 8.9. Include the guard bits, perform rounding to nearest (only for the case of effective addition), and determine if there is an exponent overflow. The representation is IEEE Standard single precision, with significands of 10 bits instead of 24 bits. Indicate the outputs of each module in Figure 8.9.

Operation	X	Y
Add	000110001001111000	001001100100011101
Add	000110001001111000	101001100100011101
Sub	000110001001111000	000110001001110111
Sub	011111110111100011	111111100001010101

Module	Delay (ns)
Exponent difference	$0.3\lceil\log_2 e\rceil + 0.5$
Swap (includes buffer for control)	0.5
R1-shifter	0.5
Compare	$0.2\lceil\log_2 m\rceil$
R-shifter	$0.2\lceil\log_2 m\rceil$
Bit-invert control (includes buffer)	0.5
Conditional bit invert	0.3
Bit invert	0.1
Two's complement compound adder	$0.3\lceil\log_2 m\rceil + 0.6$
LZA	$0.2\lceil\log_2 m\rceil$
Two's complement adder	$0.3\lceil\log_2 m\rceil + 0.3$
L1/R1-shifter (includes buffer)	0.7
L-shifter (includes buffer)	$0.2\lceil\log_2 m\rceil + 0.2$
Round and norm overflow	$0.3\lceil\log_2 m\rceil$
MUX	0.5
Add, Norm, and Round	$0.3\log_m +2.0$

TABLE 8.7 Delay of modules.

8.24 **[Calculating delays of single-path and double-path implementations]** For the module delays in Table 8.7, determine the maximum clock rate for the pipelined adders of Figure 8.11 for double precision. To account for clock skew and other delays, the maximum stage delay should not be larger than 80% of the clock cycle.

In the single-path implementation, how would you change the positioning of the stage latches to reduce the clock cycle?

In the double-path case, modify the implementation so that no swap is needed in the CLOSE path. Determine the new clock cycle.

Floating-Point Multiplication

8.25 **[Example of execution for basic implementation]** For the following floating-point operands in the IEEE Standard single-precision representation (10-bit significand, instead of 24), perform the multiplication using the basic implementation. Show the four rounding modes. Verify the correctness of your result.

X	Y
001010101010110011	101111111101110011
110011110101110010	111000111011111100

8.26 [**Delay of basic implementation**] Give an estimate of the delay (in inverter delays with a load of four) of the floating-point multiplication implementation of Figure 8.12 for single- and double-precision formats. Make reasonable estimates (and justify them) of the delay of each module.

Propose a pipelining structure so that the delay of a stage is about the delay of 20 inverters.

8.27 [**Exceptions and specials**] Show an example of a pair of operands in the IEEE Standard single-precision representation (10-bit significand, instead of 24) that produce an underflow in floating-point multiplication. Indicate the representation of the result.

8.28 [**Denormals**] Indicate which of the two operands (in IEEE Standard single-precision representation, with 10-bit significand) is denormal and perform the floating-point multiplication with rounding to nearest. Verify the correctness of your result

X	Y
001010101010110011	000000000000101010

8.29 [**Calculation of sticky**] Determine the sticky bit in the floating-point multiplication for the following operands (in IEEE Standard single-precision representation, with 10-bit significand). Use the two methods described in Section 8.5.5.

X	Y
001010101010111000	001010101010010000
010000000001100000	001010101011000000

8.30 [**Example(s) of execution for alternative implementation**] For the following floating-point operands in the IEEE Standard single-precision representation (10-bit significand, instead of 24) perform the multiplication using the implementation of Section 8.5.5. Show the four rounding modes. Verify the correctness of your result.

X	Y
001010101010110011	101111111101110011
110011110101110010	111000111011111100

8.31   **[Other rounding modes]** Extend the alternative implementation to perform

   (a)   round to zero
   (b)   round to plus infinity
   (c)   round to minus infinity

8.32   **[Comparing delay of basic and alternative implementations]** Give an estimate of the delay (in inverter delays with load of four) of the alternative floating-point implementation of Figure 8.13 for single- and double-precision formats. Estimate the reduction in delay of this implementation with respect to the basic implementation of Figure 8.12.

   Make reasonable estimates (and justify them) of the delay of each module.

   Propose a pipelining structure so that the delay of a stage is about the delay of 20 inverters.

8.33   **[Selection when c_m not in critical path]** Determine the control of selection in Figure 8.18 using the four bits composing Σ.

Floating-Point Multiply-Add Fused

8.34   **[Example of execution for basic implementation]** For the following floating-point operands in the IEEE Standard single-precision representation (10-bit significand, instead of 24) perform the multiply-add using the basic implementation. Show the four rounding modes. Verify the correctness of your result.

X	Y	W
001010101010110011	101111111101110011	110011110101110010

8.35   **[Delay of basic implementation]** Give an estimate of the delay (in inverter delays with a load of four) of the floating-point MAF implementation of Figure 8.19 for single- and double-precision formats. Make reasonable estimates (and justify them) of the delay of each module.

   Propose a pipelining structure so that the delay of a stage is about the delay of 20 inverters.

8.36 [**Adder output realignment**] Determine the amount of left shift needed to realign the adder output when the product is unnormalized, the exponents and the signs of the addends are equal, and

(a) there is no overflow in addition
(b) there is an overflow

8.37 [**Exponent updating**] Describe the updating of the result exponent in floating-point MAF operation.

Floating-Point Division

8.38 [**Example of execution for digit recurrence method**] For the following floating-point operands in the IEEE Standard single-precision representation (10-bit significand, instead of 24), perform the division using the radix-2 digit recurrence method. Show the round-to-nearest mode. Verify the correctness of your result.

X	D
001010101011010011	101111111110110011
110011110001011010	111000111101011101

8.39 [**Rounding with approximation from below**] Consider an approximation of the quotient from below with an error less than $2^{-(f+1)}$. Show that a direct algorithm for rounding to nearest involves a comparison with $2^{-(f+1)}d$ (see notation in this chapter).

8.40 [**Example of execution for iterative method**] For the following floating-point operands in the IEEE Standard single-precision representation (10-bit significand, instead of 24), perform the division using the NR iterative method with an initial approximation of 4 bits (computed using 5 bits of the divisor). Show the round-to-nearest mode. Verify the correctness of your result.

X	D
001010101011010011	101111111110110011
110011110001011010	111000111101011101

8.41 [**Example of execution for the multiplicative method**] Repeat Exercise 8.40 for the direct (multiplicative) division method described in Section 7.3.2.

8.42   **[Combined quotient conversion, correction, normalization and rounding]** For the following floating-point operands in the IEEE Standard single-precision representation (10-bit significand, instead of 24), perform the division using the digit recurrence method and show the bit-vectors used in the combined scheme for quotient conversion, correction, normalization, and rounding for the round-to-nearest mode. Verify the correctness of your result.

X	D
001010101011010011	101111111110110011

8.43   **[Conversion and rounding]** Consider the on-the-fly conversion and rounding of the quotient in the digit recurrence method. Develop an algorithm for updating of Q, QN, and QP when $a \leq r - 2$. Show that QP is needed for the rounding but not for the updating of registers when $a = r - 2$. Consequently, no register QP is required. Show that for $a < r - 2$, QP is not required at all.

8.44   **[Overflow after rounding]** Show that in floating-point division overflow after round to nearest cannot occur.

8.45   **[Normalization control]** In the algorithm given for rounding to nearest for floating-point division for the digit recurrence case, two different signals are used to determine whether the result is normalized: the signal *norm* before rounding and the bit $MM_q[0]$ after rounding. Show that *norm* can be used for both situations; that is, show that the situation "unnormalized" is not changed by the rounding. Indicate the advantage in delay of this approach.

8.46   **[Rounding to plus infinity]** Describe an algorithm for rounding toward plus infinity in floating-point division for the digit recurrence case.

8.9   Further Readings

There are several general treatments of floating-point arithmetic (Sterbenz 1974; Kulisch 1977; Kulisch and Miranker 1981; Knuth 1998; Goldberg 1991; Overton 2001). The choice of base and its effect on range and relative accuracy for a given machine precision is discussed in Brown and Richman (1969). Brent (1973) reports on the precision attainable with various floating-point number systems. Statistical properties of floating-point addition obtained from program traces, presented in Sweeney (1965), led to the adoption of base 16 floating-point format in the IBM S/360 systems. Underflow and the denormalized numbers are discussed in Coonen (1981). Loss of significance in floating-point subtraction and addition is discussed in Feldstein and Goodman (1982). Algorithms for arbitrary precision floating-point arithmetic are presented in Priest (1991). Bohlender et al. (1991) present semantics for exact floating-point operations.

Floating-Point Representation: Rounding

A classic on rounding errors in algebraic processes is Wilkinson (1963). Kuck et al. (1977) discuss the basic measures and characteristics of errors. Statistical studies of the accuracy and static and dynamic numerical characteristics of floating-point arithmetic are reported in Kuki and Cody (1973) and Cody (1973). Early work on rounding in floating-point arithmetic is presented in Yohe (1973). An axiomatic approach to rounding is discussed in Kulisch and Miranker (1981).

Floating-Point Standards

The IEEE Floating-Point Standard and its implementation aspects are discussed in Coonen (1980). Cody et al. (1984) describe a standard independent of radix and word length. An analysis of proposals for floating-point standards is described in Cody (1987). The reasons for rounding to even if tie are discussed in Reiser and Knuth (1975). The issues and status of the standard are presented in Kahan (1996).

Implementation of Floating-Point Unit: General

Design and implementation of floating-point arithmetic are the subject of numerous articles. Early designs are described in Bucholz (1962), Anderson et al. (1967), and Gosling (1971). More recent design issues are presented in Oberman (1996), Oberman and Flynn (1996b, 1997), and Even and Paul (2000). There are many descriptions of specific designs and implementations in the literature (Ware et al. 1982; Benschneider et al. 1989; Montoye et al. 1990; Darley et al. 1990; Dobberpuhl et al. 1992; Dao-Trong and Helwig 1992; Ide et al. 1993; Nicks et al. 1994; Flynn et al. 1995; Bannon and Keller 1995; Hunt 1995; Williams et al.

1995; Schwarz et al. 1999; Gerwig and Kroener 1999; Sharangpani and Arora 2000; Naini et al. 2001).

Implementation of Floating-Point Adder

Implementations of floating-point adders are presented in Vassiliadis et al. (1989), Beaumont-Smith et al. (1999), Seidel and Even (2001), and Bruguera and Lang (2001) among others. The FAR/CLOSE path scheme was proposed in Farmwald (1981), and its implementations are reported in Greenlay et al. (1995) and Oberman et al. (1999). A variable-latency adder is discussed in Oberman and Flynn (1998). Nielsen et al. (2000) propose a packet-forwarding adder to reduce the stage delay. Schemes for fast detection of leading one/zero are developed in Hokenek and Montoye (1990), Oklobdzija (1994), Suzuki et al. (1995), and Bruguera and Lang (1999).

Implementation of Floating-Point Multiplier

Designs of floating-point multipliers are described in many articles (Uya et al. 1984; Yu and Zyner 1995). Specific details of rounding schemes for multiplication have been presented in Santoro et al. (1989), Kabuo et al. (1994), Yu and Zyner (1995), Park et al. (1999), and Even and Seidel (2000).

Implementation of Floating-Point Multiply-Add Fused

Multiply-add fused designs are discussed in Hokenek et al. (1990), Jessani and Putrino (1998), and Chen et al. (2001).

Implementation of Floating-Point Division and Square Root

Soderquist and Leeser (1996) present area and performance trade-offs in floating-point division and square root implementations. Floating-point division/square root schemes using multiplicative approach are presented in Anderson et al. (1967), Oberman (1999), Clouser et al. (1999), Horel and Lauterbach (1999), and Agarwal et al. (1999). The digit recurrence schemes for division and square root described in Chapter 5 are applicable to floating-point operations and have been used frequently in practice (Prabhu and Zyner 1995; Yeager 1996; Inui et al. 1999). A self-timed floating-point divider is reported in Williams et al. (1995) and Suzuki et al. (1997). Rounding for digit recurrence and convergence division/square root are discussed in Ercegovac and Lang (1992) and in Markstein (1990), Kabuo et al. (1994), Schwarz (1995), Oberman and Flynn (1996a), and Markstein (2000), respectively. Design and implementation aspects of rounding units are presented in Burgess and Knowles (1999). Bounds on the number of bits of result required to perform correct rounding for division, square root, reciprocal, and square root reciprocal are presented in Iordache and Matula (1999) and Lang and Mullar (2001).

Verification and Testing

Verification of floating-point implementations supporting the IEEE Standard 754 is presented in Chen et al. (1996), Rusinoff (1998), Moore et al. (1998), and Cornea-Hasegan et al. (1999). An approach to the verifiable design of floating-point units is proposed in Even and Paul (2000). Benchmarks for floating-point arithmetic are presented in Karpinsky (1985). A number-theoretic approach to testing of rounding modes is developed in Parks (2000). A tool for testing of floating-point implementations is discussed in Verdonk et al. (2001).

8.10 Bibliography

Agarwal, R. C., F. G. Gustavson, and M. S. Schmookler (1999). Series approximation methods for divide and square root in the Power3 Microprocessor. In *Proceedings of the 14th IEEE Symposium on Computer Arithmetic*, pages 116–23.

Anderson, S. F., J. G. Earle, R. E. Goldschmidt, and D. M. Powers (1967). The IBM 360/370 model 91: floating-point execution unit. *IBM Journal of Research and Development*, pages 34–53.

ANSI and IEEE (1985). IEEE standard for binary floating-point arithmetic. *ANSI/IEEE Standard, Std* 754-1985, New York.

Bannon, P., and J. Keller (1995). Internal architecture of Alpha 21164 microprocessor. In *Digest of Papers COMPCON '95*, pages 79–87.

Beaumont-Smith, A., N. Burgess, D. Lefrere, and C. C. Lim (1999). Reduced latency IEEE floating-point standard adder architecture. In *Proceedings of the 14th IEEE Symposium on Computer Arithmetic*, pages 35–42.

Benschneider, B. J., W. J. Bowhill, E. M. Copper, M. N. Gavrielov, P. E. Gronowski, V. K. Maheshwari, V. Peng, J. D. Pickholtz, and S. Samudrala (1989). A pipelined 50 MHz CMOS 64-bit floating-point arithmetic processor. *IEEE Journal of Solid-State Circuits*, SC-24(5):1317–23.

Bohlender, G., P. Kornerup, D. W. Matula, and W. Walter (1991). Semantics for exact floating-point operations. In *Proceedings of the 10th IEEE Symposium on Computer Arithmetic*, pages 22–26.

Brent, R. P. (1973). On the precision attainable with various floating point number systems. *IEEE Transactions on Computers*, C-22(6):601–7.

Brown, W. S., and P. L. Richman (1969). The choice of base. *Communications of the ACM*, 12(10):560–61.

Bruguera, J. D., and T. Lang (1999). Leading-one prediction with concurrent position correction. *IEEE Transactions on Computers*, 48(10):1083–97.

Bruguera, J. D., and T. Lang (2001). Using the reverse-carry approach for double datapath floating-point addition. In *Proceedings of the 15th IEEE Symposium on Computer Arithmetic*, pages 203–10.

Bucholz, W. (1962). *Planning a New Computer System: Project STRETCH* (Chapter 14, page 210). Wiley & Sons, Inc., New York.

Burgess, N., and S. Knowles (1999). Efficient implementation of rounding units. In *Conference Record of the 33rd Asilomar Conference on Signals, Systems, and Computers*, volume 2, pages 1489–93.

Chen, C., L.-A. Chen, and J.-R. Cheng (2001). Architectural design of a fast floating-point multiplication-add fused unit using signed-digit addition. In *Proceedings Euromicro Symposium on Digital Systems Design*, pages 346–53.

Chen, Y.-A., E. Clarke, P.-H. Ho, Y. Hoskote, T. Kam, M. Khaira, J. O'Leary, and X. Zhao (1996). Verification of all circuits in a floating-point unit using word-level model checking. In *Proceedings of International Conference on Formal Methods in Computer-Aided Design (FMCAD '96)*, pages 19–33.

Clouser, J., M. Matson, R. Badeau, R. Dupcak, S. Samudrala, R. Allmon, and N. Fairbanks (1999). A 600-MHz superscalar floating-point processor. *IEEE Journal of Solid-State Circuits*, 34(7):1026–29.

Cody, W. J. (1973). Static and dynamic numerical characteristics of floating-point arithmetic. *IEEE Transactions on Computers*, C-22(6):598–601.

Cody, W. J. (1987). Analysis of proposals for the floating-point standard. *Computer*, 20(3):63–68.

Cody, W. J., J. T. Coonen, D. M. Gay, K. Hanson, D. Hough, W. Kahan, R. Karpinski, J. Palmer, F. N. Ris, and D. Stevenson (1984). A proposed radix-and-word-length-independent standard for floating-point arithmetic. *IEEE MICRO*, 4(4):86–100.

Coonen, J. T. (1980). An implementation guide to a proposed standard for floating-point arithmetic. *Computer*, 13(1):68–79.

Coonen, J. T. (1981). Underflow and the denormalized numbers. *Computer*, 14(3):75–87.

Cornea-Hasegan, M. A., R. A. Golliver, and P. Markstein (1999). Correctness proofs outline for Newton-Raphson based floating-point divide and square root algorithms. In *Proceedings of the 14th IEEE Symposium on Computer Arithmetic*, pages 96–105.

Dao-Trong, S., and K. Helwig (1992). A single-chip IBM System/390 floating-point processor in CMOS. *IBM Journal of Research and Development*, 36(4):733–48.

Darley, M., B. Kronlage, D. Bural, B. Churchill, D. Pulling, P. Wang, R. Iwamoto, and L. Yang (1990). The TMS390C602A floating-point coprocessor for Sparc systems. *IEEE Micro*, 10(3):36–47.

Dobberpuhl, D. W., R. T. Witek, R. Allmon, R. Anglin, D. Bertucci, S. Britton, L. Chao, R. A. Conrad, D. E. Dever, B. Gieseke, S. M. N. Hassoun, G. W. Hoeppner, K. Kuchler, M. Ladd, B. M. Leary, L. Madden, E. J. McLellan, D. R. Meyer, and J. Montanaro (1992). A 200-MHz 64-b dual-issue CMOS microprocessor. *IEEE Journal of Solid-State Circuits*, 27(11):1555–64.

Ercegovac, M. D., and T. Lang (1992). On-the-fly rounding. *IEEE Transactions on Computers*, 41(12):1497–1503.

Even, G., and W. J. Paul (2000). On the design of IEEE compliant floating-point units. *IEEE Transactions on Computers*, 49(5):398–413.

Even, G., and P.-M. Seidel (2000). A comparison of three rounding algorithms for IEEE floating-point multiplication. *IEEE Transactions on Computers*, 49(7):638–50.

Farmwald, M. P. (1981). *On the Design of High-Performance Digital Arithmetic Units*. PhD thesis, Stanford University.

Feldstein, A., and R. Goodman (1982). Loss of significance in floating-point subtraction and addition. *IEEE Transactions on Computers*, C-31:328–35.

Flynn, M. J., K. Nowka, G. Bewick, E. M. Schwarz, and N. Quach (1995). The SNAP project: towards sub-nanosecond arithmetic. In *Proceedings of the 12th Symposium on Computer Arithmetic*, pages 75–82.

Gerwig, G., and M. Kroener (1999). Floating-point unit in standard cell design with 116 bit wide dataflow. In *Proceedings of the 14th IEEE Symposium on Computer Arithmetic*, 266–73.

Goldberg, D. (1991). What every computer scientist should know about floating-point arithmetic. *ACM Computing Surveys*, 23(1):5–48.

Gosling, J. B. (1971). Design of large high-speed floating-point arithmetic units. In *IEE Proceedings*, volume 118, pages 493–98.

Greenley, D., et al. (1995). UltraSPARC: the next generation superscalar 64-bit SPARC. In *Digest of Papers. COMPCON '95. Technologies for the Information Superhighway*, pages 442–51.

Hokenek, E., and R. K. Montoye (1990). Leading zero anticipator (LZA) in the IBM Risc System/6000 floating-point execution unit. *IBM Journal of Research and Development*, 34(1):71–77.

Hokenek, E., R. K. Montoye, and P. W. Cook (1990). Second-generation RISC floating point with multiply-add fused. *IEEE Journal of Solid-State Circuits*, 25(5):1207–13.

Horel, T., and G. Lauterbach (1999). UltraSPARC-III: designing third-generation 64-bit performance. *IEEE Micro*, 19(3):73–85.

Hunt, D. (1995). Advanced performance features of the 64-bit PA-8000. In *Digest of Papers COMPCON '95*, pages 23–28.

Ide, N., H. Fukuhisa, Y. Kondo, T. Yoshida, M. Nagamatsu, J. Mori, I. Yamazaki, and K. Ueno (1993). A 320-MFLOPS CMOS floating-point processing unit for superscalar processors. *IEEE Journal of Solid-State Circuits*, SC-28(3):352–61.

Inui, S., T. Uesugi, H. Saito, Y. Hagihara, A. Yoshikawa, M. Nishida, and M. Yamashina (1999). A 250 MHz CMOS floating-point divider with operand pre-scaling. In *Symposium on VLSI Circuits*, pages 17–18.

Iordache, C., and D. W. Matula (1999). On infinitely precise rounding for division, square root, reciprocal and square root reciprocal. In *Proceedings of the 14th IEEE Symposium on Computer Arithmetic*, pages 233–40.

Jessani, R. M., and M. Putrino (1998). Comparison of single- and dual-pass multiply-add fused floating-point units. *IEEE Transactions on Computers*, 47(9):927–37.

Kabuo, H., T. Taniguchi, A. Miyoshi, H. Yamashita, M. Urano, H. Edamatsu, and S. Kuninobu (1994). Accurate rounding scheme for the Newton-Raphson method using redundant binary representation. *IEEE Transactions on Computers*, 43(1):43–51.

Kahan, W. (1996). Lecture Notes on the Status of IEEE 754. Technical Report http://http.cs.berkeley.edu/~wkahan/ieee754status/ieee754.ps, University of California, Berkeley.

Karpinsky, R. (1985). PARANOIA: A floating-point benchmark. *BYTE*, 10(2):223–35.

Knuth, D. E. (1998). *The Art of Computer Programming: Seminumerical Algorithms*. Addison Wesley, Reading, Massachusetts, 3rd edition.

Kuck, D. J., S. Parker, and A. Sameh (1977). Analysis of rounding methods in floating-point arithmetic. *IEEE Transactions on Computers*, C-26:643–50.

Kuki, H., and W. J. Cody (1973). A statistical study of the accuracy of floating point number systems. *Communications of the ACM*, 16(14):223–30.

Kulisch, U. W. (1977). Mathematical foundation of computer arithmetic. *IEEE Transactions on Computers*, C-26(7):610–21.

Kulisch, U. W., and W. L. Miranker (1981). *Computer Arithmetic in Theory and Practice*. Academic Press, New York.

Lang, T., and J.-M. Muller (2001). Bounds on runs of zeros and ones for algebraic functions. In *Proceedings of the 15th IEEE Symposium on Computer Arithmetic*, pages 13–20.

Markstein, P. (2000). *IA-64 and Elementary Functions: Speed and Precision*. Hewlett-Packard Professional Books. Prentice Hall.

Markstein, P. W. (1990). Computation of elementary functions on IBM RISC System/6000 processor. *IBM Journal of Research and Development*, pages 111–19.

Montoye, R. K., E. Hokonek, and S. L. Runyan (1990). Design of the floating-point execution unit of the IBM RISC System/6000. *IBM Journal of Research and Development*, 34(1):59–70.

Moore, J. S., T. W. Lynch, and M. Kaufmann (1998). A mechanically checked proof of the AMD5$_k$86 floating-point division program. *IEEE Transactions on Computers*, 47(9):913–26.

Naini, A., A. Dhablania, W. James, and D. Das Sarma (2001). 1 GHz HAL SPARC64^r dual floating point unit with RAS features. In *Proceedings of the 15th IEEE Symposium on Computer Arithmetic*, pages 173–83.

Nicks, T. N., R. E. Fry, and P. E. Harvey (1994). POWER2 floating-point unit: Architecture and implementation. *IBM Journal of Research and Development*, 38(5):525–36.

Nielsen, A. M., D. W. Matula, C. N. Lyu, and G. Even (2000). An IEEE compliant floating-point adder that conforms with the pipelined packet-forwarding paradigm. *IEEE Transactions on Computers*, 49(1):33–47.

Oberman, S. F. (1996). *Design Issues in High Performance Floating Point Arithmetic Units*. PhD thesis, Department of Electrical Engineering, Stanford University.

Oberman, S. F. (1999). Floating-point division and square root algorithms and implementation in the AMD-K7 microprocessor. In *Proceedings of the 14th IEEE Symposium on Computer Arithmetic*, pages 106–15.

Oberman, S. F., G. Favor, and F. Weber (1999). AMD 3DNow technology: architecture and implementations. *IEEE Micro*, 19(2):37–48.

Oberman, S. F., and M. J. Flynn (1996a). Fast IEEE rounding for division by functional iteration. Technical Report CSL-TR-96-700, Computer Systems Laboratory, Department of Electrical Engineering and Computer Science, Stanford University.

Oberman, S. F., and M. J. Flynn (1996b). Implementing division and other floating-point operations: A system perspective. In *Scientific Computing and Validated Numerics (Proceedings of SCAN'95)*, pages 18–24.

Oberman, S. F., and M. J. Flynn (1997). Design issues in division and other floating-point operations. *IEEE Transactions on Computers*, 46(2):154–61.

Oberman, S. F., and M. J. Flynn (1998). Reducing the mean latency of floating-point addition. *Theoretical Computer Science*, 196(1–2):201–14.

Oklobdzija, V. G. (1994). An algorithmic and novel design of a leading zero detector circuit: comparison with logic synthesis. *IEEE Transactions on Very Large Scale Integration (VLSI) Systems*, 2(1):124–28.

Overton, M. A. (2001). *Numerical Computing with IEEE Floating-Point Arithmetic*. SIAM.

Park, W.-C., T.-D. Han, S.-D. Kim, and S.-B. Yang (1999). A floating point multiplier performing IEEE rounding and addition in parallel. *Journal of Systems Architecture*, 45(14):1195–1207.

Parks, M. (2000). Number-theoretic test generation for directed roundings. *IEEE Transactions on Computers*, 49(7):651–58.

Prabhu, J. A., and G. B. Zyner (1995). 167 MHz radix-8 divide and square root using overlapped radix-2 stages. In *Proceedings of the 12th IEEE Symposium on Computer Arithmetic*, pages 155–62.

Priest, D. M. (1991). Algorithms for arbitrary precision floating point arithmetic. In *Proceedings of the 10th IEEE Symposium on Computer Arithmetic (Arith-10)*, pages 132–44.

Reiser, J. F., and D. E. Knuth (1975). Evading the drift in floating-point addition. *Information Processing Letters*, 3(3):84–87.

Rusinoff, D. (1998). A mechanically checked proof of IEEE compliance of a register-transfer-level specification of the AMD-k7 floating-point multiplication, division, and square root instructions. *LMS Journal of Computation and Mathematics*, 1:148–200.

Santoro, M. R., G. Bewick, and M. A. Horowitz (1989). Rounding algorithms for IEEE multipliers. In *Proceedings of the 9th Symposium on Computer Arithmetic*, pages 176–83.

Schwarz, E. M. (1995). Rounding for quadratically converging algorithms for division and square root. In *Conference Record of the 29th Asilomar Conference on Signals, Systems and Computers*, volume 1, pages 600–3.

Schwarz, E. M., R. M. Smith, and C. A. Krygowski (1999). The S/390 G5 floating-point unit supporting hex and binary architectures. In *Proceedings of the 14th IEEE Symposium on Computer Arithmetic*, pages 258–65.

Seidel, P.-M., and G. Even (2001). On the design of fast IEEE floating-point adders. In *Proceedings of the 15th IEEE Symposium on Computer Arithmetic*, pages 184–94.

Sharangpani, H., and K. Arora (2000). Itanium processor microarchitecture. *IEEE Micro*, 20(5):24–43.

Soderquist, P., and M. Leeser (1996). Area and performance tradeoffs in floating-point division and square root implementations. *ACM Computing Surveys*, 28(3):518–64.

Sterbenz, P. H. (1974). *Floating Point Computation*. Prentice-Hall, Englewood Cliffs, New Jersey.

Suzuki, H., H. Makino, K. Mashiko, and H. Hamano (1997). A floating-point divider using redundant binary circuits and an asynchronous clock scheme. In *Proceedings of the International Conference on Computer Design: VLSI in Computers and Processors*, pages 685–89.

Suzuki, H., Y. Nakase, H. Makino, H. Morinaka, and K. Mashiko (1995). Leading-zero anticipatory logic for high-speed floating point addition. In *Proceedings of the IEEE 1995 Custom Integrated Circuits*, pages 589–92.

Sweeney, D. W. (1965). An analysis of floating-point addition. *IBM Systems Journal*, 4:31–42.

Uya, M., K. Kaneko, and J. Yasui (1984). A CMOS floating-point multiplier. *IEEE Journal of Solid-State Circuits*, SC-19(5):697–702.

Vassiliadis, S., D. S. Lemon, and M. Putrino (1989). S/370 sign-magnitude floating-point adder. *IEEE Journal of Solid-State Circuits*, 24:1062–70.

Verdonk, B., A. Cuyt, and D. Verschaeren (2001). A precision- and range-independent tool for testing floating-point arithmetic. i. Basic operations, square root, and remainder. *ACM Transactions on Mathematical Software*, 27(1):92–118.

Ware, F. A., W. McAllister, J. R. Carlson, D. K. Sun, and R. J. Vlach (1982). 64 bit monolithic floating-point processors. *IEEE Journal of Solid-State Circuits*, SC-17(5):898–907.

Wilkinson, J. H. (1963). *Rounding Errors in Algebraic Processes*. Prentice-Hall, Englewood Cliffs, New Jersey.

Williams, T., N. Patkar, and G. Shen (1995). SPARC64: a 64-b 64-active-instruction out-of-order-execution MCM processor. *IEEE Journal of Solid-State Circuits*, 30(11):1215–26.

Yeager, K. C. (1996). The Mips R10000 superscalar microprocessor. *IEEE Micro*, 16(2):28–41.

Yohe, J. M. (1973). Roundings in floating-point arithmetic. *IEEE Transactions on Computers*, C-22(6):577–86.

Yu, R. Y., and G. B. Zyner (1995). 167 MHz radix-4 floating point multiplier. In *Proceedings of the 12th IEEE Symposium on Computer Arithmetic*, pages 149–54.

IN THIS CHAPTER WE PRESENT AND DISCUSS THE FOLLOWING TOPICS:

- Modes of operation and algorithm and implementation models
- Least-significant-digit-first (LSDF) arithmetic: addition, subtraction, and multiplication
- Most-significant-digit-first (MSDF) online arithmetic: addition, subtraction. General design method: application to multiplication and division. Multioperation and composite algorithms

CHAPTER 9

Digit-Serial Arithmetic

9.1 Introduction

In the previous chapters we described algorithms and implementations for arithmetic modules that have the inputs applied all at once (in parallel) and deliver the results in the same way. However, since the numerical values are represented by digit vectors, it is possible to apply the inputs and deliver the output one digit at a time (serially), so that all digits of the same numerical operands/results share the same digit lines. The system is usually clocked so that one digit is applied/delivered per clock cycle. This serial alternative is the topic of this chapter. We consider the case in which all operands and results are serial, although it is possible to have a mixed system in which some inputs and outputs are serial and others parallel. The methods to design these mixed systems can be devised from those for parallel and serial systems.

The main reason for having serial input/output is to reduce the number of signal lines connecting modules and to simplify their interface, since these connections and interfaces influence both area and energy dissipation. The drawback is the time (number of cycles) required to receive the inputs and to deliver the results. This delay can be compensated for by overlapping the execution of successive operations (even if dependent), since the successor operation can begin when a few digits of the operands have been received.

Although the algorithm to perform the operation, as well as the implementation of the module, is affected by the serial characteristic of the signals, it is important to distinguish between the characteristics of the input/output signals and that of the algorithm and implementation. For instance, one possible algorithm for serial input/outputs is to collect all the input digits before beginning the operation and to produce the result in parallel, before delivering it in a serial manner. However, this would add the delay of collecting and delivering to the time to perform the operation; consequently, a reduction of delay is achieved if an

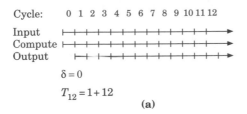

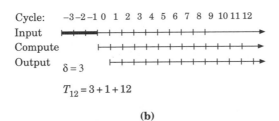

FIGURE 9.1 Timing characteristics of serial operation with $n = 12$. (a) With $\delta = 0$. (b) With $\delta = 3$.

algorithm is devised that takes into account the serial nature of the signals. Similarly, in parallel input/output systems some algorithms use operands and produce results in a digit-serial manner; examples of this are the sequential multiplication algorithm and the digit recurrence division algorithm.[1]

9.1.1   Modes of Operation and Algorithm and Implementation Models

We consider the case in which the numerical values are represented in a radix-r number system. In some cases, we use conventional representations, while in others redundant representations are preferable.

A serial signal is a numerical input or output with one digit per clock cycle. Figure 9.1 shows typical timing diagrams for a serial operation, in which in each cycle one digit of each operand is applied and one digit of the output is delivered. Note that by convention we denote as cycle 1 the cycle in which the

---

1. Although in this case the digits produced are in a signed-digit representation, so that a conventional representation can only be delivered when all digits have been obtained.

first digit of the output is delivered. The total execution time is the sum of two components:

- The *initial delay* δ, which corresponds to the additional number of operand digits required to determine the first result digit. That is, the first output digit is delivered $\delta + 1$ cycles after the application of the first input digits. So, as shown in Figure 9.1(a), $\delta = 0$ corresponds to the case in which the first output digit is delivered one cycle after the application of the first input digits. Figure 9.1(b) shows a case in which the first output is delivered in the cycle after four input digits have been applied ($\delta = 3$).
- The *time to deliver the n output digits*. Since one digit is delivered per cycle, for an output of n digits, this time is equal to n cycles.

Consequently, the *execution time* is

$$T_n = \delta + 1 + n \qquad\qquad 9.1$$

Serial Modes

Two serial modes are typical:

1. *Least-significant digit first* (LSDF) mode. The digits of the operands (result) are applied serially starting from the least-significant digit. This mode is also known as right-to-left mode and, since it was the first serial mode, typically this mode is implied when the term "serial arithmetic" is used.

    Because of the order of the digits, the indexing is simplified if right-to-left indexing is used, as in the representation of integers, namely,

$$x = \sum_{i=0}^{n-1} x_i r^i \qquad\qquad 9.2$$

2. *Most-significant digit first* (MSDF) mode. The digits are applied starting from the most-significant digit (left-to-right mode). Arithmetic performed in this mode is known as *online arithmetic*, and the corresponding initial delay is called *online delay*.

    The indexing is simplified here by using left-to-right indexing, as in the representation of fractions, that is,

$$x = \sum_{i=1}^{n} x_i r^{-i} \qquad\qquad 9.3$$

Algorithm and Implementation Model

We now describe a general model for a serial algorithm and its implementation. Consider an operation with two n radix-r digit operands, x and y, and one result z. The input-output model is described as follows.

In cycle j the result digit z_{j+1} is computed. Consequently the cycles are labeled from $-\delta, \ldots, 0, 1, \ldots, n$ so that in cycle j the operand digits $x_{j+1+\delta}$ and $y_{j+1+\delta}$ are received, output digit z_{j+1} is computed, and output digit z_j is delivered (Figure 9.2(a)). To conform with both serial modes, in LSDF (MSDF) mode digits are counted from the least-significant (most-significant) side.

The algorithm consists of recurrences on numerical values. In each of the $n + \delta$ iterations, one digit of the operands is introduced (for the last δ iterations the input digits are set to zero), an internal state w (also called a residual) is updated, and one digit of the result is produced (zero for the first δ cycles).[2] An additional cycle is needed to deliver the last result digit.

Calling $x[j], y[j]$, and $z[j]$ the numerical values of the corresponding signals when the representation consists of the first $j + \delta$ digits for the operands and j digits for the result, iteration j is described by

$$x[j + 1] = (x[j], x_{j+1+\delta})$$

$$y[j + 1] = (y[j], y_{j+1+\delta})$$

$$z_{j+1} = F(w[j], x[j], x_{j+1+\delta}, y[j], y_{j+1+\delta}, z[j]) \qquad \textbf{9.4}$$

$$z[j + 1] = (z[j], z_{j+1})$$

$$w[j + 1] = G(w[j], x[j], x_{j+1+\delta}, y[j], y_{j+1+\delta}, z[j], z_{j+1})$$

Figure 9.2(b) depicts the serial algorithm and implementation model.

The initial delay δ depends on the serial mode and on the specific operation (Table 9.1). As can be seen from the table, for the MSDF mode all basic operations can be performed with a small and fixed (independent of the precision) initial delay. On the other hand, for the LSDF mode, only addition and multiplication have a small initial delay, whereas division, square root, and max/min have an initial delay $O(n)$, which means that this mode is not suitable for these

---

2. When more than n digits of the result are required, such as in some multiplications in which all $2n$ digits are delivered, the number of iterations is increased correspondingly.

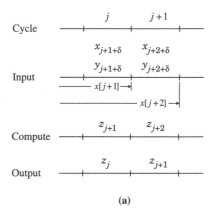

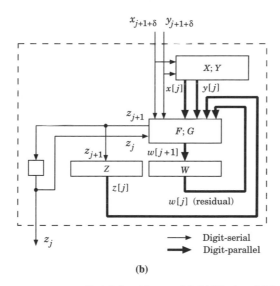

FIGURE 9.2 Serial algorithm model: (a) Timing. (b) Implementation.

operations. Moreover, the initial delay is also $O(n)$ for multiplication if only the most-significant half of the product is required (see Figure 9.3(a)).

As seen in Figure 9.3(b), online arithmetic is well-suited for variable precision computations: once a desired precision is obtained, the operation can terminate.

Operation	LSDF	MSDF
Addition	0	2 ($r = 2$)
		1 ($r \geq 4$)
Multiplication	0	3 ($r = 2$)
		2 ($r = 4$)
Only MS half of product	n	
Division	$2n$*	4
Square root	$2n$*	4
Max/min	n	0

* The result digits delivered LS first.

TABLE 9.1 Initial delay (δ).

LSDF mode

n-digit addition:

Cycle: 0 1 2 ...

      LSD   MSD

  Inputs: x x x x x x x x x
  Output: x x x x x x x x x

n by $n \rightarrow 2n$ multiplication:

      LSD   MSD

  Inputs: x x x x x x x x x
  Output: x x x x x x x x x x x x x x x x x x

                        ‾‾‾‾‾‾‾‾‾‾‾
                        MS half

           (a)

MSDF mode

  Cycle: $\overline{2}\,\overline{1}$ 0 1 2 ...
n-digit operation:

      MSD   LSD

  Inputs: x x x x x x x x x
  Output: x x x x x x x x x

        ‾‾
      online delay $= 2$

           (b)

FIGURE 9.3 (a) LSDF and (b) MSDF modes.

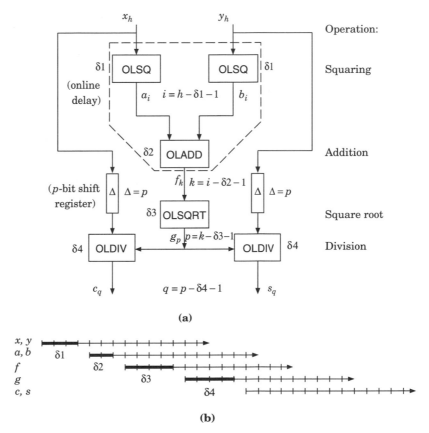

FIGURE 9.4 Online computation in 2D vector normalization: (a) Network. (b) Timing diagram.

Composite Algorithm

Since the execution time of a serial operation can be high, it is convenient to develop composite algorithms in which the execution of successive (dependent) operations overlap; that is, a successor operation can begin as soon as the result digits of its predecessors are available. This is illustrated in the following example where a sequence of operations is implemented by a network of digit-serial (online) arithmetic modules. The network in Figure 9.4(a) implements the expressions for the 2D vector normalization.[3]

$$c = \frac{x}{\sqrt{x^2 + y^2}} \qquad s = \frac{y}{\sqrt{x^2 + y^2}} \qquad\qquad \textbf{9.5}$$

The corresponding timing diagram is given in Figure 9.4(b).

The online delay of the network is the sum of online delays of the operations on the longest path. For $r = 2$, we obtain from Table 9.1

$$\Delta_{norm} = \delta 1 + \delta 2 + \delta 3 + \delta 4 = 3 + 2 + 4 + 4 = 13 \qquad\qquad \textbf{9.6}$$

The total execution time for the composite operation is $D_{norm} = \Delta_{norm} + 4 + n$.

The more levels there are in a sequence of operations and the longer the precision, the more advantageous is the online approach.

To reduce further the execution time, the three modules in the dashed box in Figure 9.4(a) can be merged into a single online module, called a *composite module*, with a shorter online delay than the sum of the online delays of the dependent components.

The latency in the case of LSDF arithmetic is obtained in a similar manner (Exercise 9.1).

9.2   LSDF Arithmetic

We consider now the algorithms and implementation for addition/subtraction and for multiplication for the LSDF mode. As discussed before, these are the basic operations that result in a small initial delay. This is true for multiplication if all result digits are required.

9.2.1   LSDF Addition and Subtraction

In addition/subtraction the internal state corresponds to the carry. Consequently, the initial delay is $\delta = 0$ and the radix-2^k implementation consists of a k-bit adder with a carry flip-flop (or a latch), as illustrated in Figure 9.5(a). Subtraction is performed by adding the two's complement of operand y. This is done by (bit) complementing input y_i and initializing the carry flip-flop to 1. Overflow is detected as in a bit-parallel adder (see Chapter 1).

---

3. Ercegovac and Lang (1988b, 1999).

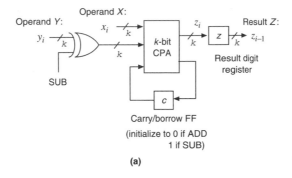

(a)

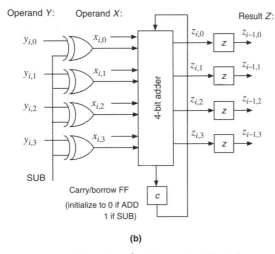

(b)

FIGURE 9.5 (a) Radix-2^k digit-serial adder/subtractor. (b) Radix-16 digit-serial adder/subtractor.

The cycle delay is

$$t_{LSDFadd\text{-}k} = t_{CPA(k)} + t_{FF} \qquad\qquad \textbf{9.7}$$

and the total time for n-bit addition is

$$T_{LSDFadd\text{-}n} = \left(\frac{n}{k} + 1\right) t_{LSDFadd\text{-}k} \qquad\qquad \textbf{9.8}$$

The cost is one k-bit CPA, k XOR gates, one flip-flop, and one k-bit output register.

A radix-16 adder/subtractor is shown in Figure 9.5(b).

9.2.2　LSDF Multiplication

There are many schemes for performing LSDF multiplication, differing in the treatment of inputs and outputs and in the design of basic cells. We concentrate on the two most commonly used digit-serial multipliers for radix-2 and two's complement representation:

1.　Serial-serial (LSDF-SS) multiplier, with both operands used in digit-serial form.

2.　Serial-parallel (LSDF-SP) multiplier, in which one operand is first converted to parallel form. It is similar to the sequential multiplier discussed in Chapter 4.

In both types of multipliers the output is produced digit-serially. Since there are n input digits and $2n$ product digits with the most-significant half obtained in cycles $n+1$ to $2n+1$, the operation cannot be completed during the input of the operands.

Serial-Serial Multiplier (Radix 2)

We define an internal state (residual)

$$w[j] = 2^{-(j+1)}(x[j] \times y[j] - p[j])\qquad\qquad 9.9$$

where

$$x[j] = \sum_{i=0}^{j} x_i 2^i$$

and similarly for $y[j]$ and $p[j]$. Since now both operands are used in serial form, the recurrence is

$$
\begin{aligned}
w[j+1] &= 2^{-(j+2)}(x[j+1] \times y[j+1] - p[j+1]) \\
&= 2^{-(j+2)}((x[j]+x_{j+1}2^{j+1})(y[j]+y_{j+1}2^{j+1}) - (p[j]+p_{j+1}2^{j+1})) \\
&= 2^{-1}(w[j]+y[j+1]x_{j+1}+x[j]y_{j+1}-p_{j+1})\qquad\qquad 9.10
\end{aligned}
$$

Calling

$$v[j] = w[j]+y[j+1]x_{j+1}+x[j]y_{j+1}\qquad\qquad 9.11$$

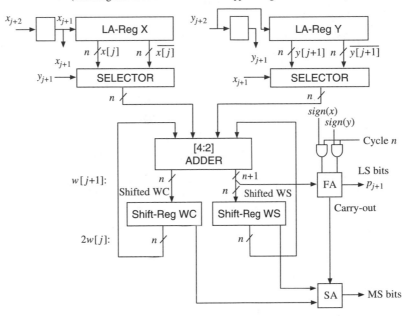

(Shift-register for load control in left-append registers not shown)

FIGURE 9.6 Serial-serial two's complement radix-2 multiplier.

to keep $w[j+1]$ an integer and $p_{j+1} \in \{0, 1\}$ we make

$$w[j+1] = \lfloor 2^{-1}v[j] \rfloor$$
$$p_{j+1} = v[j] \bmod 2$$

9.12

With a carry-save form of the residual $w[j]$ (two bit-vectors) and adding to it two multiples ($y[j+1]x_{j+1}$ and $x[j]y_{j+1}$), the addition to produce $v[j]$ is implemented by a [4:2] adder of n positions, as shown in Figure 9.6. The bit-vectors $x[j]$ and $y[j+1]$ are generated using two left-appending registers with load controlled by a "moving" 1 shift-register. The input latches for x_{j+1} and y_{j+1} are used to avoid left-appending in both $X(Y)$ registers and the corresponding selectors at the expense of one extra cycle.

The residual is produced by a (wired) shift right of one position. The least-significant digit is peeled off as the result digit.

This recurrence is performed n times. After that, since the input digits are 0, we obtain $w[j+1] = 2^{-1}w[j] - z_{j+1}$, so that the rest of the result digits are obtained by shifting right the residual.

For two's complement representation, the operand bits x_{n-1} and y_{n-1} have negative weights and the last two multiples are possibly negative. Instead of subtraction, addition of their two's complements is used. This is performed as usual by taking ones' complement and adding a 1 as carry-in. The two carry-ins are incorporated in a FA stage as shown in Figure 9.6. During the cycles 0 to $n-1$ the FA stage simply transmits the LS product bit. During the nth cycle the FA stage produces product bit p_{n-1} and a carry-out as the result of adding x_{n-1}, y_{n-1} (carry-ins for two's complement) and the LS sum bit of the [4:2] adder. The carry-out initializes the carry-in FF in the serial adder SA, which produces the remaining product bits $p_n, \ldots, p_{2n-1}$.

The total execution time of the operation is

$$T_{SSMULT} = 2nt_{cyc} \qquad\qquad \textbf{9.13}$$

where the delay of the critical path in a cycle is

$$t_{cyc} = t_{SEL} + t_{[4:2]} + t_{FF} \qquad\qquad \textbf{9.14}$$

The cost is one n-bit [4:2] adder, 5 n-bit registers, and gates to form multiples. Compared to a serial-parallel sequential multiplier, a serial-serial multiplier has a longer t_{cyc} and requires more circuits. Its main justification is the ability to begin producing product bits while inputting the operands.

Serial-Parallel Multiplier (Radix 2)

The core of this multiplier is similar to the sequential multiplier discussed in Chapter 4. To that core it is necessary to add modules to (1) convert one of the operands to parallel form and (2) deliver the result in serial form. One possible implementation is to perform the operation in $3n$ cycles split into three phases:

- Phase 1: Serial input and conversion of one operand to parallel form—not necessary if one operand is constant.[4] The second operand can be entered together with the first and stored in a register, or it can be entered during Phase 2, maybe using the same digit lines as the first operand.

---

4. This case with a constant operand is frequent in signal-processing applications, such as FIR filters (Oppenheim et al. 1999). In such a case, this type of multiplier is especially convenient.

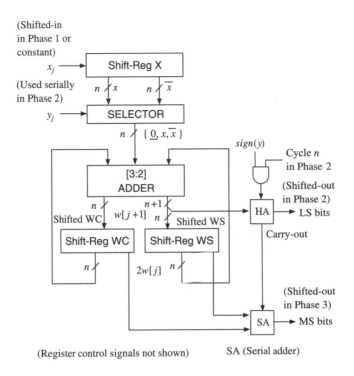

FIGURE 9.7 Three-phase serial multiplier.

- Phase 2: Serial-parallel processing and output of the LS half of the product.
- Phase 3: Serial output of the MS half of the product.

The multiplier in Phase 2 corresponds to the sequential multiplier discussed in Chapter 4. The difference here is the serial delivery of the result. Figure 9.7 shows a block diagram of the implementation.

If the multiplier is negative, a subtraction is performed in the last cycle of multiplication by adding the two's complement of the multiplicand. A half-adder is used to add a carry-in in cycle n. This assures that one product bit is produced per cycle.

Since the least-significant bit of the product is produced in the first cycle of Phase 2, the initial delay is n for the case of nonconstant operands (cycles of Phase 1) and 0 if one of the operands is constant. On the other hand, for fractional operands and n-bit result (most-significant n bits), the LS bits are suppressed and only the MS half is delivered. This increases the initial delay by n. To obtain a rounded MS half of the product, a 1 is inserted in the least-significant bit of the initial carry-save partial product.

The critical path in a cycle is

$$t_{cyc} = t_{SEL} + t_{CSA} + t_{FF} \qquad \qquad \textbf{9.15}$$

The delay of the LSDF-SP multiplier (from time of LS bit of operand to MS bit of product) is $T_{SPrnd} = 3n \times t_{cyc}$. The cost is similar to the cost of a sequential multiplier.

This design can be extended to radix 4 with recoding using the scheme described in Chapter 4, Section 4.1.

This approach is suitable for systems where a high throughput is the primary objective: the phases can be used as pipeline stages so that up to three multiplications can be in progress.

9.3   MSDF: Online Arithmetic

As indicated in Section 9.1, online arithmetic algorithms operate in a digit-serial MSDF mode. Moreover, as shown in Figure 9.8 there is an online delay δ so that to compute the first digit of the result, $\delta + 1$ digits of the input operands are needed. Thereafter, for each new digit of the operands, an extra digit of the

Cycle	-2	-1	0	1	2	$\cdots$
Input	x_1	x_2	x_3	x_4	x_5	$\cdots$
Compute			z_1	z_2	z_3	$\cdots$
Output				z_1	z_2	$\cdots$

$$\delta = 2$$

FIGURE 9.8 Timing in online arithmetic.

result is obtained. As indicated in the figure, another cycle is needed to output the computed result digit.

The left-to-right mode of computation requires a flexibility in computing output digits on the basis of partial information about inputs. This is achieved by the use of *redundancy* in the number representation system, which allows several representations of a given value. The main redundant representation systems are signed-digit and carry-save, as presented in Chapter 2. With these representations, there is a flexibility in choosing an output digit so that, if necessary, a compensation can be introduced in the following iterations. In online arithmetic the most-frequent representation used is signed-digit, with both symmetric $\{-a, \ldots, a\}$ and asymmetric $\{b, \ldots, c\}$ digit sets. Overredundant digit sets are also useful.

Since different redundant representations are possible, an integrated approach for a complex computation can use heterogeneous representations to optimize the implementation.

In addition to redundancy in the representation of the serial signals, to have fast addition operations some internal signals also use a redundant representation, as discussed in Chapter 2.

In some cases, conversion from redundant to conventional representation is needed; this conversion in parallel arithmetic requires a carry-propagate addition. In the online approach, the conversion can be performed efficiently without carry-propagate addition using an on-the-fly conversion method discussed in Chapter 5.

Algorithms have been developed for most of the basic arithmetic operations, as well as for certain composite operations. Of significance is the larger set of operations for which an online implementation has a small initial delay, in contrast with the LSDF approach.

We first describe the algorithms and implementations for online addition and subtraction, which can be obtained directly from the parallel counterparts. Then we develop a general method of designing online algorithms and implementations and apply the method to multiplication and division.

9.3.1   Addition/Subtraction

The online addition/subtraction algorithm can be obtained from the serialization of a redundant adder (carry-save or signed-digit, see Chapter 2). As indicated there, for radices larger than 2 with $a > r/2$, redundant addition allows a transfer digit that propagates only to the adjacent more significant digit. Consequently, the

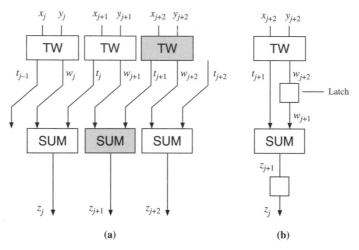

FIGURE 9.9 (a) A segment of radix-$r > 2$ signed-digit parallel adder. (b) Radix-$r > 2$ online adder. All latches cleared at start.

corresponding online adder, shown in Figure 9.9, has an online delay of 1 and corresponds to the following expressions:

$$(t_{j+1}, w_{j+2}) = \begin{cases} (0, x_{j+2} + y_{j+2}) & \text{if } |x_{j+2} + y_{j+2}| \leq a - 1 \\ (1, x_{j+2} + y_{j+2} - r) & \text{if } x_{j+2} + y_{j+2} \geq a \\ (-1, x_{j+2} + y_{j+2} + r) & \text{if } x_{j+2} + y_{j+2} \leq -a \end{cases} \qquad \textbf{9.16}$$

and

$$z_{j+1} = w_{j+1} + t_{j+1} \qquad \textbf{9.17}$$

where $x_j, y_j, z_j \in \{-a, \dots, a\}$.

EXAMPLE 9.1   In Table 9.2, we illustrate a radix-4 online addition with $a = 3$ and operands

$$x = (.12\bar{3}30\bar{1})$$
$$y = (.2\bar{1}3322)$$

The result is $z = (1.\bar{1}0\bar{1}221)$.

j	x_{j+2}	y_{j+2}	t_{j+1}	w_{j+2}	w_{j+1}	z_{j+1}	z_j
-1	1	2	1	-1	0*	1	0*
0	2	-1	0	1	-1	-1	1
1	-3	-3	-1	-2	1	0	-1
2	3	3	1	2	-2	-1	0
3	0	2	0	2	2	2	-1
4	-1	2	0	1	2	2	2
5	0	0	0	0	1	1	2
6	0	0	0	0	0	0	1

* Latches initialized to 0.

TABLE 9.2 Example of radix-4 online addition.

Note that during cycle 0 the result digit $z_0 = 1$ is produced. Although this might be interpreted as an overflow, the range of the result remains less than 1 since the next digit is –1. See also Exercise 9.10. ■

The cycle time corresponds to the delay of one digit radix-r signed-digit adder plus the loading of the register.

For $r = 2$, the condition $a > r/2$ is not satisfied, so that in the corresponding signed-digit adder output digit z_j depends on input digits up to index $j + 2$. Consequently, Figure 9.10 illustrates how a digit-parallel radix-2 signed-digit adder (Chapter 2, Section 2.12.2) is converted into a radix-2 online adder with online delay $\delta = 2$. In this implementation, a signed-digit $x_i \in \{-1, 0, 1\}$ is represented by a pair of binary variables (x_i^+, x_i^-) such that

$$x_i = x_i^+ - x_i^- \qquad \text{9.18}$$

The cycle time is

$$t_{cyc} = 2t_{FA} + t_{FF} \qquad \text{9.19}$$

and the operation time

$$T_{OLADD\text{-}2} = (2 + n + 1)t_{cyc} \qquad \text{9.20}$$

The cost is 2 FAs and 5 FFs.

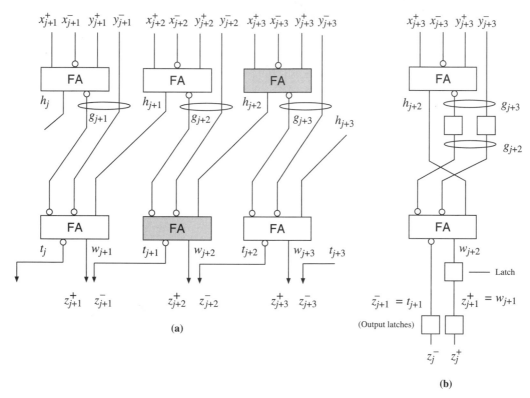

FIGURE 9.10 (a) A segment of radix-2 signed-digit parallel adder. (b) Radix-2 online adder. All latches cleared at start.

EXAMPLE 9.2   We illustrate radix-2 online addition with operands

$$x = (.010\bar{1}110\bar{1})$$
$$y = (.10\bar{1}01\bar{1}\,\bar{1}0)$$

The result is $z = (1.\bar{1}0100\bar{1}01)$. The signals in Table 9.3 correspond to Figure 9.10(b) and signed bits are encoded using 9.18.

Note that during cycle 0 the result digit $z_0 = 1$ is produced. Although this might be interpreted as an overflow, the range of the result remains less than 1 since the next digit is -1. See also Exercise 9.10.   ■

j	x_{j+3}	y_{j+3}	$x^+_{j+3}x^-_{j+3}$	$y^+_{j+3}y^-_{j+3}$	h_{j+2}	g_{j+3}	g_{j+2}	$t_{j+1}w_{j+2}$	$z^+_{j+1}z^-_{j+1}$	z_j
-2	0	1	00	10	1	10	00*	01	—	—
-1	1	0	10	00	1	10	10	00	10	—
0	0	-1	00	01	0	01	10	11	01	1
1	-1	0	01	00	0	10	01	11	11	-1
2	1	1	10	10	1	00	10	00	10	0
3	1	-1	10	01	1	11	00	01	00	1
4	0	-1	00	01	0	01	11	10	11	0
5	-1	0	01	00	0	10	01	11	01	0
6	0	0	00	00	0	00	10	11	11	-1
7	0	0	00	00	0	00	00	00	10	0
8	0	0	00	00	0	00	00	00	00	1

* g latches initialized to 00.

TABLE 9.3 Example of radix-2 online addition.

9.3.2   A Method for Developing Online Algorithms

We now describe a method to develop online algorithms and implementations. This method is a generalization of the method presented in Chapter 5 for digit recurrence division; we assume that the reader is familiar with that material and will consult some of the notation and definitions there. The method consists of two parts. Part 1 defines the residual and the corresponding digit recurrence. Part 2 determines the output-digit selection function. There are several possibilities in defining the selection function, the main ones being (1) selection using selection constants, similar to the quotient-digit selection derivation described in Chapter 5, and (2) selection by other methods such as rounding and truncation of the residual to produce the output digit. We discuss both selection techniques, leaving the choice open.

In later sections we illustrate the method by the operations of multiplication and division. Other operations, such as square root, sum of squares, and maximum, can be developed in the same manner.

In terms of the components described in Section 9.1, Part 1 consists of the development of the recurrence on the residual (internal state) $w[j]$ such that

$$w[j+1] = G(w[j], x[j], x_{j+1+\delta}, y[j], y_{j+1+\delta}, z[j], z_{j+1}) \qquad 9.21$$

for $-\delta \leq j \leq n-1$ where

$$x[j] = \sum_{i=1}^{j+\delta} x_i r^{-i}, \quad y[j] = \sum_{i=1}^{j+\delta} y_i r^{-i}, \quad z[j] = \sum_{i=1}^{j} z_i r^{-i} \qquad 9.22$$

are the online forms of the operands and the result, respectively. Moreover, the bounds of the residual are determined.

In Part 2 the result digit is obtained as

$$z_{j+1} = F(w[j], x[j], x_{j+1+\delta}, y[j], y_{j+1+\delta}, z[j]) \qquad 9.23$$

Part 1: Residual and Its Recurrence

- **Step 1.** Describe the online operation by the bound on the error after j digits have been computed. For an operation f with operands x and y and result z, this bound has the form (for simplicity we consider the case $\rho = 1$)

$$|f(x[j], y[j]) - z[j]| < r^{-j} \qquad 9.24$$

- **Step 2.** Transform expression (9.24) so that it can be used to develop a recurrence with only primitive operations, such as multiplication by r (shift), addition/subtraction, and multiplication by a single digit. The form of the resulting expression is

$$\underline{B} < G(f(x[j], y[j]) - z[j]) < \overline{B} \qquad 9.25$$

where G represents the required transformation and $\underline{B}$ and $\overline{B}$ are the transformed bounds.

For example, a division error expression

$$|x[j]/y[j] - z[j]| < r^{-j}$$

is transformed into

$$|x[j] - z[j] \cdot y[j]| < |r^{-j} y[j]| \qquad 9.26$$

to avoid the use of division. Similarly, for square root the error expression

$$|x[j]^{1/2} - z[j]| < r^{-j}$$

is transformed into

$$-2z[j]r^{-j} + r^{-2j} < x[j] - z[j]^2 < 2z[j]r^{-j} + r^{-2j} \qquad 9.27$$

- **Step 3.** Define a scaled residual (called in the sequel just "residual") as follows:[5]

$$w[j] = r^j (G(f(x[j], y[j]) - z[j]))$$ 9.28

with the bound

$$\underline{\omega} = r^j \underline{B} < w[j] < r^j \overline{B} = \overline{\omega}$$ 9.29

and the initial condition $w[-\delta] = 0$. The values $\underline{\omega}$ and $\overline{\omega}$ are the actual bounds to be determined in Step 6.

- **Step 4.** Determine a recurrence on $w[j]$ of the form

$$w[j+1] = r w[j] + r^{j+1}(G(f(x[j+1], y[j+1]) - z[j+1]) \\ - G(f(x[j], y[j]) - z[j])))$$ 9.30

- **Step 5.** For purposes of selection of the result digit z_{j+1}, express the recurrence as

$$w[j+1] = r w[j] + H_1 + H_2(z_{j+1})$$ 9.31

so that H_1 is independent of z_{j+1}. This leads to the following decomposition:

$$v[j] = r w[j] + H_1$$ 9.32

$$w[j+1] = v[j] + H_2(z_{j+1})$$ 9.33

Note that H_1 depends on the online delay δ, the radix r, and the redundancy factor ρ. Moreover, to reduce the recurrence delay, redundant adders are used, resulting in redundant representations for $v[j]$ and $w[j]$.

- **Step 6.** Determine the bounds of $w[j+1]$ in terms of H_1 and H_2. From (9.31) we obtain[6]

$$\overline{\omega} = r \overline{\omega} + \max(H_1) + H_2(a)$$ 9.34

resulting in

$$\overline{\omega} = -\frac{\max(H_1) + H_2(a)}{r-1}$$ 9.35

---

5. The scaling is done to have a bound that is not multiplied by r^{-j}.
6. For simplicity we consider the case in which the bound of $w[j]$ is independent of j. If this is not the case (for example, for square root), the derivation has to be modified accordingly (see Chapter 6).

Similarly,

$$\underline{\omega} = -\frac{\min(H_1) + H_2(-a)}{r - 1} \qquad \textbf{9.36}$$

Part 2a: Selection Function with Selection Constants

The selection function produces the result digit z_{j+1} so that $w[j+1]$ is bounded according to (9.35) and (9.36). In the method with selection constants it is described by the the selection constants[7] m_k such that

$$z_{j+1} = k \text{ if } m_k \leq \widehat{v}[j] < m_{k+1} \qquad \textbf{9.37}$$

where $\widehat{v}[j]$ is an estimate of $v[j]$. In the type of implementations considered here, this estimate is obtained by truncating the redundant representation of $v[j]$ to t fractional bits.

To produce a correct selection function, the selection constants need to satisfy

$$\max(\widehat{L}_k) \leq m_k \leq \min(\widehat{U}_{k-1}) \qquad \textbf{9.38}$$

where $[\widehat{L}_k, \widehat{U}_k]$ is the selection interval of the estimate $\widehat{v}[j]$, which we determine now. The max and min operators relate to variables on which the selection interval depend, such as the divisor in division and the result in square root.

- **Step 7.** Determine $[\widehat{L}_k, \widehat{U}_k]$, the limits of the selection intervals of $\widehat{v}[j]$. We begin by obtaining $[L_k, U_k]$, the selection intervals on $v[j]$, and then restrict these intervals to take into account the effect of using the estimate $\widehat{v}[j]$.

  From the relation between $w[j+1]$ and $v[j]$ (expression (9.33)) we have

$$\overline{\omega} = U_k + H_2(k) \qquad \underline{\omega} = L_k + H_2(k) \qquad \textbf{9.39}$$

Substituting $\overline{\omega}$ and $\underline{\omega}$ from (9.35) and (9.36), we get the selection intervals for $v[j]$,

$$U_k = -\frac{\max(H_1) + H_2(a)}{r - 1} - H_2(k)$$

$$\qquad \qquad \qquad \qquad \qquad \qquad \qquad \textbf{9.40}$$

$$L_k = -\frac{\min(H_1) + H_2(-a)}{r - 1} - H_2(k)$$

---

7. To simplify the description we consider the case in which there is only one selection constant for each k; if this is not possible, a staircase function has to be developed, as described in Section 5.4.

Now we restrict the intervals because of the use of the estimate $\widehat{v}[j]$. The estimate introduces an error such that

$$e_{min} \leq v[j] - \widehat{v}[j] \leq e_{max} \qquad\qquad \textbf{9.41}$$

producing the error-restricted selection interval $[L_k^*, U_k^*]$ with

$$U_k^* = U_k - e_{max} \qquad L_k^* = L_k + |e_{min}| \qquad\qquad \textbf{9.42}$$

Specifically, as shown in Section 5.4, for a redundant representation truncated to t fractional bits, the errors are as follows:

For carry-save representation, $e_{max} = 2^{-t+1} - ulp$ and $e_{min} = 0$.

For signed-digit representation, $e_{max} = 2^{-t} - ulp$ and $e_{min} = -(2^{-t} - ulp)$.

Since the estimate is obtained by assimilating t fractional bits of $v[j]$, the errors are multiples of 2^{-t}. Consequently, the actual selection interval boundaries and the selection constants have a granularity of 2^{-t}. Let $\widehat{U}$ and $\widehat{L}$ denote the actual (grid-restricted) selection intervals. As shown in Section 5.4,

$$\widehat{U}_{k-1} = \lfloor U_{k-1}^* + 2^{-t} \rfloor_t$$
$$\widehat{L}_k = \lceil L_k^* \rceil_t \qquad\qquad \textbf{9.43}$$

where $\lfloor x \rfloor_t$ and $\lceil x \rceil_t$ indicate x values truncated to t fractional bits. The choice of constants m_k is illustrated in Figure 9.11.

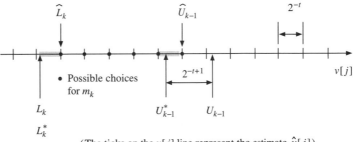

(The ticks on the $v[j]$ line represent the estimate $\widehat{v}[j]$)

FIGURE 9.11 The choices of selection constant m_k.

- **Step 8.** Determination of t and δ. To be able to determine m_k, from (9.38) we need

$$\min(\widehat{U}_{k-1}) - \max(\widehat{L}_k) \geq 0 \qquad\qquad \textbf{9.44}$$

This gives a relation between t and δ that is used to choose suitable values.

- **Step 9.** Determination of the selection constants m_k using expression (9.38). Moreover, determine the range of $\widehat{v}[j]$ as

$$\lfloor r\underline{\omega} + \min(H_1) - e_{max}\rfloor_t \leq \widehat{v}[j] \leq \lfloor r\overline{\omega} + \max(H_1) + |e_{min}|\rfloor_t \qquad \textbf{9.45}$$

Part 2b: Other Selection Methods

In algorithms using a higher radix ($r > 4$), implementing a selection function based on selection constants becomes quickly impractical. In such a case there are other methods for selecting output digits in online algorithms. We present here a selection method based on rounding of the residual part $v[j]$.

Selection by Rounding   Consider the residual expression (9.33)

$$w[j+1] = rw[j] + H_1 + H_2(z_{j+1}) = v[j] + H_2(z_{j+1}) \qquad \textbf{9.46}$$

In the rounding method, the result digit is obtained as

$$z_{j+1} = \left\lfloor v[j] + \frac{1}{2} \right\rfloor \qquad\qquad \textbf{9.47}$$

with $|v[j]| < r - \frac{1}{2}$ to avoid overredundant output digit. Replacing this selection in the recurrence, we get

$$w[j+1] = v[j] + H_2\left(\left\lfloor v[j] + \frac{1}{2}\right\rfloor\right) \qquad \textbf{9.48}$$

This next residual has to satisfy the bounds for convergence. This limits the direct application of the approach to some operations, while for others some transformations are required. If applicable, for a high radix this type of selection is far simpler to implement than a selection function using constants. Its implementation depends on the representation of the output digit, and in the case of a two's complement conventional representation of the digits, it corresponds to a short CPA.

For residuals in redundant form, the rounding is performed on an estimate $\widehat{v}[j]$ defined in the expression (9.41). Since the selection by rounding is equivalent to the selection using selection constants

$$m_k = \frac{2k - 1}{2}$$

we can use a similar procedure in determining the necessary precision of the estimate as presented in Step 8.

Specifically, for an estimate of the residual in carry-save form of t fractional bits, the estimate error is $e_{max} = 2^{-t+1} - ulp$. When $\widehat{v}[j] = m_k - 2^{-t}$ it must be possible to choose $z_{j+1} = k - 1$. Consequently, to have a correct selection it is necessary that

$$m_k - 2^{-t} + e_{max} = \frac{2k - 1}{2} + 2^{-t} \le \widehat{U}_{k-1} \qquad \textbf{9.49}$$

9.3.3 Generic Form of Execution and Implementation

We now describe a generic execution of an online algorithm and present the components of an implementation.

The execution corresponds to $n + \delta$ iterations of the recurrence, each corresponding to one clock cycle. The iterations (cycles) are labeled from $-\delta$ to $n - 1$. One digit of each input is introduced during cycles $-\delta$ to $n - 1 - \delta$ and digits value 0 thereafter. The result digits are 0 for cycles $-\delta$ to -1, and z_1 is produced in cycle 0. Finally, the result digit z_j is output in cycle j. Consequently, one additional cycle is required to output z_n.

For an operation with two operands x and y and one output z, the execution in cycle j consists of the following actions:

- Input $x_{j+1+\delta}$ and $y_{j+1+\delta}$.
- Update $x[j + 1] = (x[j], x_{j+1+\delta})$ and $y[j + 1] = (y[j], y_{j+1+\delta})$ by appending the input digits.
- Compute $v[j] = rw[j] + H_1$.
- Determine z_{j+1} using the selection function.
- In some algorithms, update $z[j + 1] = (z[j], z_{j+1+\delta})$ by appending the result digits.
- Compute the next residual $w[j + 1] = v[j] + H_2(z_{j+1})$.

In addition, result digit z_j is output.

Due to this similar structure of the algorithms, they are all implemented using the same basic components, such as the following:

1. Registers to store operands, results, and residual vectors

2. Multiplication of vector by digit

3. Append units to append a new digit to a vector

4. Two-operand and multioperand redundant adders, such as signed digit adders, [3:2] carry-save adders, and their generalization to [4:2] and [5:2] adders

5. Converters from redundant representations (i.e., signed-digit and carry-save) to conventional representations

6. Carry-propagate adders of limited precision (3 to 6 bits) to produce estimates of the residual functions

7. Digit-selection schemes to obtain output digits

An online algorithm implementation is similar to implementation of the digit recurrence algorithms discussed in Chapter 5 and consists of a linear array of digit slices, as shown in Figure 9.12. The number of digit slices depends on the operation implemented.

9.3.4   Algorithms and Implementations

We now develop algorithms for online multiplication and division and give examples of these for radix 2.

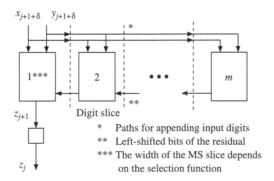

FIGURE 9.12 A typical digit slice organization of an online arithmetic unit.

Multiplication

Using the design method discussed in Section 9.3.2, we develop a radix-r online multiplication algorithm for n-digit signed operands x and y, and product p in the range $(-1, 1)$ represented with n signed digits from the set $\{-a, \ldots, a\}$. Let the operands and the result at cycle j be

$$x[j] = \sum_{i=1}^{j+\delta} x_i r^{-i}, \quad y[j] = \sum_{i=1}^{j+\delta} y_i r^{-i}, \quad p[j] = \sum_{i=1}^{j} p_i r^{-i} \qquad \textbf{9.50}$$

The error bound at cycle j is

$$|x[j] \cdot y[j] - p[j]| < r^{-j} \qquad \textbf{9.51}$$

The corresponding residual is defined as

$$w[j] = r^j (x[j] \cdot y[j] - p[j]) \qquad \textbf{9.52}$$

with the bound $|w[j]| < \omega$.

The resulting recurrence is

$$w[j+1] = r w[j] + (x[j]y_{j+1+\delta} + y[j+1]x_{j+1+\delta})r^{-\delta} - p_{j+1} \qquad \textbf{9.53}$$

This is decomposed into

$$v[j] = r w[j] + (x[j]y_{j+1+\delta} + y[j+1]x_{j+1+\delta})r^{-\delta}$$
$$w[j+1] = v[j] - p_{j+1} \qquad \textbf{9.54}$$

resulting in

$$H_1 = (x[j]y_{j+1+\delta} + y[j+1]x_{j+1+\delta})r^{-\delta} \qquad H_2 = -p_{j+1} \qquad \textbf{9.55}$$

and the bound (from 9.35)

$$\overline{\omega} = -\underline{\omega} = \omega = \rho(1 - 2r^{-\delta}) \qquad \textbf{9.56}$$

The selection intervals are

$$U_k = -\frac{2ar^{-\delta} - a}{r-1} + k = \rho(1 - 2r^{-\delta}) + k$$
$$L_k = -\frac{-2ar^{-\delta} + a}{r-1} + k = -\rho(1 - 2r^{-\delta}) + k \qquad \textbf{9.57}$$

Radix	ρ	t	δ	Initial Number of Bits/Operand
2	1	2	3	6
4	1	2	2	4
	$\frac{2}{3}$	3	3	6
8	$\frac{2}{3}$	2	3	9

TABLE 9.4 Examples of relations between r, ρ, t, and δ based on (9.60).

Using a carry-save representation for $w[j]$ and $v[j]$, the grid-restricted intervals are

$$\widehat{U}_k = \lfloor \rho(1 - 2r^{-\delta}) + k - 2^{-t} \rfloor_t$$
$$\widehat{L}_k = \lceil -\rho(1 - 2r^{-\delta}) + k \rceil_t \qquad\qquad \textbf{9.58}$$

The expression to determine t and δ is[8]

$$\lfloor \rho(1 - 2r^{-\delta}) + k - 1 - 2^{-t} \rfloor_t - \lceil -\rho(1 - 2r^{-\delta}) + k \rceil_t \geq 0 \qquad \textbf{9.59}$$

This results in

$$\lfloor \rho(1 - 2r^{-\delta}) \rfloor_t \geq 2^{-1}(1 + 2^{-t}) \qquad\qquad \textbf{9.60}$$

Several examples of relations based on (9.60) between ρ, t, and δ for radices 2, 4, and 8 are presented in Table 9.4.

The selection constants are determined using the selection interval (9.58). The range of the estimate $\widehat{v}[j]$, using expression (9.45), is

$$\lfloor -r\rho + 2r^{-\delta}(r\rho - 1) - e_{max} \rfloor_t \leq \widehat{v}[j] \leq \lfloor r\rho - 2r^{-\delta}(r\rho - 1) + |e_{min}| \rfloor_t \quad \textbf{9.61}$$

which is simplified to

$$\lfloor -\rho(r - 2r^{-\delta}) - 2^{-t+1} \rfloor_t \leq \widehat{v}[j] \leq \lfloor \rho(r - 2r^{-\delta}) \rfloor_t \qquad \textbf{9.62}$$

EXAMPLE 9.3   We now present a radix-2 online multiplication algorithm and its implementation for the carry-save representation of the residual. From Table 9.4, the online delay is $\delta = 3$ and $t = 2$.

---

8. For multiplication the selection interval does not depend on another variable, so the min and max operators in the general description are not needed.

The selection constants m_k's are obtained from

$$\widehat{L}_k \leq m_k \leq \widehat{U}_{k-1} \qquad\qquad \textbf{9.63}$$

where

$$\begin{aligned}
\widehat{U}_k &= \lfloor 1 - 2^{-2} + k - 2^{-2} \rfloor_2 = k + 2^{-1} \\
\widehat{L}_k &= \lceil -1 + 2^{-2} + k \rceil_2 = k - 3 \times 2^{-2}
\end{aligned} \qquad\qquad \textbf{9.64}$$

So $\widehat{U}_{k-1} = k - 2^{-1}, \widehat{L}_k = k - 3 \times 2^{-2}$, so that $m_k = k - 2^{-1}$ is acceptable. Therefore, the selection constants are

$$m_0 = -2^{-1}, \quad m_1 = 2^{-1} \qquad\qquad \textbf{9.65}$$

The range of $\widehat{v}[j]$ is

$$-2 \leq \widehat{v}[j] \leq \frac{7}{4} \qquad\qquad \textbf{9.66}$$

The corresponding selection function $SELM(\widehat{v}[j])$ is

$$p_{j+1} = SELM(\widehat{v}[j]) = \begin{cases} 1 & \text{if } \frac{1}{2} \leq \widehat{v}[j] \leq \frac{7}{4} \\ 0 & \text{if } -\frac{1}{2} \leq \widehat{v}[j] \leq \frac{1}{4} \\ -1 & \text{if } -2 \leq \widehat{v}[j] \leq -\frac{3}{4} \end{cases} \qquad\qquad \textbf{9.67}$$

This selection function has a simple implementation. The assimilated estimate $\widehat{v}$ is (v_{-1}, v_0, v_1, v_2). Since the selection constants have one fractional bit, bit v_2 of the estimate is not used. The product digit p_{j+1} is coded with two bits (pp, pn) as follows:

p_{j+1}	pp	pn
1	1	0
0	0	0
−1	0	1

Using this coding and the fact that the estimate bit v_2 is not used, the selection function is described by Table 9.5.

$\hat{v}$	$v_{-1}v_0.v_1$	p_{j+1}
$\frac{3}{2}$	01.1	1
1	01.0	1
$\frac{1}{2}$	00.1	1
0	00.0	0
$-\frac{1}{2}$	11.1	0
-1	11.0	-1
$-\frac{3}{2}$	10.1	-1
-2	10.0	-1

TABLE 9.5 Radix-2 multiplication selection function.

The corresponding switching expressions are

$$pp = v'_{-1}(v_0 + v_1), \quad pn = v_{-1}(v'_0 + v'_1) \qquad \textbf{9.68}$$

The algorithm is shown in Figure 9.13, and its implementation in Figure 9.14(a). The latches LX and LY are the output latches of the predecessor online units. The carries c_x and c_y correspond to the signs of x_{j+4} and y_{j+4}, respectively. The module V produces the estimate of $v[j]$. The calculation of $2w[j+1]$ is illustrated in Figure 9.14(b). The subtraction of $v[j] - p_{j+1}$ can be implemented by simply complementing the estimate bit v_0 if $p_{j+1} \neq 0$ (Exercise 9.11). The critical path consists of a SELECTOR (2-input MUX), a [4:2] adder, a 4-bit CPA, the SELM module, and an XOR for complementing v_0.

In Table 9.6, we illustrate radix-2 online multiplication with operands

$$x = (.110\bar{1}10\bar{1}1)$$

$$y = (.101\bar{1}\bar{1}110)$$

For simplicity we show v and w in nonredundant form.

The computed product is $p = (.10\bar{1}01\bar{1}10)$. The true double-precision product is $p^* = (.0110010110000010)$. The absolute error with respect to the true product truncated to 8 bits is $|p - p^*_{tr}| = 2^{-8}$. Note that $p[8] + w[8]2^{-8} = p^*$.   ∎

1. [*Initialize*]

   $x[-3] = y[-3] = w[-3] = 0$

   for $j = -3, -2, -1$

   $x[j + 1] \leftarrow CA(x[j], x_{j+4}); \ y[j + 1] \leftarrow CA(y[j], y_{j+4})$

   $v[j] = 2w[j] + (x[j]y_{j+4} + y[j + 1]x_{j+4})2^{-3}$

   $w[j + 1] \leftarrow v[j]$

   end for   .

2. [*Recurrence*]

   for $j = 0 \ldots n - 1$

   $x[j + 1] \leftarrow CA(x[j], x_{j+4}); \ y[j + 1] \leftarrow CA(y[j], y_{j+4})$

   $v[j] = 2w[j] + (x[j]y_{j+4} + y[j + 1]x_{j+4})2^{-3}$

   $p_{j+1} = SELM(\widehat{v[j]});$

   $w[j + 1] \leftarrow v[j] - p_{j+1}$

   $P_{out} \leftarrow p_{j+1}$

   end for

where

- The residual is in redundant form, represented by the pseudosum *WS* and stored-carry *WC* bit-vectors. For simplicity, we use $w[j]$ in the description.
- n is the precision in bits.
- The online delay $\delta = 3$; the estimate $\widehat{v}[j]$ is computed with $t = 2$.
- $SELM(\widehat{v}[j])$ is the product-digit selection function. Since the selection constants are $\pm\frac{1}{2}$, the second fractional bit of the estimate is not used.
- CA is on-the-fly conversion/appending function producing the online operands in the conventional representation (discussed in Section 5.2.3).
- P_{out} is the product digit register.

FIGURE 9.13 Radix-2 online multiplication algorithm.

Online Division

Using the design method discussed in Section 9.3.2, we develop a radix-r online division algorithm for n-digit signed operands x and y, and quotient q in the range $(-1, 1)$ represented with n signed digits from the set $\{-a, \ldots, a\}$. Let the

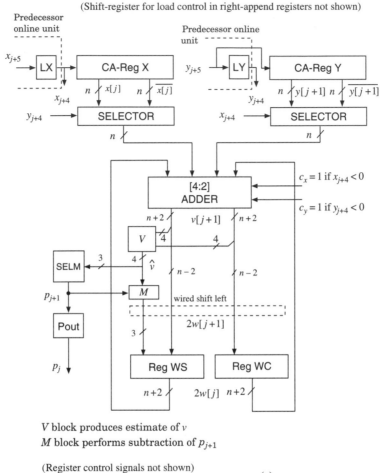

V block produces estimate of v

M block performs subtraction of p_{j+1}

(Register control signals not shown)

(a)

$$v[j] \begin{vmatrix} vs_{-1}vs_0 . vs_1 \ vs_2 \ vs_3 \ vs_4 \ \dots \\ vc_{-1}vc_0 . vc_1 \ vc_2 \ vc_3 \ vc_4 \ \dots \end{vmatrix}$$

Estimate of $v[j]$ $v_{-1} \ v_0 \ \cdot v_1 \ v_2$

$$2w[j+1] \begin{vmatrix} v_0^* \quad v_1 \cdot v_2 \ vs_3 \ vs_4 \ \dots \\ \qquad\qquad\quad vc_3 \ vc_4 \ \dots \end{vmatrix}$$

$v_0^* = v_0 \, \mathrm{XOR} \, |p_{j+1}|$

(b)

FIGURE 9.14 (a) Implementation of radix-2 online multiplier. (b) Calculation of $2w[j+1]$.

j	x_{j+4}	y_{j+4}	$x[j+1]$	$y[j+1]$	$v[j]$	p_{j+1}	$w[j+1]$
−3	1	1	.1	.1	00.0001	0	00.0001
−2	1	0	.11	.10	00.00110	0	00.00110
−1	0	1	.110	.101	00.011110	0	00.011110
0	−1	−1	.1011	.1001	00.1100011	1	11.1100011
1	1	−1	.10111	.10001	11.10000111	0	11.10000111
2	0	1	.101110	.100011	11.001001010	−1	00.001001010
3	−1	1	.1011011	.1000111	00.0100111101	0	00.0100111101
4	1	0	.10110111	.10001110	00.10110000010	1	11.10110000010
5	0	0	.10110111	.10001110	11.0110000010	−1	00.0110000010
6	0	0	.10110111	.10001110	00.110000010	1	11.110000010
7	0	0	.10110111	.10001110	11.10000010	0	11.10000010

TABLE 9.6 Example of radix-2 online multiplication.

operands and the result at cycle j be

$$x[j] = \sum_{i=1}^{j+\delta} x_i r^{-i}, \quad y[j] = \sum_{i=1}^{j+\delta} y_i r^{-i}, \quad q[j] = \sum_{i=1}^{j} q_i r^{-i} \qquad \textbf{9.69}$$

The error bound at cycle j is

$$|x[j] - q[j]d[j]| < d[j]r^{-j} \qquad \textbf{9.70}$$

The residual is

$$w[j] = r^j(x[j] - q[j]d[j]) \qquad \textbf{9.71}$$

with the bound $|w[j]| < \omega \le d[j]$.

The residual recurrence is

$$w[j+1] = rw[j] + x_{j+1+\delta}r^{-\delta} - q[j]d_{j+1+\delta}r^{-\delta} - d[j+1]q_{j+1} \quad \textbf{9.72}$$

which is decomposed as

$$v[j] = rw[j] + x_{j+1+\delta}r^{-\delta} - q[j]d_{j+1+\delta}r^{-\delta}$$
$$w[j+1] = v[j] - d[j+1]q_{j+1} \qquad \textbf{9.73}$$

In terms of the notation of Section 9.3.1,

$$H_1 = x_{j+1+\delta}r^{-\delta} - q[j]d_{j+1+\delta}r^{-\delta} \qquad H_2 = -d[j+1]q_{j+1} \qquad \textbf{9.74}$$

The bound of $w[j]$ is

$$\omega = -\frac{2ar^{-\delta} - ad[j+1]}{r-1} = \rho(d[j+1] - 2r^{-\delta}) \qquad \textbf{9.75}$$

The selection intervals on $v[j]$ are

$$U_k = \rho(d[j+1] - 2r^{-\delta}) + kd[j+1]$$
$$L_k = -\rho(d[j+1] - 2r^{-\delta}) + kd[j+1] \qquad \textbf{9.76}$$

Using a carry-save representation for $w[j]$ and $v[j]$, the grid-restricted intervals are

$$\widehat{U}_k = \lfloor \rho(d[j+1] - 2r^{-\delta}) + kd[j+1] - 2^{-t} \rfloor_t$$
$$\widehat{L}_k = \lceil -\rho(d[j+1] - 2r^{-\delta}) + kd[j+1] \rceil_t \qquad \textbf{9.77}$$

The expression to determine t and δ is

$$d[j+1]_{min}(\lfloor \rho(d[j+1] - 2r^{-\delta}) + (k-1)d[j+1] - 2^{-t} \rfloor_t)$$
$$-d[j+1]_{max}(\lceil -\rho(d[j+1] - 2r^{-\delta}) + kd[j+1] \rceil_t) \geq 0 \qquad \textbf{9.78}$$

Using $d[j+1]_{max} = 1$ and $d[j+1]_{min} = \frac{1}{2}$, the expression has a valid solution only for $r = 2$. We consider only this case here; for higher radices it is necessary to divide the range of $d[j+1]$ into intervals and develop a staircase selection function as was done in Section 5.4.

For $r = 2$ ($\rho = 1$), we get

$$\lfloor 2^{-1} - 2^{-\delta+1} + 2^{-1}(k-1) - 2^{-t} \rfloor_t - \lceil -1 + 2^{-\delta+1} + k \rceil_t \geq 0 \qquad \textbf{9.79}$$

The worst case is for $k = 1$, resulting in

$$\lfloor 2^{-1} - 2^{-\delta+1} \rfloor_t - \lceil 2^{-\delta+1} \rceil_t \geq 2^{-t} \qquad \textbf{9.80}$$

A solution to this is $t = 3$ and $\delta = 4$.

The selection constants are obtained using the selection intervals of (9.77).

The range of $\widehat{v}[j]$ is

$$\lfloor -r\rho(1 - 2r^{-\delta}) - 2ar^{-\delta} - 2^{-t+1} \rfloor_t \leq \widehat{v}[j] \leq \lfloor r\rho(1 - 2r^{-\delta}) - 2ar^{-\delta} \rfloor_t \qquad \textbf{9.81}$$

Since this is the same expression as for multiplication, it can be likewise simplified to

$$\lfloor -\rho(r - 2r^{-\delta}) - 2^{-t+1} \rfloor_t \leq \widehat{v}[j] \leq \lfloor \rho(r - 2r^{-\delta}) \rfloor_t \qquad \textbf{9.82}$$

EXAMPLE 9.4   We now present a radix-2 online division algorithm and its implementation for the carry-save representation of the residual. Using $t = 3$, $\delta = 4$, and min/max values[9] of $d[j + 1]$, we obtain

$$\min \widehat{U}_0 = \widehat{U}_0[d[j + 1] = 1/2] = 2^{-1} - 2^{-3} + 0 - 2^{-3} = 2^{-2}$$
$$\max \widehat{L}_1 = \widehat{L}_1[d[j + 1] = 1] = -1 + 2^{-3} + 1 = 2^{-3} \qquad \textbf{9.83}$$

resulting in $m_1 = 2^{-2}$

$$\min \widehat{U}_{-1} = \widehat{U}_{-1}[d[j + 1] = 1] = 1 - 2^{-3} - 1 - 2^{-3} = -2^{-2}$$
$$\max \widehat{L}_0 = \widehat{L}_0[d[j + 1] = 1/2] = -2^{-1} + 2^{-3} = -3 \times 2^{-3} \qquad \textbf{9.84}$$

so that $m_0 = -2^{-2}$.

The range of $\widehat{v}[j]$ is

$$-2 \leq \widehat{v}[j] \leq \frac{15}{8} \qquad \textbf{9.85}$$

The corresponding quotient-selection function $SELD(\widehat{v}[j])$ is

$$q_{j+1} = SELD(\widehat{v}[j]) = \begin{cases} 1 & \text{if } \frac{1}{4} \leq \widehat{v}[j] \leq \frac{15}{8} \\ 0 & \text{if } -\frac{1}{4} \leq \widehat{v}[j] \leq \frac{1}{8} \\ -1 & \text{if } -2 \leq \widehat{v}[j] \leq -\frac{1}{2} \end{cases} \qquad \textbf{9.86}$$

The assimilated estimate $\widehat{v}$ is $(v_{-1}, v_0, v_1, v_2, v_3)$. Since the selection constants have two fractional bits, bit v_3 of the estimate is not used. The implementation of the selection function is left as Exercise 9.18. In summary, the radix-2 algorithm is given in Figure 9.15.

The corresponding implementation is shown in Figure 9.16. The latches LX and LD are the output latches of the predecessor online units. The carries c_d and c_q are determined by the signs of the divisor d and digit

---

9. Actually for $r = 2$, since $\rho = 1$, $\widehat{L}_1$ and $\widehat{U}_{-1}$ do not depend on $d[j + 1]$ so only $d[j + 1] = \frac{1}{2}$ matters in the derivation of the selection constants.

1. [*Initialize*]

$$x[-4] = d[-4] = w[-4] = q[0] = 0$$

for $j = -4, \ldots, -1$

$$d[j + 1] \leftarrow CA(d[j], d_{j+5})$$

$$v[j] = 2w[j] + x_{j+5}2^{-4}$$

$$w[j + 1] \leftarrow v[j]$$

end for

2. [*Recurrence*]

for $j = 0 \ldots n - 1$

$$d[j + 1] \leftarrow CA(d[j], d_{j+5})$$

$$v[j] = 2w[j] + x_{j+5}2^{-4} - q[j]d_{j+5}2^{-4}$$

$$q_{j+1} = SELD(\widehat{v}[j]);$$

$$w[j + 1] \leftarrow v[j] - q_{j+1}d[j + 1]$$

$$q[j + 1] \leftarrow CA(q[j], q_{j+1})$$

$$Q_{out} \leftarrow q_{j+1}$$

end for

where

- The residual is in redundant form, represented by the pseudosum *WS* and stored-carry *WC* bit-vectors. For simplicity, we use $w[j]$ in the description.
- n is the precision in bits.
- The online delay $\delta = 4$; the estimate $\widehat{v}[j]$ is computed with $t = 3$.
- $SELD(\widehat{v}[j])$ is the quotient-digit selection function. Since the selection constants are $\pm 1/4$, the third fractional bit of the estimate is not used.
- *CA* is on-the-fly conversion/appending function producing the online operands in the conventional representation (discussed in Section 5.2.3).
- Q_{out} is the quotient digit register.

FIGURE 9.15 Radix-2 online division algorithm.

d_{j+5}, and the quotient q and digit q_{j+1}, respectively. Block U combines the dividend digit $x_{j+5}2^{-4}$ and the sign extension bits of $q[j]d_{j+5}2^{-4}$. This allows the use of a [3:2] adder in computing $v[j]$. The design details of module U are considered in Exercise 9.19. The critical path consists of a SELECTOR

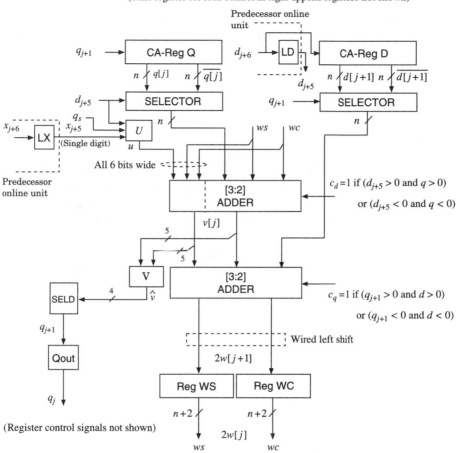

(Shift-register for load control in right-append registers not shown)

FIGURE 9.16 Block diagram of radix-2 online divider.

(2-input MUX), module U, a [3:2] adder, estimate module V, selection module SELD, a SELECTOR, and a [3:2] adder.

In comparison with conventional digit recurrence division using residuals in carry-save form, the online division is somewhat more complex and has a longer cycle time. For example, while the conventional division uses a [3:2] adder, 1 fractional bit estimate of the partial remainder, one 2-input multiplexer, and 5 registers (1 for the nonredundant divisor, 2 for the redundant partial remainder, and 2 for the redundant quotient), the online algorithm uses two [3:2] adders, a 3-bit (fractional) estimate, 6 registers (2 for the redundant divisor, 2 for the redundant residual, and 2 for the redundant quotient), two 2-input multiplexers, two on-the-fly converters (using the divisor and the quotient registers), and two appending networks. The implementation of the selection function has similar complexity as that of multiplication.   ∎

The Reduction of Digit Slices in Online Implementations

The number of bit slices required in an implementation of an online algorithm is smaller than that required in serial-parallel implementation. As illustrated in Figure 9.17(a), n is the number of bits in the result, ib is the number of integer bits of $v[j]$, and t the number of fractional bits in its estimate $\widehat{v}[j]$. Let $p < n$ be the number of fractional bit slices in the implementation. In the first p cycles the computation of the residual is exact. Beginning with step $p + 1$, an error in the residual due to "truncation" of the fractional bit slices $p, \ldots, n + \delta$ is introduced

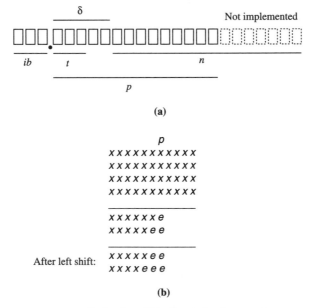

FIGURE 9.17 Reduction of bit slices in implementation.

and propagated to the left by one position due to the term $2w[j]$. For residuals formed using a [4:2] adder, the error in cycle $p + 1$ (after left shift) affects bits in positions $p - 2$, $p - 1$, p (Figure 9.17(b)). After cycles $p + 1, \ldots, p + h$ are performed, the truncation error has affected all bit positions up to and including bit position $p - 2h + \delta$ (in the last δ cycles the input digits are 0 and the error propagation is caused by the left shift of the residual). To have a valid selection using an estimate of t fractional bits,

$$p - 2h + \delta \geq t \qquad\qquad\qquad \textbf{9.87}$$

Since $p + h = n + \delta$, we obtain

$$p = \left\lceil \frac{2n + \delta + t}{3} \right\rceil \qquad\qquad\qquad \textbf{9.88}$$

and the total number of bit slices is $ib + p$.

For example, the number of bit slices for 32-bit radix-2 online multiplication is

$$2 + \left\lceil \frac{2 \times 32 + 3 + 2}{3} \right\rceil = 2 + 23 = 25 \qquad\qquad\qquad \textbf{9.89}$$

compared to 34 in an implementation without slice reduction.

Multioperation and Composite Online Algorithms

To reduce the overall online delay of a group of operations, it is often advantageous and feasible to combine several operations into a single *multioperation online algorithm*. As an example,[10] below we show an online algorithm for sum of squares $x^2 + y^2 + z^2$, which is used in 3-D normalization. The inputs are in the range $[\frac{1}{2}, 1)$ and the output in the range $[\frac{1}{4}, 3)$ and its online delay $\delta_{ss} = 0$ when the output digit is overredundant. This is in contrast with the delay of $(3 + 2 + 2 = 7)$ of the corresponding network consisting of three online multipliers and two adders. This reduction in delay is partially due to the overredundant output digit. The algorithm is given in Figure 9.18 and the corresponding implementation in Figure 9.19(a).

As shown in Figure 9.19(b) the selection function $s_{j+1} = csint(v[j])$ produces an output digit in the range from 0 to 8. If it can be used in this form in the next operation, which is the case in the 3-D normalization, no recoding to the digit set $\{-1, 0, 1\}$ is necessary.

---

10. Ercegovac and Lang (1999).

---

1. [*Initialize*]

   $w[0] = x[0] = y[0] = z[0] = 0$

2. [*Recurrence*]

   for $j = 0 \ldots n - 1$

   $v[j] = 2w[j] + (2x[j] + x_{j+1}2^{-j-1})x_{j+1} + (2y[j] + y_{j+1}2^{-j-1})y_{j+1}$
   $\qquad + (2z[j] + z_{j+1}2^{-j-1})z_{j+1}$

   $w[j+1] \leftarrow csfract(v[j])$

   $s_{j+1} \leftarrow csint(v[j])$

   $x[j+1] \leftarrow (x[j], x_{j+1}); \, y[j+1] \leftarrow (y[j], y_{j+1}); \, z[j+1] \leftarrow (z[j], z_{j+1})$

   $S_{out} \leftarrow s_{j+1}$

   end for

   where

   - n is the precision in bits.
   - $csfract(v)$ and $csint(v)$ correspond to the fractional and integer parts of v obtained in carry-save form (see Figure 9.19(b)).
   - S_{out} is the result-digit register.

FIGURE 9.18 Radix-2 online sum-of-squares algorithm.

For more complicated algorithms that cannot be implemented by a single online module, an interconnection of modules is required. A modular approach would be to use the standard online implementation of primitive operations as components. However, this might lead to suboptimal implementations with respect to area and online delay. An alternative is to develop an integrated approach and develop one *composite algorithm*. As an illustration we show in Figure 9.20 the use of the sum-of-squares with overredundant output digit in the set {0, ..., 8}, and a square root algorithm[11] developed for this input digit set to perform

$$d = \sqrt{(x^2 + y^2 + z^2)}$$

with an overall online delay of 5. This is part of an online unit to compute the 3-D normalization. A network of standard online modules would have an online delay of 11.

---

11. Ercegovac and Lang (1999).

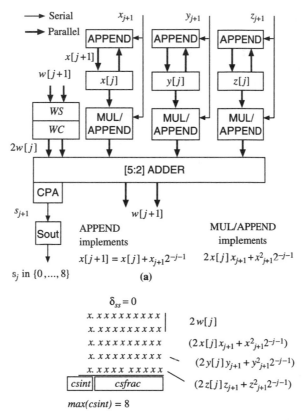

FIGURE 9.19 (a) Radix-2 online unit for computing sum of squares. (b) Carry-save operation: obtaining residual and output digit.

Online Implementation of Recursive Algorithms

An important characteristic of MSDF/online arithmetic is its capability to reduce the latency between successive recurrence evaluations, independent of the precision of computation. These recursive algorithms are frequently used in

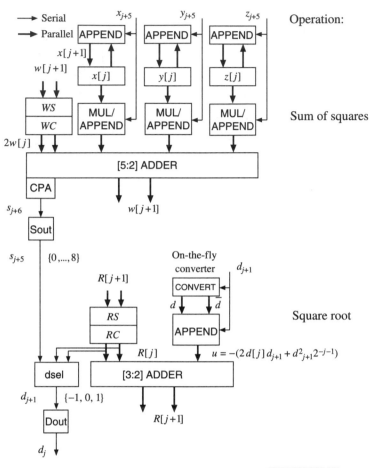

FIGURE 9.20 Composite scheme for computing $d = \sqrt{(x^2 + y^2 + z^2)}$.

signal processing applications, such as recursive filtering.[12] As an illustration of the potential benefits of online arithmetic in such applications, we consider a second-order IIR filter (Figure 9.21(a)) characterized by the output expression

$$y[k] = a_1 y[k-1] + a_2 y[k-2] + bx[k] \qquad\qquad \textbf{9.90}$$

---

12. Oppenheim et al. (1999).

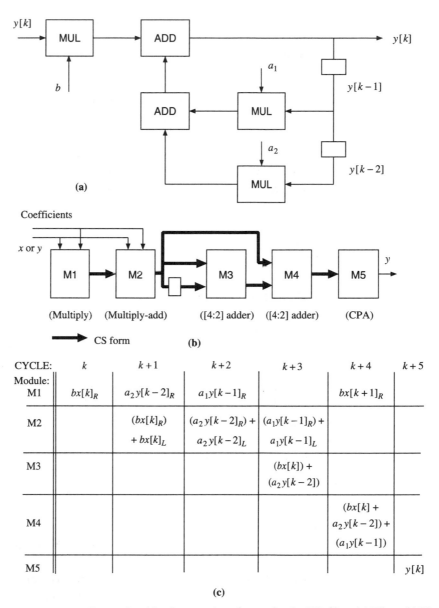

FIGURE 9.21 Conventional implementation of second-order IIR filter: (a) Filter. (b) Five-module network. (c) Schedule.

Without attempting any optimizations, several filter implementation alternatives are analyzed, assuming (i) parallel-in/parallel-out interface, (iii) coefficients in parallel form, and (iii) fixed-point format with n bits. Alternatives 2 and 3 use digit-serial arithmetic internally.

1. **Conventional parallel arithmetic** implementation may, for example, use a five-module network: Module M1 is a $n \times n/2$ multiplier producing a carry-save product; module M2 is a multiply-add unit producing the $2n$-bit product in carry-save form; modules M3 and M4 are [4:2] adders; and module M5 is a carry-propagate adder (Figure 9.21(b)). As indicated in the schedule (Figure 9.21(c)), the time to obtain $y[k]$ is $T_{CONV} = 6t_{module}$. We assume that the longest critical path is in module M5 corresponding to a $2n$-bit CPA with $t_{module} \approx 6t_{FA}$ for $n \leq 31$, where t_{FA} is the delay of a full-adder. Since the next computation can begin in cycle $k + 4$, the rate of filter computation using this implementation is $R_{CONV} \approx 1/(4 \times 6t_{FA})$.

2. **LSDF serial arithmetic** implementation uses three serial multipliers and two serial adders. Since it takes n clocks to begin producing the most-significant half of the product, the time to produce $y[k]$ is $T_{LSDF} \approx nt_{FA}$. The rate of generating outputs in this alternative is $R_{LSDF} \approx 1/(n \times t_{FA})$. The input/output format conversions are not in the critical path.

3. **Online arithmetic** implementation uses online multioperation module M shown in Figure 9.22(a). The module M consists of the following components: (i) one online multiplier (one operand in parallel form) with online delay of 2 and (ii) two online multiply-add modules computing $vu + w$, where v is in parallel form while u and w are in online form. This module has an online delay of 3. The cycle time of module M is $t_M \approx 3t_{FA}$. A filter consisting of a single M module has a rate of one n-bit output every n cycles.

   To produce a higher throughput we can use the fact that online algorithms operate in the MSDF mode, so that several successive computations can be overlapped. As indicated in the timing diagram of Figure 9.22(c), consecutive y outputs can be produced digit-serially $\Delta_{iter} = 3 + 1 = 4$ cycles apart. Since $t_M \approx 3t_{FA}$, the rate is $R_{OL} = 1/(\Delta_{iter} \times t_M) \approx 1/(12t_{FA})$, which, for $n > 12$, is better than rates achievable with

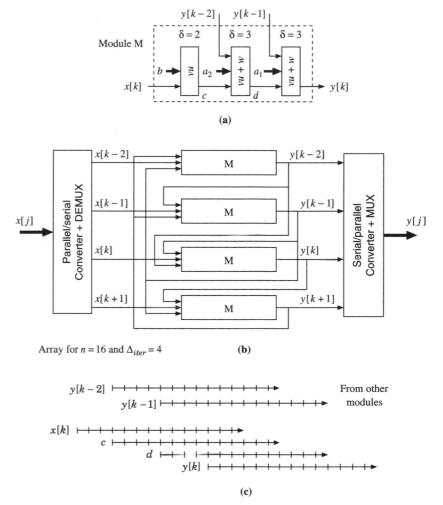

FIGURE 9.22 Online implementation of second-order IIR filter: (a) The online multioperation module. (b) Network for higher throughput. (c) Timing diagram.

approaches 1 and 2. Moreover, the throughput in the online approach is independent of the working precision. However, achieving this throughput requires $\lceil n / \Delta_{iter} \rceil$ multioperation operators, as illustrated in Figure 9.22(b). The implementation cost would be comparable to the cost of that many conventional serial-parallel carry-save multipliers (without carry-propagate adders). The number of bit slices in online units is reduced as discussed on page 526. Moreover, the modules are interconnected serially. The parallel input is serialized and demultiplexed to the online modules. The serial outputs are converted to parallel form using on-the-fly conversion and multiplexed to obtain the filter output.

This example indicates that significant speedups might be possible using the online alternative.

9.4   Concluding Remarks

Digit-serial arithmetic is attractive in implementations where the area and the interconnection width between the modules should be minimized while the increased latency is acceptable. It allows appropriate choice of radix, digit set, and precision to satisfy design needs. This approach is well-suited to the design of massively parallel implementations. Two modes of computation are considered: right-to-left (a conventional LSD first approach), and left-to-right (MSDF or online approach). These modes have different characteristics in terms of delays, cost, and applicability. The MSDF mode allows overlapping successive operations after a few cycles, which is important in implementing recursive algorithms. The discussion in this chapter focused on the basic concepts and algorithms: more advanced developments of digit-serial arithmetic and its applications are covered in the references at the end.

9.5   Exercises

9.1   Show a block diagram for performing 2D vector normalization, similar to the one in Figure 9.4, for LSDF arithmetic. Compare the total number of cycles to the online scheme discussed in the text.

LSDF Addition/Subtraction

9.2   Write a recurrence for radix-r LSDF addition.

LSDF Multiplication

9.3 A serial algorithm is described by the following expressions:

$$x[j] = x[j-1] + 2^j x_j$$
$$y[j] = y[j-1] + 2^j y_j$$
$$w[j] = \lfloor (1/2)(w[j-1] + x[j-1]y_j + y[j]x_j) \rfloor$$
$$z_j = (w[j-1] + x[j-1]y_j + y[j]x_j) \bmod 2$$
$$z[j] = z[j-1] + 2^j z_j$$

Show $x[j]$, $y[j]$, $z[j]$ for $0 \le j \le 6$ for the following input sequence:

$$j \quad 0123456$$
$$x[j] = 1011010$$
$$y[j] = 1110011$$

Is this an LSDF or MSDF algorithm ?

9.4 Place latches in Figure 9.6 to have a pipelined implementation.

9.5 For a serial-serial multiplier show a timing diagram with the contents of all registers for the 5×5 bit two's complement multiplication of $x = 01011$ by $y = 10001$.

9.6 For each of the 3×3 serial-parallel LSDF multipliers shown in Figure 9.23:
  (a) Give a timing diagram (schedule).
  (b) Determine the number of cycles to produce the 6-bit product.
  (c) Determine the critical path in a cycle.

9.7 Develop an LSDF algorithm for squaring of unsigned and two's complement n-bit integers. Design a bit slice and show the network for $n = 8$.

MSDF Addition/Subtraction

9.8 Perform the radix-4 online addition of $x = 0.2\bar{3}1\bar{2}$ and $y = 0.223\bar{1}$.

9.9 Perform the radix-2 online addition of $x = 0.10\bar{1}\bar{1}01\bar{1}0$ and $y = 0.10101101\bar{1}$.

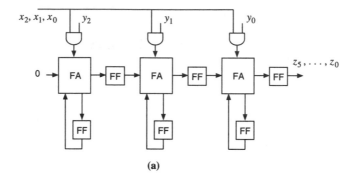

(a)

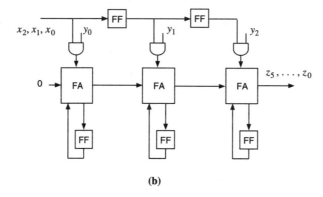

(b)

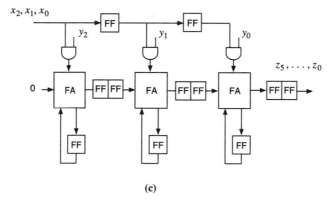

(c)

FIGURE 9.23 Serial-parallel multipliers (Exercise 9.6).

9.10   Consider radix-2 online addition with operands satisfying $x + y < 1$.

(a)   Perform the addition algorithm to show that if $z_0 = 1$, then the next nonzero result digit must have a value -1.

(b)   Devise a modification to the online addition algorithm to produce z_0 to 0 and develop the corresponding design.

MSDF Multiplication

9.11   Show that the subtraction of $v[j] - p_{j+1}$ can be implemented by complementing the estimate bit v_0 if $p_{j+1} \neq 0$.

9.12   Show the execution of the radix-2 online multiplication of $0.1\bar{1}1$ by $0.11\bar{1}$.

9.13   Develop an algorithm and a design to perform on-the-fly conversion and appending for radix-2 online multiplication.

9.14   Design a radix-2 online multiplier using a signed-digit adder. A signed digit $x_i \in \{-1, 0, 1\}$ is encoded as $x_i = x_{i1} - x_{i0}, x_{i1}, x_{i0} \in \{0, 1\}$. Compare with the implementation in Figure 9.14.

9.15   Develop a radix-4 online multiplication algorithm with digit selection by rounding. Show the execution of the algorithm to multiply $x = 0.2\bar{2}1$ by $y = 0.\bar{1}2\bar{1}$.

9.16   Develop a radix-2 online algorithm for computing $s = x^2, x \in (-1, 1), n$ bits of precision.

9.17   Derive an MSDF multiplication algorithm for radix 2, multiplicand x in parallel two's complement form, and online multiplier y and product p in signed-bit form. The residual is in carry-save form. Determine the online delay δ and the number of fractional bits t of the estimate $\widehat{v}[j]$. Show a block diagram of implementation, and compare it with the online multiplication implementation shown in Figure 9.14 with respect to the critical path and the modules used.

MSDF Division

9.18   Implement the selection function for radix-2 online division.

9.19   Module U of Figure 9.16 is specified as follows:

Output u

	$x_{j+5}; d_{j+5}$	-1	$0.$	1	2	3	4
1.	1 1	0	0	0	0	0	0
2.	0	0	0	0	0	0	1
3.	-1	0	0	0	0	0	1
4.	0 1	1	1	1	1	1	1
5.	0	0	0	0	0	0	0
6.	-1	0	0	0	0	0	0
7.	-1 1	1	1	1	1	1	0
8.	0	1	1	1	1	1	1
9.	-1	1	1	1	1	1	1

If $q < 0$, rows 1 and 3, 4 and 6, 7 and 9 are swapped. The input x_{j+5} and d_{j+5} are coded as (xp, xn) and (dp, dn). xz and dz denote zero digit values. qs is the sign of the quotient.

(a)   Show that the output of module U is

$$u_i = 1, \; i = -1, \ldots, 3 \; \text{if} \; xn + xz(dp \cdot \overline{qs} + dn \cdot qs) = 1$$

$$u_4 = 1 \; \text{if} \; \overline{xz} \cdot dz + \overline{xz}(dp \cdot qs + dn \cdot \overline{qs}) + xz(dp \cdot \overline{qs} + dn \cdot qs) = 1$$

(b)   Show that the given table and expressions are correct.

(c)   Show a gate network implementing the module. Compare its delay with the delay of a 6-bit [4:2] adder.

9.20   Perform radix-2 online division for $x = 0.10110010$ and $d = 0.11011101$.

9.21   Derive an MSDF division algorithm for radix 2, dividend x in parallel two's complement form, and online divisor d and quotient q in signed-bit form. The residual is in carry-save form. Determine the online delay δ and the number of fractional bits t of the estimate $\widehat{v}[j]$. Show a block diagram of implementation, and compare it with the online division implementation shown in Figure 9.16 with respect to the critical path and the modules used.

MSDF Other Operations

9.22 Develop a radix-2 multiply-add online algorithm using the method in Section 9.3.2.

9.23 Develop a radix-2 square root online algorithm using the method in Section 9.3.2.

9.24 Develop an online algorithm for $z = \max(x, y)$ with a minimum online delay for the following cases:

(a)   Signed-digit inputs and output
(b)   Nonredundant magnitude inputs and output
(c)   Nonredundant two's complement inputs and output

Compare the two algorithms with respect to online delay, cycle time, and cost.

Number of Slices

9.25 In a manner similar to that of Figure 9.17, determine the bits affected by implementing $p < n$ bit slices to update a residual using

(a)   [3:2] reduction
(b)   [5:2] reduction
(c)   [6:2] reduction

MSDF Composite

9.26 (a)   Develop a composite online algorithm to compute $ab + cd$.
(b)   Show a design at the level of Figure 9.14.
(c)   Identify the critical path.
(d)   Compare online delay, clock cycle, and cost with respect to a scheme that uses two online multipliers and one online adder.

MSDF Multimodule

9.27 Show a block diagram for the execution of the following operation using several online modules. Give a timing diagram of the operation, using the initial delays of Table 9.1.

$$z = \frac{\sqrt{x^2 + y^2}}{\sqrt{w^2 + v^2}}$$

Compare the execution time with the case in which each operation requires the operands to be provided in parallel form. Assume that the operands x, y, w, v are provided in parallel form and that the result is required in parallel form. Make any reasonable assumption on the execution time of the operations.

9.6   Further Readings

Digit-serial arithmetic is the subject of several books and many articles (Denyer and Renshaw 1985; Smith and Denyer 1988; Hartley and Corbett 1990; Hartley and Parhi 1995).

LSDF Arithmetic—General

The least-significant digit first (LSDF) algorithms and implementations for addition are often covered in standard texts on digital systems. LSDF multiplication is discussed in Lyon (1976), Chen and Willoner (1979), Danielson (1984), Gnanasekaran (1985), Dadda (1989), and Ienne and Viredez (1994). Design issues in digit-serial signal processors are discussed in Irwin and Owens (1989, 1990). An architecture and implementation of digit-serial processor are presented in Owens et al. (1993).

MSDF Arithmetic—General

A discussion of the MSDF approach and its application to the evaluation of polynomial and rational functions is presented in Ercegovac (1975, 1977) (see more details in Chapter 10). Variations of MSDF and LSDF bit-serial arithmetic are discussed in Sips (1984). An overview of online arithmetic is given in Ercegovac (1984), and a method for the design of online algorithms appears in Ercegovac and Lang (1988a). The design of the corresponding selection functions is discussed in Tu (1990). The properties of functions computable in online arithmetic are studied in Muller (1994).

MSDF Algorithms

Online division and multiplication algorithms are introduced in Trivedi and Ercegovac (1977), and variations and implementations are reported in Irwin

(1977), Trivedi and Rusnak (1978), Gorji-Sinaki (1981), Lin and Sips (1987), Tu and Ercegovac (1989), Guyot et al. (1989), Tu and Ercegovac (1991), and Tenca and Ercegovac (1999). Other algorithms and implementations appear in Owens (1980), Irwin and Owens (1987), and Bajard et al. (1994). Online square root is discussed in Ercegovac (1978), Oklobdzija and Ercegovac (1982), and Tu (1990). Online algorithms for evaluation of elementary functions are discussed in Kla et al. (1991).

The use of high-radix online arithmetic for accurate computing is investigated in Lynch and Schulte (1995) and Daumas et al. (1997).

VLSI implementations of online arithmetic algorithms are discussed in Irwin and Owens (1983) and Tullsen and Ercegovac (1986).

MSDF Recursive Computations

The use of online arithmetic in recursive computations and the development of algorithms and implementations are discussed in a number of places (Brackert 1988; Knowles et al. 1989; Brackert et al. 1989; Ercegovac 1991; Ercegovac and Lang 1992; Fernando 1993; Fernando and Ercegovac 1994, 1997). A method for designing MSDF algorithms for recursive filters is discussed in McQuillan and McCanny (1995).

MSDF Floating-Point

Floating-point online arithmetic and its implementation are presented in Watanuki (1981), Lin and Sips (1987), Tu (1990), Tu and Ercegovac (1991), and Duprat et al. (1991). Error analysis of floating-point online arithmetic is discussed in Watanuki and Ercegovac (1983).

MSDF Complex Number Arithmetic

Online arithmetic algorithms and implementations for operations on complex numbers are presented in Nielsen (1997) and McIlhenny (2002).

MSDF Variable Precision

Variable-precision algorithms and implementations using online arithmetic are reported in Tenca (1998) and Tenca and Ercegovac (1999).

MSDF FPGA-Based Algorithms and Implementations

Implementations of online arithmetic algorithms in FPGAs are reported in Daumas et al. (1994), Tenca et al. (1999), Tisserand et al. (1999), and Tenca and Hussaini (2001).

MSDF Various Applications

Online arithmetic has been applied to CORDIC (Ercegovac and Lang 1990; Lin and Sips 1990; Osorio et al. 1995), 2D DCT (Bruguera and Lang 1995), signal processing (Galli and Tenca 2001), digital communications (Rajagopal and Cavallaro 2001), digital control of real-time systems (Dimmler 1999; Dimmler et al. 1999), and neural networks (Girau and Tisserand 1996).

Composite online algorithms and implementations for various applications including various matrix computations such as triangularization, singular value decomposition, and 2-D and 3-D normalization are presented in Ercegovac and Lang (1987, 1988b), Tu (1990), Ercegovac and Tu (1991), Ercegovac and Lang (1999), and Huang and Ercegovac (2001).

9.7   Bibliography

Bajard, J. C., J. Duprat, S. Kla, and J.-M. Muller (1994). Some operators for on-line radix 2 computations. *Journal of Parallel and Distributed Computing*, 22(2):336–45.

Brackert, R. H. (1988). *Design and Implementation of Recursive Filters Using On-line Arithmetic*. PhD thesis, University of California, Los Angeles.

Brackert, R. H., M. D. Ercegovac, and A. N. Willson (1989). Design of an on-line multiply-add module for recursive digital filters. In *Proceedings of the 9th IEEE Symposium on Computer Arithmetic*, pages 34–41.

Bruguera, J., and T. Lang (1995). 2-D DCT using online arithmetic. In *International Conference on Acoustics, Speech, and Signal Processing*, volume 5, pages 3275–78.

Chen, I. N., and R. Willoner (1979). An o(n) parallel multiplier with bit-sequential input and output. *IEEE Transactions on Computers*, C-28(10):721–27.

Dadda, L. (1989). On serial input multipliers for two's complement numbers. *IEEE Transactions on Computers*, 38:1341–45.

Danielsson, P. E. (1984). Serial/parallel convolvers. *IEEE Transactions Computers*, C-33(7):652–67.

Daumas, M., J.-M. Muller, and A. Tisserand (1997). Very high radix on-line arithmetic for accurate computations. In *15th IMACS World Congress on Scientific Computation, Modelling and Applied Mathematics*, Berlin, Germany.

Daumas, M., J.-M. Muller, and J. Vuillemin (1994). Implementing on-line arithmetic on PAM. In *4th International Workshop on Field-Programmable Logic and Applications*.

Denyer, P., and D. Renshaw (1985). *VLSI Signal Processing: A Bit-Serial Approach*. Addison-Wesley, Reading, Massachusetts.

Dimmler, M. (1999). *Digital Control of Micro-Systems Using On-line Arithmetic*. PhD thesis, Ecole Polytechnique Federal de Lausanne.

Dimmler, M., A. Tisserand, U. Holmbeg, and R. Longchamp (1999). On-line arithmetic for real-time control of microsystems. *IEEE/ASME Transactions on Mechatronics*, 4(2):213–17.

Duprat, J., M. Fiallos, J.-M. Muller, and H. J. Yeh (1991). Delays of on-line floating-point operators in borrow-save representation. In *IFIP Workshop on Algorithms and Parallel VLSI Architectures*.

Ercegovac, M. D. (1975). *A General Hardware-Oriented Method for Evaluation of Functions and Computations in a Digital Computer*. PhD thesis, Department of Computer Science, University of Illinois at Urbana-Champaign. (Technical Report UIUCDCS-R-750.)

Ercegovac, M. D. (1977). A general hardware-oriented method for evaluation of functions and computations in a digital computer. *IEEE Transactions on Computers*, C-26(7):667–80.

Ercegovac, M. D. (1978). An on-line square rooting algorithm. In *4th IEEE Symposium on Computer Arithmetic*. IEEE Computer Society Press, Los Alamitos, California.

Ercegovac, M. D. (1984). On-line arithmetic: An overview. In *Proceedings of the SPIE, Real Time Signal Processing VII*, pages 86–93.

Ercegovac, M. D. (1991). On-line arithmetic for recurrence problems. In *Proceedings of the SPIE: Advanced Signal Processing Algorithms, Architectures, and Implementations II*.

Ercegovac, M. D., and T. Lang (1987). On-line scheme for computing rotation factors. In *Proceedings of the 8th IEEE Symposium on Computer Arithmetic*, pages 196–203.

Ercegovac, M. D., and T. Lang (1988a). On-line arithmetic: A design methodology and applications. In R. W. Brodersen and H. S. Moscovitz, editors, *VLSI Signal Processing, III*, Chapter 24. IEEE Press, New York.

Ercegovac, M. D., and T. Lang (1988b). On-line scheme for computing rotation factors. *Journal of Parallel and Distributed Computing*, Special Issue on Parallelism in Computer Arithmetic (5).

Ercegovac, M. D., and T. Lang (1990). Redundant and on-line CORDIC: Application to matrix triangularization and SVD. *IEEE Transactions on Computers*, 39(6):725–40.

Ercegovac, M. D., and T. Lang (1992). Fast arithmetic for recursive computations. In *Proceedings of the IEEE Workshop on VLSI Signal Processing*, pages 14–28.

Ercegovac, M. D., and T. Lang (1999). On-line scheme for normalizing a 3-D vector. In *Proceedings of the 33rd Asilomar Conference on Signals, Systems and Computers*, pages 1460–64.

Ercegovac, M. D., and P. K. G. Tu (1991). Application of on-line arithmetic algorithms to the SVD computation: preliminary results. In *Proceedings of the 10th IEEE Symposium on Computer Arithmetic*, pages 246–55.

Fernando, J. S. (1993). *Design Alternatives for Recursive Digital Filters Using On-line Arithmetic*. PhD thesis, University of California, Los Angeles.

Fernando, J. S., and M. D. Ercegovac (1994). Conventional and on-line arithmetic designs for high-speed recursive digital filters. *Journal of VLSI Signal Processing*, 7:189–97.

Fernando, J. S., and M. D. Ercegovac (1997). A method of eliminating oscillations in high-speed recursive digital filters. *IEEE Transactions on Circuits and Systems—II: Analog and Digital Signal Processing*, 44(10):861–64.

Galli, R., and A. F. Tenca (2001). Design and evaluation of on-line arithmetic for signal processing applications on FPGAs. In *Proceedings of the SPIE—Advanced Signal Processing Algorithms, Architectures, and Implementations XI*, volume 4474, pages 134–44.

Girau, B., and A. Tisserand (1996). On-line arithmetic-based reprogrammable hardware implementation of multilayer perceptron back-propagation. In *Proceedings of the 5th International Conference on Microelectronics for Neural Networks and Fuzzy Systems. MicroNeuro'96*, pages 168–75.

Gnanasekaran, R. (1985). A fast serial-parallel binary multiplier. *IEEE Transactions on Computers*, C-34:741–44.

Gorji-Sinaki, A. (1981). *Error-Coded Algorithms for On-line Arithmetic*. PhD thesis, University of California, Los Angeles.

Guyot, A., Y. Herreros, and J.-M. Muller (1989). Janus, an on-line multiplier/ divider for manipulating large numbers. In *Proceedings of the 9th IEEE Symposium on Computer Arithmetic*, pages 106–11. IEEE Computer Society Press, Los Alamitos, California.

Hartley, R., and P. Corbett (1990). Digit-serial processing techniques. *IEEE Transactions on Circuits and Systems*, 37(6):707–19.

Hartley, R., and K. K. Parhi (1995). *Digit-Serial Computation*. Kluwer Academic Publishers.

Huang, Z., and M. D. Ercegovac (2001). FPGA implementation of pipelined on-line scheme for 3-D vector normalization. In *IEEE Symposium on Field-Programmable Custom Computing Machines*, pages 1–4.

Icnne, P., and M. A. Viredez (1994). Bit-serial multipliers and squarers. *IEEE Transactions Computers*, 43(12):1445–50.

Irwin, M. J. (1977). *An Arithmetic Unit for On-line Computation*. PhD thesis, Department of Computer Science, University of Illinois at Urbana-Champaign. (Technical Report UIUCDCS-R-77-873.)

Irwin, M. J., and R. M. Owens (1983). Fully digit on-line networks. *IEEE Transactions on Computers*, C-32(4):402–6.

Irwin, M. J., and R. M. Owens (1987). Digit-pipelined arithmetic as illustrated by the Paste-up system: a tutorial. *IEEE Computer*, 20(4):61–73.

Irwin, M. J., and R. M. Owens (1989). Design issues in digit-serial signal processors. In *Proceedings of ISCAS'89*, volume 1, pages 441–44.

Irwin, M. J., and R. M. Owens (1990). A case for digit serial VLSI signal processors. *Journal of VLSI Signal Processing*, 1(4):321–34.

Kla, S., C. Mazenc, X. Merrheim, and J.-M. Muller (1991). New algorithms for on-line computation of elementary functions. In *Proceedings of the SPIE: Advanced Signal Processing Algorithms, Architectures, and Implementations II*, volume 1566, pages 275–85.

Knowles, S. C., R. F. Woods, J. McWirther, and J. McCanny (1989). Bit-level systolic architectures for high-performance IIR filtering. *Journal of VLSI Signal Processing*, 1(1):9–24.

Lin, H., and H. J. Sips (1987). A novel floating-point on-line division algorithm. In *Proceedings of the 8th IEEE Symposium on Computer Arithmetic*, pages 188–95.

Lin, H., and H. J. Sips (1990). On-line CORDIC algorithms. *IEEE Transactions on Computers*, 39(8):1038–52.

Lynch, T., and M. J. Schulte (1995). A high radix on-line arithmetic for credible and accurate computing. *Journal of Universal Computer Science*, 1(7):439–53.

Lyon, R. F. (1976). Two's complement pipeline multipliers. *IEEE Transactions on Communication*, pages 418–25.

McIlhenny, R. D. (2002). *Complex Number On-line Arithmetic for Reconfigurable Hardware: Algorithms, Implementations, and Applications*. PhD thesis, University of California, Los Angeles.

McQuillan, S., and J. V. McCanny (1995). A systematic methodology for the design of high performance recursive digital filters. *IEEE Transactions on Computers*, 44(8):971–82.

Muller, J.-M. (1994). Some characterizations of functions computable in on-line arithmetic. *IEEE Transactions on Computers*, 43(6):752–55.

Nielsen, A. M. (1997). *Number Systems and Digit Serial Arithmetic*. PhD thesis, Department of Mathematics and Computer Science, Odense University, Denmark.

Oklobdzija, V. G., and M. D. Ercegovac (1982). An on-line square root algorithm. *IEEE Transactions on Computers*, C-31:70–75.

Oppenheim, A. V., R. W. Schafer, and J. R. Buck (1999). *Discrete-Time Signal Processing*. Prentice-Hall, Upper Saddle River, New Jersey.

Osorio, R. R., E. Antelo, J. D. Bruguera, J. Villalba, and E. L. Zapata (1995). Digit on-line large radix CORDIC rotator. In *Proceedings of ASAP-95* (Strasbourg, France), pages 246–57.

Owens, R. M. (1980). *Digit On-line Algorithms for Pipeline Architectures*. PhD thesis, Department of Computer Science, Pennsylvania State University, University Park. (Technical Report CS-80-21.)

Owens, R. M., T. P. Kelliher, M. J. Irwin, M. Vishwanath, R. S. Bajwa, and W.-L. Yang (1993). The design and implementation of the Arithmetic Cube II, a VLSI signal processing system. *IEEE Transactions on Very Large Scale Integration (VLSI) Systems*, 1(4):491–502.

Rajagopal, S., and J. R. Cavallaro (2001). On-line arithmetic for detection in digital communication receivers. In *Proceedings of the 15th IEEE Symposium on Computer Arithmetic*, pages 257–65.

Sips, H. J. (1984). Bit-sequential arithmetic for parallel processors. *IEEE Transactions on Computers*, C-33(1):7–20.

Smith, S. G., and P. Denyer (1988). *Serial-Data Computation*. Kluwer Academic Publishers.

Tenca, A. F. (1998). *Variable Long-Precision Arithmetic (VLPA) for Reconfigurable Coprocessor Architectures*. PhD thesis, University of California, Los Angeles.

Tenca, A. F., and M. D. Ercegovac (1999). On the design of high-radix on-line division for long precision. In *Proceedings of the 14th IEEE Symposium on Computer Arithmetic*, pages 44–51.

Tenca, A. F., M. D. Ercegovac, and M. E. Louie (1999). Fast on-line multiplication units using LSA organization. In *Proceedings of the SPIE—Advanced Signal Processing Algorithms, Architectures, and Implementations IX*, volume 3807, pages 74–83.

Tenca, A. F., and S. U. Hussaini (2001). A design of radix-2 on-line division using LSA organization. In *Proceedings of the 15th IEEE Symposium on Computer Arithmetic*, pages 266–73.

Tisserand, A., P. Marchal, and C. Piguet (1999). FPOP: field-programmable on-line operators. In *Proceedings of the SPIE—Advanced Signal Processing Algorithms, Architectures, and Implementations IX*, volume 3807, pages 31–42.

Trivedi, K. S., and M. D. Ercegovac (1977). On-line algorithms for division and multiplication. *IEEE Transactions on Computers*, C-26(7):161–67.

Trivedi, K. S., and J. G. Rusnak (1978). Higher radix on-line division. In *Proceedings of the 4th IEEE Symposium on Computer Arithmetic*, pages 164–74.

Tu, P. K.-G. (1990). *On-line Arithmetic Algorithms for Efficient Implementations*. PhD thesis, University of California, Los Angeles.

Tu, P. K.-G., and M. D. Ercegovac (1989). Design of on-line division unit. In *Proceedings of the 9th IEEE Symposium on Computer Arithmetic*, pages 42–49.

Tu, P. K.-G., and M. D. Ercegovac (1991). Gate array implementation of on-line algorithms for floating-point operations. *Journal of VLSI Signal Processing*, 3(4):307–17.

Tullsen, D. M., and M. D. Ercegovac (1986). Design and implementation of an on-line algorithm. In *Proceedings of the SPIE Real Time Signal Processing IX*, volume 698, pages 92–99.

Watanuki, O. (1981). *Floating-Point On-line Arithmetic for Highly Concurrent Digit-Serial Computation: Application to Mesh Problems*. PhD thesis, University of California, Los Angeles.

Watanuki, O., and M. D. Ercegovac (1983). Error analysis of certain floating-point on-line algorithms. *IEEE Transactions on Computers*, C-32:352–58.

IN THIS CHAPTER WE PRESENT AND DISCUSS THE FOLLOWING TOPICS:

- Polynomial approximations and piecewise interpolation
- Reduction, approximation, and reconstruction
- Rational approximation
- Linear convergence method:
  - Multiplicative normalization: Logarithm function
  - Additive normalization: Exponential function

The evaluation of functions is an important part of many numerical computations. The set of functions we consider includes logarithm, exponential, and various trigonometric functions. The computation of these functions can be performed in software, using the standard floating-point instructions. For this, there are libraries that are suitable for particular processors. On the other hand, these functions can be computed by hardware/firmware implementations. These implementations might be specific for one particular function or for a set of functions. Since the hardware/firmware implementations can customize the primitives used and different types of parallelism can be available, the algorithms suited for this type of implementation might be different than those used for evaluation by software. In this chapter we concentrate on hardware/firmware implementations.

In general, these functions cannot be computed exactly with a finite number of arithmetic operations. Consequently, they have to be approximated. Moreover, the argument, coefficients, intermediate variables, and result are represented by finite-precision digit vectors. Therefore, the accuracy is determined by the error of the approximation and by the roundoff errors that occur during the evaluation of the approximation.

The choice of a method and a particular implementation depends on the requirements, such as delay, throughput, area, and energy. Of particular significance are the number of bits of the argument and the result, as well as the accuracy required. In this respect, requirements vary widely, from low-precision fixed-point representations to double-precision or quad-precision floating-point representations. Later we give comments on the domain of applicability of each method.

In particular for floating-point representations, as discussed in Chapter 8, the IEEE Standard specifies four rounding modes for the basic operations. No such requirement is specified for the functions because of the difficulty in obtaining correctly rounded results; this difficulty is described as the Table Maker's Dilemma.

However, as we comment later, for limited precision, such as single-precision representation, it is possible to have practical implementations that produce correctly rounded results. Moreover, in several applications it is convenient to have other numerical properties, such as monotonicity.

Usually, approximation methods are applicable only for a limited domain of the argument. Consequently, it is typical to include an initial step of domain reduction and a final step of reconstruction.

The most-direct evaluation method is a table lookup; this traditional method has become more practical recently because of the possibility of building larger tables. As a consequence, it might be the most-effective method for precisions of up to, say, 12 bits. For larger precisions, the resulting table is too large for practical implementation so that other methods have to be used.

Implementation of suitable approximation algorithms should utilize only basic operations, such as additions, multiplications, and table lookups. Because of this, the approximations we consider fall into two classes. The first class uses an approximating polynomial and can be used for any continuous function. We discuss variations of this method that reduce the degree of the polynomial by incorporating also table lookup. An interesting recent approach is to use only tables and adders. The second class consists in forming a recurrence that converges to the value of the function.[1] This recurrence depends on the function being evaluated; as a consequence, this approach is only useful for some functions. To have a simple implementation, the operations in the recurrence are limited to multiplication by the radix (shifts), multiplication by a radix-r digit, additions, and table lookups. Because of this, these are called *shift-and-add algorithms*, although for high radices a rectangular multiplier is also required.

In some instances a rational approximation that consists of the quotient of two polynomials can be used. The curve-fitting ability of a rational approximation consisting of polynomials of degree M and N roughly corresponds to that of a polynomial of degree $M + N$. Moreover, the two polynomials can be evaluated in parallel, reducing the evaluation time. Rational approximations are preferable to polynomial approximations for functions with a pole, such as $\tan(x)$, or an asymptote, such as $\arctan(x)$. On the other hand, the drawback is the division required.

---

1. Algorithms of this type have been considered in Chapter 7 for reciprocal, reciprocal square root, and square root.

10.1   Argument Range Reduction

Approximation methods can usually be applied only for a limited range of the argument. Consequently, before applying the approximating algorithm it is necessary to perform a transformation that reduces the range of the argument. Moreover, after obtaining the approximation another transformation produces the final value. The specific transformations depend on the function and on the approximating method.

Calling x_r the reduced argument, the most-used reduction methods are either

- additive, in which $x_r = x - kC$. This type is used, for instance, for trigonometric functions, where $C = \pi/4$.
- multiplicative, in which $x_r = xC^k$. This is used, for example, for the logarithm function.

In the case of floating-point representations, the reduction is applied to the representation consisting of sign, exponent, and significand. For instance:[2]

- For the logarithm (base 2) function, there is no need to perform explicit range reduction, since it is possible to approximate directly the significand and then add the exponent. A reduction step is performed only for zero argument exponent, to avoid leading zeros that result in a loss of accuracy.
- For the exponential (base 2) function with floating-point argument with exponent E and significand M, the reduction step results in

$$M_r = M \times 2^E - \lfloor M \times 2^E \rfloor \quad E_r = \lfloor M \times 2^E \rfloor \qquad \textbf{10.1}$$

In general, the reduction step should not result in a loss of accuracy. This might require that x_r be represented with additional precision.[3]

10.2   Correct Rounding and Monotonicity

The rounding modes for the basic floating-point operations are defined in Chapter 8. Moreover, methods for obtaining correctly rounded results are described for addition, multiplication, and division. In particular, the methods for division[4] are

---

2. For additional details see Schulte and Swartzlander (1994).
3. For details see Muller (1997) and Daumas et al. (1995).
4. And other algebraic functions, such as square root and reciprocal square root.

based on the calculation of the remainder produced by the rounded approximation. These methods are not applicable for nonalgebraic functions, such as those discussed in this chapter. For these functions, the following approaches have been used:

- For implementations using table lookup only, the correctly rounded values can be stored. As indicated, these implementations are practical for low-precision cases.
- For implementations based on piecewise interpolations, the coefficients can be tuned so as to produce correctly rounded results. Since the tuning is done by exhaustively examining all argument values (or at least all values of the significand), this method is practical for medium-precision cases.[5]
- As discussed in Chapter 8, the cases that are problematic for rounding correspond to those in which the infinite-precision result has a large number of consecutive zeros or ones after the rounding bit. Consequently, if a bound p on the maximum number of zeros (or ones) is known, for correct rounding to bit n, the approximation should be computed with an error less than $2^{-(n+1+p)}$. Although in general these bounds are not known, they have been obtained for some values of n by selective searches.[6]

When the result is not correctly rounded, it is convenient to preserve some properties of the function, such as monotonicity. This preservation means that if $f(x + ulp) > f(x)$, then the approximation F should satisfy $F(x + ulp) \geq F(x)$, and similarly if $f(x + ulp) < f(x)$. For the basic elementary functions sin, tan, arctan, $\log_2$, and exp, monotonicity is preserved (in specific intervals) if the approximation has an additional accuracy of a few bits.[7]

10.3 Polynomial Approximations and Interpolations

The approximation of a function by a polynomial has the advantages of being general, since any continuous function can be approximated in this way, and that the implementation consists of multiplications and additions. Because of

---

5. For details see Schulte and Swartzlander (1994).
6. For details see Muller (1997).
7. For details see Ferguson and Brightman (1991).

this, the implementation can accommodate a family of functions, where only the polynomial coefficients determine which function is being computed.

Different types of polynomials are possible. The most appropriate depends on the domain of the function, the error objective, and the implementation requirements. With respect to the error there are methods to obtain the optimal polynomial to minimize the maximum absolute error (called *minimax approximation*) or the average error (called *least-squares approximation*).[8] For hardware implementation usually the minimax case is considered, or an easily implemented (although not optimal) approximation is used and the required error is obtained by adapting the degree of the polynomial. The total error is obtained by the approximation error plus the roundoff errors arising from the use of finite-precision arithmetic in the evaluation.

For high accuracy and for a large argument domain, a high-degree polynomial is required. Two related alternatives are used to reduce the degree of the polynomial:

1. Partition into subranges and perform piecewise interpolation. This requires table lookup in addition to the polynomial evaluation.

2. Range reduction, polynomial approximation, and range recovery. This also requires table lookup and is suitable only for some functions, in which the recovery is simple.

We consider these alternatives now.

10.3.1   Polynomial Approximations

The most direct polynomials that can be used to approximate a function are obtained from a truncated Taylor or Maclaurin series. Although, these polynomials are effective to approximate a function in one point, they do not produce the minimum error for approximation in a range. Consequently, they are used when the range is small (maybe as part of a piecewise interpolation or together with range reduction).

The Taylor series of $f(x)$ about x_0 is given by

$$f(x) = f(x_0) + \sum_{i=1}^{\infty} \frac{f^{(i)}(x_0)}{i!}(x - x_0)^i \qquad \textbf{10.2}$$

---

8. See, for example, Davis (1990).

where $f^{(i)}(x_0)$ is the ith derivative of $f(x)$ evaluated at x_0. The Maclaurin series is the special case for $x_0 = 0$.

These series converge for an interval of values of x, which depends on the function. The absolute error (also called the *Lagrange remainder*) when the series is truncated at term k (that is, all terms for $i > k$ are omitted) is

$$\epsilon_k(x) = \frac{f^{(k+1)}(\alpha)(x - x_0)^{k+1}}{(k+1)!} \qquad \textbf{10.3}$$

where α is an unknown value such that $x_0 < \alpha < x$.

EXAMPLE 10.1   Consider the evaluation of $y = \sin x$ for $0 \leq x \leq \frac{1}{2}$ with an absolute error less than 2^{-32}. The Taylor series expansion about $x = 0$ is

$$\sin x = x - \frac{x^3}{3!} + \frac{x^5}{5!} - \frac{x^7}{7!} + \cdots \qquad \textbf{10.4}$$

Because of the alternating signs, a bound for the error when using a k-term approximation is

$$|\epsilon_k(x)| < \frac{x^{2k+1}}{(2k+1)!} \qquad \textbf{10.5}$$

This error is maximum for $x = \frac{1}{2}$. Consequently,

$$\frac{2^{-(2k+1)}}{(2k+1)!} < 2^{-32} \qquad \textbf{10.6}$$

which is satisfied for $k = 5$.

A better approximation is obtained if the expansion is about the middle point of the interval. However, in such a case the approximation includes all powers of x, so it might be more expensive to evaluate. ∎

For a polynomial of the same degree, a significantly smaller maximum error than that of using a truncated Taylor series is obtained by using Chebyshev polynomials of the first kind. For details, see the references at the end of the chapter.

Another method to obtain a polynomial approximation is by interpolation. In this method, a polynomial of degree N is obtained by making its value coincide with the function at $N + 1$ points (breakpoints). The most direct way to obtain the coefficients c_j is to solve the set of $N + 1$ linear equations

$$\sum_{j=0}^{N} c_j X_i^j = Y_i \quad 0 \le i \le N \qquad \textbf{10.7}$$

where (X_i, Y_i) are the $N + 1$ breakpoints. The resulting polynomial is then

$$p_N(x) = \sum_{j=0}^{N} c_j x^j \qquad \textbf{10.8}$$

Instead of solving these equations, there are several direct methods to obtain the coefficients (see the references at the end of the chapter).

Implementation

The evaluation of the polynomial approximation computation requires the coefficients, which can be hardwired or stored in memory, and multiplier/accumulator units.

The scheduling of the operations depends on the characteristics and the number of multiplier/accumulator units.

If one nonpipelined unit is available, a sequential algorithm is required. For this, it is convenient to factor the polynomial as follows (called *Horner's rule*):

$$p_N(x) = c_0 + x(c_1 + x(c_2 + x(\ldots x(c_{N-1} + xc_N)\ldots))) \qquad \textbf{10.9}$$

Then the evaluation of the polynomial results in the following recurrence:

$$R[i - 1] = c_{i-1} + x R[i], \quad i = N, \ldots, 1 \qquad \textbf{10.10}$$

with the initial condition $R[N] = c_N$ and the result $p_N(x) = R[0]$. The execution time corresponds to N multiply/adds.

The presented approach is not the best when the multiplier/accumulator unit is pipelined or when several units are available, since in those cases a parallel algorithm is required. For instance, for a polynomial of degree 7, we can write

$$p_7(x) = x^4(x^2(c_7x + c_6) + c_5x + c_4) + x^2(c_3x + c_2) + (c_1x + c_0) \qquad \textbf{10.11}$$

This can be performed in three multiply/accumulate steps as shown in Figure 10.1. An implementation is illustrated in Figure 10.2. It requires four multiplier/accumulator units and one squarer.

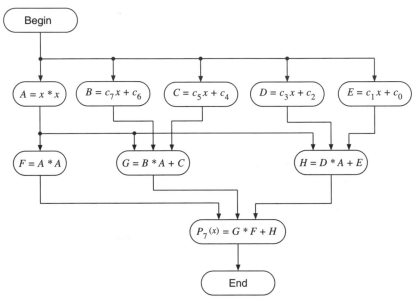

FIGURE 10.1 Concurrent execution graph for $P_7(x)$.

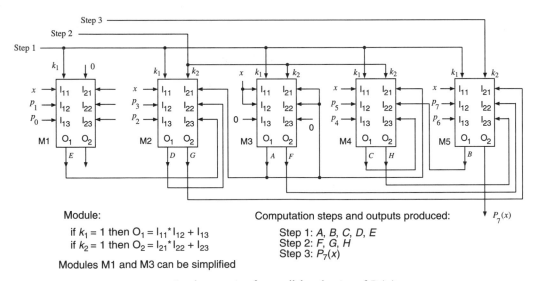

Module:

if $k_1 = 1$ then $O_1 = I_{11}*I_{12} + I_{13}$
if $k_2 = 1$ then $O_2 = I_{21}*I_{22} + I_{23}$

Modules M1 and M3 can be simplified

Computation steps and outputs produced:

Step 1: A, B, C, D, E
Step 2: F, G, H
Step 3: $P_7(x)$

FIGURE 10.2 Implementation for parallel evaluation of $P_7(x)$.

Cycle	1	2	3	4	5	6	7	8	9	10	11
Stage 1	$A1$	$B1$	$C1$	$D1$	$E1$	$F1$	$G1$	$H1$	$P1$		
Stage 2		$A2$	$B2$	$C2$	$D2$	$E2$	$F2$	$G2$	$H2$	$P2$	
Stage 3			A	B	C	D	E	F	G	H	P

FIGURE 10.3 Evaluation of $P_7(x)$ on a three-stage pipelined multiplier/accumulator.

For a polynomial of degree N, the number of steps is $\lceil \log_2 N + 1 \rceil$ and the number of multiplier/accumulator units is $(N + 1)/2$.

The operations can also be scheduled on a pipelined multiplier/accumulator. For instance, a scheduling for a three-stage unit is shown in Figure 10.3. In this unit, Stages 1 and 2 perform partial product reductions, producing a redundant product, and Stage 3 performs the accumulation.

If some of the polynomial coefficients are zero, a different decomposition might be preferable. For instance, if in $p_7(x)$ the c_i are zero for i even, we can get

$$p_7(x) = c_7x^7 + c_5x^5 + c_3x^3 + c_1x = x^4(x^2(c_7x) + c_5x) + x(x^2c_3 + c_1)$$

10.12

The accuracy of the result depends on the error of the approximation and the error introduced by the finite-precision coefficients, intermediate variables, and result.

10.3.2 Piecewise Interpolation

An alternative to fitting a polynomial of degree N through $N + 1$ breakpoints is to have different polynomials (of lower degree) through subsets of the breakpoints. This is called *piecewise interpolation*. So, if the breakpoints are sufficiently close, it might be accurate enough to do a linear interpolation, fitting a straight line between adjacent points. The polynomial for the linear interpolation between breakpoints i and $i + 1$, illustrated in Figure 10.4, is

$$p_1^{(i)}(x) = Y_i + \frac{Y_{i+1} - Y_i}{X_{i+1} - X_i}(x - X_i)$$

10.13

Consequently, for each breakpoint two values are required, namely, Y_i and $(Y_{i+1} - Y_i)/(X_{i+1} - X_i)$.

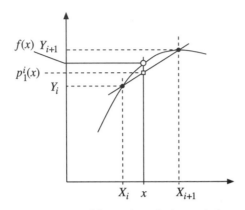

FIGURE 10.4 Linear piecewise interpolation.

EXAMPLE 10.2   Consider an approximation of $f(x) = x^{1/3}$ in the domain $\frac{1}{2} \leq x < 1$ by piecewise interpolation with linear interpolation. As an illustration we only consider four intervals, namely, $X_i = \frac{1}{2} + (1/8)i$ with $i = 0, 1, 2, 3$. The following table contains the required constants:

i	Y_i	$(Y_{i+1} - Y_i)/(X_{i+1} - X_i)$
0	0.7937	0.4904
1	0.8550	0.4288
2	0.9086	0.3824
3	0.9564	0.3488

For instance, for $x = 0.788$ we obtain $i = 2$ and
$$p(x) = 0.9086 + 0.3824(0.788 - 0.75) = 0.9231$$ ■

Implementation

If the X_i are equally spaced and are multiples of 2^{-k}, for an n-bit argument x we can write

$$x = X_i + x_r 2^{-k} \qquad \textbf{10.14}$$

with x_r integer. That is, X_i is obtained as x truncated at fractional bit k. Moreover, the rest of x corresponds to $(x - X_i)$. Consequently, for an x of n bits, f of which

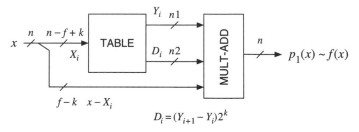

$$D_i = (Y_{i+1} - Y_i)2^k$$

k, $n1$ and $n2$ are determined by desired accuracy of the result

FIGURE 10.5 A generic implementation of linear interpolation.

are fractional, X_i has $n - f + k$ bits and $x - X_i$ has $f - k$ bits. The corresponding implementation consists of

- a module (table) that stores the function values Y_i. The input to this module has $n - f + k$ bits (or one less if x is normalized so that its most-significant bit is always 1). Moreover, it could store the values $(Y_{i+1} - Y_i)/2^{-k}$, or this difference can be computed.
- a multiply-add unit. The multiplier has $f - k$ bits.

The value k and the width of the table and of the multiplicand depend on the accuracy required.

A generic implementation is shown in Figure 10.5.

Error

The error is composed of the error of the interpolation and of the error due to the roundoff of the intermediate values that are formed during the evaluation. The first component depends on the particular function as well as on the number and position of the breakpoints. For equally spaced breakpoints, their number determines the number of inputs to the module storing the values (the size of the table).

To reduce the error it is possible to use higher-degree polynomials and/or more breakpoints. Although this requires more constants because of the reduced error, for the same accuracy it requires smaller tables than the linear interpolation. However, it requires more multiplication-adds. Special attention has been given recently to hardware implementations of second-order (quadratic) interpolation.[9] An implementation of a quadratic interpolation is illustrated in Figure 10.6.

---

9. For details see Cao et al. (2001).

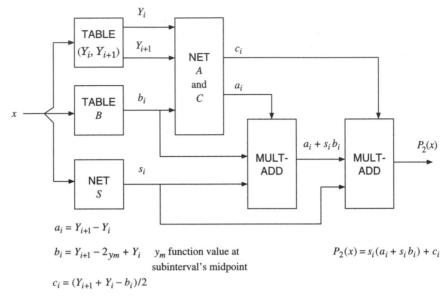

$$a_i = Y_{i+1} - Y_i$$

$b_i = Y_{i+1} - 2y_m + Y_i$ y_m function value at
subinterval's midpoint $P_2(x) = s_i(a_i + s_i b_i) + c_i$

$$c_i = (Y_{i+1} + Y_i - b_i)/2$$

FIGURE 10.6 Implementation of a quadratic interpolator. Adapted from Cao et al. (2001).

10.3.3   Reduction, Approximation, and Reconstruction

This method is a variation of the piecewise interpolation method and also uses table lookup in addition to the polynomial evaluation. Since the approximation by a polynomial is more accurate for smaller domains, this method reduces first the domain,[10] then performs the approximation, and finally reconstructs the approximation of the function. It consists of a series of N breakpoints X_i and a table storing approximations of $f(X_i)$. To compute $f(x)$ the method consists of the following three steps:

1.   Reduction: Select the breakpoint X_i closest to x and apply the reduction transformation producing r such that

$$r = R(x, X_i)$$

The function R is chosen so as to simplify the remaining steps.

---

10. This is a second-level reduction, which is applied in addition to the initial range reduction, discussed before.

2. Approximation: Calculate an approximation to $g(r)$ by using a polynomial $p(r)$.

3. Reconstruction: The value $f(x)$ is obtained from $g(r)$ and $Y_i \approx f(X_i)$ by the function

$$f(x) = S(g(r), Y_i)$$

The location of the breakpoints is selected so that the reconstruction function is simple. Consequently, the location of these points is different for each function. Also the method might not be applicable to functions for which the reconstruction is complicated.

EXAMPLE 10.3   Compute $\ln(x)$ on $[1, 2]$.

1. Reduction: Find the breakpoint $X_i = 1 + i/64, i = 0, 1, \ldots, 64$ such that $|x - X_i| \le 1/128$. Obtain

$$r = 2(x - X_i)/(x + X_i), \quad |r| \le 1/128$$

2. Approximation: Approximate $\ln(x/X_i)$ by a polynomial $p(r)$. Since

$$\ln\left(\frac{x}{X_i}\right) = \ln\left(\frac{1 - 2^{-1}r}{1 + 2^{-1}r}\right)$$

the polynomial is of the form

$$p(r) = r + p_1 r^3 + p_2 r^5 + \cdots + p_N r^{2N+1}$$

3. Reconstruction: Reconstruct $\ln(x)$ using the following relations

$$\ln(x) = \ln(X_i) + \ln(x/X_i)$$
$$\approx \ln(X_i) + p(r)$$
$$\approx Y_i + p(r)$$

where $Y_i \approx \ln(X_i), i = 0, 1, \ldots, 64$ are stored in a table.   ∎

As in the piecewise interpolation method the error depends mainly on the number of breakpoints and on the degree of the polynomial.[11]

---

11. For additional examples and error analysis see Tang (1991).

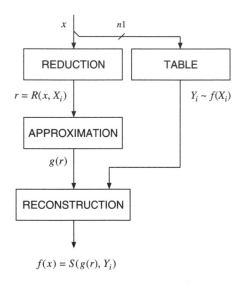

FIGURE 10.7 Block diagram of reduction, approximation, and reconstruction.

Implementation

As for piecewise approximation the implementation consists of a table and a polynomial evaluator. However, since the domain is first reduced to a common subdomain, only one polynomial is required. On the other hand, additional modules are required for the reduction and reconstruction, and these depend on the function being computed. Moreover, as indicated above, the locations of the breakpoints also depend on the function, and this influences the bits of the argument used to access the table. Figure 10.7 illustrates implementation of reduction, approximation, and reconstruction.

10.4   Bipartite and Multipartite Table Method

This method uses table lookup and additions and therefore reduces the size of required table(s), with respect to the table-only method. Moreover, it does not require the multipliers used in the polynomial methods.

A bipartite formula to approximate $f(x)$ is obtained as follows. Split the n-bit argument x into three parts[12] as

$$x = x_1 + x_2 2^{-k} + x_3 2^{-2k} \qquad\qquad \textbf{10.15}$$

where $k = n/3$ and $0 \le x_i \le 1 - 2^{-k}$.

The Taylor series expansion of $f(x)$ at $x_1 + x_2 2^{-k}$ is

$$f(x) = f(x_1 + x_2 2^{-k}) + x_3 2^{-2k} f^{(1)}(x_1 + x_2 2^{-k}) + \epsilon_1 \qquad\qquad \textbf{10.16}$$

where

$$\epsilon_1 = \frac{1}{2} x_3^2 2^{-4k} f^{(2)}(\alpha) \qquad\qquad \textbf{10.17}$$

and $\alpha \in [x_1 + x_2 2^{-k}, x]$.

The derivative $f^{(1)}(x_1 + x_2 2^{-k})$ is approximated by $f^{(1)}(x_1)$, resulting in

$$f(x) = f(x_1 + x_2 2^{-k}) + x_3 2^{-2k} f^{(1)}(x_1) + \epsilon_1 + \epsilon_2 \qquad\qquad \textbf{10.18}$$

where $\epsilon_2 = x_2 x_3 2^{-3k} f^{(2)}(\xi)$ and $\xi \in [x_1, x_1 + x_2 2^{-k}]$.

Therefore, the bipartite formula is

$$f(x) \approx F_0(x_1, x_2) + F_1(x_1, x_3) \qquad\qquad \textbf{10.19}$$

with an error $\epsilon \approx 2^{-3k} f_{max}^{(2)}$.

10.4.1 Implementation

F_0 and F_1 are precomputed and stored in tables $T0$ and $T1$. As illustrated in Figure 10.8, for an input x the corresponding values are obtained from the tables and added to produce the approximation of $f(x)$.

The $T0$ table stores a value of the function on the domain segments defined by (x_1, x_2), while table $T1$ stores the "offset" values defined by (x_1, x_3) to be added to the segment values. This is illustrated in Figure 10.9. To minimize the error, the values in table $T0$ are at the middle points of the segment.

---

12. In general, the split need not be into equal parts.

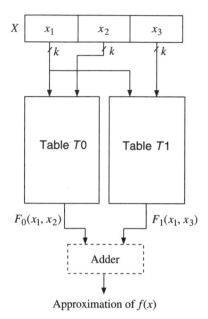

$$F_0(x_1, x_2) \qquad F_1(x_1, x_3)$$

FIGURE 10.8 Bipartite method.

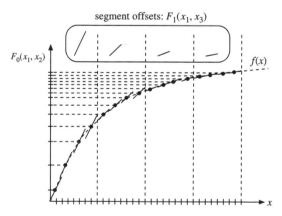

FIGURE 10.9 Segment and offset values.

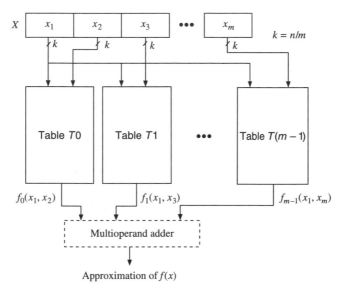

FIGURE 10.10 Multipartite method of function approximation.

10.4.2 Comparison

For an argument of n bits, an approximation using a single table requires 2^n words, while a bipartite method requires two tables of $2^{2n/3}$ words each—a significant saving. For example, for $n = 16$, a bipartite method uses two tables of 2^{11} words compared to a direct method using one table of 2^{16} words. Note that since the offset is small (many leading zeros), the width of the second table is also small. Moreover, the output of the two tables corresponds to a carry-save representation of the sum, so that no carry-propagate adder is required if this carry-save representation can be used in the operations that follow.

10.4.3 Multipartite Table Approach

The bipartite table approach can be generalized by subdividing x into m parts $(x_1, x_2, \ldots, x_m)$ and having one offset for each pair (x_1, x_p), $p = 2, \ldots, m$. As shown in Figure 10.10, this results in a reduction in the total number of table bits, but requires more additions.[13]

---

13. For additional details see Stine and Schulte (1999) and Muller (1999).

10.5   Rational Approximation

Any continuous elementary function can be approximated by a polynomial of degree L, $P_L(x)$, or by a rational function

$$R_{M,N}(x) = \frac{P_M(x)}{Q_N(x)}$$ **10.20**

As mentioned before, in many instances rational approximations are more accurate than the polynomial approximations using the same number of coefficients. Moreover, rational functions have a higher degree of parallelism in execution. A disadvantage is the need for a divider.

The coefficients of a rational approximation $R_{M,N}(x)$ for a function $f(x)$ are determined so as to minimize the maximum relative error

$$\max_{[a, b]} \left[\frac{R_{M,N}(x) - f(x)}{f(x)} \right]$$ **10.21**

over an interval $[a, b]$. Such an approximation is unique. The coefficients can be obtained using methods discussed in the literature.[14]

EXAMPLE 10.4   The following rational function[15] approximates $\tan(\frac{\pi}{4}x)$ in the interval $x \in [0, 1]$ with an absolute error less than 10^{-8} (not including roundoff errors):

$$\tan\left(\frac{\pi}{4}x\right) \approx x R_{1,2}(x^2) = x \frac{P_1(x^2)}{Q_2(x^2)} = x \frac{p_1 x^2 + p_0}{q_2 x^4 + q_1 x^2 + q_0}$$ **10.22**

The coefficients of the P and Q polynomials are

$$p_1 = -0.125288887278448 \times 10^2$$
$$p_0 = 0.211849369664121 \times 10^3$$
$$q_2 = 1.0 \times 10^1$$ **10.23**
$$q_1 = -0.714145309347748 \times 10^2$$
$$q_0 = 0.269735013121412 \times 10^3$$

---

14. For example, consult Hart et al. (1978).
15. From Hart et al. (1978), pages 119 and 216, TAN 4142.

In comparison, a polynomial approximation of $\tan(\frac{\pi}{4}x)$ in the same interval with a similar absolute error[16] is

$$\tan\left(\frac{\pi}{4}x\right) \approx x\,P_6(x^2) = x(c_6 x^{12} + c_5 x^{10} + c_4 x^8 + c_3 x^6 + c_2 x^4 + c_1 x^2 + c_0)$$

10.24

where the coefficients are

$$c_6 = 0.4443199695 \times 10^{-3}$$
$$c_5 = 0.951307678 \times 10^{-4}$$
$$c_4 = 0.2931842304 \times 10^{-2}$$
$$c_3 = 0.97543639755 \times 10^{-2}$$
$$c_2 = 0.398891627332 \times 10^{-1}$$
$$c_1 = 0.1614868943266$$
$$c_0 = 0.7853982781345$$

10.25

The computational graphs of these two approximations are illustrated in Figure 10.11. The rational approximation requires four multiplications, two multiply-adds, one addition, and one division. If implemented with three multiply-add units, an adder, and a divider, the critical path corresponds, roughly, to multiplication, multiply-add, addition, and division operation. On the other hand, the polynomial approximation requires six multiply-adds and four multiplications. Implemented with four multiply-add units, it has a critical path of two multiplications and three multiply-adds. The choice of the method depends on the number and relative delays of functional units as well as on the design objectives. ∎

10.5.1   MSDF Polynomial/Rational Function Evaluator

We now discuss an approach for evaluation of polynomial and rational functions suitable for hardware implementation. The approach is also of interest since it eliminates the use of explicit division in evaluation of certain rational functions. The approach uses most-significant-digit-first (MSDF) serial arithmetic, discussed in Chapter 9.

---

16. See Hart et al. (1978), pages 119 and 215, TAN 4225.

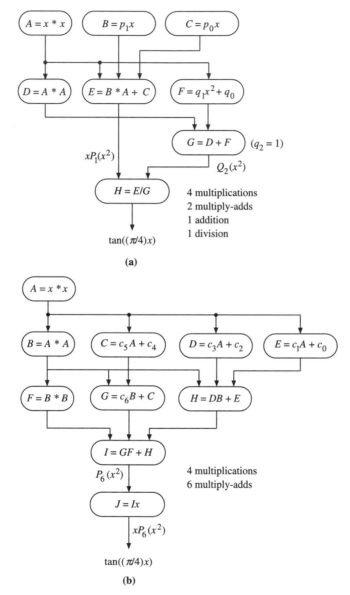

FIGURE 10.11 Computational graphs for computing $\tan(\frac{\pi}{4}x)$: (a) Rational approximation. (b) Polynomial approximation.

To introduce the method we illustrate a correspondence between a solution to a system of linear equations and a rational function. Consider the system

$$
\begin{aligned}
y_1 &= p_0 + x \cdot y_2 \\
y_2 &= p_1 - q_1 \cdot y_1 + x \cdot y_3 \\
y_3 &= p_2 - q_2 \cdot y_1 + x \cdot y_4 \\
y_4 &= -q_3 \cdot y_1
\end{aligned}
\qquad \textbf{10.26}
$$

Solving for y_1, we obtain

$$
y_1 = \frac{p_2 x^2 + p_1 x + p_0}{q_3 x^3 + q_2 x^2 + q_1 x + 1}
\qquad \textbf{10.27}
$$

or, $y_1 = R_{2,3}(x)$. That is, a rational function R can be evaluated by solving a system of linear equations similar to system (10.26).

Clearly, solving the system (10.26) by a direct method such as the Gaussian elimination, would not be attractive. Instead, we solve the system iteratively using MSDF serial arithmetic. The coefficients p's, q's and the argument x are in parallel form while y'_ks are produced and used digit-by-digit in MSDF manner. Each y_k of system (10.26) is evaluated on a separate module that uses digit $\times$ digit-vector multiplication, addition, and output digit selection to perform MSDF multiply-add operation. To obtain one digit of each y_k per iteration step, the coefficients and the argument x are bounded as discussed later. The solution is in the $(-1, 1)$ range, i.e., the MS digit of each y_k is 0, to allow initialization of the iterative process. In step j the network of modules produces the $j + 1$-st digit dk_{j+1} of each y_k using digits dk_j produced in the previous step. In m steps, the result of m radix-r digits is obtained. In other words, the iterative method used is linearly convergent. Note that division required by the rational function is not explicitly performed. As discussed shortly, the iterative method used is a generalization of a scalar digit-recurrence division to a vector by matrix division. The network for solving system (10.26) is shown in Figure 10.12. The result in digit-parallel form can be obtained during the computation using on-the-fly conversion.

We now give a general formulation of the MSDF method for evaluating polynomials and rational functions.[17] As mentioned above, the corresponding

---

17. Details of the method are in Ercegovac (1977).

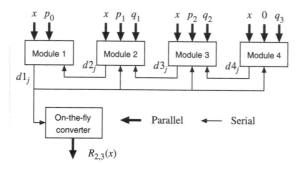

FIGURE 10.12 MSDF network for evaluation of rational function $R_{2,3}(x)$.

implementations have a delay linearly proportional to the number of digits in the result. The approach is (1) to transform a polynomial or a rational function into a system of linear equations, and (2) to solve the system using digit-recurrence division generalized to matrices and vectors in which the coefficient matrix corresponds to the divisor and the right-hand side vector to the dividend. The quotient is the solution vector. The elements of the solution vector are, as expected, obtained starting with the most-significant digits. Like in scalar division, redundancy in the quotient and residual representation is used to reduce the delay and simplify the selection of result digits.

In the following discussion boldface letters denote matrices and vectors.

First, map a function $f(x)$ (rational function or a polynomial)

$$f(x) \Rightarrow L : \mathbf{A} \cdot \mathbf{y} = \mathbf{b} \qquad \textbf{10.28}$$

such that $y_1 = f(x)$.

For example, a rational function $R_{2,3}(x)$, discussed above (10.26) is mapped to the matrix/vector form as follows:

$$
\begin{bmatrix}
1 & -x & 0 & 0 \\
q_1 & 1 & -x & 0 \\
q_2 & 0 & 1 & -x \\
q_3 & 0 & 0 & 1
\end{bmatrix}
\begin{bmatrix}
y_1 \\
y_2 \\
y_3 \\
y_4
\end{bmatrix}
=
\begin{bmatrix}
p_0 \\
p_1 \\
p_2 \\
0
\end{bmatrix}
\qquad \textbf{10.29}
$$

Solving the system produces **y** such that

$$y_1 = R_{3,2}(x) = \frac{p_2 x^2 + p_1 x + p_0}{q_3 x^3 + q_2 x^2 + q_1 x + 1} \qquad \textbf{10.30}$$

Similarly, a polynomial $P_3(x)$ is mapped to the following system

$$\begin{bmatrix} 1 & -x & 0 & 0 \\ 0 & 1 & -x & 0 \\ 0 & 0 & 1 & -x \\ 0 & 0 & 0 & 1 \end{bmatrix} \begin{bmatrix} y_1 \\ y_2 \\ y_3 \\ y_4 \end{bmatrix} = \begin{bmatrix} p_0 \\ p_1 \\ p_2 \\ p_3 \end{bmatrix} \qquad \textbf{10.31}$$

such that $y_1 = P_3(x) = p_3 x^3 + p_2 x^2 + p_1 x + p_0$

Then, as mentioned above, the system L is solved by a digit recurrence division algorithm applied to a divisor that is a matrix (**A**) and a dividend (**b**) that is a vector. The solution vector **y** is computed most-significant digit first, producing m significant digits in m steps.

$$\mathbf{y} = \frac{\mathbf{b}}{\mathbf{A}} \qquad \textbf{10.32}$$

In general, for a solution to exist, the matrix **A** must be nonsingular. Moreover, for a digit recurrence method to be applicable, the matrix must be diagonally dominant. That is, for each row, the sum of absolute values of off-diagonal coefficients must be smaller that the diagonal element. Since the matrices considered here have 1s on the diagonal, a necessary condition for convergence is

$$\sum_{j \neq i} |a_{i,j}| < 1 \qquad \textbf{10.33}$$

For radix r and the quotient-digit selection by rounding, the condition (10.33) is more restricted and requires prescaling—as in the high-radix division with selection by rounding.[18] For simplicity we consider here only radix 2. The algorithm for radix 2 is summarized in Figure 10.13.

In the algorithm we use the following notation:

- Matrices and vectors of elements are in boldface: the coefficient matrix **A** of order N; the solution vector $\mathbf{y} = (y_1, \dots, y_N)$; the right-hand side vector $\mathbf{b} = (b_1, \dots, b_N)$.

---

18. Ercegovac (1975, 1977).

1. [*Initialize*]

   $\mathbf{w}[0] = \mathbf{b}; \ \mathbf{d}[0] = \mathbf{0};$

2. [*Recurrence*]

   for $j = 0 \ldots m - 1$

   $\mathbf{v}[j] = 2(\mathbf{w}[j] - \mathbf{Ad}[j]);$

   $\mathbf{d}[j+1] \leftarrow SEL(\widehat{\mathbf{v}}[j]);$

   $\mathbf{w}[j+1] \leftarrow \mathbf{v}[j];$

   $y_1[j+1] \leftarrow CONVERT(y_1[j], SEL(\widehat{\mathbf{v}}[j]))$

   end for

3. [*Result*]

   $y_1[m] \approx f(x)$

where

- Each residual is in redundant form, represented by the pseudosum *WS* and stored-carry *WC* bit-vectors. For simplicity, we use $wk[j]$ in the description.
- m is the precision in bits.
- *SEL* is the digit selection function

$$dk_j = SEL(\widehat{vk[j]}) = \begin{cases} 1 & \text{if } \widehat{vk[j]} \geq 0.5 \\ 0 & \text{if } -0.5 \leq \widehat{vk[j]} \leq 0 \\ -1 & \text{if } \widehat{wk[j]} \leq -1 \end{cases}$$

where $\widehat{vk[j]}$ is the estimate of $vk[j] = 2(wk[j] - dk_j - q_i d1_j + xd(k+1)_j)$ truncated to one fractional bit.

FIGURE 10.13 Radix-2 MSDF algorithm for evaluating polynomial and rational functions.

- The residual vector at step j:

$$\mathbf{w}[j] = (w1[j], \ldots, wN[j]) \qquad \text{10.34}$$

- The result digit-vector at step j:

$$\mathbf{d}[j] = (d1_j, \ldots, dN_j) \qquad \text{10.35}$$

where digit $dk_j \in \{-1, 0, 1\}$ is the jth digit of

$$y_k = \sum_{j=1}^{m} dk_j 2^{-j}$$

Note that the multiplications in the term $\mathbf{A} \times \mathbf{d}[j]$ are implemented as digit-vector by digit multipliers.

The convergence of the algorithm requires the following conditions to be satisfied:

$$|y_i| \leq 1$$
$$\max_i |b_i| \leq \frac{3}{4} \qquad \textbf{10.36}$$
$$\max_i \left(\sum_{j \neq i} |a_{ij}| \right) \leq \frac{1}{8}$$

The mapping onto a linear system L in the case of rational functions requires that $q_0 = 1$. This may require recalculation of the coefficients by dividing P and Q by q_0.

EXAMPLE 10.5 We present an implementation for evaluation of the rational function $R_{3,4}(x)$ as an approximation to $\sinh(x)$.[19]

To satisfy the bounds (10.36) and to have $a_{1,1} = 1$, the original coefficients are divided by q_0. Moreover, we restrict the argument x to $[0, \frac{1}{8}]$ and divide all normalized coefficients of P by 2 to make them $\leq \frac{3}{4}$. This scaling requires one additional iteration.

We illustrate the algorithm for $m = 12$. The normalized coefficients, rounded to 12 bits, are shown in hexadecimal:

$$p_3 = 0.0d\,8$$
$$p_2 = 0.000$$
$$p_1 = 0.800$$
$$p_0 = 0.000$$
$$q_4 = 0.007 \qquad \textbf{10.37}$$
$$q_3 = 0.000$$
$$q_2 = -0.0fa$$
$$q_1 = 0.000$$
$$q_0 = 1.000$$

---

19. The coefficients are obtained from rational function approximation of $\sinh(x)$ in the interval $x \in [0, \frac{1}{6}]$ with a relative error less than 10^{-13}; see Hart et al. (1978), pages 104 and 182, SINH 2002.

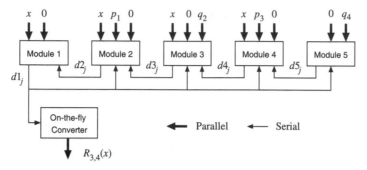

FIGURE 10.14 Implementation of rational function evaluator for $R_{3,4}(x)$. (The initial values correspond to the rational approximation for $\sinh(x)$.)

As shown in Figure 10.14, there are five modules performing the following residual recurrences:

$$w1[j+1] = 2(w1[j] - d1_j + x \cdot d2_j)$$
$$w2[j+1] = 2(w2[j] - d2_j - q_1 \cdot d1_j + x \cdot d3_j)$$
$$w3[j+1] = 2(w3[j] - d3_j - q_2 \cdot d1_j + x \cdot d4_j) \qquad \textbf{10.38}$$
$$w4[j+1] = 2(w4[j] - d4_j - q_3 \cdot d1_j + x \cdot d5_j)$$
$$w5[j+1] = 2(w5[j] - d5_j - q_4 \cdot d1_j)$$

The digits dk_j are selected using the selection function defined in Figure 10.13. The initial residuals are

$$(w1[0], w2[0], w3[0], w4[0], w5[0]) = (0, p_1, 0, p_3, 0)$$

A parallel form of the result can be obtained using on-the-fly conversion.

The evaluation of $R_{3,4}(x)$ for $x = 0.000110100001$ with 12-bit precision, showing nonredundant next residual $v1$ (for simplicity), is illustrated in Table 10.1. Other residuals are not shown. ∎

Implementation

An implementation consists of one module per row of the system L: the number of rows (the order of the system) $N = \max(degree(P), degree(Q)) + 1$. In a

j	$v1[j]$	$d\,1_{j+1}$	$d\,2_{j+1}$	$d\,3_{j+1}$	$d\,4_{j+1}$	$d\,5_{j+1}$	$y_1[j+1]^*$
0	0.000000000000	0	1	0	0	0	0.000000000000
1	0.001101000010	0	0	0	0	0	0.000000000000
2	0.011010000100	0	0	0	0	0	0.000000000000
3	0.110100001000	1	0	0	1	0	0.001000000000
4	−0.010111110000	0	0	0	0	0	0.001000000000
5	−0.101111100000	−1	0	1	−1	0	0.000110000000
6	0.100001000000	1	0	−1	1	0	0.000111000000
7	−0.111110000000	−1	0	0	−1	0	0.000110100000
8	0.000100000000	0	0	0	1	0	0.000110100000
9	0.001000000000	0	1	1	0	0	0.000110100000
10	0.011101000010	0	0	−1	0	0	0.000110100100
11	0.111010000100	1	−1	1	0	0	0.000110100010
12	−0.011000111010	0	1	0	0	−1	0.000110100010

TABLE 10.1 Evaluation of sinh(0.10197) using rational approximation and radix-2 generalized division algorithm. The error $|\sinh(x) - y_1[13]| < 2^{-12}$. $y_1[13]$ is computed to compensate for the initial scaling of p coefficients by 2.

radix-2 implementation for a rational function, a module, shown in Figure 10.15, has a [4:2] adder, four registers, two multiplexers with complementers, and a digit selection. The modules for row 1 and row N are simpler, using a [3:2] adder, three registers, and one multiplexer. The modules are initialized in a bit-parallel manner. During the evaluation steps, only single digits are passed between the modules. The result produced serially by Module 1 is converted into a conventional bit-parallel form using an on-the-fly converter. In the case of a polynomial evaluation, all modules but the last one use [3:2] adders. The module corresponding to the last row simply shifts out serially the coefficient p_{N-1}.

The delay for a rational function evaluation implementation is

$$T = (t_{sel} + t_{MUX} + t_{[4:2]} + t_{REG})m \qquad \textbf{10.39}$$

Comparison with a conventional implementation for evaluating a rational function consisting of multiply-add modules and a divider is left as Exercise 10.11.

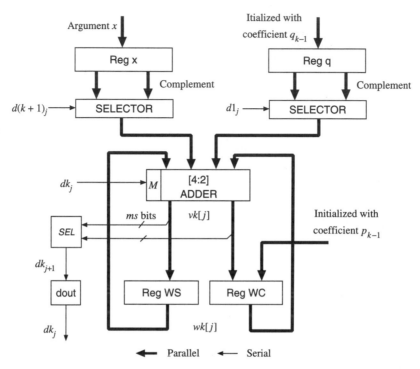

Argument x

Itialized with
coefficient q_{k-1}

Reg x

Reg q

Complement

Complement

$d(k+1)_j$ —— SELECTOR

$d1_j$ —— SELECTOR

dk_j ——

M | [4:2] ADDER

ms bits

$vk[j]$

Initialized with
coefficient p_{k-1}

SEL

dk_{j+1}

dout

dk_j

Reg WS

Reg WC

$wk[j]$

⟵ Parallel   ⟵ Serial

SEL block produces estimate and
performs selection

M block performs subtraction of dk_j

(register control signals not shown)

FIGURE 10.15   Implementation of MSDF module used in evaluation of rational functions.

10.6   Linear Convergence Method

In this method a sequence of approximations is constructed that converge to the
function. To control the sequence, an auxiliary sequence is used, which converges
either to one or to zero. In the first case, the method is called *multiplicative
normalization*,[20] and in the second *additive normalization*.

---

20. Note that the term "normalization" is used with a different meaning in Chapter 8.

In Chapter 7, multiplicative normalization has been used for reciprocal and for square root. In that case, the main primitive operation is multiplication and the convergence is quadratic. Here we consider algorithms with linear convergence. This replaces full multiplications by multiplication by one digit and also allows the computation of some functions, such as logarithm, because the recurrences include functions that are difficult to compute, but can be stored in tables of reasonable size for the case of linear convergence.

In these linear convergence algorithms the primitives are multiplication by powers of the radix (shifts), multiplication by a radix-r digit, and additions. Because of this, this class of algorithms is called *shift-and-add*, although for high radices rectangular multipliers are also needed.

10.6.1   Multiplicative Normalization

This approach has been used for several functions, such as reciprocal, division, square root, reciprocal square root, and logarithm. In all cases, there is a sequence that converges toward one, and this controls the convergence of another sequence toward the result. Because of this, we first consider the convergence of the first sequence and then apply this to the logarithm function.

Multiplicative Convergence toward One

The iterative algorithm consists of determining a sequence $P[j]$ such that the sequence $x[j]$ converges to one, where

$$x[j + 1] = x[j]P[j] \qquad \textbf{10.40}$$

with $x[0] = x$. For linear convergence we make

$$P[j] = (1 + s_j\, r^{-j}) \qquad \textbf{10.41}$$

where r is the radix of the algorithm and s_j is a radix-r digit. Note that the multiplicative normalization produces a *continued product* representation of the reciprocal of x, that is,

$$\frac{1}{x} \approx \prod_{j=0}^{n} P[j] = \prod_{j=0}^{n}(1 + s_i\, r^{-i}) \qquad \textbf{10.42}$$

That is, this normalization can be used to produce an approximation of the reciprocal function.

Selection Function and Residual

The specific selection function depends on the radix, the digit set, and the representation of $x[j]$. The design of a selection function follows the same method as discussed for division and square root (Chapters 5 and 6). With respect to the radix, since the selection function now depends only on one variable (instead of on $w[j]$ and d as in division), higher radices, such as radix 16, are practical. In addition, because of the convergence toward one, selection by rounding is possible and might be appropriate for high radices.[21]

As for digit recurrence division, we can have restoring and nonrestoring algorithms (with nonredundant digit set). Also possible is to use a redundant digit set. Moreover, in this latter case, it is possible to use a nonredundant adder (CPA) or a redundant adder (CSA or signed-digit adder). Since the use of a redundant digit set and a redundant adder results in a faster iteration, we concentrate on that case.

To simplify the selection, we define a (scaled) residual

$$w[j] = r^j(1 - x[j]) \qquad\qquad \textbf{10.43}$$

and since $x[j + 1] = (1 + s_j r^{-j})x[j]$, we obtain the recurrence

$$w[j + 1] = r(w[j] - x[j]s_j) = r(w[j] - s_j + s_j w[j]r^{-j}) \qquad \textbf{10.44}$$

with initial condition

$$w[0] = 1 - x \qquad\qquad \textbf{10.45}$$

The digit s_j is selected so that the residual is bounded. Calling $\overline{B[j]}$ the upper bound in iteration j, and using a signed-digit set $-a \leq s_j \leq a$, from the recurrence we get

$$\overline{B[j + 1]} = r(\overline{B[j]} - a + \overline{B[j]}ar^{-j}) \qquad\qquad \textbf{10.46}$$

The solution to this recurrence is complicated. Assuming that $B[j + 1] = B[j]$ (which is not the case), we get

$$\overline{B[j]} = \frac{r\rho}{1 + \rho r^{-j+1}} \qquad\qquad \textbf{10.47}$$

---

21. See Chapter 9.

If this bound is used, we get $w[j + 1] = \overline{B[j]}$, which makes the algorithm converge since $w[j + 1] < \overline{B[j + 1]}$.

Similarly, for the lower bound (assuming the bound is independent of j) we get

$$\underline{B[j]} = -\frac{r\rho}{1 - \rho r^{-j+1}} \qquad \textbf{10.48}$$

However, in this case using this bound the algorithm would not converge since we get $|w[j + 1]| > |\underline{B[j + 1]}|$. A solution is to use

$$\underline{B[j]} = -\frac{r\rho}{1 - \rho r^{-(j+2)}} \qquad \textbf{10.49}$$

The selection interval of $w[j]$ is then[22]

$$U_k = \frac{\overline{B[j + 1]} + rk}{r(1 + kr^{-j})} = \frac{\rho + k(1 + \rho r^{-j})}{(1 + \rho r^{-j})(1 + kr^{-j})} \qquad \qquad \textbf{10.50}$$

$$L_k = \frac{-\rho + k(1 - \rho r^{-(j+3)})}{(1 - \rho r^{-(j+3)})(1 + kr^{-j})}$$

For the case of carry-save representation of the residual and an estimate corresponding to the assimilation up to fractional bit t, the selection function is described by the selection constants m_k, such that

$$\max(\widehat{L_k}) \leq m_k \leq \min(\widehat{U}_{k-1}) \qquad \textbf{10.51}$$

where

$$\widehat{L}_k = \lceil L_k \rceil_t, \ \ \widehat{U}_{k-1} = \lfloor U_{k-1} - 2^{-t} \rfloor_t \qquad \textbf{10.52}$$

EXAMPLE 10.6 We describe a radix-2 multiplicative normalization algorithm for $x \in [\frac{1}{2}, 1]$. The recurrence becomes

$$w[j + 1] = 2(w[j] - s_j + w[j]s_j 2^{-j}) \qquad \textbf{10.53}$$

The selection intervals ($\rho = 1, j \geq 1$) become

$$U_1 = \frac{1 + (1 + 2^{-j})}{(1 + 2^{-j})(1 + 2^{-j})} \geq 1.11 \quad L_1 = \frac{-1 + (1 - 2^{-(j+3)})}{(1 + 2^{-(j+3)})(1 + 2^{-j})} \leq -0.04$$

---

22. See Chapter 5 for definitions and detailed method.

$$U_0 = \frac{1}{1 + 2^{-j}} \geq 0.66 \quad L_0 = \frac{-1}{1 - 2^{-(j+3)}} \leq -1$$

$$U_{-1} = \frac{1 - (1 + 2^{-j})}{(1 + 2^{-j})(1 - 2^{-j})} \geq -0.67 \quad L_{-1} = \frac{-1 - (1 - 2^{-(j+3)})}{(1 - 2^{-(j+3)})(1 - 2^{-j})} \leq -4.1$$

For $j = 0$, the selection intervals are defined as for $j \geq 1$ except that L_{-1} and U_{-1} are not defined. Since $s_0 \neq -1$, this presents no problem.

If $t = 2$ we get

$$\min(\widehat{U}_0) = \lfloor 1.11 - 0.25 \rfloor_2 = 0.75 \quad \max(\widehat{L}_1) = \lceil -0.04 \rceil_2 = 0$$

$$\min(\widehat{U}_{-1}) = \lfloor -0.67 - 0.25 \rfloor_2 = -1 \quad \max(\widehat{L}_0) = \lceil -1 \rceil_2 = -1$$

The corresponding selection constants are

$$m_1 = 0, \quad m_0 = -1 \qquad\qquad \textbf{10.54}$$

This selection function produces $s_0 = 1$, which is a valid choice for the range of $1/x$. ∎

An implementation of multiplicative normalization for radix 2 is illustrated in Figure 10.16. Note the variable shifter required for this algorithm.

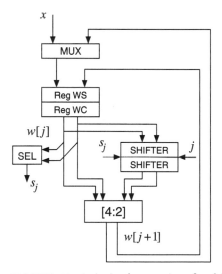

FIGURE 10.16 An implementation of multiplicative normalization.

Selection by Rounding for (Very) High Radices

Convergence to 1 permits selection by rounding and therefore a high-radix algorithm. However, in the first iterations selection by rounding is not possible, so that an initial approximation of $1/x$ by other means is required. This can be done by table lookup or by a linear approximation (see Section 7.1.3).

In selection by rounding, the digit is obtained by rounding the carry-save residual (truncated to fractional bit t). Calling this truncated residual $\widehat{w}[j]$, we get

$$s_j = \left\lfloor \widehat{w}[j] + \frac{1}{2} \right\rfloor \tag{10.55}$$

The next residual is then

$$w[j+1] = r\left(w[j] - \left\lfloor \widehat{w}[j] + \frac{1}{2} \right\rfloor + s_j w[j] r^{-j} \right) \tag{10.56}$$

Since

$$-\frac{1}{2} \le \left(w[j] - \left\lfloor \widehat{w}[j] + \frac{1}{2} \right\rfloor \right) \le \frac{1}{2} + 2^{-t} \tag{10.57}$$

the worst case is the upper bound. Since for selection by rounding

$$w[j] < a + \frac{1}{2} \tag{10.58}$$

we get

$$w[j+1] \le r\left(\frac{1}{2} + 2^{-t} + a\left(a + \frac{1}{2} \right) r^{-j} \right) \tag{10.59}$$

Making $a = (r-1)$ for simplicity, there is convergence if

$$2^{-t} < \frac{(r-1)}{2r}(1 - 2r^{-j+2} - r^{-j+1}) \tag{10.60}$$

Consequently, there is convergence for $j \ge 3$. The s_j digits for $j = 1$ and $j = 2$ can be obtained from an approximation of the reciprocal of x using only its most-significant bits.

In such a case, for $j \ge 3$,

$$2^{-t} < \frac{(r-1)}{2r}(1 - 2/r - 1/r^2) = \frac{1}{4} + \epsilon \tag{10.61}$$

resulting in $t \ge 2$.

Actually, instead of obtaining an approximation corresponding to the first two digits, the algorithm converges if the approximation has h fractional bits with $\log_2(r) < h < 2\log_2(r)$ (that is, more than one radix-r digit but less than two). This reduces the requirement on the precision of the approximation and simplifies the calculation of the initial residual using a rectangular multiplier, as discussed now.

The algorithm is then as follows:[23]

1. Obtain an initial approximation of the reciprocal with h fractional bits. Call this approximation A and

$$P[-1] = A \qquad\qquad\qquad \textbf{10.62}$$

2. Obtain the initial values as follows:

$w[0] = 2^h(1 - P[-1]x)$ requires an $h + 1$ by n rectangular multiplication

$y[0] = P[-1]$

$$\qquad\qquad\qquad\qquad\qquad\qquad\qquad \textbf{10.63}$$

3. **Begin** iteration

$$s_j = \left\lfloor \widehat{w}[j] + \frac{1}{2} \right\rfloor$$
$$w[j+1] = r(w[j] - s_j + s_j w[j]2^{-h}r^{-j})$$
$$P[j] = (1 + s_j 2^{-h}r^{-j}) \qquad\qquad \textbf{10.64}$$
$$y[j+1] = y[j]P[j]$$

End iteration

EXAMPLE 10.7 We now give a numerical example. For simplicity, we use a nonredundant representation of the residual (which makes $t = 0$) and the relatively low radix 16. We give the result of the execution in radix-16 representation.

Consider the calculation of the reciprocal of $x = (0.A3B6)_{16} \approx 0.6395_{10}$.

1. Initial approximation of reciprocal with $h = 6$ fractional bits:

$$A = 1.100100_2 = 1.90_{16} = 1.5625_{10}$$

---

23. The algorithm computes the values $P[j]$ as well as the approximation to the reciprocal $y[j]$. For the application in function evaluation the sequence $P[j]$ is used.

2. Initial values:

$$P[-1] = 1.90$$

$$w[0] = 2^6(1 - 0.ffcc\,60) = 0.0ce\,8$$

$$y[0] = 1.90$$

3. Iterations:

$$s_0 = 0$$

$$P[0] = 1$$

$$w[1] = 16(0.0ce\,8) = 0.ce\,8$$

$$y[1] = 1.90$$

$$s_1 = 1$$

$$P[1] = (1 + 2^{-10}) = 1.004$$

$$w[2] = 16(0.ce\,8 - 1 + 0.ce\,8 \times 2^{-10}) = -3.14c6$$

$$y[2] = 1.90(1.004) = 1.9064$$

$$s_2 = -3$$

$$P[2] = (1 - 3 \times 2^{-14}) = 0.fff4$$

$$y[3] = 1.9064 \times 0.fff4 = 1.90513b5$$

$$1/x \approx (1.9050100)_{16}$$ ∎

Input Domain

The reciprocal of x is represented by the product of the $P[j]$. Consequently for a digit set $s_j \in \{-a, \ldots, a\}$,

$$\frac{1}{\prod_{i=0}^{m}(1 + ar^{-i})} \le x \le \frac{1}{\prod_{i=0}^{m}(1 - ar^{-i})} \qquad \textbf{10.65}$$

For instance, for $r = 2$ (eliminating the factor for $i = 0$ in the right product) the domain is $0.21 \le x \le 3.45$, and for $r = 4$ and $a = 2$ the domain is $0.19 \le x \le 2.38$.

Logarithm

We want to compute $y = \ln(x)$. Since the multiplicative normalization produces

$$\frac{1}{x} \approx \prod_{j=0}^{m} P[j] = \prod_{j=0}^{m}(1 + s_j 2^{-j}) \qquad \textbf{10.66}$$

we get

$$\ln(x) \approx -\sum_{j=0}^{m} \ln\, P[j] = -\sum_{i=0}^{m} \ln\,(1 + s_j 2^{-j}) \qquad \textbf{10.67}$$

Consequently, the s_i obtained from the multiplicative normalization are used to obtain $\ln(1 + s_j 2^{-j})$ from a table and these values are then added. So, in addition to the recurrence for multiplicative normalization, we have the recurrence

$$y[j+1] = y[j] - \ln(P[j]) \qquad \textbf{10.68}$$

The result is $y[m+1] \approx y[0] + \ln(x)$.

Error

The absolute error[24] is

$$E = \ln(x) - y[m+1] \qquad \textbf{10.69}$$

However,

$$\ln(x) = \ln\left(x \frac{\prod P[j]}{\prod P[j]}\right) = \ln\left(x \prod_{j=0}^{m} P[j]\right) - \sum_{j=0}^{m} \ln(P[j])$$

$$= \ln(x[m+1]) + y[m+1] \qquad \textbf{10.70}$$

resulting in

$$E = \ln(x[m+1]) \qquad \textbf{10.71}$$

Since the power series expansion of $\ln(z)$ is

$$\ln(z) = (z-1) - \frac{(z-1)^2}{2} + \cdots \quad (0 < z \le 2) \qquad \textbf{10.72}$$

---

24. This error is produced by the convergent algorithm. In a particular implementation, the contributions due to finite-precision representations have to be included.

making $z = x[m + 1]$, the error is bounded by

$$|E| \leq |x[m + 1] - 1| + \frac{(x[m + 1] - 1)^2}{2} \qquad \textbf{10.73}$$

From this expression we see that a more accurate approximation than (10.68) is obtained as

$$\ln(x) \approx y[m + 1] + x[m + 1] - 1 \qquad \textbf{10.74}$$

The resulting error is bounded by

$$|E| \leq \frac{(x[m + 1] - 1)^2}{2} \qquad \textbf{10.75}$$

For instance, if the approximation of $1/x$ has an error bound of 2^{-n}, then the basic algorithm would also have an error of 2^{-n}, whereas the modified approximation would have an error bound of 2^{-2n-1}. However, this reduced error might result in an increased computation cost since $x[m + 1]$ has to be computed, while for the basic algorithm only $w[j]$ is required.

Since the power series expansion used is valid for $0 < x[m + 1] \leq 2$, the analysis is valid for $x \geq \frac{1}{2}$.

Algorithm for $\ln(x)$

We summarize in Figure 10.17 the radix-2 algorithm for computing $\ln(x)$ with multiplicative normalization, using the approximation $\ln(x) \approx y[m + 1]$ with an absolute error of 2^{-m}.

The evaluation of $\ln(0.631)$ with 12-bit precision showing nonredundant residuals (for simplicity) is illustrated in Table 10.2.

Implementation

As shown in Figure 10.18, the overall implementation requires two variable shifters, one [4:2] adder, one [3:2] adder, one CPA, the selection function module, two multiplexers, a module with a table for generating L_j constants, and four registers.

The delay is

$$T_{LN} = [\max((\max(t_{sel}, t_{shift}) + t_{4-2}), (t_{sel} + t_{table} + t_{CSA})) + t_{REG}]m + t_{CPA}$$

$$\textbf{10.76}$$

1. *[Initialize]*

$$y[0] = 0; \ w[0] = 1 - x$$

2. *[Recurrence]*

   for $j = 0 \ldots m$

   $$s_j = SEL(\widehat{w[j]});$$
   $$w[j + 1] \leftarrow 2(w[j] - s_j + s_j w[j]2^{-j})$$
   $$y[j + 1] \leftarrow y[j] - L_j$$

   end for

3. *[Result]*

$$y[m + 1] \approx ln(x)$$

where

- The residual is in redundant form, represented by the pseudosum WS and stored-carry WC bit-vectors. For simplicity, we use $w[j]$ in the description.
- m is the precision in bits.
- SEL is the continued-product digit selection function defined by

$$s_j = SEL(\widehat{w[j]}) = \begin{cases} 1 & \text{if } \widehat{w[j]} \geq 0 \\ 0 & \text{if } -1 \leq \widehat{w[j]} \leq 0.25 \\ -1 & \text{if } \widehat{w[j]} \leq -0.75 \end{cases}$$

where $\widehat{w[j]}$ is an estimate of the residual $w[j]$ with $t = 2$ fractional bits.

- The constants L_j are defined as

$$L_j = \begin{cases} \ln(1 + 2^{-j}) & \text{if } s_j = 1 \text{ and } j \leq m/2 \\ \ln(1 - 2^{-j}) & \text{if } s_j = -1 \text{ and } j \leq m/2 \\ 0 & \text{if } s_j = 0 \text{ and } j \leq m/2 \\ s_j 2^{-j} & \text{if } j > m/2 \end{cases}$$

and the constants $L_j = \ln(1 \pm 2^{-j})$ are stored in a table.

FIGURE 10.17 Radix-2 algorithm for $\ln(x), x \in [1/2, 1)$.

j	$w[j]$	s_j	L_j	$y[j]$
0	0.010111100111	1	0.101100010111	0.000000000000
1	−0.100001100010	0	0.000000000000	−0.101100010111
2	−1.000011000101	−1	−0.010010011010	−0.101100010111
3	0.011011011001	1	0.000111100010	−0.011001111101
4	−1.000010011000	−1	−0.000100001000	−0.100001011111
5	0.000011100100	1	0.000001111110	−0.011101010110
6	−1.111000101010	−1	−0.000001000000	−0.011111010101
7	−1.101101100011	−1	−0.000000100000	−0.011110010100
8	−1.011001011000	−1	−0.000000010000	−0.011101110100
9	−0.110010000100	0	0.000000000000	−0.011101100100
10	−1.100100000111	−1	−0.000000000100	−0.011101100100
11	−1.001000000010	−1	−0.000000000010	−0.011101100000
12	−0.001111111111	0	0.000000000000	−0.011101011110
13	−0.011111111110			−0.011101011110

TABLE 10.2 Evaluation of $\ln(0.631)$ using radix-2 multiplicative normalization. The error $|\ln(0.631) - y[13])| < 0.0001 < 2^{-12}$.

where m is the number of iterations. The table L contains $m/2 \times 2$ constants. The access to the table can be removed from the critical path (see Exercise 10.18).

10.6.2   Exponential by Additive Normalization

The function $y = e^x$ can be computed by an additive normalization. To do this we obtain a sequence of $\{b_j\}$ so that

$$x - \sum_{j=1}^{m} \ln(b_j) \to 0 \quad \text{(normalizes to 0)} \qquad \textbf{10.77}$$

Then

$$e^x \approx \prod_{j=1}^{m} b_j \qquad \textbf{10.78}$$

Although the method allows unrestricted values for b_j, to have an implementation with only additions and multiplications with a one-digit multiplier, the b_j are

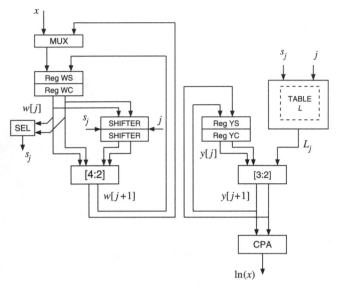

FIGURE 10.18 Implementation of radix-2 algorithm for computing $\ln(x)$.

restricted to be of the form

$$b_j = 1 + s_j r^{-j}$$

with s_j a radix-r digit. This also permits the use of a table lookup for $\ln(b_j)$.

As in other digit recurrence algorithms, a residual is defined as

$$w[j] = r^j \left(x - \sum_{i=1}^{j} \ln(1 + s_j r^{-j}) \right) \qquad \textbf{10.79}$$

resulting in the recurrence

$$w[j + 1] = r(w[j] - r^j \ln(1 + s_j r^{-j})) \qquad \textbf{10.80}$$

The exponential is obtained by the recurrence

$$y[j + 1] = y[j](1 + s_j r^{-j}) \qquad \textbf{10.81}$$

where $y[0] = 1$.

Selection Function

The selection function is determined in the same manner as for logarithm. For $s_j \in \{-a, \ldots, a\}$, the convergence bound is

$$r^j \sum_{i=j+1}^{\infty} \ln(1 - ar^{-i}) \leq w[j] \leq r^j \sum_{i=j+1}^{\infty} \ln(1 + ar^{-i}) \qquad \textbf{10.82}$$

We give next a selection function for radix 2 and leave the derivation as an exercise (see Exercise 10.20):

$$s_j = SEL(\widehat{w[j]}) = \begin{cases} 1 & \text{if} \quad \widehat{w[j]} \geq 0.5 \\ 0 & \text{if} \quad -0.5 \leq \widehat{w[j]} \leq 0.25 \\ -1 & \text{if} \quad \widehat{w[j]} \leq -0.75 \end{cases} \qquad \textbf{10.83}$$

where $\widehat{w[j]}$ is an estimate of the residual $w[j]$ (in carry-save form) with $t = 2$ fractional bits.

Algorithm for e^x

Figure 10.19 summarizes the radix-2 algorithm for computing e^x using additive normalization. Let the input argument be $x \in (-\ln(2), \ln(2))$.[25] Since $L_0 > U_{-1}$ for $j = 0$, an additional transformation is applied to allow the use of the same selection function in all steps. This initial transformation makes $y[0] = e^{-0.5}$ and $w[0] = x + 2^{-1}$. Since the residual is in carry-save form, no addition is needed to initialize $w[0]$.

EXAMPLE 10.8 The evaluation of $\exp(-0.437)$ with 12-bit precision, showing nonredundant residuals (for simplicity), is illustrated in Table 10.3. ∎

---

25. This range is obtained from an argument x_{in} by using the transformation

$$e^{x_{in}} = e^{x_{in}(\log_2 e)(\log_e 2)} = e^{(I+f)\ln(2)}$$

where I is an integer and $-1 < f < 1$. Therefore,

$$e^{x_{in}} = e^{I\ln(2)}e^{f\ln(2)} = 2^I e^x$$

where $x = f\ln(2) \in (-\ln(2), \ln(2))$.

---

1. [*Initialize*]

   $$y[0] = e^{-0.5}; \quad w[0] = x + 0.5$$

2. [*Recurrence*]

   for $j = 0 \ldots m$

   $s_j = SEL(\widehat{w[j]})$;

   $w[j+1] \leftarrow 2(w[j] - L_j 2^j)$

   $y[j+1] \leftarrow y[j] + y[j]s_j 2^{-j}$

   end for

3. [*Result*]

   $$y[m+1] \approx e^x$$

where

- The residual is in redundant form, represented by the pseudosum *WS* and stored-carry *WC* bit-vectors. For simplicity, we use $w[j]$ in the description.
- m is the precision in bits.
- *SEL* is the continued-sum digit selection function defined by expression (10.83). $\widehat{w[j]}$ is the estimate of the residual truncated to two fractional bits.
- The constants L_j are defined as

$$L_j = \begin{cases} \ln(1 + 2^{-j}) & \text{if } s_j = 1 \text{ and } j \leq m/2 \\ \ln(1 - 2^{-j}) & \text{if } s_j = -1 \text{ and } j \leq m/2 \\ 0 & \text{if } s_j = 0 \text{ and } j \leq m/2 \\ s_j 2^{-j} & \text{if } j > m/2 \end{cases}$$

where constants $L_j = \ln(1 \pm 2^{-j})$ are stored in a table.

FIGURE 10.19 Radix-2 algorithm for e^x, $x \in (-\ln(2), \ln(2))$.

Implementation

As shown in Figure 10.20, the overall implementation is similar to that of $\ln(x)$: it uses two variable shifters, one [4:2] adder, one [3:2] adder, and one CPA, the selection function module, one multiplexer, a module for generating the L_j

j	$w[j]$	s_j	L_j	$y[j]$
0	0.000100000010	0	0.000000000000	0.100110110100
1	0.001000000100	0	0.000000000000	0.100110110100
2	0.010000001000	0	0.000000000000	0.100110110100
3	0.100000010000	1	0.000111100010	0.100110110100
4	−0.111000001100	−1	−0.000100001000	0.101011101011
5	0.010011111110	0	0.000000000000	0.101000111100
6	0.100111111111	1	−0.000000111111	0.101000111100
7	−0.101111000110	−1	−0.000000100000	0.101001100101
8	0.100001110100	1	0.000000010000	0.101001010000
9	−0.111100011000	−1	−0.000000001000	0.101001011010
10	0.000111010000	0	0.000000000000	0.101001010101
11	0.001110100000	0	0.000000000000	0.101001010101
12	0.011101000000	0	0.000000000000	0.101001010101
13	0.111010000001			0.101001010101

TABLE 10.3 Evaluation of $\exp(-0.437)$ using radix-2 additive normalization. The error $|\exp(-0.437) - y[13])| < 2^{-13}$.

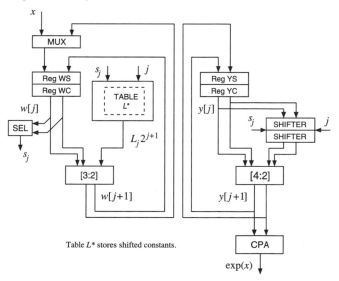

Table L^* stores shifted constants.

FIGURE 10.20 Implementation of radix-2 algorithm for computing $\exp(x)$.

constants, and four registers. The table stores shifted constants $L_j 2^{j+1}$. If the table is to be shared with implementation for $\ln(x)$, a left shifter is used.

The delay is

$$T_{EXP} = [\max((\max(t_{sel}, t_{shift}) + t_{4-2}), (t_{sel} + t_{table} + t_{CSA})) + t_{REG}]m + t_{CPA}$$

$$10.84$$

where m is the number of iterations. The table L contains $m/2 \times 2$ constants.

Error

The error in the approximation

$$e^x \approx \prod_{i=1}^{m} b_i = y[m+1] \qquad 10.85$$

is

$$E_{exp} = |e^x - y[m+1]| = y[m+1]|e^{x[m+1]} - 1| \qquad 10.86$$

where

$$x[m+1] = x - \sum_{i=1}^{m} \ln(b_i)$$

Since[26]

$$|e^{x[m+1]} - 1| < |x[m+1]|e^{|x[m+1]|} \qquad 10.87$$

we have that

$$E_{exp} < |x[m+1]|e^{|x[m+1]|}y[m+1] \qquad 10.88$$

The linear additive normalization guarantees that

$$|x[m+1]| = |x - \sum_{i=1}^{m} \ln(b_i)| < r^{-m} \qquad 10.89$$

---

26. Note that $e^{x[m+1]} - 1 = x[m+1] + x[m+1]^2/2 + \cdots = x[m+1](1 + x[m+1]/2 + x[m+1]^2/6 + \cdots) < x[m+1]e^{x[m+1]}$.

Consequently, $e^{|x[m+1]|} < e^{r^{-m}} < 1 + r^{-m} + r^{-2m}$. Therefore,

$$E_{exp} = |e^x - y[m+1]| < r^{-m}(1 + r^{-m} + r^{-2m})y[m+1] \approx r^{-m}y[m+1]$$

10.90

Since $y[m+1]$ is a close approximation to e^x, we have

$$y[m+1] \leq \frac{e^x}{1 - r^{-m}} \leq \frac{2}{1 - r^{-m}}$$

10.91

for $x < \ln(2)$. Therefore,

$$y[m+1] < 2(1 + r^{-m})$$

and

$$E_{exp} < 2r^{-m} + O(r^{-2m})$$

10.92

10.6.3 Trigonometric and Inverse Trigonometric Functions

Implementation of these functions using the linear convergence technique is commonly based on the CORDIC method described in Chapter 11.

10.7 Concluding Remarks

In this chapter we presented several methods for evaluation of functions that are suitable for implementation in hardware. These include general polynomial and rational function approximations, and linear convergence methods based on shift-and-add algorithms applicable to particular functions. The methods discussed require lookup tables of varying sizes, some of the standard components such as adders, multipliers, and dividers, and special components such as variable shifters and digit-by-vector multipliers. In general there is a tradeoff between complexity of tables and of computation involved. As the VLSI technology progresses, the use of increasingly larger tables is becoming feasible and attractive. We also discussed general issues such as argument range transformations and rounding. Hardware-oriented evaluation of functions is also covered in other chapters (reciprocal in

Chapter 5 and square root in Chapter 6). As mentioned above, trigonometric and inverse trigonometric functions are discussed in Chapter 11.

10.8   Exercises

Argument Range Reduction

10.1   Apply the reduction described in expression (10.1) to the floating-point argument $x = 1.5325 \times 2^{22}$. Describe the reconstruction required after obtaining an approximation of $y_r = e^{x_r}$.

Rounding

10.2   Develop a table for a correctly rounded to nearest $y = \sin(x)$ for 6-bit x and y and $0 \le x \le \pi/4$.

Polynomial Approximation and Interpolations

10.3   Using a Taylor series expansion, show an implementation of an approximation of the function $\sin x$ for $-\pi \le x \le \pi$ with operand and result of 16 bits in two's complement representation and an absolute error of less than one ulp. Determine the width of each variable and the latency and throughput. Show the execution for the computation of $\sin(1.25)$.

Consider the following three cases:

(a)   Use one nonpipelined multiplier-accumulator and store the coefficients of the polynomial in a table. The access to the table plus a multiply-accumulate takes four cycles.

(b)   Use one pipelined multiplier-accumulator with four stages and store the coefficients in a table. The access to the table takes one cycle and each stage of the multiplier accumulator corresponds to one cycle. Show the scheduling of the operations.

(c)   Use as many multiplier accumulators as required for minimum latency. Pipeline these modules for increased throughput. The coefficients are hardwired. Determine the width of each multiplier accumulator.

10.4 Determine an interpolating polynomial of degree 4 for the function $\tan(x)$ for $0 \leq x \leq \pi/4$. Obtain an approximation of $\tan(0.5)$ with 16 bits and compare with that produced by a Taylor series expansion of the same degree.

Piecewise Interpolation

10.5 Show an implementation of a piecewise interpolation of $x^{1/3}$ for $\frac{1}{2} \leq x < 1$, 16-bit input and output. Use four intervals and linear interpolation.

Determine the degree of the polynomial based on Taylor series required for the same error.

10.6 Obtain a formula for a polynomial for quadratic piecewise interpolation. Show an implementation and compare with the implementation for linear interpolation.

Reduction, Approximation, and Reconstruction

10.7 Show an implementation for the computation of $\log(x)$ according to the procedure described in Example 10.3.

Bipartite Table Method

10.8 Develop the tables required for the computation of an approximation of $1/x$ for $\frac{1}{2} \leq x < 1$ with 9-bit precision using the bipartite table method. Use the tables to compute an approximation of the reciprocal of 0.100110010.

10.9 For the approximation of a function using an operand of 16 bits, compare the implementation using piecewise interpolation with that of using the bipartite table method. Consider the modules required, the latency, and the possibility of pipelining.

MSDF Polynomial and Rational Function Evaluation

10.10 Evaluate the following polynomial using the algorithm in Figure 10.13:

$$P_2(x) = -0.39x^2 + 0.15x + 0.18$$

for $x = 0.23$ and $m = 8$.

Show an implementation and compare it with a corresponding conventional parallel polynomial evaluation with respect to cost and delay.

10.11 Compare the cost and delay of a rational evaluation MSDF scheme and a conventional scheme using multiply-add and a radix-4 divider. Compare delays assuming similar costs.

10.12 Modify the algorithm in Figure 10.13 to accept the argument x MSDF serially.

10.13 Show that the solution of the following linear system satisfies $y_i = x^{5-i}$, $i = 1, \ldots, 4$:

$$
\begin{bmatrix}
1 & -x & 0 & 0 \\
0 & 1 & -x & 0 \\
0 & 0 & 1 & -x \\
0 & 0 & 0 & 1
\end{bmatrix}
\begin{bmatrix}
y_1 \\
y_2 \\
y_3 \\
y_4
\end{bmatrix}
=
\begin{bmatrix}
0 \\
0 \\
0 \\
x
\end{bmatrix}
$$

Determine the cost and delay for $m = 24$ and radix 2. Compare this scheme for generating integer powers of x with a corresponding conventional implementation with respect to cost and delay.

Convergence Methods

10.14 Compute an approximation of the reciprocal of $x = (0.10101111)_2$ using the radix-2 multiplicative normalization algorithm (with carry-save adder).

10.15 Develop a radix-4 multiplicative normalization algorithm for reciprocal.

10.16 Compute an approximation of the reciprocal of $x = (0.AF636456)_{16}$ using a radix-16 multiplicative normalization algorithm with selection by rounding and carry-save adder. For $j \leq 2$ compute a suitable initial approximation.

10.17 Compute a 12-bit approximation of $y = \ln(0.625)$ using the radix-2 multiplicative normalization algorithm. Use the table of $\ln(1 + 2^{-j})$ given in the Appendix.

10.18 Consider reducing the delay of the implementation of the radix-2 algorithm for computing $\ln(x)$. For this, design a network for generating the constants L_j such that the access to the table is not in the critical path. Compare the delay with the implementation described in the text.

10.19 An algorithm to compute

$$y = \log_2(x) = \sum_{j=0}^{n} y_j 2^j$$

is as follows:

$$w[n] = x$$

$$w[n - j - 1] = w[n - j]2^{-y_{k-j}2^{k-j}}$$

$$y_{k-j} = 1 \text{ if } w[n - j] \geq 2^{2^{k-j}}$$

Show that the algorithm is correct. Determine k for x integer. Compute $y = \log_2(0.625)$ with $n = 8$ bits precision.

Compare the implementation of this algorithm with that of linear multiplicative normalization.

10.20 Derive the selection function for exponential additive normalization algorithm in radix 2 defined by expression (10.83).

10.21 Calculate $y = e^{0.75}$ with 12 fractional bits using a radix-2 algorithm (with carry-save adder). Determine the error of the resulting approximation.

10.22 General exponentiation can be described as $y = x^v = (e^{\ln(x)})^v = e^{v \ln(x)}$. Show an implementation using normalization. Compute $y = 0.75^{1.25}$ with 12 bits.

10.23 Describe an implementation of $y = x^v$ with v positive integer using the operations of squaring and multiplication.

10.24 Show an algorithm for the computation of reciprocal square root and square root by linear multiplicative normalization.

10.25 Compare the computation of $\ln(x)$ and of e^x by Taylor series expansion and by normalization methods. Give expressions for the delay and list the modules required. Give some reasonable conclusions.

10.9   Further Readings

Books and Surveys

The theoretical foundations and algorithms for evaluation of mathematical functions suitable for hardware design are covered in a comprehensive manner in Muller (1997). Approximation theory useful in deriving algorithms is discussed in standard books on numerical methods such as Dahlquist and Bjorck (1974) and Mathews (1992). Cheney (1966) and Davis (1990) are classics on the function approximation theory. Early work on approximations for software implementation is found in Hart et al. (1978). Practical polynomial and rational approximations are surveyed in Cody (1970) and Cody and Waite (1980).

Argument Reduction

A comprehensive discussion of several methods for reducing argument range is presented in Muller (1997) (Chapter 9). Specifc reduction methods are described in Tang (1991), Daumas et al. (1994), Schulte and Swartzlander (1994), and Ferguson (1995).

Correct Rounding and Monotonicity

Problems and approaches to correct rounding are considered in Schulte and Swartzlander (1993, 1994), Muller (1997), and Lefèvre et al. (1998). A technique for obtaining approximations with monotonicity property for some transcendental functions is introduced in Ferguson and Brightman (1991). How to get some transcendentals correctly rounded in double-precision is shown in Lefèvre et al. (1998). An analysis of worst cases for correct rounding in double precision for elementary functions is described in Lefevre and Muller (2001).

Hardware Polynomial Evaluators

Pipelined combinational networks for polynomial evaluation are developed in Tung and Avizienis (1970). Ercegovac (1977) describes an MSDF scheme for polynomial evaluation. A pipelined scheme for evaluating elementary functions with Chebyshev polynomials is presented in Hwang et al. (1987). Duprat and Muller (1988) and Corbaz et al. (1991) propose hardware polynomial evaluators. An online polynomial evaluation scheme is discussed in Merrheim et al. (1993). Schemes for parallel and MSDF evaluation suitable for FPGA implementation are presented in Ercegovac et al. (1995). Ercegovac and Muller (1998) propose a MSDF scheme for polynomial evaluation at regularly spaced points. Burleson (1990) proposes a scheme for polynomial evaluation using distributed arithmetic.

Lookup Tables and Interpolation

An overview of table-based function evaluation methods is presented in Muller (1998). Methods using small table lookups followed by polynomial/rational approximation evaluation suitable for general-purpose systems are presented in Tang (1989, 1990, 1991, 1992). Approaches based on interpolating polynomials using table lookups and multipliers have been frequently considered with the aim of reducing sizes of tables and multipliers. Noetzel (1989) presents the design of an interpolating memory for evaluation of function approximations with Lagrange interpolating polynomials. An error analysis is also given. This approach is followed later by Lewis (1994) among others. Jain and Lin (1995, 1997) describe an interpolation technique based on matched interpolating polynomials for double-precision computation of reciprocals, square root, sine, and arctangent functions. Das Sarma and W. Matula (1997) discuss the use of interpolation in reciprocal tables. Das Sarma and Matula (1994) present an analysis of accuracy in ROM tables for reciprocals. Cao et al. (2001) describe a design for evaluation of functions in single precision using interpolation with second-order polynomials and optimized tables. A VLSI implementation of second-order polynomial interpolation with unequal subintervals for sine/cosine evaluation is presented in Paliouras et al. (2000). Farmwald (1981) describes a design for evaluation of functions based on the Taylor series implemented with large tables and fast multipliers. Wong and Goto (1994) present a technique based on the evaluation of the Taylor series using a difference method. It is implemented with adders and large tables. Lefèvre and Muller (1999) describe a table-based method for evaluating the exponential function in double precision. A table lookup method for 100-bit precision is described in Daumas et al. (2000). A method for evaluating functions using tables and small multipliers is described in Ercegovac et al. (2000).

Bipartite and Multipartite Table Methods

Introduced in Das Sarma and Matula (1995), the bipartite tables and their variations have been reported on frequently. Symmetric bipartite tables are discussed in Schulte and Stine (1997a, 1997b, 1999) and Stine and Schulte (1999). Muller (1999) discusses a generalization to multipartite tables. De Dinechin and Tisserand (2001) present a unified approach to the previously reported bipartite and multipartite tables leading to smaller tables. Hassler and Takagi (1995) present a function evaluation using table lookup and addition similar to the bipartite method.

Rational Function Evaluation

Koren and Zinaty (1990) develop a coprocessor implementation for evaluating rational approximations in extended double-precision format. An MSDF approach to rational function evaluation without explicit division is introduced in Ercegovac (1975, 1977).

Linear Convergence Method

Specker (1965) and Linhardt and Miller (1969) discuss multiplicative and additive algorithms of the shift-and-add type for computing logarithm, exponential, and trigonometric functions. A systematic study of radix-2 shift-and-add algorithms with $\{-1, 0, 1\}$ digit set and nonredundant residuals is presented in DeLugish (1970). A radix-16 extension of DeLugish's approach with digit selection using rounding is reported in Ercegovac (1973). The use of higher radix 2^k and predictive techniques in the multiplicative normalization has been considered by Baker (1973, 1975). Further developments of this type of algorithms are discussed in Zurawski (1980) and Rodrigues et al. (1981). The computation of log and exp are related to the CORDIC algorithm, which is also of the shift-and-add type, described in the next chapter; see references there, especially for very-high-radix algorithms and implementations. Chen (1972) provides another approach to function evaluation resulting in shift-and-add algorithms.

Complex Function Hardware Evaluation

Bajard et al. (1994) discuss a shift-and-add method for function evaluation in the complex domain.

Function Evaluation in Processors

Agarwal et al. (1986) discuss scalar and vector elementary functions for the IBM System 370. Markstein (1990) describes computation of elementary functions on the IBM RISC system/6000 processor. Rauchwerger and Farmwald (1990) discuss evaluation of polynomials on a multiple floating-point coprocessor architecture. Transcendental function evaluation for Intel IA-64 is described in Harrison et al. (1999) and Story and Tang (1999), and for AMD K5 processor in Lynch et al. (1995).

10.10 Bibliography

Agarwal, R. C., J. C. Cooley, F. G. Gustavson, J. B. Shearer, G. Slishman, and B. Tuckerman (1986). New scalar and vector elementary functions for the IBM system/370. *IBM Journal of Research and Development*, 30(2):126–44.

Bajard, J.-C., S. Kla, and J.-M. Muller (1994). BKM: A new hardware algorithm for complex elementary functions. *IEEE Transactions on Computers*, 43(8):955–63.

Baker, P. W. (1973). Predictive algorithms for some elementary functions in radix 2. *Electronics Letters*, 9(21):493–94.

Baker, P. W. (1975). Parallel multiplicative algorithms for some elementary functions. *IEEE Transactions on Computers*, C-24(3):321–24.

Burleson, W. P. (1990). Polynomial evaluation in VLSI using distributed arithmetic. *IEEE Transaction on Circuits and Systems*, 37(10):1299–304.

Cao, J., B. W. Y. Wei, and J. Cheng (2001). High-performance architectures for elementary function generation. In *Proceedings of the 15th IEEE Symposium on Computer Arithmetic*, pages 136–44.

Chen, T. C. (1972). Automatic computation of exponentials, logarithms, ratios, and square roots. *IBM Journal of Research and Development*, pages 380–89.

Cheney, E. W. (1966). *Introduction to Approximation Theory*. International Series in Pure and Applied Mathematics. McGraw Hill, New York.

Cody, W. J. (1970). A survey of practical rational and polynomial approximation of functions. *SIAM Review*, 12(3):400–423.

Cody, W. J., and W. Waite (1980). *Software Manual for the Elementary Functions*. Prentice-Hall, Englewood Cliffs, New Jersey.

Corbaz, G., J. Duprat, B. Hochet, and J.-M. Muller (1991). Implementation of a VLSI polynomial evaluator for real-time applications. In *Proceedings of ASAP91*, pages 13–24.

Dahlquist, G., and A. Bjorck (1974). *Numerical Methods*. Prentice-Hall, Englewood Cliffs, New Jersey.

Das Sarma, D., and D. W. Matula (1994). Measuring the accuracy of ROM reciprocal tables. *IEEE Transactions on Computers*, 43(8):932–40.

Das Sarma, D., and D. W. Matula (1995). Faithful bipartite ROM reciprocal tables. In *Proceedings of the 12th IEEE Symposium on Computer Arithmetic*, pages 17–28.

Das Sarma, D., and D. W. Matula (1997). Faithful interpolation in reciprocal tables. In *Proceedings of the 13th IEEE Symposium on Computer Arithmetic*, pages 82–91.

Daumas, M., C. Finot, and J.-M. Muller (2000). Table based implementation of elementary functions for hundred-bit precision. In *16th IMACS World Congress on Computational and Applied Mathematics*.

Daumas, M., C. Mazenc, X. Merrheim, and J.-M. Muller (1994). Fast and accurate range reduction for computation of the elementary functions. In *Proceedings of the 14th IMACS World Congress on Computational and Applied Mathematics*, pages 1196–98.

Daumas, M., C. Mazenc, X. Merrheim, and J.-M. Muller (1995). Modular range reduction: A new algorithm for fast and accurate computation of the elementary functions. *Journal of Universal Computer Science*, 1(3):162–75.

Davis, P. J. (1990). *Interpolation and Approximation*. Dover Publications, New York.

de Dinechin, F., and A. Tisserand (2001). Some improvements on multipartite table methods. In *Proceedings of the 15th IEEE Symposium on Computer Arithmetic*, pages 128–35.

DeLugish, B. G. (1970). *A Class of Algorithms for Automatic Evaluation of Certain Elementary Functions in a Binary Computer*. PhD thesis, Department of Computer Science, University of Illinois at Urbana-Champaign. (Technical Report UIUCDCS-R-399.)

Duprat, J., and J.-M. Muller (1988). Hardwired polynomial evaluation. *Journal of Parallel and Distributed Computing*, 5(3):291–309.

Ercegovac, M. D. (1973). Radix-16 evaluation of certain elementary functions. *IEEE Transactions on Computers*, C-22(6):561–66.

Ercegovac, M. D. (1975). *A General Hardware-Oriented Method for Evaluation of Functions and Computations in a Digital Computer*. PhD thesis, Department of Computer Science, University of Illinois at Urbana-Champaign. (Technical Report UIUCDCS-R-750.)

Ercegovac, M. D. (1977). A general hardware-oriented method for evaluation of functions and computations in a digital computer. *IEEE Transactions on Computers*, C-26(7):667–80.

Ercegovac, M. D., T. Lang, J.-M. Muller, and A. Tisserand (2000). Reciprocation, square root, inverse square root, and some elementary functions using small multipliers. *IEEE Transactions on Computers*, 49(7):628–37.

Ercegovac, M. D., and J.-M. Muller (1998). Fast evaluation of functions at regularly-spaced points. In *SPIE International Symposium on Optical Science, Engineering, and Instrumentation*, vol. 3461, pages 555–66.

Ercegovac, M. D., J.-M. Muller, and A. Tisserand (1995). FPGA implementation of polynomial evaluation algorithms. In *SPIE Photonics East '95 Conference Proceedings*, vol. 2607, pages 177–88.

Farmwald, P. M. (1981). High bandwidth evaluation of elementary functions. In *Proceedings of the 5th IEEE Symposium on Computer Arithmetic*, pages 139–42.

Ferguson, W. (1995). Exact computation of a sum or difference with applications to argument reduction. In *Proceedings of the 12th IEEE Symposium on Computer Arithmetic*, pages 216–21.

Ferguson, W., and T. Brightman (1991). Accurate and monotone approximations of some transcendental functions. In *Proceedings of the 10th IEEE Symposium on Computer Arithmetic*, pages 237–44.

Harrison, J., T. Kubaska, S. Story, and P. T. P. Tang (1999). The computation of transcendental functions on the IA-64 architecture. *Intel Technology Journal*, Q4.

Hart, J. F., E. W. Cheney, C. L. Lawson, H. J. Maehly, C. K. Mesztenyi, J. R. Rice, H. G. Thacher, and C. Witzgall (1978). *Computer Approximations*. Robert E. Krieger Publishing Company, Florida.

Hassler, H., and N. Takagi (1995). Function evaluation by table look-up and addition. In *Proceedings of the 12th IEEE Symposium on Computer Arithmetic*, pages 10–16.

Hwang, K., H. C. Wang, and Z. Xu (1987). Evaluating elementary functions with chebyshev polynomials on pipeline nets. In *Proceedings of the 8th IEEE Symposium on Computer Arithmetic*, pages 121–28.

Jain, V. K., and L. Lin (1995). High-speed double precision computation of non-linear functions. In *Proceedings of the 12th IEEE Symposium on Computer Arithmetic*, pages 107–14.

Jain, V. K., and L. Lin (1997). Complex-argument universal nonlinear cell for rapid prototyping. *IEEE Transactions on Very Large Scale Integration (VLSI) Systems*, 5(1):15–27.

Koren, I., and O. Zinaty (1990). Evaluating elementary functions in a numerical coprocessor based on rational approximations. *IEEE Transactions on Computers*, 39(8):1030–37.

Lefèvre, V., and J. Muller (1999). Table methods for the elementary functions. In *SPIE Symposium on Optical Science and Technology*, vol. 3807, pages 43–9.

Lefevre, V., and J.-M. Muller (2001). Worst cases for correct rounding of the elementary functions in double precision. In *Proceedings of the 15th IEEE Symposium on Computer Arithmetic*, pages 111–18.

Lefèvre, V., J.-M. Muller, and A. Tisserand (1998). Toward correctly rounded transcendentals. *IEEE Transactions on Computers*, 47(11):1235–43.

Lewis, D. M. (1994). Interleaved memory function interpolators with application to an accurate lns arithmetic unit. *IEEE Transactions on Computers*, 43(8): 974–82.

Linhardt, R. J., and H. S. Miller (1969). Digit-by-digit transcendental function computation. *RCA Review*, 30:209–47.

Lynch, T., A. Ahmed, M. J. Schulte, T. Callaway, and R. Tisdale (1995). The K5 transcendental functions. In *Proceedings of the 12th IEEE Symposium on Computer Arithmetic*, pages 163–70.

Markstein, P. W. (1990). Computation of elementary functions on IBM RISC System/6000 processor. *IBM Journal of Research and Development*, pages 111–19.

Mathews, J. H. (1992). *Numerical Methods for Mathematics, Science, and Engineering, Second Edition*. Prentice-Hall, Englewood Cliffs, New Jersey.

Merrheim, X., J.-M. Muller, and H. J. Yeh (1993). Fast evaluation of polynomials and inverses of polynomials. In *Proceedings of the 11th IEEE Symposium on Computer Arithmetic*, pages 186–92.

Muller, J.-M. (1997). *Elementary Functions, Algorithms and Implementation*. Birkhauser, Boston.

Muller, J.-M. (1999). A few results on table-based methods. *Reliable Computing*, 5(3):279–88.

Noetzel, A. S. (1989). An interpolating memory unit for function evaluation: analysis and design. *IEEE Transactions on Computers*, 38(3):377–84.

Paliouras, V., K. Karagianni, and T. Stouraitis (2000). A floating-point processor for fast and accurate sine/cosine evaluation. *IEEE Transactions on Circuits and Systems II: Analog and Digital Signal Processing*, 47(5):441–51.

Rauchwerger, L., and P. M. Farmwald (1990). A multiple floating point coprocessor architecture. In *Proceedings of the 23rd Annual Workshop and Symposium*, pages 216–22.

Rodrigues, M. R. D., J. H. P. Zurawski, and J. B. Gosling (1981). Hardware evaluation of mathematical functions. *IEE Proceedings E (Computers and Digital Techniques)*, 128(4):155–64.

Schulte, M. J., and J. E. Stine (1997a). Accurate function evaluation by symmetric table lookup and addition. In *Proceedings of the IEEE International*

Conference on Application-Specific Systems, Architectures and Processors, pages 144–53.

Schulte, M. J., and J. E. Stine (1997b). Symmetric bipartite tables for accurate function approximation. In *Proceedings of the 13th IEEE Symposium on Computer Arithmetic*, pages 175–83.

Schulte, M. J., and J. E. Stine (1999). Approximating elementary functions with symmetric bipartite tables. *IEEE Transactions on Computers*, 48(8):842–47.

Schulte, M. J., and E. E. Swartzlander (1993). Exact rounding of certain elementary functions. In *Proceedings of the 11th IEEE Symposium on Computer Arithmetic*, pages 138–45.

Schulte, M. J., and E. E. Swartzlander (1994). Hardware designs for exactly rounded elementary functions. *IEEE Transactions on Computers*, 43(8):964–73.

Specker, W. H. (1965). A class of algorithms for $\ln(x)$, $\exp(x)$, $\sin(x)$, $\cos(x)$, $\tan^{-1}(x)$ and $\cot^{-1}(x)$. *IEEE Transactions on Electronic Computers*, EC-14: 85–86.

Stine, J. E., and M. J. Schulte (1999). The symmetric table addition method for accurate function approximation. *Journal of VLSI Signal Processing*, 21:167–77.

Story, S., and P. T. P. Tang (1999). New algorithms for improved transcendental functions on IA-64. In *Proceedings of the 14th IEEE Symposium on Computer Arithmetic*, pages 4–11.

Tang, P. T. P. (1989). Table-driven implementation of the exponential function in IEEE floating-point arithmetic. *ACM Transactions on Mathematical Software*, 15(2):144–57.

Tang, P. T. P. (1990). Table-driven implementation of the logarithm function in IEEE floating-point arithmetic. *ACM Transactions on Mathematical Software*, 16(4):378–400.

Tang, P. T. P. (1991). Table lookup algorithms for elementary functions and their error analysis. In *Proceedings of the 10th IEEE Symposium on Computer Arithmetic*, pages 232–36.

Tang, P. T. P. (1992). Table-driven implementation of the expm1 function in IEEE floating-point arithmetic. *ACM Transactions on Mathematical Software*, 18(2):211–22.

Tung, C., and A. Avizienis (1970). Combinational arithmetic systems for the approximation of functions. In *AFIPS Conference Proceedings 1970 Spring Joint Computer Conference*, pages 95–107.

Wong, W. F., and E. Goto (1994). Fast hardware-based algorithms for elementary function computations using rectangular multipliers. *IEEE Transactions on Computers*, 43(3):278–94.

Zurawski, J. H. P. (1980). *High Performance Evaluation of Division and Other Elementary Functions*. PhD thesis, University of Manchester, England.

IN THIS CHAPTER WE PRESENT AND DISCUSS THE FOLLOWING TOPICS:

- CORDIC method
- Rotation and vectoring modes
- Convergence, precision, and range
- Compensation of scaling factor
- Implementations: word-serial and pipelined
- Extension to hyperbolic and linear coordinates
- Unified description
- Redundant addition and high radix

CHAPTER 11

CORDIC Algorithm and Implementations

In this chapter we consider the CORDIC algorithm and its implementation.[1] This algorithm permits the realization of rotations, the calculation of trigonometric functions, such as sin and cosine, of the inverse trigonometric function $\tan^{-1}(a/b)$, and of $\sqrt{a^2 + b^2}$. Moreover, it has been extended to hyperbolic functions and multiplication and division. In addition, minor modifications allow the calculation of other functions such as square root, exponential, and logarithm. The algorithm is attractive because of its generality, as well as its efficiency for some calculations, such as rotations. It has been used for applications in signal and image processing, in robotics, and in 3D graphics. Application-specific versions are being used for linear transforms, digital filters, and solution of linear systems.

The algorithm is based on the rotation of a vector on the plane.[2] As shown in Figure 11.1, the vector (terminating in point) x_{in}, y_{in} is rotated by the angle θ, producing the vector (terminating in) x_R, y_R. This rotation is described by the expressions

$$x_R = M_{in} \cos(\beta + \theta) = x_{in} \cos\theta - y_{in} \sin\theta$$
$$y_R = M_{in} \sin(\beta + \theta) = x_{in} \sin\theta + y_{in} \cos\theta$$

11.1

where M_{in} is the modulus of the vector and β is the initial angle. This rotation can be expressed in matrix form as

$$\begin{bmatrix} x_R \\ y_R \end{bmatrix} = \begin{bmatrix} \cos\theta & -\sin\theta \\ \sin\theta & \cos\theta \end{bmatrix} \begin{bmatrix} x_{in} \\ y_{in} \end{bmatrix} = ROT(\theta) \begin{bmatrix} x_{in} \\ y_{in} \end{bmatrix}$$

11.2

---

1. This corresponds to the class of linear convergence algorithms discussed in Chapter 10.
2. It has been extended to more dimensions, but here we will restrict ourselves to the two-dimensional case. Moreover, this description is for circular coordinates; the extension to hyperbolic and linear coordinates is discussed in Section 11.5

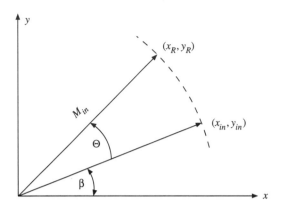

FIGURE 11.1 Vector rotation.

This is called a *perfect rotation* (or just *rotation*) because the modulus of the vector is preserved. Its direct implementation requires the evaluation of $\cos\theta$ and $\sin\theta$, four multiplications, and two additions.

The CORDIC algorithm performs rotations by a sequence of (micro) rotations by elementary angles. For this, define the sequence of elementary rotation angles α_j and decompose the angle θ as a sum of elementary angles[3]

$$\theta = \sum_{j=0}^{\infty} \alpha_j \qquad 11.3$$

Consequently,

$$ROT(\theta) = \prod_{j=0}^{\infty} ROT(\alpha_j) \qquad 11.4$$

and $ROT(\alpha_j)$ is described by the following equations:[4]

$$x_R[j+1] = x_R[j]\cos(\alpha_j) - y_R[j]\sin(\alpha_j)$$

$$y_R[j+1] = x_R[j]\sin(\alpha_j) + y_R[j]\cos(\alpha_j) \qquad 11.5$$

---

3. Now we consider the theoretical case in which the number of microrotations is infinite and later determine the error introduced in the function by performing a finite number.
4. We use the subscript R so as not to confuse with the $x[j]$, $y[j]$ of the CORDIC microrotation introduced later.

This microrotation is still complex to implement since it requires multiplications. The multiplications are avoided by the following:

1. Decomposing the rotation into a scaling operation and a rotation-extension (also called a *similarity*) by factoring the term $\cos(\alpha_j)$. The result is

$$x_R[j+1] = \cos(\alpha_j)(x_R[j] - y_R[j]\tan(\alpha_j))$$

$$y_R[j+1] = \cos(\alpha_j)(y_R[j] + x_R[j]\tan(\alpha_j))$$

11.6

2. Choosing as elementary angles the sequence[5]

$$\alpha_j = \tan^{-1}(\sigma_j(2^{-j})) = \sigma_j\tan^{-1}(2^{-j})$$

11.7

with $\sigma_j \in \{-1, 1\}$.

With this choice the rotation-extension becomes

$$x[j+1] = x[j] - \sigma_j 2^{-j}y[j]$$

$$y[j+1] = y[j] + \sigma_j 2^{-j}x[j]$$

11.8

which is implemented using only additions and shifts.

This rotation-extension scales the modulus $M[j]$ so that

$$M[j+1] = K[j]M[j] = \frac{1}{\cos\alpha_j}M[j] = \left(1 + \sigma_j^2 2^{-2j}\right)^{1/2}M[j]$$

$$= (1 + 2^{-2j})^{1/2}M[j]$$

11.9

The CORDIC algorithm consists in applying a sequence of rotation-extensions. The total scaling factor is then

$$K = \prod_{j=0}^{\infty}(1 + 2^{-2j})^{1/2} \approx 1.6468$$

11.10

Note that, because $\sigma_j \in \{-1, 1\}$, the scaling factor is constant, independent of the angle being rotated.

Moreover, to decompose the angle θ or to accumulate it, depending on the operation mode as discussed later, the following third recurrence is used:

$$z[j+1] = z[j] - \alpha_j = z[j] - \sigma_j\tan^{-1}(2^{-j})$$

11.11

---

5. This results in a radix-2 algorithm; the higher-radix case is considered in Section 11.5.

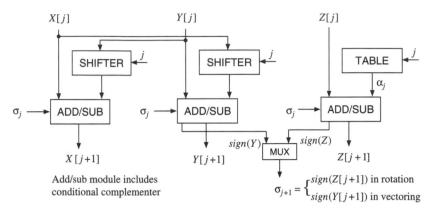

FIGURE 11.2 Implementation of one iteration.

The resulting iteration, called a CORDIC microrotation,[6] is

$$x[j + 1] = x[j] - \sigma_j 2^{-j} y[j]$$
$$y[j + 1] = y[j] + \sigma_j 2^{-j} x[j] \qquad \textbf{11.12}$$
$$z[j + 1] = z[j] - \sigma_j \tan^{-1}(2^{-j})$$

An implementation of one iteration is as shown in Figure 11.2. It consists of shifters and adders and a table to contain the angles $\tan^{-1}(2^{-j})$. The signals $sign(Y)$ and $sign(Z)$ are used to determine the value of σ_j, as discussed below.

11.1   Rotation and Vectoring Modes

The CORDIC algorithm is used in two modes: rotation and vectoring. We consider these modes now.

11.1.1   Rotation Mode

In this mode, an initial vector (x_{in}, y_{in}) is rotated by an angle θ. As shown in Figure 11.3, to do this, the angle is decomposed into the primitive angles (using the recurrence z) and the vector is rotated by these angles. To decompose the

---

6. Although this is a rotation-extension, to simplify the discussion it is called a microrotation.

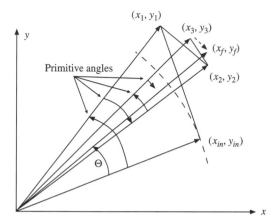

FIGURE 11.3 Rotating a vector using microrotations.

angle, the initial value of z is made equal to θ, and σ_j (the direction of rotation) is selected so that the final angle is zero. That is,

$$z[0] = \theta$$

$$\sigma_j = \begin{cases} 1 & \text{if } z[j] \geq 0 \\ -1 & \text{if } z[j] < 0 \end{cases} \qquad \textbf{11.13}$$

Then, this value of σ_j is used in the microrotation to produce $x[j+1]$, $y[j+1]$, and $z[j+1]$. The initial condition is

$$(x[0], \ y[0]) = (x_{in}, \ y_{in}) \qquad \textbf{11.14}$$

The final values are[7]

$$x_f = K(x_{in} \cos\theta - y_{in} \sin\theta)$$
$$y_f = K(x_{in} \sin\theta + y_{in} \cos\theta) \qquad \textbf{11.15}$$
$$z_f = 0$$

To obtain a perfect rotation it is necessary to compensate for the scaling factor K. The methods for performing this compensation are discussed in Section 11.3.

---

7. We use the subscript f for the final values of the rotation-extension to contrast with the subscript R for the rotation.

j	$z[j]$	σ_j	$x[j]$	$y[j]$
0	1.1693	1	1.0	0.125
1	0.3839	1	0.875	1.125
2	−0.0796	−1	0.3125	1.1562
3	0.1653	1	0.7031	1.4843
4	0.0409	1	0.5175	1.5722
5	−0.0214	−1	0.4193	1.6046
6	0.0097	1	0.4694	1.5915
7	−0.0058	−1	0.4445	1.5988
8	0.0019	1	0.4570	1.5953
9	−0.0019	−1	0.4508	1.5971
10	0.0000	1	0.4539	1.5962
11	−0.0009	−1	0.4524	1.5967
12	−0.0004	−1	0.4531	1.5965
13			0.4535	1.5963

TABLE 11.1 Example of vector rotation.

EXAMPLE 11.1   Table 11.1 illustrates rotation of a vector ($x_{in} = 1$, $y_{in} = 0.125$) by an angle of 67° using $n = 12$ microrotations. The expected coordinates of the rotated vector are $x_R = 0.2756$, $y_R = 0.9693$.

After performing compensation of the scaling factor $K = 1.64676$, the coordinates are $x[13]/K = 0.2753$ and $y[13]/K = 0.9693$, with errors smaller than 2^{-12}.   ∎

As a special case, to compute $\cos\theta$ and $\sin\theta$ the initial conditions are $x[0] = 1/K$ and $y[0] = 0$. More in general, if a and b are constants, $a\cos\theta - b\sin\theta$ and $a\sin\theta + b\cos\theta$ are computed by setting the initial conditions to $x[0] = a/K$ and $y[0] = b/K$.

11.1.2   Vectoring Mode

In this mode, the initial vector (x_{in}, y_{in}) is rotated until the y component is zero. Moreover, the corresponding rotation angle is accumulated in z. To accomplish this rotation, for an initial vector in the first quadrant, the direction of rotation is selected as

$$\sigma_j = \begin{cases} 1 & \text{if } y[j] < 0 \\ -1 & \text{if } y[j] \geq 0 \end{cases} \qquad \textbf{11.16}$$

For the initial values $(x[0], \, y[0]) = (x_{in}, \, y_{in})$ and $z[0] = z_{in}$, the final values are

$$x_f = K\left(x_{in}^2 + y_{in}^2\right)^{1/2}$$
$$y_f = 0 \qquad \textbf{11.17}$$
$$z_f = z_{in} + \tan^{-1}\left(\frac{y_{in}}{x_{in}}\right)$$

The compensation of the scale factor is again discussed in Section 11.3.

EXAMPLE 11.2 Table 11.2 illustrates the vectoring mode. A vector $(x_{in} = 0.75, \, y_{in} = 0.43)$ is rotated clockwise to force the y component to zero. We perform $n = 12$

j	$y[j]$	σ_j	$x[j]$	$z[j]$
0	0.43	-1	0.75	0.0
1	-0.32	1	1.18	0.7853
2	0.27	-1	1.34	0.3217
3	-0.065	1	1.4075	0.5667
4	0.1109	-1	1.4156	0.4423
5	0.0224	-1	1.4225	0.5047
6	-0.0219	1	1.4232	0.5360
7	0.0002	-1	1.4236	0.5204
8	-0.0108	1	1.4236	0.5282
9	-0.0053	1	1.4236	0.5243
10	-0.0025	1	1.4236	0.5223
11	-0.0011	1	1.4236	0.5213
12	-0.0004	1	1.4236	0.5208
13			1.4236	0.5206

TABLE 11.2 Example of vectoring.

microrotations. The expected coordinates of the rotated vector are

$$x_R = \sqrt{x_{in}^2 + y_{in}^2} = 0.8645, \quad y_R = 0.0$$

and the rotated angle

$$z_f = \tan^{-1}\left(\frac{0.43}{0.75}\right) = 0.5205$$

The accumulated angle $z[13] = 0.5206$. After performing compensation of the scaling factor $K = 1.64676$, we obtain $x[13]/K = 0.864$. The errors are smaller than 2^{-12}. ∎

11.2 Convergence, Precision, and Range

In this section we check the convergence of the algorithm, determine the precision obtained with n iterations, and the range of the rotation angle. We first consider the rotation mode.

11.2.1 Convergence

The condition of convergence of the algorithm is that the residual angle to rotate after iteration j is not greater than the maximum angle than can be rotated in the remaining iterations. That is,

$$|z[j]| \leq \sum_{i=j}^{\infty} \tan^{-1}(2^{-i}) \qquad\qquad \textbf{11.18}$$

From this expression we obtain the maximum value of the rotation angle, namely,

$$\theta_{max} = z[0]_{max} = \sum_{j=0}^{\infty} \tan^{-1}(2^{-j}) \approx 1.7433 \ (99.88°) \qquad\qquad \textbf{11.19}$$

For this angle all $\sigma_j = 1$ and all $z[j] > 0$.

Now consider an angle $|\theta| < \theta_{max}$. In this case, as shown in Figure 11.4, there is an iteration i for which $z[i]$ is negative. The maximum negative $z[i]$ occurs when $z[i-1] = 0$. Since the rotation angle in the iteration to produce $z[i]$

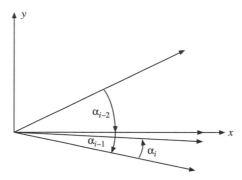

FIGURE 11.4 Convergence condition: the maximum negative case.

is $\tan^{-1}(2^{-(i-1)})$ we obtain

$$|z[i]| \leq \tan^{-1}\left(2^{-(i-1)}\right) \tag{11.20}$$

Consequently, the convergence condition requires that

$$\tan^{-1}\left(2^{-(i-1)}\right) \leq \sum_{j=i}^{\infty} \tan^{-1}(2^{-j}) \tag{11.21}$$

which is equivalent to

$$\tan^{-1}(2^{-i}) \leq \sum_{j=i+1}^{\infty} \tan^{-1}(2^{-j}) \tag{11.22}$$

Since this condition is satisfied for all i, the algorithm converges.[8] In conclusion, the CORDIC algorithm converges as long as the rotation angle is not larger than θ_{max}.

Since the maximum angle for convergence is somewhat larger than $\pi/2$, it might be necessary to do a preprocessing (argument range reduction) to achieve a larger angle. For instance, to achieve a range of $[-\pi, \pi]$, when the magnitude of the angle is larger than $\pi/2$, an initial rotation by $\pi/2$ is performed, which consists in an interchange of x and y and a sign change.

---

8. This is an instance of the more general condition, which states that an algorithm converges if the bases satisfy $2\alpha_{j+1} \geq \alpha_j$.

11.2.2   Range and Error for n Iterations and Truncation

Up to now we have considered the theoretical case in which the sequence of iterations is infinite. In practice, of course, the sequence is finite. Moreover, all variables are represented by a finite number of bits. This affects the range and produces an error in the result.

As shown in (11.20), the residual angle after n iterations $z[n]$ is bounded by

$$|z[n]| \leq \tan^{-1}\left(2^{-(n-1)}\right) \qquad\qquad \textbf{11.23}$$

Moreover,

$$2^{-n} < \tan^{-1}\left(2^{-(n-1)}\right) < 2^{-(n-1)} \qquad\qquad \textbf{11.24}$$

This means that the angle after n iterations has an error bound of $2^{-(n-1)}$.

With respect to the maximum angle for convergence, the expression for n iterations is

$$\theta_{max}(n) = \sum_{j=0}^{n-1} \tan^{-1}(2^{-j}) + 2^{-n+1} \qquad\qquad \textbf{11.25}$$

where 2^{-n+1} is the maximum residual angle. Note that the maximum angle does not change significantly with the number of iterations, after a reasonable number of iterations.

For the vectoring mode we have $\tan^{-1}(y_{in}/x_{in}) \leq \theta_{max}$, and the conclusions about the maximum angle, precision, and range remain the same as in the rotation mode.

Truncation errors

The error bound of expression (11.24) assumes variables of infinite precision. The representation with a finite number of bits requires roundoff (usually truncation) and introduces additional errors. These errors are of two (interrelated) types:

- Accumulation of the roundoff errors.
- Error in the determination of σ_j. This produces a rotation in the wrong direction.

Error analysis have been performed to bound the total error.[9] This analysis is used to determine the number of bits required to obtain a desired precision.

11.3   Compensation of Scaling Factor

If a perfect rotation is required (in rotation mode) or if the modulus is required (in vectoring mode), it is necessary to compensate for the scaling factor K. The following compensation methods have been proposed:

- The most direct method is to multiply by $1/K$. Moreover, since K is a constant, the multiplication can be simplified by taking advantage of the zeros in the representation of $1/K$. Recoding can be used to increase the number of zeros. This method requires additional hardware to perform the multiplication.

- Another method of compensation is to approximate $1/K$ by a product of factors of the form (1 ± 2^{-i}). For an acceptable approximation error, the number of factors is between $n/3$ and $n/4$. This product can be implemented by a sequence of scaling iterations of the form

$$x_s = x \pm x(2^{-i})$$

Consequently, these iterations can use the same hardware as the CORDIC iterations.

- A related method (which can be combined with scaling iterations) is to use repetitions of CORDIC iterations, that is, to perform more than one CORDIC iteration for some index values. This can be done without changing the convergence condition (of course, in each iteration, the value of σ has to be determined so that the algorithm converges). This is correct since the condition

$$|z[i + 1]| \leq \tan^{-1}(2^{-i})$$

applies also for the case with repetitions.

Since the repetitions also produce a scaling of the modulus, they can be used together with the scaling iterations to compensate for the original

---

9. For details see Hu (1992) and Antelo, Bruguera, et al. (1997).

Scaling iterations	$(-1)(+2)(-5)(+10)(+16)(+19)(+22)$
Scalings + repetitions	$(-2)(+16)(+17)$ $1, 3, 5, 6$

TABLE 11.3 Scale-factor compensation for $n = 24$.

scaling factor. The problem consists then in finding the minimum number of scaling iterations plus repetitions so that the scale factor is compensated. For instance, for $n = 24$, Table 11.3 shows a set of scaling iterations and of scaling plus repetitions.

Although for the $n = 24$ case there is no difference in the number of iterations among both methods, in some cases (for instance, in the use of redundant adders as described later) repetitions are necessary for convergence. In such cases there is some flexibility in the position of these repetitions. Consequently, they can be used also as part of the compensation of the scaling factor.

The scale-factor compensation introduces additional errors because of the approximate method used and because of the truncations. These errors have to be included in the total error.

11.4   Implementations

The implementation can be word-serial or pipelined. We describe these alternatives now.

11.4.1   Word-Serial Implementation

In the word-serial implementation the hardware for one iteration is reused. Consequently, this implementation is a sequential system in which each microrotation corresponds to one clock cycle. This implementation is shown in Figure 11.5. Note the variable shifters required for the multiplication by 2^{-j} in iteration j. The critical path delay is the sum of the delays of the shifter, the conditional complementer (for the subtraction), the adder, and the register.

The same hardware is also used for the scale-factor compensation, for the case in which the compensation is done by scaling iterations and the repetition

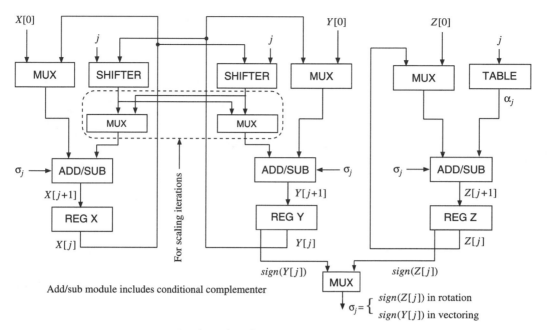

FIGURE 11.5 Word-serial implementation.

of CORDIC iterations. Note the MUX required to implement the scaling iterations.

11.4.2 Pipelined Implementation

In the pipelined implementation the iterations are unfolded so that each microrotation uses its own hardware (Figure 11.6). From the point of view of latency this has the advantage that the shift amount in each iteration is constant, so that the corresponding shifters are implemented just by the suitable wiring. Consequently, the delay of one iteration is now only the sum of the delays of the conditional complementer and the adder.

Additional hardware is required in this case for the scale-factor compensation.

Moreover, because the iterations are unfolded, this allows the execution of several CORDIC operations in a pipelined fashion, resulting in a high throughput. In this case, the delay of the latches should be included in the latency.

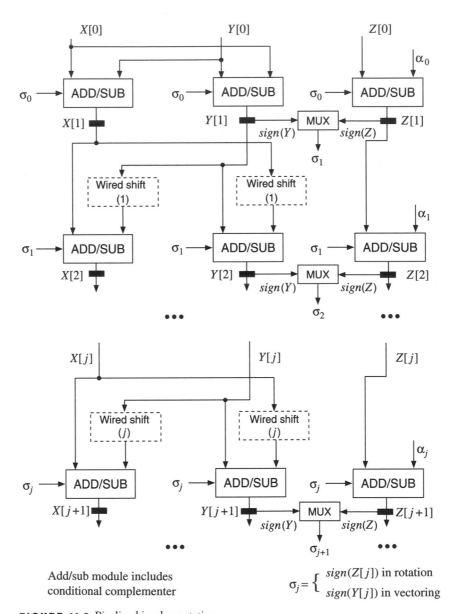

FIGURE 11.6 Pipelined implementation.

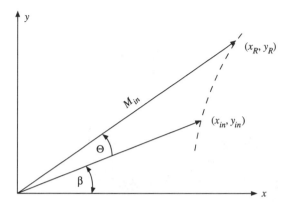

FIGURE 11.7 Rotation in hyperbolic coordinate system.

11.5 Extension to Hyperbolic and Linear Coordinates

The algorithm described in the previous sections is for circular coordinates. We now consider its extension to hyperbolic coordinates and linear coordinates.

11.5.1 Hyperbolic Coordinates

Similarly as for circular coordinates, as shown in Figure 11.7, an hyperbolic rotation by angle θ is described by

$$\begin{bmatrix} x_R \\ y_R \end{bmatrix} = \begin{bmatrix} \cosh\theta & \sinh\theta \\ \sinh\theta & \cosh\theta \end{bmatrix} \begin{bmatrix} x_{in} \\ y_{in} \end{bmatrix} \qquad \textbf{11.26}$$

Notice the change in sign in the upper-right element with respect to the circular case. Consequently, the corresponding CORDIC microrotation is

$$\begin{aligned} x[j+1] &= x[j] + \sigma_j 2^{-j} y[j] \\ y[j+1] &= y[j] + \sigma_j 2^{-j} x[j] \\ z[j+1] &= z[j] - \sigma_j \tanh^{-1}(2^{-j}) \end{aligned} \qquad \textbf{11.27}$$

The scaling factor in iteration j is[10]

$$K_h[j] = (1 - 2^{-2j})^{1/2} \qquad \textbf{11.28}$$

---

10. We use the subscript h to differentiate with the factor K for circular coordinates.

Since $\tanh^{-1} 2^0 = \infty$ (and $K_h[0] = 0$), for hyperbolic coordinates it is necessary to begin from iteration $j = 1$.

Moreover, in this hyperbolic case, a complication is that the algorithm does not converge with the sequence of angles $\tanh^{-1}(2^{-j})$ since

$$\sum_{j=i+1}^{\infty} \tanh^{-1}(2^{-j}) < \tanh^{-1}(2^{-i}) \qquad \textbf{11.29}$$

A solution is to repeat some iterations. Since

$$\sum_{i=j+1}^{\infty} \tanh^{-1}(2^{-i}) < \tanh^{-1}(2^{-j}) < \sum_{i=j+1}^{\infty} \tanh^{-1}(2^{-i}) + \tanh^{-1}\left(2^{-(3j+1)}\right)$$

$$\textbf{11.30}$$

repeating iterations 4, 13, 40, . . . , k, $3k+1$, . . . results in a convergent algorithm. Including these repetitions, we get

$$K_h \approx 0.82816$$
$$\theta_{max} = 1.11817$$

For these coordinates we can also have the rotation and vectoring modes, with the same expressions for the calculation of σ_j as in the circular coordinate case. The final values are, for rotation mode,

$$x_f = K_h(x_{in} \cosh \theta + y_{in} \sinh \theta)$$
$$y_f = K_h(x_{in} \sinh \theta + y_{in} \cosh \theta) \qquad \textbf{11.31}$$
$$z_f = 0$$

and for vectoring mode,

$$x_f = K_h\left(x_{in}^2 - y_{in}^2\right)^{1/2}$$
$$y_f = 0 \qquad \textbf{11.32}$$
$$z_f = z_{in} + \tanh^{-1}\left(\frac{y_{in}}{x_{in}}\right)$$

Similar considerations as those given for the circular mode with respect to errors and to scale-factor compensation apply also to the hyperbolic case. Moreover, the implementations of Figures 11.5 and 11.6 can be adapted, either to include both types of coordinates or just for the hyperbolic case.

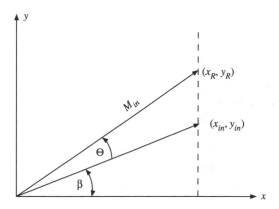

FIGURE 11.8 Rotation in linear coordinate system.

11.5.2   Linear Coordinates

The rotation for linear coordinates is shown in Figure 11.8. That is,

$$x_R = x_{in}$$
$$y_R = y_{in} + x_{in}z_{in}$$

11.33

Consequently, the elementary angles are of the form 2^{-i}, and the corresponding microrotation is

$$x[j+1] = x[j]$$
$$y[j+1] = y[j] + \sigma_j 2^{-j} x[j]$$
$$z[j+1] = z[j] - \sigma_j(2^{-j})$$

11.34

The scaling factor is 1.

For the vectoring mode the final values are

$$x_f = x_{in}$$
$$z_f = z_{in} + \frac{y_{in}}{x_{in}}$$

11.35

From the expressions we see that this linear mode can be used to perform multiply-add and divide-add.

11.5.3   Unified Description

From the previous development it is possible to describe the algorithm in the three coordinate systems in a unified manner by defining the parameter m so that

- $m = 1$ for circular coordinates
- $m = -1$ for hyperbolic coordinates
- $m = 0$ for linear coordinates

In that case, the unified microrotation is[11]

$$x[j+1] = x[j] - m\sigma_j 2^{-j} y[j]$$
$$y[j+1] = y[j] + \sigma_j 2^{-j} x[j]$$

$$z[j+1] = \begin{cases} z[j] - \sigma_j \tan^{-1}(2^{-j}) & \text{if } m = 1 \\ z[j] - \sigma_j \tanh^{-1}(2^{-j}) & \text{if } m = -1 \\ z[j] - \sigma_j(2^{-j}) & \text{if } m = 0 \end{cases}$$

11.36

and the scaling factor is

$$K_m[j] = (1 + m2^{-2j})^{1/2}$$

11.37

Table 11.4 summarizes the two modes in the three coordinate systems.

11.5.4   Other Functions

The functions shown in Table 11.4 are obtained directly from the application of the CORDIC algorithm. Additional functions can be obtained with suitable initial values; some of these functions are shown in Table 11.5.

11.6   Redundant Addition and High Radix

We now describe two of the many modifications that have been proposed for the CORDIC algorithm and its implementation. The main objective of these modifications is to reduce the latency and/or to increase the throughput. The modifications considered here are applicable to a unified implementation for rotation and vectoring. Modifications that are applicable only to one mode are discussed in the next section.

---

11. The z recurrence is also written as $z[j+1] = z[j] - \sigma_j m^{-1/2} \tan^{-1}(m^{1/2} 2^{-j})$.

Coordinates	Rotation mode $\sigma_j = sign(z[j])$	Vectoring mode $\sigma_j = -sign(y[j])$
Circular ($m = 1$) $\alpha_j = \tan^{-1}(2^{-j})$ Initial $j = 0$ $j = 0, 1, 2, \ldots, n-1$ $K_1 \approx 1.64676$ $\theta_{max} \approx 1.74329$	$x_f = K_1(x_i \cos(z_i) - y_i \sin(z_i))$ $y_f = K_1(x_i \sin(z_i) + y_i \cos(z_i))$ $z_f = 0$	$x_f = K_1\left(x_i^2 + y_i^2\right)^{1/2}$ $y_f = 0$ $z_f = z_i + \tan^{-1}\left(\dfrac{y_i}{x_i}\right)$
Linear ($m = 0$) $\alpha_j = 2^{-j}$ Initial $j = 0$ $j = 0, 1, 2, \ldots, n-1$ $K_0 = 1$ $\theta_{max} = 2 - 2^{-n}$	$x_f = x_i$ $y_f = y_i + x_i z_i$ $z_f = 0$	$x_f = x_i$ $y_f = 0$ $z_f = z_i + \dfrac{y_i}{x_i}$
Hyperbolic ($m = -1$) $\alpha_j = \tanh^{-1}(2^{-j})$ Initial $j = 1$ $j = 1, 2, 3, 4, 4, 5, \ldots, 13,$ $13, \ldots, n$ $K_{-1} \approx 0.82816$ $\theta_{max} \approx 1.11817$	$x_f = K_{-1}(x_i \cosh(z_i) + y_i \sinh(z_i))$ $y_f = K_{-1}(x_i \sinh(z_i) + y_i \cosh(z_i))$ $z_f = 0$	$x_f = K_{-1}\left(x_i^2 - y_i^2\right)^{1/2}$ $y_f = 0$ $z_f = z_i + \tanh^{-1}\left(\dfrac{y_i}{x_i}\right)$

Note: $sign(a) = 1$ if $a \geq 0$, $sign(a) = -1$ if $a < 0$. x_i, y_i, z_i are the initial values.

TABLE 11.4 Unified CORDIC.

Although these modifications are applicable to both modes and all coordinates, the details given here are limited to the circular case.

11.6.1 Redundant Representation

As can be seen from the description and evaluation of its implementation, the main delay in the critical path of the CORDIC iteration is that of the adder, even when a fast adder is used. An evident way of reducing this delay is to use one of the redundant adders (see Chapter 2). This results in a redundant representation of the variables.

m	Mode	Initial values			Functions	
		x_{in}	y_{in}	z_{in}	x_R	y_R or z_R
1	rotation	1	0	θ	$\cos\theta$	$y_R = \sin\theta$
-1	rotation	1	0	θ	$\cosh\theta$	$y_R = \sinh\theta$
-1	rotation	a	a	θ	ae^θ	$y_R = ae^\theta$
1	vectoring	1	a	$\pi/2$	$\sqrt{a^2+1}$	$z_R = \cot^{-1}(a)$
-1	vectoring	a	1	0	$\sqrt{a^2-1}$	$z_R = \coth^{-1}(a)$
-1	vectoring	$a+1$	$a-1$	0	$2\sqrt{a}$	$z_R = 0.5\ln(a)$
-1	vectoring	$a+\frac{1}{4}$	$a-\frac{1}{4}$	0	$\sqrt{a}$	$z_R = \ln\left(\frac{1}{4}a\right)$
-1	vectoring	$a+b$	$a-b$	0	$2\sqrt{ab}$	$z_R = 0.5\ln\left(\frac{a}{b}\right)$

Note: The final values x_R and y_R are obtained after compensation of the scale factor.

TABLE 11.5 Some additional functions.

The main problem with this approach is the need to detect the sign to obtain σ_j. This sign detection might be done by converting to a conventional representation, but this would defeat the purpose of using a redundant adder. It might also be done by a sign detection module; however, the sign depends on all the bits so that the sign detection delay is intrinsically of the same order as the conversion. The solution to this, used in other digit recurrences such as division (see Chapter 5), is to obtain σ_j from an estimate of the sign. However, to assure convergence some modification is required for the case in which the estimate is not correct.

One possibility is to use a redundant digit set for σ_j. In particular, use the value 0 in addition to ± 1. This is the solution adopted for division. However, for CORDIC it has the disadvantage that the introduction of 0 makes the scaling factor variable, dependent on the angle. Two approaches have been proposed to handle this situation:

- Calculate the variable scaling factor and perform the corresponding compensation. This can be done by evaluating the scaling factor using a recurrence (and the compensation by a division), or by calculating the logarithm of the scaling factor and then using an exponential function for the compensation (see Chapter 10 for the algorithms to compute logarithm and exponential). Note that for the required precision, only the first half of the σs affect the scaling factor.

* Modify the recurrence to keep a constant scaling factor. The corresponding iteration has been called a *double rotation*.

Another possibility is to maintain the digit set ± 1 to have a constant scale factor. The possibility of an incorrect estimate producing a nonconvergent algorithm can be handled by the following proposals:

* Introduce additional CORDIC iterations (called *correcting iterations*) to correct any possible error. To maintain the constant scaling factor it is necessary to include these iterations at fixed points, irrespective of whether an error occurred.
* Use two CORDIC modules, called the *plus* and the *minus module*. Whenever an estimate is inconclusive, initiate the operation in both modules and determine later which of the two is correct.

We now discuss at a hight level the double-rotation and the correcting-iterations approaches.

Double-Rotation Approach

In this approach the set of values of σ_j is $\{-1, 0, 1\}$. To maintain the constant scale factor the corresponding rotations are performed by a double rotation, as follows:

* $\sigma_j = 1$. Both rotations are by angle $\tan^{-1}(2^{-(j+1)})$.
* $\sigma_j = 0$. The two rotations are by the angles $\tan^{-1}(2^{-(j+1)})$ and $-\tan^{-1}(2^{-(j+1)})$.
* $\sigma_j = -1$. Both rotations are by the angle $-\tan^{-1}(2^{-(j+1)})$.

Consequently, the scaling factor is constant and has value

$$K = \prod_{j=1}^{n}(1 + 2^{-2j})$$

The elementary angles for this algorithm are $\alpha_j = 2\tan^{-1}(2^{-(j+1)})$ (instead of the $\tan^{-1}(2^{-j})$ of the conventional CORDIC). The algorithm converges for these elementary angles because $2\alpha_{j+1} \geq \alpha_j$.

The double rotation is incorporated into a single iteration, resulting in the following recurrences:

$$x[j+1] = x[j] - q_j 2^{-j} y[j] - p_j 2^{-2j-2} x[j]$$
$$y[j+1] = y[j] + q_j 2^{-j} x[j] - p_j 2^{-2j-2} y[j] \qquad \qquad \textbf{11.38}$$
$$z[j+1] = z[j] - q_j \left(2 \tan^{-1}\left(2^{-(j+1)}\right)\right)$$

The two control variables (q_j, p_j) take values $(1, 1)$ for $\sigma_j = 1$; $(0, -1)$ for $\sigma_j = 0$; and $(-1, 1)$ for $\sigma_j = -1$. The value of σ_j is determined essentially from an estimate of the sign of the corresponding variable ($z[j]$ for rotation and $y[j]$ for vectoring); since the variable converges to 0, the estimate of the sign uses the bits $j - 1$, j, and $j + 1$ of the carry-save representation of $z[j]$.

As indicated, the algorithm uses a redundant representation and produces a constant scaling factor. However, the recurrence is more complicated than the conventional CORDIC because of the three terms required to produce $x[j + 1]$ and $y[j + 1]$.

Correcting-Iterations Approach

In this approach the values of σ_j are kept as ± 1, which results in a constant scale factor. However, because of the redundant representation, it is not possible to determine accurately the sign of the corresponding variable. Consequently, an estimate of the sign is obtained by examining a limited number of digits. In case the sign estimation is incorrect, the rotation is in the incorrect direction so that the algorithm might not converge. To assure convergence, some additional iterations (called *repetitions* because they correspond to repeating the rotation with the same elementary angle) are introduced at predetermined intervals.

The interval between these iterations depends on the maximum error committed when the estimation is incorrect, and this is influenced by the number of digits used to estimate the sign. Consequently, there is a trade-off between the complexity of the sign detection and the number of repetitions. It can be shown that when m digits are used for the estimation, the distance between repetitions is about m iterations for the rotation mode and $m - 2$ iterations for the vectoring mode.[12]

Since the scale factor of the required precision is affected only by the first half of the σ's, for the second half it is possible to use the redundant digit set

---

12. For details see Takagi et al. (1991) and Lee and Lang (1992).

$\{-1, 0, 1\}$, in which case no repetitions are needed in these iterations. Moreover, beginning in $n/4$ the scale factor can be approximated by linear terms of the form $1 + 2^{-2j-1}$; consequently, a constant scale factor is obtained when using the digit set $\{-1, 0, 1\}$ if a scaling iteration is performed when $\sigma_j = 0$ (this is performed with the same hardware of the CORDIC iteration).[13]

In summary, in this approach the iteration is as in the conventional CORDIC, but additional iterations are required to compensate the errors produced by using an estimate of the sign in the first $n/2$ or $n/4$ iterations. As indicated before, these correcting iterations can be combined with other CORDIC repetitions and with scaling iterations to compensate the constant scaling factor.

11.6.2 Higher Radix

To reduce the number of iterations, it is possible to extend the algorithm to a higher radix. The corresponding recurrence is

$$
\begin{aligned}
x[j+1] &= x[j] - \sigma_j r^{-j} y[j] \\
y[j+1] &= y[j] + \sigma_j r^{-j} x[j] \\
z[j+1] &= z[j] - \tan^{-1}(\sigma_j r^{-j})
\end{aligned}
\qquad \textbf{11.39}
$$

and the value of σ_j is a signed radix-r digit (nonredundant or redundant). Note that the σ_j is now part of the argument of the elementary angle; this is in contrast to the radix-2 case, in which the σ_j has values ± 1, so that it just determines the direction of rotation.

A selection function determines the value of σ_j; this function is obtained using a method similar to that of division (see Chapter 5). Because, as in division, the complexity of the selection function increases with the radix, direct implementations are practical for radix 4.[14]

In the high-radix algorithm the scale factor is

$$
K = \prod \left(1 + \sigma_j^2 r^{-2j}\right)^{1/2}
\qquad \textbf{11.40}
$$

Consequently, this scale factor is variable: it is necessary to compute it and then to compensate. A method to do this is to compute the logarithm of the scale factor

---

13. In a pipelined implementation an additional wired shift of $2j + 1$ positions is required.
14. For details see Antelo, Villalba, et al. (1997) and Villalba et al. (1998b).

and then to compensate by multiplying by the exponential of this logarithm. This multiplication can be done with a recurrence similar to CORDIC, as described in Chapter 10.

As stated before, the scale factor is affected only by the first half bits. Consequently, a constant scale factor can be achieved if the first half is done radix-2 and the rest high radix. Of course, this increases the number of iterations with respect to the case in which the whole algorithm is high radix.

The iteration can be performed using nonredundant or redundant adders. As in division, the use of a redundant adder complicates somewhat the selection function.

11.6.3   Example: 24-Bit Unit

We now describe the implementation of a 24-bit unit for circular coordinates, using the enhancements described in this section. Specifically:

1. The iteration is performed using redundant adders. To be specific, we select a signed-digit (radix-2) adder.

2. The scale factor is constant. This is achieved by the following:

   - Iterations 0 to 6 use a σ_j with values ± 1. The selection function uses the four most-significant signed digits of the corresponding variable to estimate the sign. This requires two correcting iterations for convergence; the position of these iterations is determined so that they also contribute to the scale-factor compensation (see below).

   - Iterations 7 to 12 use a σ_j with values $-1, 0, 1$. To maintain the constant scaling factor, scaling by $(1 + 2^{-(2j+1)})$ is required when $\sigma_j = 0$. The selection function is the same as for the first case, except that $\sigma_j = 0$ is selected when the value of the four digits is 0.

   - Iterations 13 to 18 are radix 4. To simplify the implementation, σ_j is in the set $\{-2, -1, 0, 1, 2\}$. The selection function is especially simple for the rotation mode, since in these iterations $\tan^{-1}(2^{-j})$ can be approximated by 2^{-j}.

3. The scale-factor compensation is by repetitions and scalings. The repetitions are chosen so as to include those required for convergence and to minimize the total number of iterations.

11.7   Application-Specific Variations

In the previous sections we have considered a CORDIC unit applicable to all the coordinates and operation modes. In some applications this is not required, so that particular optimizations are possible. Moreover, in some cases a vectoring operation is followed by a rotation, so that the rotation angle does not have to be computed explicitly in conventional representation. We briefly summarize some of these cases now. More details are given in the references at the end of the chapter.

11.7.1   Only Rotation

If only the rotation mode is required, then the following optimizations have been considered:

Obtaining the σ_j Values Directly from the Angle

Especially when using redundant adders in a pipelined implementation, the critical path is affected by the delay of the selection of σ_j. In rotation mode, these σs depend on the value of the angle to rotate. In fact, the set of σs is just a different representation of the angle. Because of this, there have been attempts to obtain the σs directly from the conventional representation of the angle. This is direct for small angles because in this case the $\tan^{-1}\theta$ can be approximated by θ. More precisely, from the Taylor expansion we get

$$\tan^{-1}x = x - \frac{x^3}{3} + \cdots$$

Consequently, for a precision of n bits and $x < 2^{-n/3}$ we have that

$$\tan^{-1}(2^{-i}) \approx 2^{-i}$$

This indicates that after the iteration with $j = \lceil n/3 \rceil$ the σs can be obtained directly from $z[\lceil n/3 \rceil]$. For the initial $n/3$ σs the standard z iteration can be performed. Alternatively, the following approaches have been proposed:

- Obtain the σs from a table, using as address some most-significant bits of $z[0]$.
- Obtain the σs directly as the most-significant bits of $z[0]$. Since an error is produced, include correcting iterations.

Rotation by Predefined Angles

Some applications, such as the computation of transforms (Fourier, cosine, etc.), are performed by a sequence of rotations of predefined angles. In such cases, the representation of the angles by σs can be predetermined and the sequences of σs stored. Therefore, it is not necessary to perform the z iteration, resulting in a reduction of the delay and the area.

However, in this case, since the rotation angles are known, the sine and cosine can be stored and the rotation implemented by multipliers and adders.

11.7.2   Vectoring Followed by Rotation

In some applications, such as normalization, matrix triangularization, and SVD, it is necessary first to compute an angle and then perform rotations by that angle. In such cases, the angle can be computed by a vectoring (plus some other operations, such as additions) and then the angle is used for the rotation. Consequently, the following improvements can be used:

- The angle can be kept in its representation by the σs and used in this form for the rotation. In this way, the z recurrences are avoided, both in the vectoring and in the rotation.
- The rotation can be initiated as soon as the first σ is produced. That is, the vectoring and the rotation can be overlapped, so that the overall delay is essentially equal to that of the rotation.

11.8   Concluding Remarks

The CORDIC algorithm is part of the class of shift-and-add linear convergence algorithms discussed in Chapter 10. In circular coordinates it can be used directly to compute several trigonometric functions, such as cosine, sine, and arctan. However, its greatest potential is to compute directly functions of several variables, such as the rotation of a vector by a specified angle (three variables), the modulus of a vector, and $\arctan(y/x)$ (two variables each). This can compete favorably with other multioperation algorithms for these functions.

The algorithm has been extended to hyperbolic and linear coordinates so that the corresponding unified implementation is very versatile. Moreover, because of its very regular nature, it can be pipelined for high-throughput applications. This is particularly true when using low-latency redundant adders.

Although the basic algorithm is radix 2, it has been extended to higher radices. The most-direct extension is to radix 4, which results in half the number of iterations as for radix 2. However, in the high-radix cases, the scaling factor is not constant, so additional hardware is required to compute the scaling factor. Consequently, an alternative approach is to perform the first iterations radix 2 and the last iterations radix 4, since the scaling factor is affected only by the first iterations.

Very-high-radix implementations have been proposed using selection by rounding (see the references at the end of the chapter). These implementations reduce the number of iterations, but require rectangular multipliers, larger tables, and some special initial iterations.

Many other variations have been proposed, mainly to reduce the delay and versatility of the implementations. Moreover, the implementations have been adapted and combined with other modules for specific applications (see Section 11.10).

The basic CORDIC algorithm and implementations are for fixed-point representations. As for other algorithms, this limits the dynamic range of the operands and result, requiring range reductions and adjustments to reduce loss of precision. To overcome these limitations, algorithms with floating-point representations have been proposed. If the operations, mainly additions, are performed in floating point, the implementation is inefficient; however, it is possible to perform the operations in fixed point and to add preprocessing and postprocessing steps to convert from and to floating point. Additional complications occur if the angle is represented in floating point (see the references at the end of the chapter).

11.9 Exercises

Circular CORDIC: Rotation and Vectoring

11.1 Compute $\sin(30°)$ and $\cos(30°)$ to a precision of seven bits using the CORDIC algorithm.

(a) Utilize a datapath width of 7 fractional bits. Determine the error.

(b) Utilize a datapath width of 10 fractional bits and truncate the final result to 7 fractional bits. Compare the error with that of part (a).

(c) Determine an angle for which the error difference between part (a) and part (b) is large.

11.2   Perform the rotation of the vector $(x_{in}, y_{in}) = (1, 1)$ by an angle $\theta = 2\pi/3$ using the CORDIC algorithm. Perform a range reduction before the CORDIC algorithm. Perform the scale-factor compensation by multiplication.

(a)   Utilize a datapath width of 7 fractional bits. Determine the error.

(b)   Utilize a datapath width of 10 fractional bits and truncate the final result to 7 fractional bits. Compare the error with that of part (a).

(c)   Determine an angle for which the error difference between part (a) and part (b) is large.

11.3   Compute $\tan^{-1}(2.13/3.25)$ and the modulus of the vector $(3.25, 2.13)$, using the CORDIC algorithm. Perform the scale-factor compensation by multiplication.

(a)   Utilize a datapath width of 7 fractional bits. Determine the error.

(b)   Utilize a datapath width of 10 fractional bits and truncate the final result to 7 fractional bits. Compare the error with that of part (a).

(c)   Determine a vector for which the error difference between part (a) and part (b) is large.

Convergence, Precision, Range

11.4   Consider an algorithm that produces a variable A as a sum of the form

$$A = \sum_{i=0}^{\infty} s_i \alpha_i$$

where the α_i are the basis elements and $s_i \in \{0, 1\}$. Show that a recurrent algorithm that produces s_i in iteration i converges if $2\alpha_{i+1} \geq \alpha_i$.

Determine whether the same condition is valid for $s_i \in \{-1, 1\}$.

Apply this condition to the CORDIC algorithm.

11.5   Show the convergence of the CORDIC algorithm for vectoring mode.

11.6   Perform the rotation of $(x_{in}, y_{in}) = (1.0, 1.0)$ by an angle of 2 radians using the CORDIC algorithm directly, without argument range reduction. Does the algorithm converge?

Compensation of Scaling Factor

11.7   Compute the value of $1/K$ obtained by performing the scalings and the repetitions and scaling iterations indicated by Table 11.3.

Implementations

11.8   The CORDIC recurrences for x and y require the use of two shifters. Redefine the recurrences in the vectoring mode so that only one shifter (possibly different from the original one) is used. What are the implications of the modified recurrences on the implementation? Why is this alternative not reasonable for rotation?

11.9   Determine the value of i, for $j \geq i$, such that $|\tan^{-1}(2^{-j}) - 2^{-j}| \leq 2^{-n}$ (consider the Taylor series expansion). Indicate how this can be used to simplify the implementation of the CORDIC algorithm.

Other Functions

11.10   Show that the functions described in Table 11.5 are obtained as indicated. Determine the range of the argument for convergence.

11.11   Compute $0.5e^{0.76}$ using the CORDIC algorithm with a datapath width of eight fractional bits. Perform six iterations and initialize so that no scale-factor compensation is required.

11.12   Compute $\ln(0.17)$ using the CORDIC algorithm with a datapath of eight fractional bits. Perform six iterations.

11.13   Compute $\tan(0.7)$ using two passes through a unified CORDIC unit. Use a datapath of eight bits and perform six iterations in each pass.

Is it possible to use a radix-4 CORDIC module, without need of scale-factor compensation?

Redundant Addition

11.14   Show that to compute a variable scaling factor with an error of 2^{-n} it is only necessary to consider the σs for $j \leq n/2$.

11.15   Using the CORDIC implementation with redundant adder (signed digit) perform the computation of $\sin(\pi/4)$ with 8-bit precision. Use a selection function with an estimate of the sign with two digits and introduce the required repetitions.

Compute the resulting scaling factor and compensate using a multiplication.

11.16   Repeat the previous exercise for vectoring of the vector $(x_{in}, y_{in}) = (1, 1)$.

Radix 4

11.17 Compute $\sin(\pi/4)$ with a radix-4 CORDIC algorithm using an implementation with carry-save adders and the following selection function:

$$
\sigma_0 = \begin{cases}
2 & \text{if } \frac{5}{8} \le \widehat{w}[0] \le \frac{14}{8} \\
1 & \text{if } \frac{3}{8} \le \widehat{w}[0] \le \frac{4}{8} \\
0 & \text{if } -\frac{4}{8} \le \widehat{w}[0] \le \frac{2}{8} \\
-1 & \text{if } -\frac{7}{8} \le \widehat{w}[0] \le -\frac{5}{8} \\
-2 & \text{if } -\frac{15}{8} \le \widehat{w}[0] \le -1
\end{cases}
$$

and for $j > 0$,

$$
\sigma_j = \begin{cases}
2 & \text{if } \frac{12}{8} \le \widehat{w}[j] \le \frac{21}{8} \\
1 & \text{if } \frac{4}{8} \le \widehat{w}[j] \le \frac{11}{8} \\
0 & \text{if } -\frac{4}{8} \le \widehat{w}[j] \le \frac{3}{8} \\
-1 & \text{if } -\frac{12}{8} \le \widehat{w}[j] \le -\frac{5}{8} \\
-2 & \text{if } -\frac{23}{8} \le \widehat{w}[j] \le -\frac{13}{8}
\end{cases}
$$

where $w[j] = 2^j z[j]$ and $\widehat{w}[j]$ is the carry-save $w[j]$ truncated to 3 fractional bits.

Perform three iterations and determine the error. Determine the (variable) scaling factor and perform the compensation by multiplication.

11.10 Further Readings

General

The CORDIC algorithm was first presented (for circular coordinates) in Volder (1959) (see also Volder 2000) and extended to hyperbolic and linear coordinates in Walther (1971) (see also Walther 2000), where there is also a discussion of the conditions for convergence and of the variety of functions that can be performed. Extensions to hyperbolic and linear coordinates are also described in Linhardt and Miller (1969). Delosme (1989) presents a theory and further extensions. Schmid (1974) discusses decimal CORDIC algorithms and implementations.

Scale-Factor Calculation and Compensation

The different techniques for the compensation of the constant scale factor are discussed in Haviland and Tuszinsky (1980), Delosme (1986), Timmermann et al. (1991a), and Villalba et al. (1998a).

Range Extension

Range extension of CORDIC algorithms is discussed in Hu et al. (1991) and Hahn et al. (1994).

σ Prediction in Rotation Mode

To reduce the critical path in rotation mode, several authors have proposed methods to obtain the values of the rotation directions directly from the rotation angle. Since this is not exact, especially for the first iterations, different correction techniques are included (Baker 1976; Naseem 1984; Naseem and Fisher 1985; Timmermann et al. 1992; Hu and Naganathan 1993; Antelo et al. 1995; Wang et al. 1997; Kwak et al. 2000).

Redundant Representation

In the last decade several proposals have dealt with the reduction of the latency by the use of redundant addition. Ercegovac and Lang (1990) propose to use the digit set $\{1, 0, -1\}$, resulting in a variable scaling factor; the computation of this scaling factor as well as the compensation are done online. Two variations for constant scaling factor are proposed for rotation in Takagi et al. (1991) and for vectoring in Lee and Lang (1992). Redundant CORDIC is the subject of Lee (1990). A branching algorithm is presented in Duprat and Muller (1993) and extended in Phatak (1998a, 1998b), and the differential CORDIC algorithm in Dawid and Meyr (1996). CORDIC with carry-save representation is considered in Kunemund et al. (1990).

Online Algorithms

Online arithmetic (see Chapter 9) has been applied to the CORDIC algorithm. In Ercegovac and Lang (1987) it has been used for the computation of cosine/sine, and in Lin and Sips (1990) to perform the general CORDIC algorithm. In Ercegovac and Lang (1990) the calculation of the variable scale factor and its compensation is performed with online modules. An online large radix CORDIC rotator is presented in Osorio et al. (1995).

High-Radix Algorithms

A combined radix-2 and radix-4 implementation, with the first half of the iterations in radix 2 and the second half in radix 4, is used in Lee and Lang (1992); this reduces the number of iterations but results in a constant scaling factor. This method has been generalized to the unified CORDIC in Antelo et al. (1996). Completely radix-4 algorithms are described in Antelo, Bruguera, et al. (1997) and Villalba et al. (1998b). Radix 2^k algorithms for some elementary functions are proposed in Baker (1975). Recent work on higher-radix CORDIC includes Antelo et al. (2000a, 2000b) and Lewis (1999).

Error Analysis

Error analysis of the CORDIC algorithm when implemented with finite-width datapaths is presented in Hu (1992b) and Hu and Bass (1993). In Antelo, Bruguera, et al. (1997), error in the argument is included and approaches are proposed to reduce the overall error. Bekooij et al. (2000) deal with numerical accuracy of Fast Fourier Transforms with CORDIC. The quantization effects are studied in Hu (1992c). Numerical accuracy and hardware trade-offs are discussed in Kota and Cavallaro (1993).

Multidimensional CORDIC

Multidimensional CORDIC has been presented in Delosme (1989). It is also investigated in Hsiao (1993). Its use is suggested for zeroing out several components of a vector (multidimensional vectoring) and applying the corresponding σs to rotate other vectors. Alternative multidimensional algorithms are presented in Delosme and Hsiao (1990), and Hsiao and Delosme (1995), and its application to complex SVD in Hsiao et al. (2000). The use of the 3D CORDIC as part of an algorithm for 3D rotation of rigid bodies is presented in Lang and Antelo (2001).

Floating-Point CORDIC

Floating-point applications in which the vector is in floating-point representation, but the CORDIC algorithm is in fixed-point representation, with preprocessing and postprocessing stages, are discussed in Ercegovac and Lang (1990), de Lange and Deprettere (1991), and Timmermann et al. (1994). This is extended in Hekstra and Deprettere (1993) to the case in which the angle is also in floating-point representation. A floating-point vectoring is discussed in der Kolk et al. (2000). Cavallaro and Luk (1988a, 1988b) present floating-point CORDIC for matrix computations. Floating-point CORDIC implementations are described in Ahmed (1982), Metafas and Goutis (1995), and Hekstra (1998).

Extensions

The CORDIC algorithm has been modified so that other functions can be performed. In particular for $\cos^{-1}$ and $\sin^{-1}$, see Mazenc et al. (1993), Krieger and Hosticka (1996), and Lang and Antelo (2000). This latter has been generalized to perform vectoring with an arbitrary target in Lang and Antelo (1998). An extension to perform CORDIC for interval arithmetic is presented in Hormigo et al. (1999).

Applications and Implementations

Applications and implementations of CORDIC have been frequently described in the literature. We present here some of that work. CORDIC applications fall roughly into signal processing, graphics, and robotics areas. Common to all are the needs for fast and efficient evaluation of transcendental functions, matrix computations, and operations on vectors and angles. Implementations are typically application-specific, although there are several general CORDIC processors, such as Haviland and Tuszinsky (1980), König and Böhme (1990), an Deprettere et al. (1990). Signal processing applications and implementations CORDIC are surveyed in Hu (1992a), which provides an extensive bibliogra up to 1992. Cavallaro and Luk (1988a, 1988b), Hu et al. (1993), and Hemk and Cavallaro (1994) present a CORDIC approach to matrix compu' Zou and Kornerup (1995) and Bruguera et al. (1996) deal with varior forms. There are numerous discussions of various implementations of ' algorithms for digital signal processing applications (Despain 1974; Ahmed et al. 1982; Deprettere et al. 1984; Wald and Despain 1984; Sung et al. 1986; Timmermann et al. 1991b; Hekstra 2000). The literature contains many presentations of various aspects of the design and implementation of CORDIC processors (Ahmed 1982; Cavallaro 1988; Harber 1989; Hu 1989; Wang 1998; Hekstra 1998; Kwak 2000). A bit-serial floating-point CORDIC processor is described in Bass et al. (1991). A radix-4 pipelined CORDIC processor is presented in Bruguera et al. (1993). Mencer et al. (2000) discuss implementation of CORDIC with reconfigurable arrays. Numerical accuracy and hardware trade-offs are discussed in Kota and Cavallaro (1993). Graphics and robotics applications are the subject of Yang et al. (1987), Yoshimura et al. (1989), Krieger and Hosticka (1996), and Lang and Antelo (2001). CORDIC approach to decimal-binary conversion is presented in Daggett (1959).

11.11 Bibliography

Ahmed, H. M. (1982). *Signal Processing Algorithms and Architectures*. PhD thesis, Stanford University.

Ahmed, H. M., J.-M. Delosme, and M. Morf (1982). Highly concurrent computing structures for matrix arithmetic and signal processing. *Computer*, 15(1):65–82.

Antelo, E., J. D. Bruguera, T. Lang, and E. L. Zapata (1997). Error analysis and reduction for angle calculation using the CORDIC algorithm. *IEEE Transactions on Computers*, 46(11):1264–71.

Antelo, E., J. D. Bruguera, J. Villalba, and E. L. Zapata (1995). Redundant CORDIC rotator based on parallel prediction. In *Proceedings of the 12th IEEE Symposium on Computer Arithmetic*, pages 172–79.

Antelo, E., J. Bruguera, and E. Zapata (1996). Unified mixed radix-2-4 redundant CORDIC processor. *IEEE Transactions on Computers*, 45(9):1086–73.

Antelo, E., T. Lang, and J. D. Bruguera (2000a). Very-high radix circular CORDIC: vectoring and unified rotation/vectoring. *IEEE Transactions on Computers*, 49(7):727–39.

Antelo, E., T. Lang, and J. D. Bruguera (2000b). Very-high radix CORDIC rotation based on selection by rounding. *Journal of VLSI Signal Processing Systems*, 25(2):141–54.

Antelo, E., J. Villalba, J. D. Bruguera, and E. L. Zapata (1997). High-performance rotation architectures based on the radix-4 CORDIC algorithm. *IEEE Transactions on Computers*, 46(8):855–70.

Baker, P. W. (1975). Parallel multiplicative algorithms for some elementary functions. *IEEE Transactions on Computers*, C-24(3):321–24.

Baker, P. W. (1976). Suggestion for a fast binary sine/cosine generator. *IEEE Transactions on Computers*, C-25(11):1134–36.

Bass, S. C., G. M. Butler, R. L. Williams, F. Barlos, and D. R. Miller (1991). A bit-serial, floating point CORDIC processor in VLSI. In *Proceedings of ICASSP-91*, pages 1165–68.

Bekooij, M., J. Huisken, and K. Nowak (2000). Numerical accuracy of fast fourier transforms with CORDIC arithmetic. *Journal of VLSI Signal Processing Systems*, 25(2):187–93.

Bruguera, J. D., E. Antelo, and E. L. Zapata (1993). Design of a pipelined radix-4 cordic processor. *Journal of Parallel Computing*, 19(7):729–44.

Bruguera, J. D., N. Guil, T. Lang, J. Villalba, and E. Zapata (1996). CORDIC based parallel/pipelined architecture for the Hough transform. *Journal of VLSI Signal Processing*, pages 207–21.

Cavallaro, J. R. (1988). *VLSI CORDIC Processor Architectures for the Singular Value Decomposition*. PhD thesis, Cornell University.

Cavallaro, J. R., and F. T. Luk (1988a). CORDIC arithmetic for an SVD processor. In *Proceedings of the 8th IEEE Symposium on Computer Arithmetic*, pages 113–20.

Cavallaro, J. R., and F. T. Luk (1988b). Floating-point CORDIC for matrix computations. In *Proceedings of the 1988 IEEE International Conference on Computer Design*, pages 40–42.

Daggett, D. H. (1959). Decimal-binary conversion in CORDIC. *IRE Transactions on Electronic Computers*, EC-8(3):335–39.

Dawid, H., and H. Meyr (1996). The differential CORDIC algorithm: Constant scale factor redundant implementation without correcting iterations. *IEEE Transactions on Computers*, 45(3):307–18.

de Lange, A. A., and E. F. A. Deprettere (1991). Design and implementation of a floating-point quasi systolic general purpose CORDIC rotator for high-rate parallel data and signal processing. In *Proceedings of the 10th IEEE Symposium on Computer Arithmetic*, pages 272–81.

Delosme, J.-M. (1986). The matrix exponential approach to elementary operations. In *Proceedings of the SPIE—Advanced Algorithms and Architectures for Signal Processing*, volume 696, pages 188–95.

Delosme, J.-M. (1989). CORDIC algorithms: theory and extensions. In *SPIE Advanced Algorithms and Architectures for Signal Processing IV*, volume 1152, pages 131–45.

Delosme, J.-M., and S. H. Hsiao (1990). CORDIC algorithms in four dimensions. In *SPIE Advanced Signal-Processing Algorithms, Architectures and Implementations*, volume 1348, pages 349–60.

Deprettere, E. F. A., A. de Lange, and P. Dewilde (1990). The synthesis and implementation of signal processing applications specific VLSI CORDIC arrays. In *Proceedings of ISCAS'90*, pages 974–77.

Deprettere, E. F. A., P. Dewilde, and R. Udo (1984). Pipelined CORDIC architectures for fast VLSI filtering and array processing. In *Proceedings of ICASSP '84*, pages 41.A.6.1–41.A.6.4.

der Kolk, K. J. V., J. A. Lee, and E. F. A. Deprettere (2000). A floating-point vectoring algorithm based on fast rotations. *Journal of VLSI Signal Processing Systems*, 25(2):125–40.

Despain, A. M. (1974). Fourier transform computers using CORDIC iterations. *IEEE Transactions on Computers*, C-23(10):993–1001.

Duprat, J., and J.-M. Muller (1993). The CORDIC algorithm: New results for fast VLSI implementation. *IEEE Transactions on Computers*, 42(2):168–78.

Ercegovac, M. D., and T. Lang (1987). Fast cosine/sine implementation using on-line CORDIC. In *Proceedings of the 21st Asilomar Conference Signals, Systems, Computers*, pages 222–26.

Ercegovac, M. D., and T. Lang (1990). Redundant and on-line CORDIC: Application to matrix triangularization and SVD. *IEEE Transactions on Computers*, 39(6):725–40.

Hahn, H., D. Timmermann, B. J. Hosticka, and B. Rix (1994). A unified and division-free CORDIC argument reduction method with unlimited convergence domain including inverse hyperbolic functions. *IEEE Transactions on Computers*, 43(11):1339–44.

Harber, R. G. (1989). *VLSI Design of Systems of CORDIC Processors*. PhD thesis, Purdue University.

Haviland, G. H., and A. A. Tuszinsky (1980). A CORDIC arithmetic processor chip. *IEEE Transactions on Computers*, C-29(2):68–79.

Hekstra, G. J. (1998). *CORDIC for High-Performance Numerical Computation*. PhD thesis, Technical University of Delft, The Netherlands.

Hekstra, G. J. (2000). Evaluation of fast rotation methods. *Journal of VLSI Signal Processing Systems*, 25(2):113–24.

Hekstra, G. J., and E. F. A. Deprettere (1993). Floating-point CORDIC. In *Proceedings of the 11th IEEE Symposium on Computer Arithmetic*, pages 130–37.

Hemkumar, N. D., and J. R. Cavallaro (1994). Redundant and on-line CORDIC for unitary transformations. *IEEE Transactions on Computers*, 43(8):941–54.

Hormigo, J., J. Villalba, and E. L. Zapata (1999). Interval sine and cosine functions computation based on Variable-precision CORDIC algorithm. In *Proceedings of the 14th IEEE Symposium on Computer Arithmetic*, pages 186–93.

Hsiao, S.-F. (1993). *Multi-Dimensional CORDIC Algorithms*. PhD thesis, Yale University.

Hsiao, S.-F., and J.-M. Delosme (1995). Householder CORDIC algorithms. *IEEE Transactions on Computers*, 44(8):990–1000.

Hsiao, S.-F., C.-Y. Lau, and J.-M. Delosme (2000). Redundant constant-factor implementation of multi-dimensional CORDIC and its application to complex SVD. *Journal of VLSI Signal Processing Systems*, 25(2):155–66.

Hu, X. (1989). *A Silicon Compiler for Dedicated Mathematical Systems Based on CORDIC Arithmetic Processors*. PhD thesis, Purdue University.

Hu, X., and S. C. Bass (1993). A neglected error source in the CORDIC algorithm. In *Proceedings of ISCAS'93*, volume 1, pages 766–69.

Hu, X., S. C. Bass, and R. G. Harber (1993). An efficient implementation of singular value decomposition rotation transformations with CORDIC processors. *Journal of Parallel and Distributed Computing*, 17:360–62.

Hu, X., G. Harber, and S. Bass (1991). Expanding the range of the cordic algorithm. *IEEE Transactions on Computers*, 40(1):13–21.

Hu, Y. H. (1992a). CORDIC-based VLSI architectures for digital signal processing. *IEEE Signal Processing Magazine*, pages 16–35.

Hu, Y. H. (1992b). The quantization effects of the CORDIC algorithm. *IEEE Transactions on Signal Processing*, 40(4):834–44.

Hu, Y. H. (1992c). The quantization effects of the CORDIC algorithm. *IEEE Transactions on Signal Processing*, 40(4):834–45.

Hu, Y. H., and S. Naganathan (1993). An angle recoding method for CORDIC algorithm implementation. *IEEE Transactions on Computers*, 42(1):99–102.

König, D., and J. F. Böhme (1990). Optimizing the CORDIC algorithm for processors with pipeline architectures. In *Signal Processing V: Theories and Applications*. Elsevier Science, Amsterdam, The Netherlands.

Kota, K., and J. R. Cavallaro (1993). Numerical accuracy and hardware tradeoffs for CORDIC arithmetic for special-purpose processors. *IEEE Transactions on Computers*, 42(7):769–79.

Krieger, C., and B. Hosticka (1996). Inverse kinematics computations with modified CORDIC iterations. *IEE Proceedings Computer Digital Techniques*, 143(1):87–92.

Kunemund, R., H. Soldner, S. Wohlleben, and T. Noll (1990). CORDIC processor with carry-save architecture. In *Proceedings of the 16th European Solid-State Circuit Conference*, pages 193–96.

Kwak, J.-H. (2000). *High Speed CORDIC Processor Design: Algorithms, Architectures, and Applications*. PhD thesis, University of Texas at Austin.

Kwak, J.-H., J. H. Choi, and E. E. Swartzlander (2000). High-speed CORDIC based on an overlapped architecture and a novel σ-prediction method. *Journal of VLSI Signal Processing Systems*, 25(2):167–78.

Lang, T., and E. Antelo (1998). CORDIC vectoring with arbitrary target value. *IEEE Transactions on Computers*, 47(7):736–49.

Lang, T., and E. Antelo (2000). CORDIC-based computation of ArcCos and $\sqrt{1 - t^2}$. *Journal of VLSI Signal Processing*, 25:19–38.

Lang, T., and E. Antelo (2001). High-throughput 3D rotations and normalizations. In *Conference Record of the 35th Asilomar Conference on Signals, Systems and Computers*, volume 1, pages 846–51.

Lee, J.-A. (1990). *Redundant CORDIC: Theory and Its Application to Matrix Computations*. PhD thesis, University of California, Los Angeles.

Lee, J. A., and T. Lang (1992). Constant-factor redundant CORDIC for angle calculation and rotation. *IEEE Transactions on Computers*, 41(8):1016–25.

Lewis, D. (1999). Complex logarithmic number system arithmetic using high-radix redundant CORDIC algorithms. In *Proceedings of the 14th IEEE Symposium on Computer Arithmetic*, pages 194–203.

Lin, H., and H. J. Sips (1990). On-line CORDIC algorithms. *IEEE Transactions on Computers*, 39(8):1038–52.

Linhardt, R. J., and H. S. Miller (1969). Digit-by-digit transcendental function computation. *RCA Review*, 30:209–47.

Mazenc, C., X. Merrheim, and J.-M. Muller (1993). Computing functions $\cos^{-1}$ and $\sin^{-1}$ using CORDIC. *IEEE Transactions on Computers*, 42(1):118–22.

Mencer, O., L. Semeria, M. Morf, and J.-M. Delosme (2000). Application of reconfigurable CORDIC architectures. *Journal of VLSI Signal Processing*, 24(2-3):211–21.

Metafas, D. E., and C. E. Goutis (1995). A floating-point advanced CORDIC processor. *Journal of VLSI Signal Processing*, 10:53–65.

Naseem, A. (1984). *Implementation of Parallel Computational Algorithms on a Modified CORDIC Arithmetic Logic Unit*. PhD thesis, Michigan State University.

Naseem, A., and P. D. Fisher (1985). The modified CORDIC algorithm. In *Proceedings of the 7th IEEE Symposium on Computer Arithmetic*, pages 144–52.

Osorio, R. R., E. Antelo, J. D. Bruguera, J. Villalba, and E. L. Zapata, (1995). Digit on-line large radix CORDIC rotator. In *Proceedings of ASAP-95 (Strasbourg, France)*, pages 246–57.

Phatak, D. S. (1998a). Comments on Duprat and Muller's branching CORDIC paper. *IEEE Transactions on Computers*, 47(9):1037–40.

Phatak, D. S. (1998b). Double step branching CORDIC: a new algorithm for fast sine and cosine generation. *IEEE Transactions on Computers*, 47(5):587–602.

Schmid, H. (1974). *Decimal Computation*. John Wiley & Sons, New York.

Sung, T. Y., Y. H. Hu, and H. J. Yu (1986). Doubly pipelined CORDIC array for digital signal processing. In *IEEE International Conference on ASSP*, pages 1169–72.

Takagi, N., T. Asada, and S. Yajima (1991). Redundant CORDIC methods with a constant scale factor. *IEEE Transactions on Computers*, 40(9):989–95.

Timmermann, D., H. Hahn, and B. J. Hosticka (1992). Low latency time CORDIC algorithms. *IEEE Transactions on Computers*, 41(8):1010–15.

Timmermann, D., H. Hahn, B. J. Hosticka, and B. Rix (1991a). A new addition scheme and fast scaling factor compensation methods for CORDIC algorithms. *INTEGRATION, The VLSI Journal*, 11:85–100.

Timmermann, D., H. Hahn, B. J. Hosticka and G. Schmidt (1991b). A programmable CORDIC chip for digital signal processing applications. *IEEE Journal of Solid-State Circuits*, 26(9):1317–21.

Timmermann, D., B. Rix, H. Hahn, and B. J. Hosticka (1994). A CMOS floating-point vector-arithmetic unit. *IEEE Journal of Solid-State Circuits*, 29(5):634–39.

Villalba, J., T. Lang, and E. L. Zapata (1998a). Parallel compensation of scale factor for the CORDIC algorithm. *Journal of VLSI Signal Processing*, 19:227–41.

Villalba, J., E. L. Zapata, E. Antelo., and J. D. Bruguera (1998b). Radix-4 vectoring CORDIC. *Journal of VLSI Signal Processing*, 19:127–47.

Volder, J. (1959). The CORDIC computing techniquie. *IRE Transactions on Electronic Computers*, EC-8(3):330–34.

Volder, J. E. (2000). The birth of CORDIC. *Journal of VLSI Signal Processing Systems*, 25(2):101–5.

Walther, J. S. (1971). A unified algorithm for elementary functions. In *Proceedings of the Spring Joint Computer Conference*, pages 379–85.

Walther, J. S. (2000). The story of unified CORDIC. *Journal of VLSI Signal Processing Systems*, 25(2):107–12.

Wang, S. (1998). *A CORDIC Arithmetic Processor*. PhD thesis, University of Texas at Austin.

Wang, S., V. Piuri, and E. E. Swartzlander (1997). Hybrid CORDIC algorithms. *IEEE Transactions on Computers*, 46(11):1202–7.

Wold, E. H. and A. M. Despain (1984). Pipeline and parallel-pipeline FFT processors for VLSI implementations. *IEEE Transactions on Computers*, C-33(5):414–26.

Yang, B., D. Timmermann, J. F. Bome, H. Hahn, B. J. Hosticka, G. Schmidt, and G. Zimmer (1987). Special computers: graphics, robotics. In *Proceedings of VLSI Computer, COMPEURO*, pages 727–30.

Yoshimura, H., T. Nakanishi, and H. Yamauchi (1989). A 50-MHz CMOS geometrical mapping processor. *IEEE Transactions on Circuits and Systems*, 36(10):1360–63.

Zou, F., and P. Kornerup (1995). High speed DCT/IDCT using a pipelined CORDIC algorithm. In *Proceedings of the 12th IEEE Symposium on Computer Arithmetic*, pages 180–87.

Bibliography

Abu-Khater, I. S., A. Bellaouar, and M. I. Elmasry (1996). Circuit techniques for CMOS low-power high-performance multipliers. *IEEE Journal of Solid-State Circuits*, 31(10):1535–46.

Agarwal, R. C., J. C. Cooley, F. G. Gustavson, J. B. Shearer, G. Slishman, and B. Tuckerman (1986). New scalar and vector elementary functions for the IBM system/370. *IBM Journal of Research and Development*, 30(2):126–44.

Agarwal, R. C., F. G. Gustavson, and M. S. Schmookler (1999). Series approximation methods for divide and square root in the Power3 Microprocessor. In *Proceedings of the 14th IEEE Symposium on Computer Arithmetic*, pages 116–23.

Agrawal, D. P. (1979). High-speed arithmetic arrays. *IEEE Transactions on Computers*, C-28(3):215–24.

Agrawal, D. P., and T. R. N. Rao (1978). On multiple operand addition of signed binary numbers. *IEEE Transactions on Computers*, C-27(11):1068–70.

Ahmed, H. M. (1982). *Signal Processing Algorithms and Architectures*. PhD thesis, Stanford University.

Ahmed, H. M., J.-M. Delosme, and M. Morf (1982). Highly concurrent computing structures for matrix arithmetic and signal processing. *Computer*, 15(1):65–82.

Al-Twaijry, H. A. (1997). *Area and Performance Optimized CMOS Multipliers*. PhD thesis, Stanford University.

Anderson, S. F., J. G. Earle, R. E. Goldschmidt, and D. M. Powers (1967). The IBM 360/370 model 91: floating-point execution unit. *IBM Journal of Research and Development*, pages 34–53.

ANSI and IEEE (1985). IEEE standard for binary floating-point arithmetic. *ANSI/IEEE Standard, Std* 754-1985, New York.

Antelo, E., J. D. Bruguera, T. Lang, and E. L. Zapata (1997). Error analysis and reduction for angle calculation using the CORDIC algorithm. *IEEE Transactions on Computers*, 46(11):1264–71.

Antelo, E., J. D. Bruguera, J. Villalba, and E. L. Zapata (1995). Redundant CORDIC rotator based on parallel prediction. In *Proceedings of the 12th IEEE Symposium on Computer Arithmetic*, pages 172–79.

Antelo, E., J. Bruguera, and E. Zapata (1996). Unified mixed radix-2-4 redundant CORDIC processor. *IEEE Transactions on Computers*, 45(9):1086–73.

Antelo, E., T. Lang, and J. D. Bruguera (2000). Very-high radix circular CORDIC: vectoring and unified rotation/vectoring. *IEEE Transactions on Computers*, 49(7):727–39.

Antelo, E., T. Lang, and J. D. Bruguera (2000). Very-high radix CORDIC rotation based on selection by rounding. *Journal of VLSI Signal Processing Systems*, 25(2):141–54.

Antelo, E., T. Lang, and J. D. Bruguera (1998). Computation of $\sqrt{x/d}$ in a very high radix combined division/square-root unit with scaling. *IEEE Transactions on Computers*, 47(2):152–61.

Antelo, E., J. Villalba, J. D. Bruguera, and E. L. Zapata (1997). High-performance rotation architectures based on the radix-4 CORDIC algorithm. *IEEE Transactions on Computers*, 46(8):855–70.

Atkins, D. E. (1968). Higher-radix division using estimates of the divisor and partial remainders. *IEEE Transactions on Computers*, C-17(10):925–34.

Atkins, D. E. (1970). Design of the arithmetic units of ILLIAC III: Use of redundancy and higher radix methods. *IEEE Transactions on Computers*, C-19(8):720–33.

Atkins, D. E. (1970). *A Study of Methods for Selection of Quotient Digits during Digital Division*. PhD thesis, Department of Computer Science, University of Illinois at Urbana-Champaign. Technical report UIUCDCS-R-397.

Atkins, D. E. (1970). Design of the arithmetic units of ILLIAC III: Use of redundancy and higher radix methods. *IEEE Transactions on Computers*, C-19(8):720–33.

Atkins, D. E. (1975). Higher radix, non-restoring division: History and recent developments. In *Proceedings of the 3rd IEEE Symposium on Computer Arithmetic*, pages 158–60.

Atkins, D. E., and S. Ong (1979). Time-component complexity of two approaches to multioperand binary addition. *IEEE Transactions on Computers*, C-28(12):918–26.

Avizienis, A. (1960). *A Study of Redundant Number Representations for Parallel Digital Computers*. PhD thesis, University of Illinois, Urbana.

Avizienis, A. (1961). Signed digit number representations for fast parallel arithmetic. *IRE Transactions on Electronic Computers*, EC-10(9):389–400.

Avizienis, A. (1962). On flexible implementation of digital computer arithmetic. In *Proc. IFIP Congress*, pages 664–70.

Avizienis, A. (1964). Binary-compatible signed-digit arithmetic. In *Proc. Fall Joint Computer Conference*, pages 663–72.

Avizienis, A. (1966). Arithmetic microsystems for the synthesis of function generators. *Proceedings of the IEEE*, 54(12):1910–19.

Avizienis, A. (1971). Digital computer arithmetic: A unified algorithmic specification. In *Proceedings of the Symposium on Computers and Automata*, pages 509–25, April 13–15.

Bajard, J.-C., J. Duprat, S. Kla, and J.-M. Muller (1994). Some operators for on-line radix 2 computations. *Journal of Parallel and Distributed Computing*, 22(2):336–45.

Bajard, J.-C., S. Kla, and J.-M. Muller (1994). BKM: A new hardware algorithm for complex elementary functions. *IEEE Transactions on Computers*, 43(8):955–63.

Baker, P. W. (1973). Predictive algorithms for some elementary functions in radix 2. *Electronics Letters*, 9(21):493–94.

Baker, P. W. (1975). Parallel multiplicative algorithms for some elementary functions. *IEEE Transactions on Computers*, C-24(3):321–24.

Baker, P. W. (1976). Suggestion for a fast binary sine/cosine generator. *IEEE Transactions on Computers*, C-25(11):1134–36.

Bannon, P., and J. Keller (1995). Internal architecture of Alpha 21164 microprocessor. In *Digest of Papers COMPCON '95*, pages 79–87.

Bass, S. C., G. M. Butler, R. L. Williams, F. Barlos, and D. R. Miller (1991). A bit-serial, floating point CORDIC processor in VLSI. In *Proceedings of ICASSP-91*, pages 1165–68.

Baugh, C. R., and B. A. Wooley (1973). A two's complement parallel array multiplication algorithm. *IEEE Transactions on Computers*, C-22:1045–47.

Bayoumi, M. A., G. A. Jullien, and W. C. Miller (1983). An area-time efficient NMOS adder. *Integration*, 1:317–34.

Beame, P., S. Cook, and H. Hoover (1986). Log depth circuits for division and related problems. *SIAM Journal on Computing*, 15:994–1003.

Beaumont-Smith, A., N. Burgess, D. Lefrere, and C. C. Lim (1999). Reduced latency IEEE floating-point standard adder architecture. In *Proceedings of the 14th IEEE Symposium on Computer Arithmetic*, pages 35–42.

Bedrij, O. J. (1962). Carry-select adder. *IRE Transactions on Electronic Computers*, EC-11(6):340–46.

Bekooij, M., J. Huisken, and K. Nowak (2000). Numerical accuracy of fast fourier transforms with CORDIC arithmetic. *Journal of VLSI Signal Processing Systems*, 25(2):187–93.

Benschneider, B. J., W. J. Bowhill, E. M. Copper, M. N. Gavrielov, P. E. Gronowski, V. K. Maheshwari, V. Peng, J. D. Pickholtz, and S. Samudrala (1989). A pipelined 50 MHz CMOS 64-bit floating-point arithmetic processor. *IEEE Journal of Solid-State Circuits*, SC-24(5):1317–23.

Bewick, G. W. (1994). *Fast Multiplication: Algorithms and Implementation*. PhD thesis, Stanford University.

Bickerstaff, K. C., E. E. Swartzlander, and M. J. Schulte (2001). Analysis of column compression multipliers. In *Proceedings of the 15th IEEE Symposium on Computer Arithmetic*, pages 33–39.

Bohlender, G., P. Kornerup, D. W. Matula, and W. Walter (1991). Semantics for exact floating-point operations. In *Proceedings of the 10th IEEE Symposium on Computer Arithmetic*, pages 22–26.

Booth, A. D. (1951). A signed binary multiplication technique. *Quarterly Journal of Mechanics and Applied Mathematics*, 4(2):236–40.

Borovec, R. T. (1968). The logical design of a class of limited carry-borrow propagation adders. Technical report no. 275, Dept. of Computer Science, University of Illinois.

Bose, B. K., L. Pei, G. S. Taylor, and D. A. Paterson (1987). Fast multiply and divide for a VLSI floating-point unit. In *Proceedings of the 8th IEEE Symposium on Computer Arithmetic*, pages 87–94.

Brackert, R. H. (1988). *Design and Implementation of Recursive Filters Using On-line Arithmetic*. PhD thesis, University of California, Los Angeles.

Brackert, R. H., M. D. Ercegovac, and A. N. Willson (1989). Design of an on-line multiply-add module for recursive digital filters. In *Proceedings of the 9th IEEE Symposium on Computer Arithmetic*, pages 34–41.

Braun, E. L. (1963). *Digital Computer Design*. Academic Press, New York.

Brent, R. P. (1970). On the addition of binary numbers. *IEEE Transactions on Computers*, C-19(8):758–59.

Brent, R. P. (1973). On the precision attainable with various floating point number systems. *IEEE Transactions on Computers*, C-22(6):601–7.

Brent, R. P., and H. T. Kung (1981). The area-time complexity of binary multiplication. *Journal of the ACM*, 28(3).

Brent, R. P., and H. T. Kung (1982). A regular layout for parallel adders. *IEEE Transactions on Computers*, C-31(3):260–64.

Briggs, W. S., and D. W. Matula (1993). A 17 × 69 bit multiply and add unit with redundant binary feedback and single cycle latency. In *Proceedings of the 11th IEEE Symposium on Computer Arithmetic*, pages 163–71.

Brown, W. S., and P. L. Richman (1969). The choice of base. *Communications of the ACM*, 12(10):560–61.

Bruguera, J., and T. Lang (1995). 2-D DCT using online arithmetic. In *International Conference on Acoustics, Speech, and Signal Processing*, volume 5, pages 3275–78.

Bruguera, J. D., E. Antelo, and E. L. Zapata (1993). Design of a pipelined radix-4 cordic processor. *Journal of Parallel Computing*, 19(7):729–44.

Bruguera, J. D., N. Guil, T. Lang, J. Villalba, and E. Zapata (1996). CORDIC based parallel/pipelined architecture for the Hough transform. *Journal of VLSI Signal Processing*, pages 207–21.

Bruguera, J. D., and T. Lang (1999). Leading-one prediction with concurrent position correction. *IEEE Transactions on Computers*, 48(10):1083–97.

Bruguera, J. D., and T. Lang (2000). Multilevel reverse-carry adder. In *Proceedings of the IEEE International Conference on Computer Design: VLSI in Computers and Processors (ICCD'00)*, pages 155–62.

Bruguera, J. D., and T. Lang (2001). Using the reverse-carry approach for double datapath floating-point addition. In *Proceedings of the 15th IEEE Symposium on Computer Arithmetic*, pages 203–10.

Bryant, R. (1996). Bit-level analysis of an SRT divider circuit. In *Proceedings of the 33rd Design Automation Conference*, pages 661–65.

Bucholz, W. (1962). *Planning a New Computer System: Project STRETCH*, Chapter 14, p. 210. Wiley and Sons, Inc., New York.

Burgess, N. (1991). Radix-2 SRT division with simple quotient digit selection. *Electronics Letters*, 27(21):1910–11.

Burgess, N. (1994). Prescaled maximally-redundant radix-4 SRT divider. *Electronics Letters*, 30(23):1926–28.

Burgess, N. (2001). Accelerated carry-skip adders with low hardware cost. In *Proceedings of the 35th Asilomar Conference on Signals, Systems and Computers*, pages 852–56.

Burgess, N., and S. Knowles (1999). Efficient implementation of rounding units. In *Conference Record of the 33rd Asilomar Conference on Signals, Systems, and Computers*, volume 2, pages 1489–93.

Burgess, N., and T. Williams (1995). Choices of operand truncation in the SRT division algorithm. *IEEE Transactions on Computers*, 44(7):933–38.

Burks, A., H. H. Goldstine, and J. Von Neumann (1946). Preliminary discussion of the logic design of an electronic computing instrument. Technical report, Institute for Advanced Study, Princeton. Reprinted in C. G. Bell, *Computer Structures, Readings and Examples*, McGraw-Hill, New York, 1971.

Burleson, W. P. (1990). Polynomial evaluation in VLSI using distributed arithmetic. *IEEE Transaction on Circuits and Systems*, 37(10):1299–304.

Bushard, L. B. (1983). A minimum table size result for higher radix nonrestoring division. *IEEE Transactions on Computers*, C-32(6):521–26.

Callaway, T. (1996). *Area, Delay, and Power Modeling of CMOS Adders and Multipliers*. PhD thesis, The University of Texas at Austin.

Callaway, T. K., and E. E. Swartzlander (1997). Power-delay characteristics of CMOS multipliers. In *Proceedings of the 13th IEEE Symposium on Computer Arithmetic*, pages 26–32.

Cao, J., B. W. Y. Wei, and J. Cheng (2001). High-performance architectures for elementary function generation. In *Proceedings of the 15th IEEE Symposium on Computer Arithmetic*, pages 136–44.

Cappa, M., and V. C. Hamacher (1973). An augmented iterative array for high-speed binary division. *IEEE Transactions on Computers*, C-22(2):172–75.

Cappello, P. R., and K. Steiglitz (1983). A VLSI layout for a pipelined Dadda multiplier. *ACM Transactions on Computer Systems*, 1(2):157–74.

Cappuccino, G., G. Cocorullo, P. Corsonello, and S. Perri (1999). High speed self-timed pipelined datapath for square rooting. *IEE Proceedings—Circuits, Devices and Systems*, 146(1):16–22.

Cappuccino, G., P. Corsonello, and G. Cocorullo (1998). High performance VLSI modules for division and square root. *Microprocessors and Microsystems*, 22(5):239–46.

Carter, T. M., and J. E. Robertson (1990). The set theory of arithmetic decomposition. *IEEE Transactions on Computers*, C-39(8):993–1005.

Carter, T. M., and J. E. Robertson (1990). Radix-16 signed-digit division. *IEEE Transactions on Computers*, C-39(12):1424–33.

Cavallaro, J. R. (1988). *VLSI CORDIC Processor Architectures for the Singular Value Decomposition*. PhD thesis, Cornell University.

Cavallaro, J. R., and F. T. Luk (1988). CORDIC arithmetic for an SVD processor. In *Proceedings of the 8th IEEE Symposium on Computer Arithmetic*, pages 113–20.

Cavallaro, J. R., and F. T. Luk (1988). Floating-point CORDIC for matrix computations. In *Proceedings of the 1988 IEEE International Conference on Computer Design*, pages 40–42.

Cavanagh, J. (1984). *Digital Computer Arithmetic*. McGraw-Hill, New York.

Chan, P. K., and M. D. F. Schlag (1990). Analysis and design of CMOS Manchester adder with variable carry-skip. *IEEE Transactions on Computers*, C-39(8): 983–92.

Chan, P. K., M. D. Schlag, C. D. Thomborson, and V. G. Oklobdzija (1992). Delay optimization of carry-skip adders and block carry-lookahead adders using multidimensional dynamic programming. *IEEE Transactions on Computers*, 41(8):920–30.

Chen, C., L.-A. Chen, and J.-R. Cheng (2001). Architectural design of a fast floating-point multiplication-add fused unit using signed-digit addition. In *Proceedings Euromicro Symposium on Digital Systems Design*, pages 346–53.

Chen, I. N., and R. Willoner (1979). An o(n) parallel multiplier with bit-sequential input and output. *IEEE Transactions on Computers*, C-28(10):721–27.

Chen, T. C. (1971). A binary multiplication scheme based on squaring. *IEEE Transactions on Computers*, C-20:678–80.

Chen, T. C. (1972). Automatic computation of exponentials, logarithms, ratios, and square roots. *IBM Journal of Research and Development*, pages 380–89.

Chen, Y.-A., E. Clarke, P.-H. Ho, Y. Hoskote, T. Kam, M. Khaira, J. O'Leary, and X. Zhao (1996). Verification of all circuits in a floating-point unit using word-level model checking. In *Proceedings of International Conference on Formal Methods in Computer-Aided Design (FMCAD '96)*, pages 19–33.

Cheney, E. W. (1966). *Introduction to Approximation Theory*. International Series in Pure and Applied Mathematics. McGraw Hill, New York.

Cheng, F.-C., S. H. Unger, and M. Theobald (2000). Self-timed carry-lookahead adders. *IEEE Transactions on Computers*, 49(7):659–72.

Cherkauer, B. S., and E. G. Friedman (1997). A hybrid radix-4/radix-8 low power signed multiplier architecture. *IEEE Transactions on Circuits and Systems—II: Analog and Digital Signal Processing*, 44(8):656–59.

Chow, C. Y., and J. E. Robertson (1978). Logical design of a redundant binary adder. In *Proceedings of the 4th IEEE Symposium on Computer Arithmetic*, pages 109–15.

Ciminiera, L., and P. Montuschi (1990). Higher radix square rooting. *IEEE Transactions on Computers*, 39(10):1220–31.

Ciminiera, L., and P. Montuschi (1996). Carry-save multiplication schemes without final addition. *IEEE Transactions on Computers*, 45(9):1050–55.

Clarke, E. M., S. M. German, and X. Zhao (1999). Verifying the SRT division algorithm using theorem proving techniques. *Formal Methods in System Design*, 14(1):7–44.

Clouser, J., M. Matson, R. Badeau, R. Dupcak, S. Samudrala, R. Allmon, and N. Fairbanks (1999). A 600-MHz superscalar floating-point processor. *IEEE Journal of Solid-State Circuits*, 34(7):1026–29.

Cocke, J., and D. W. Sweeney (1957). High speed arithmetic in a parallel device. Technical report, IBM.

Cody, W. J. (1970). A survey of practical rational and polynomial approximation of functions. *SIAM Review*, 12(3):400–423.

Cody, W. J. (1973). Static and dynamic numerical characteristics of floating-point arithmetic. *IEEE Transactions on Computers*, C-22(6):598–601.

Cody, W. J. (1987). Analysis of proposals for the floating-point standard. *Computer*, 20(3):63–68.

Cody, W. J., J. T. Coonen, D. M. Gay, K. Hanson, D. Hough, W. Kahan, R. Karpinski, J. Palmer, F. N. Ris, and D. Stevenson (1984). A proposed radix-and-word-length-independent standard for floating-point arithmetic. *IEEE MICRO*, 4(4):86–100.

Cody, W. J., and W. Waite (1980). *Software Manual for the Elementary Functions*. Prentice-Hall, Englewood Cliffs, New Jersey.

Coonen, J. T. (1980). An implementation guide to a proposed standard for floating-point arithmetic. *Computer*, 13(1):68–79.

Coonen, J. T. (1981). Underflow and the denormalized numbers. *Computer*, 14(3):75–87.

Corbaz, G., J. Duprat, B. Hochet, and J.-M. Muller (1991). Implementation of a VLSI polynomial evaluator for real-time applications. In *Proceedings of ASAP91*, pages 13–24.

Cornea-Hasegan, M. A., R. A. Golliver, and P. Markstein (1999). Correctness proofs outline for Newton-Raphson based floating-point divide and square root algorithms. In *Proceedings of the 14th IEEE Symposium on Computer Arithmetic*, pages 96–105.

Cornetta, G., and J. Cortadella (1999). A radix-16 SRT division unit with speculation of the quotient digits. In *Proceedings of the 9th Great Lakes Symposium on VLSI*, pages 74–77.

Cornetta, G., and J. Cortadella (2001). A multi-radix approach to asynchronous division. In *ASYNC 2001, Proceedings of the 7th International Symposium on Asynchronous Circuits and Systems*, pages 25–34.

Corsonello, P., S. Perri, and G. Cocorullo (2000). Performance comparison between static and dynamic CMOS logic implementations of a pipelined square-rooting circuit. *IEE Proceedings—Circuits, Devices and Systems*, 147(6):347–55.

Cortadella, J., and J. M. Llaberia (1992). Evaluation of $A + B = K$ conditions without carry propagation. *IEEE Transactions on Computers*, 41(11):1484–88.

Cortadella, J., and T. Lang (1993). Division with speculation of quotient digits. In *Proceedings of the 11th IEEE Symposium on Computer Arithmetic*, pages 87–94.

Cortadella, J., and T. Lang (1994). High-radix division and square-root with speculation. *IEEE Transactions on Computers*, 43(8):919–31.

Dadda, L. (1965). Some schemes for parallel multipliers. *Alta Frequenza*, 34:349–56.

Dadda, L. (1976). On parallel digital multipliers. *Alta Frequenza*, 45:574–80.

Dadda, L. (1989). On serial input multipliers for two's complement numbers. *IEEE Transactions on Computers*, 38:1341–45.

Dadda, L., and V. Piuri (1996). Pipelined adders. *IEEE Transactions on Computers*, 45(3):348–56.

Daggett, D. H. (1959). Decimal-binary conversion in CORDIC. *IRE Transactions on Electronic Computers*, EC-8(3):335–39.

Dahlquist, G., and A. Bjorck (1974). *Numerical Methods*. Prentice-Hall, Englewood Cliffs, New Jersey.

Danielsson, P. E. (1984). Serial/parallel convolvers. *IEEE Transactions Computers*, C-33(7):652–67.

Dantzig, T. (1954). *The Number: The Language of Science*. Free Press (Macmillan Publishing Co.), New York.

Dao, H., and V. G. Oklobdzija (2001). Application of logical effort for speed optimization and analysis of representative adders. In *Proceedings of the 35th Asilomar Conference on Signals, Systems and Computers*, pages 1666–69.

Dao-Trong, S., and K. Helwig (1992). A single-chip IBM System/390 floating-point processor in CMOS. *IBM Journal of Research and Development*, 36(4):733–48.

Darley, M., B. Kronlage, D. Bural, B. Churchill, D. Pulling, P. Wang, R. Iwamoto, and L. Yang (1990). The TMS390C602A floating-point coprocessor for Sparc systems. *IEEE Micro*, 10(3):36–47.

Das Sarma, D. (1995). *Highly Accurate Initial Reciprocal Approximations for High Performance Division Algorithms*. PhD thesis, Southern Methodist University.

Das Sarma, D., and D. W. Matula (1994). Measuring the accuracy of ROM reciprocal tables. *IEEE Transactions on Computers*, 43(8):932–40.

Das Sarma, D., and D. W. Matula (1995). Faithful bipartite ROM reciprocal tables. In *Proceedings of the 12th IEEE Symposium on Computer Arithmetic*, pages 17–28.

Das Sarma, D., and D. W. Matula (1997). Faithful interpolation in reciprocal tables. In *Proceedings of the 13th IEEE Symposium on Computer Arithmetic*, pages 82–91.

Daumas, M., C. Finot, and J.-M. Muller (2000). Table based implementation of elementary functions for hundred-bit precision. In *16th IMACS World Congress on Computational and Applied Mathematics*.

Daumas, M., C. Mazenc, X. Merrheim, and J.-M. Muller (1994). Fast and accurate range reduction for computation of the elementary functions. In *Proceedings of the 14th IMACS World Congress on Computational and Applied Mathematics*, pages 1196–98.

Daumas, M., C. Mazenc, X. Merrheim, and J.-M. Muller (1995). Modular range reduction: A new algorithm for fast and accurate computation of the elementary functions. *Journal of Universal Computer Science*, 1(3):162–75.

Daumas, M., J.-M. Muller, and A. Tisserand (1997). Very high radix on-line arithmetic for accurate computations. In *15th IMACS World Congress on Scientific Computation, Modelling and Applied Mathematics*, Berlin, Germany.

Daumas, M., J.-M. Muller, and J. Vuillemin (1994). Implementing on-line arithmetic on PAM. In *4th International Workshop on Field-Programmable Logic and Applications*.

Davis, P. J. (1990). *Interpolation and Approximation*. Dover Publications, New York.

Dawid, H., and H. Meyr (1996). The differential CORDIC algorithm: Constant scale factor redundant implementation without correcting iterations. *IEEE Transactions on Computers*, 45(3):307–18.

de Angel, E., and E. Swartzlander (1996). Low power parallel multipliers. In *Proceedings of the IEEE Workshop on VLSI Signal Processing*, pages 199–208.

de Dinechin, F., and A. Tisserand (2001). Some improvements on multipartite table methods. In *Proceedings of the 15th IEEE Symposium on Computer Arithmetic*, pages 128–35.

de Lange, A. A., and E. F. A. Deprettere (1991). Design and implementation of a floating-point quasi systolic general purpose CORDIC rotator for high-rate parallel data and signal processing. In *Proceedings of the 10th IEEE Symposium on Computer Arithmetic*, pages 272–81.

Deegan, I. (1971). Concise cellular array for multiplication and division. *Electronics Letters*, 7(23):702–4.

Delosme, J.-M. (1986). The matrix exponential approach to elementary operations. In *Proceedings of the SPIE—Advanced Algorithms and Architectures for Signal Processing*, volume 696, pages 188–95.

Delosme, J.-M. (1989). CORDIC algorithms: theory and extensions. In *SPIE Advanced Algorithms and Architectures for Signal Processing IV*, volume 1152, pages 131–45.

Delosme, J.-M., and S. H. Hsiao (1990). CORDIC algorithms in four dimensions. In *SPIE Advanced Signal-Processing Algorithms, Architectures and Implementations*, volume 1348, pages 349–60.

DeLugish, B. G. (1970). *A Class of Algorithms for Automatic Evaluation of Certain Elementary Functions in a Binary Computer*. PhD thesis, Department of Computer Science, University of Illinois at Urbana-Champaign. (Technical Report UIUCDCS-R-399.)

Dempster, A. G., and M. D. Macleod (1994). Constant integer multiplication using minimum adders. *IEE Proceedings Circuits Devices Systems*, 141(5):407–13.

Dempster, A. G., and M. D. Macleod (1995). Use of minimum-adder multiplier blocks in FIR digital filters. *IEEE Transactions on Circuits and Systems—II: Analog and Digital Signal Processing*, 42(9):569–77.

Denyer, P., and D. Renshaw (1985). *VLSI Signal Processing: A Bit-Serial Approach*. Addison-Wesley, Reading, Massachusetts.

Deprettere, E. F. A., A. de Lange, and P. Dewilde (1990). The synthesis and implementation of signal processing applications specific VLSI CORDIC arrays. In *Proceedings of ISCAS'90*, pages 974–77.

Deprettere, E. F. A., P. Dewilde, and R. Udo (1984). Pipelined CORDIC architectures for fast VLSI filtering and array processing. In *Proceedings of ICASSP '84*, pages 41.A.6.1–41.A.6.4.

der Kolk, K. J. V., J. A. Lee, and E. F. A. Deprettere (2000). A floating-point vectoring algorithm based on fast rotations. *Journal of VLSI Signal Processing Systems*, 25(2):125–40.

Despain, A. M. (1974). Fourier transform computers using CORDIC iterations. *IEEE Transactions on Computers*, C-23(10):993–1001.

Deverell, J. (1975). Pipeline iterative arithmetic arrays. *IEEE Transactions on Computers*, C-24(3):317–22.

Dimmler, M. (1999). *Digital Control of Micro-Systems Using On-line Arithmetic*. PhD thesis, Ecole Polytechnique Federal de Lausanne.

Dimmler, M., A. Tisserand, U. Holmbeg, and R. Longchamp (1999). On-line arithmetic for real-time control of microsystems. *IEEE/ASME Transactions on Mechatronics*, 4(2):213–17.

Dobberpuhl, D. W., R. T. Witek, R. Allmon, R. Anglin, D. Bertucci, S. Britton, L. Chao, R. A. Conrad, D. E. Dever, B. Gieseke, S. M. N. Hassoun, G. W. Hoeppner, K. Kuchler, M. Ladd, B. M. Leary, L. Madden, E. J. McLellan, D. R. Meyer, and J. Montanaro (1992). A 200-MHz 64-b dual-issue CMOS microprocessor. *IEEE Journal of Solid-State Circuits*, 27(11):1555–64.

Doran, R. W. (1988). Variants of an improved carry-lookahead adder. *IEEE Transactions on Computers*, C-37(9):1110–13.

Dormido, S., and M. A. Canto (1981). Synthesis of generalized parallel counters. *IEEE Transactions on Computers*, C-30(9):699–703.

Dormido, S., and M. A. Canto (1982). An upper bound for the synthesis of generalized parallel counters. *IEEE Transactions on Computers*, C-31(8): 802–5.

Duprat, J., M. Fiallos, J.-M. Muller, and H. J. Yeh (1991). Delays of on-line floating-point operators in borrow-save representation. In *IFIP Workshop on Algorithms and Parallel VLSI Architectures*.

Duprat, J., and J.-M. Muller (1988). Hardwired polynomial evaluation. *Journal of Parallel and Distributed Computing*, 5(3):291–309.

Duprat, J., and J.-M. Muller (1991). Writing numbers differently for faster calculation. *Technique et Science Informatiques*, 10(3):211–24.

Duprat, J., and J.-M. Muller (1993). The CORDIC algorithm: New results for fast VLSI implementation. *IEEE Transactions on Computers*, 42(2):168–78.

Efe, K. (1981). Multi-operand addition with conditional sum logic. In *Proceedings of the 5th IEEE Symposium on Computer Arithmetic*, pages 251–55.

Eisig, D., J. Rotstain, and I. Koren (1993). The design of a 64-bit integer multiplier/divider unit. In *Proceedings of the 11th Symposium on Computer Arithmetic*, pages 171–78.

Ercegovac, M. D. (1973). Radix-16 evaluation of certain elementary functions. *IEEE Transactions on Computers*, C-22(6):561–66.

Ercegovac, M. D. (1975). *A General Hardware-Oriented Method for Evaluation of Functions and Computations in a Digital Computer*. PhD thesis, Department of Computer Science, University of Illinois at Urbana-Champaign. (Technical Report UIUCDCS-R-750.)

Ercegovac, M. D. (1977). A general hardware-oriented method for evaluation of functions and computations in a digital computer. *IEEE Transactions on Computers*, C-26(7):667–80.

Ercegovac, M. D. (1978). An on-line square rooting algorithm. In *4th IEEE Symposium on Computer Arithmetic*. IEEE Computer Society Press, Los Alamitos, California.

Ercegovac, M. D. (1983). A higher radix division with simple selection of quotient digits. In *Proceedings of the 6th IEEE Symposium on Computer Arithmetic*, pages 94–98.

Ercegovac, M. D. (1984). On-line arithmetic: An overview. In *Proceedings of the SPIE, Real Time Signal Processing VII*, pages 86–93.

Ercegovac, M. D. (1991). On-line arithmetic for recurrence problems. In *Proceedings of the SPIE: Advanced Signal Processing Algorithms, Architectures, and Implementations II*.

Ercegovac, M. D., L. Imbert, D. W. Matula, J.-M. Muller, and G. Wei (2000). Improving Goldschmidt division, square root and square root reciprocal. *IEEE Transactions on Computers*, 49(7):759–63.

Ercegovac, M. D. and T. Lang (1985). *Digital Systems and Hardware/Firmware Algorithms*. John Wiley & Sons, Inc., New York.

Ercegovac, M. D., and T. Lang (1985). A division algorithm with prediction of quotient digits. In *Proceedings of the 7th IEEE Symposium on Computer Arithmetic*, pages 51–56.

Ercegovac, M. D., and T. Lang (1986). Alternative on-the-fly conversion of redundant into conventional representations. Technical report CSD-860027, Computer Science Department, University of California, Los Angeles.

Ercegovac, M. D., and T. Lang (1987). Fast cosine/sine implementation using on-line CORDIC. In *Proceedings of the 21st Asilomar Conference Signals, Systems, Computers*, pages 222–26.

Ercegovac, M. D., and T. Lang (1987). On-line scheme for computing rotation factors. In *Proceedings of the 8th IEEE Symposium on Computer Arithmetic*, pages 196–203.

Ercegovac, M. D., and T. Lang (1987). On-the-fly conversion of redundant into conventional representations. *IEEE Transactions on Computers*, C-36(7):895–97.

Ercegovac, M. D., and T. Lang (1987). Simple radix-4 division with divisor scaling. Technical report CSD-870015, Computer Science Department, University of California, Los Angeles.

Ercegovac, M. D., and T. Lang (1988). On-line arithmetic: A design methodology and applications. In R. W. Brodersen and H. S. Moscovitz, editors, *VLSI Signal Processing, III*, Chapter 24. IEEE Press, New York.

Ercegovac, M. D., and T. Lang (1988). On-line scheme for computing rotation factors. *Journal of Parallel and Distributed Computing*, Special Issue on Parallelism in Computer Arithmetic (5).

Ercegovac, M. D., and T. Lang (1989). Binary counter with counting period of one half adder independent of counter size. *IEEE Transactions on Circuits and Systems*, 36(6):924–26.

Ercegovac, M. D., and T. Lang (1989). Fast radix-2 division with quotient-digit prediction. *Journal of VLSI Signal Processing*, 2(1):169–80.

Ercegovac, M. D., and T. Lang (1989). Radix-4 square root without initial PLA. In *Proceedings of the 9th IEEE Symposium on Computer Arithmetic*, pages 162–68.

Ercegovac, M. D., and T. Lang (1990). Fast multiplication without carry-propagate addition. *IEEE Transactions on Computers*, 39(11):1385–90.

Ercegovac, M. D., and T. Lang (1990). Radix-4 square root without initial PLA. *IEEE Transactions on Computers*, C-39(8):1016–24.

Ercegovac, M. D., and T. Lang (1990). Redundant and on-line CORDIC: Application to matrix triangularization and SVD. *IEEE Transactions on Computers*, 39(6):725–40.

Ercegovac, M. D., and T. Lang (1990). Fast multiplication without carry-propagate addition. *IEEE Transactions on Computers*, C-39(11):1385–90.

Ercegovac, M. D., and T. Lang (1990). Simple radix-4 division with operands scaling. *IEEE Transactions on Computers*, C-39(9):1204–7.

Ercegovac, M. D., and T. Lang (1991). Module to perform multiplication, division and square root in systolic arrays for matrix computations. *Journal of Parallel and Distributed Computing*, 11(3):212–21.

Ercegovac, M. D., and T. Lang (1992). Fast arithmetic for recursive computations. In *Proceedings of the IEEE Workshop on VLSI Signal Processing*, pages 14–28.

Ercegovac, M. D., and T. Lang (1992). On-the-fly rounding. *IEEE Transactions on Computers*, 41(12):1497–1503.

Ercegovac, M. D., and T. Lang (1994). *Division and Square Root: Digit-Recurrence Algorithms and Implementations*. Kluwer Academic Publishers.

Ercegovac, M. D., and T. Lang (1996). On recoding in arithmetic algorithms. *Journal of VLSI Signal Processing*, 14:283–94.

Ercegovac, M. D., and T. Lang (1997). Effective coding for fast redundant adders using radix-2 digit set $\{0, 1, 2, 3\}$. In *Proceedings of the 31st Asilomar Conference on Signals, Systems and Computers*, pages 1163–67.

Ercegovac, M. D., and T. Lang (1999). On-line scheme for normalizing a 3-D vector. In *Proceedings of the 33rd Asilomar Conference on Signals, Systems and Computers*, pages 1460–64.

Ercegovac, M. D., T. Lang, and R. Modiri (1988). Implementation of fast radix-4 division with operands scaling. In *Proceedings of the ICCD '88 Conference*, pages 486–89, New York.

Ercegovac, M. D., T. Lang, and P. Montuschi (1993). Very high radix division with selection by rounding and prescaling. In *Proceedings of the 11th IEEE Symposium on Computer Arithmetic*, pages 112–19.

Ercegovac, M. D., T. Lang, and P. Montuschi (1994). Very-high radix division with prescaling and rounding. *IEEE Transactions on Computers*, 43(8): 909–18.

Ercegovac, M. D., T. Lang, J.-M. Muller, and A. Tisserand (2000). Reciprocation, square root, inverse square root, and some elementary functions using small multipliers. *IEEE Transactions on Computers*, 49(7):628–37.

Ercegovac, M. D., T. Lang, J. G. Nash, and L. P. Chow (1987). An area-time efficient binary divider. In *Proceedings of the ICCD '87 Conference*, pages 645–48, New York.

Ercegovac, M. D., and J.-M. Muller (1998). Fast evaluation of functions at regularly-spaced points. In *SPIE International Symposium on Optical Science, Engineering, and Instrumentation*, vol. 3461, pages 555–66.

Ercegovac, M. D., J.-M. Muller, and A. Tisserand (1995). FPGA implementation of polynomial evaluation algorithms. In *SPIE Photonics East '95 Conference Proceedings*, vol. 2607, pages 177–88.

Ercegovac, M. D., and P. K. G. Tu (1991). Application of on-line arithmetic algorithms to the SVD computation: preliminary results. In *Proceedings of the 10th IEEE Symposium on Computer Arithmetic*, pages 246–55.

Estrin, G., B. Gilchrist, and J. H. Pomerane (1956). A note on high-speed digital multiplication. *IRE Transactions on Electronic Computers*, page 140.

Even, G., and P.-M. Seidel (2000). A comparison of three rounding algorithms for IEEE floating-point multiplication. *IEEE Transactions on Computers*, 49(7):638–50.

Even, G., and W. J. Paul (2000). On the design of IEEE compliant floating-point units. *IEEE Transactions on Computers*, 49(5):398–413.

Fandrianto, J. (1987). Algorithm for high-speed shared radix-4 division and radix-4 square-root. In *Proceedings of the 8th IEEE Symposium on Computer Arithmetic*, pages 73–79.

Fandrianto, J. (1989). Algorithm for high-speed shared radix-8 division and radix-8 square root. In *Proceedings of the 9th IEEE Symposium on Computer Arithmetic*, pages 68–75.

Farmwald, M. P. (1981). *On the Design of High-Performance Digital Arithmetic Units*. PhD thesis, Stanford University.

Farmwald, P. M. (1981). High bandwidth evaluation of elementary functions. In *Proceedings of the 5th IEEE Symposium on Computer Arithmetic*, pages 139–42.

Feldstein, A., and R. Goodman (1982). Loss of significance in floating-point subtraction and addition. *IEEE Transactions on Computers*, C-31:328–35.

Fenwick, P. M. (1987). A fast-carry adder with CMOS transmission gates. *Computer Journal*, 30(1):77–79.

Ferguson, M. I., and M. D. Ercegovac (1999). A multiplier with redundant operands. In *Proceedings of the 33rd Asilomar Conference on Signals, Systems and Computers*, volume 2, pages 1322–26.

Ferguson, W. (1995). Exact computation of a sum or difference with applications to argument reduction. In *Proceedings of the 12th IEEE Symposium on Computer Arithmetic*, pages 216–21.

Ferguson, W., and T. Brightman (1991). Accurate and monotone approximations of some transcendental functions. In *Proceedings of the 10th IEEE Symposium on Computer Arithmetic*, pages 237–44.

Fernando, J. S. (1993). *Design Alternatives for Recursive Digital Filters Using On-line Arithmetic*. PhD thesis, University of California, Los Angeles.

Fernando, J. S., and M. D. Ercegovac (1994). Conventional and on-line arithmetic designs for high-speed recursive digital filters. *Journal of VLSI Signal Processing*, 7:189–97.

Fernando, J. S., and M. D. Ercegovac (1997). A method of eliminating oscillations in high-speed recursive digital filters. *IEEE Transactions on Circuits and Systems—II: Analog and Digital Signal Processing*, 44(10):861–64.

Ferrari, D. (1967). A division method using a parallel multiplier. *IEEE Transactions Electronic Computers*, EC-16(4):224–26.

Flynn, M. J. (1970). On division by functional iteration. *IEEE Transactions on Computers*, C-19(8):702–6.

Flynn, M. J., K. Nowka, G. Bewick, E. M. Schwarz, and N. Quach (1995). The SNAP project: towards sub-nanosecond arithmetic. In *Proceedings of the 12th Symposium on Computer Arithmetic*, pages 75–82.

Flynn, M. J., and S. F. Oberman (2001). *Advanced Computer Arithmetic Design*. John Wiley & Sons, Inc., New York.

Foster, C. C., and F. D. Stockton (1971). Counting responders in an associative memory. *IEEE Transactions on Computers*, C-20:1580–83.

Fowler, D. L., and J. E. Smith (1989). An accurate, high speed implementation of division by reciprocal approximation. In *Proceedings of the 9th IEEE Symposium on Computer Arithmetic*, pages 60–67.

Franklin, M. A., and T. Pan (1994). Performance comparison of asynchronous adders. In *Proceedings of the International Symposium on Advanced Research in Asynchronous Circuits and Systems*, pages 117–25.

Freiman, C. V. (1961). Statistical analysis of certain binary division algorithms. *Proceedings of IRE*, 49:91–103.

Fried, R. (1997). Minimizing energy dissipation in high-speed multipliers. In *Proceedings of 1997 International Symposium on Low Power Electronics and Design*, pages 214–19.

Gajski, D. D. (1980). Parallel compressors. *IEEE Transactions on Computers*, C-29(5):393–98.

Galli, R., and A. F. Tenca (2001). Design and evaluation of on-line arithmetic for signal processing applications on FPGAs. In *Proceedings of the SPIE—Advanced Signal Processing Algorithms, Architectures, and Implementations XI*, volume 4474, pages 134–44.

Gardiner, A. B., and J. Hont (1972). Cellular-array arithmetic unit with multiplication and division. *Proceedings of the IEE*, 119(6):559–60.

Garner, H. L. (1965). Number systems and arithmetic. In *Advances in Computers*, volume 6, pages 131–94. Academic Press, New York.

Gaviland, J., and V. C. Hamacher (1973). High-speed multiplier/divider iterative arrays. In *1973 Sagamore Computer Conference on Parallel Processing*, pages 91–100.

Gazale, M. (2000). *Number from Ahmes to Cantor*. Princeton University Press, Princeton, New Jersey.

Gerwig, G., and M. Kroener (1999). Floating-point unit in standard cell design with 116 bit wide dataflow. In *Proceedings of the 14th IEEE Symposium on Computer Arithmetic*, 266–73.

Gilchrist, B., J. H. Pomerene, and S. Y. Wong (1955). Fast carry logic for digital computers. *IRE Transactions on Electronic Computers*, EC-4:133–36.

Girau, B., and A. Tisserand (1996). On-line arithmetic-based reprogrammable hardware implementation of multilayer perceptron back-propagation. In

Proceedings of the 5th International Conference on Microelectronics for Neural Networks and Fuzzy Systems. MicroNeuro'96, pages 168–75.

Gnanasekaran, R. (1985). A fast serial-parallel binary multiplier. *IEEE Transactions on Computers*, C-34:741–44.

Goldberg, D. (1991). What every computer scientist should know about floating-point arithmetic. *ACM Computing Surveys*, 23(1):5–48.

Goldschmidt, R. E. (1964). *Applications of Division by Convergence*. Master's thesis, Massachusetts Institute of Technology.

Gorji-Sinaki, A. (1981). *Error-Coded Algorithms for On-line Arithmetic*. PhD thesis, University of California, Los Angeles.

Gosling, J. B. (1971). Design of large high-speed floating-point arithmetic units. In *IEE Proceedings*, volume 118, pages 493–98.

Gosling, J. B. (1971). Review of high-speed addition techniques. *Proceedings of IEE*, 118(1):29–35.

Gosling, J. B. (1980). *Design of Arithmetic Units for Digital Computers*. Springer-Verlag, New York.

Greenley, D., et al. (1995). UltraSPARC: the next generation superscalar 64-bit SPARC. In *Digest of Papers. COMPCON '95. Technologies for the Information Superhighway*, pages 442–51.

Guedj, D. (1996). *Numbers: The Universal Language*. Harry B. Abrams, Inc., New York.

Guild, H. H. (1969). Fully iterative fast arrays for binary multiplication and addition. *Electronic Letters*, 5:263.

Guyot, A., Y. Herreros, and J.-M. Muller (1989). Janus, an on-line multiplier/divider for manipulating large numbers. In *Proceedings of the 9th IEEE Symposium on Computer Arithmetic*, pages 106–11. IEEE Computer Society Press, Los Alamitos, California.

Guyot, A., B. Hochet, and J.-M. Muller (1987). A way to build efficient carry-skip adders. *IEEE Transactions on Computers*, C-36(10).

Guyot, A., L. Montalvo, A. Houelle, H. Mehrez, and N. Vaucher (1995). Comparison of the layout synthesis of radix-2 and pseudo-radix-4 dividers. In *Proceedings of the 8th International Conference on VLSI Design*, pages 386–91.

Guyot, A., M. Renaudin, B. El Hassan, and V. Levering (1996). Self timed division and square-root extraction. In *Ninth International Conference on VLSI Design*, pages 376–81.

Hahn, H., D. Timmermann, B. J. Hosticka, and B. Rix (1994). A unified and division-free CORDIC argument reduction method with unlimited convergence domain including inverse hyperbolic functions. *IEEE Transactions on Computers*, 43(11):1339–44.

Han, T., and D. A. Carlson (1987). Fast area-efficient VLSI adders. In *Proceedings of the 8th IEEE Symposium on Computer Arithmetic*, pages 49–56.

Harata, Y., Y. Nakamura, H. Nagase, M. Takigawa, and N. Takagi (1987). A high-speed multiplier using a redundant binary adder tree. *IEEE Journal of Solid-State Circuits*, SC-22(1):28–34.

Harber, R. G. (1989). *VLSI Design of Systems of CORDIC Processors*. PhD thesis, Purdue University.

Harris, D., S. F. Oberman, and M. H. Horowitz (1997). SRT division architectures and implementations. In *Proceeding of the 13th IEEE Symposium on Computer Arithmetic*, pages 18–25.

Harrison, J., T. Kubaska, S. Story, and P. T. P. Tang (1999). The computation of transcendental functions on the IA-64 architecture. *Intel Technology Journal*, Q4.

Hart, J. F., E. W. Cheney, C. L. Lawson, H. J. Maehly, C. K. Mesztenyi, J. R. Rice, H. G. Thacher, and C. Witzgall (1978). *Computer Approximations*. Robert E. Krieger Publishing Company, Florida.

Hartley, R., and P. Corbett (1990). Digit-serial processing techniques. *IEEE Transactions on Circuits and Systems*, 37(6):707–19.

Hartley, R., and K. K. Parhi (1995). *Digit-Serial Computation*. Kluwer Academic Publishers.

Hashemian, R. (1990). Square rooting algorithms for integer and floating-point numbers. *IEEE Transactions on Computers*, C-39(8):1025–29.

Hassler, H., and N. Takagi (1995). Function evaluation by table look-up and addition. In *Proceedings of the 12th IEEE Symposium on Computer Arithmetic*, pages 10–16.

Haviland, G. H., and A. A. Tuszinsky (1980). A CORDIC arithmetic processor chip. *IEEE Transactions on Computers*, C-29(2):68–79.

Hekstra, G. J. (1998). *CORDIC for High-Performance Numerical Computation*. PhD thesis, Technical University of Delft, The Netherlands.

Hekstra, G. J. (2000). Evaluation of fast rotation methods. *Journal of VLSI Signal Processing Systems*, 25(2):113–24.

Hekstra, G. J., and E. F. A. Deprettere (1993). Floating-point CORDIC. In *Proceedings of the 11th IEEE Symposium on Computer Arithmetic*, pages 130–37.

Hemkumar, N. D., and J. R. Cavallaro (1994). Redundant and on-line CORDIC for unitary transformations. *IEEE Transactions on Computers*, 43(8):941–54.

Hennessy, J. L. and D. A. Patterson, (1995). *Computer Architecture: A Quantitative Approach*. Morgan Kaufmann, San Francisco, 2nd edition.

Ho, I. T., and T. C. Chen (1973). Multiple addition by residue threshold functions and their representation by array logic. *IEEE Transactions on Computers*, C-22:762–67.

Hokenek, E., and R. K. Montoye (1990). Leading zero anticipator (LZA) in the IBM Risc System/6000 floating-point execution unit. *IBM Journal of Research and Development*, 34(1):71–77.

Hokenek, E., R. K. Montoye, and P. W. Cook (1990). Second-generation RISC floating point with multiply-add fused. *IEEE Journal of Solid-State Circuits*, 25(5):1207–13.

Horel, T., and G. Lauterbach (1999). UltraSPARC-III: designing third-generation 64-bit performance. *IEEE Micro*, 19(3):73–85.

Hormigo, J., J. Villalba, and E. L. Zapata (1999). Interval sine and cosine functions computation based on Variable-precision CORDIC algorithm. In *Proceedings of the 14th IEEE Symposium on Computer Arithmetic*, pages 186–93.

Hsiao, S.-F. (1993). *Multi-Dimensional CORDIC Algorithms*. PhD thesis, Yale University.

Hsiao, S.-F., and J.-M. Delosme (1995). Householder CORDIC algorithms. *IEEE Transactions on Computers*, 44(8):990–1000.

Hsiao, S.-F., C.-Y. Lau, and J.-M. Delosme (2000). Redundant constant-factor implementation of multi-dimensional CORDIC and its application to complex SVD. *Journal of VLSI Signal Processing Systems*, 25(2):155–66.

Hu, X. (1989). *A Silicon Compiler for Dedicated Mathematical Systems Based on CORDIC Arithmetic Processors*. PhD thesis, Purdue University.

Hu, X., and S. C. Bass (1993). A neglected error source in the CORDIC algorithm. In *Proceedings of ISCAS'93*, volume 1, pages 766–69.

Hu, X., S. C. Bass, and R. G. Harber (1993). An efficient implementation of singular value decomposition rotation transformations with CORDIC processors. *Journal of Parallel and Distributed Computing*, 17:360–62.

Hu, X., G. Harber, and S. Bass (1991). Expanding the range of the cordic algorithm. *IEEE Transactions on Computers*, 40(1):13–21.

Hu, Y. H. (1992). CORDIC-based VLSI architectures for digital signal processing. *IEEE Signal Processing Magazine*, pages 16–35.

Hu, Y. H. (1992). The quantization effects of the CORDIC algorithm. *IEEE Transactions on Signal Processing*, 40(4):834–44.

Hu, Y. H., and S. Naganathan (1993). An angle recoding method for CORDIC algorithm implementation. *IEEE Transactions on Computers*, 42(1):99–102.

Huang, X., W.-J. Liu, and B. W. Y. Wei (1994). A high-performance CMOS redundant binary multiplication-and-accumulation (MAC) unit. *IEEE Transactions on Circuits and Systems I: Fundamental Theory and Applications*, 41(1): 33–39.

Huang, Z., and M. D. Ercegovac (2001). FPGA implementation of pipelined on-line scheme for 3-D vector normalization. In *IEEE Symposium on Field-Programmable Custom Computing Machines*, pages 1–4.

Hunt, D. (1995). Advanced performance features of the 64-bit PA-8000. In *Digest of Papers COMPCON '95*, pages 23–28.

Hwang, K. (1978). *Computer Arithmetic Principles, Architecture and Design*. John Wiley & Sons, Inc., New York.

Hwang, K., H. C. Wang, and Z. Xu (1987). Evaluating elementary functions with chebyshev polynomials on pipeline nets. In *Proceedings of the 8th IEEE Symposium on Computer Arithmetic*, pages 121–28.

Ide, N., H. Fukuhisa, Y. Kondo, T. Yoshida, M. Nagamatsu, J. Mori, I. Yamazaki, and K. Ueno (1993). A 320-MFLOPS CMOS floating-point processing unit for superscalar processors. *IEEE Journal of Solid-State Circuits*, SC-28(3):352–61.

Ienne, P., and M. A. Viredez (1994). Bit-serial multipliers and squarers. *IEEE Transactions Computers*, 43(12):1445–50.

Ifrah, G. (1985). *From One to Zero: A Universal History of Numbers*. Viking, New York.

Inui, S., T. Uesugi, H. Saito, Y. Hagihara, A. Yoshikawa, M. Nishida, and M. Yamashina (1999). A 250 MHz CMOS floating-point divider with operand pre-scaling. In *Symposium on VLSI Circuits*, pages 17–18.

Iordache, C., and D. W. Matula (1999). On infinitely precise rounding for division, square root, reciprocal and square root reciprocal. In *Proceed-*

ings of the 14th IEEE Symposium on Computer Arithmetic, pages 233–40.

Irwin, M. J. (1977). An Arithmetic Unit for On-line Computation. PhD thesis, Department of Computer Science, University of Illinois at Urbana-Champaign. (Technical Report UIUCDCS-R-77-873.)

Irwin, M. J., and R. M. Owens (1983). Fully digit on-line networks. IEEE Transactions on Computers, C-32(4):402–6.

Irwin, M. J., and R. M. Owens (1987). Digit-pipelined arithmetic as illustrated by the Paste-up system: a tutorial. IEEE Computer, 20(4):61–73.

Irwin, M. J., and R. M. Owens (1989). Design issues in digit-serial signal processors. In Proceedings of ISCAS'89, volume 1, pages 441–44.

Irwin, M. J., and R. M. Owens (1990). A case for digit serial VLSI signal processors. Journal of VLSI Signal Processing, 1(4):321–34.

Ito, M., N. Takagi, and S. Yajima (1997). Efficient initial approximation for multiplicative division and square root by a multiplication with operand modification. IEEE Transactions on Computers, 46(4):495–98.

Iwamura, J., K. Suganama, S. Taguchi, M. Kimura, and K. Maeguchi (1982). A 16-bit CMOS/SOS multiplier-accumulator. In Proceedings of the ICCD '82 Conference, pages 151–54, New York.

Jain, V. K., and L. Lin (1995). High-speed double precision computation of nonlinear functions. In Proceedings of the 12th IEEE Symposium on Computer Arithmetic, pages 107–14.

Jain, V. K., and L. Lin (1997). Complex-argument universal nonlinear cell for rapid prototyping. IEEE Transactions on Very Large Scale Integration (VLSI) Systems, 5(1):15–27.

Jessani, R. M., and M. Putrino (1998). Comparison of single- and dual-pass multiply-add fused floating-point units. IEEE Transactions on Computers, 47(9):927–37.

Jou, J. M., and S. R. Kuang (1999). Design of low-error fixed-width multiplier for DSP applications. IEEE Transactions on Circuits and Systems—II: Analog and Digital Signal Processing, 46(6):836–42.

Kabuo, H., T. Taniguchi, A. Miyoshi, H. Yamashita, M. Urano, H. Edamatsu, and S. Kuninobu (1994). Accurate rounding scheme for the Newton-Raphson method using redundant binary representation. IEEE Transactions on Computers, 43(1):43–51.

Kahan, W. (1996). Lecture Notes on the Status of IEEE 754. Technical Report http://http.cs.berkeley.edu/~wkahan/ieee754status/ieee754.ps, University of California, Berkeley.

Kanie, Y., Y. Kubota, S. Toyoyama, Y. Iwase, and S. Suchimoto (1994). 4-2 compressor with complementary pass-transistor logic. *IEICE Transactions on Electronics*, E77-C(4):647–49.

Kantabutra, V. (1991). Designing optimum carry-skip adders. In *Proceedings of the 10th IEEE Symposium on Computer Arithmetic*, pages 146–55.

Kantabutra, V. (1993). Accelerated two-level carry-skip adders—a type of very fast adder. *IEEE Transactions on Computers*, C-42(11):1389–93.

Kantabutra, V. (1993). A recursive carry-look-ahead/carry-select hybrid adder. *IEEE Transactions on Computers*, C-42(12):1495–99.

Karatsuba, A., and Y. Ofman (1962). Multiplication of multidigit numbers on automata. *Soviet Physics-Doklaty*, 7(7):595–96.

Karp, A. H., and P. Markstein (1997). High-precision division and square root. *ACM Transactions on Mathematical Software*, 23(4):561–89.

Karpinsky, R. (1985). PARANOIA: A floating-point benchmark. *BYTE*, 10(2):223–35.

Kilburn, T., D. B. G. Edwards, and D. Aspinall (1959). Parallel addition in a digital computer—a new fast carry. *Proceedings of the IEE*, 106B:460–64.

Kilburn, T., D. B. G. Edwards, and G. F. Thomas (1956). The Manchester University Mark II Computing Machine. *Proceedings of the IEE*, pt. 103B, Suppl. 2:247–68.

King, E. J., and E. E. Swartzlander (1998). Data-dependent truncation scheme for parallel multipliers. In *Proceedings of the 31st Asilomar Conference on Signals, Systems and Computers*, volume 2, pages 1178–82.

Kinniment, D. J. (1996). An evaluation of asynchronous addition. *IEEE Transactions on VLSI Systems*, 4(1):137–40.

Kla, S., C. Mazenc, X. Merrheim, and J.-M. Muller (1991). New algorithms for on-line computation of elementary functions. In *Proceedings of the SPIE: Advanced Signal Processing Algorithms, Architectures, and Implementations II*, volume 1566, pages 275–85.

Klir, J. (1963). A note on Svoboda's algorithm for division. *Information Processing Machines (Stroje na Zpracovani Informaci)*, (9):35–39.

Knowles, S. (1999). A family of adders. In *Proceedings of the 14th IEEE Symposium on Computer Arithmetic*, pages 30–34.

Knowles, S. C., R. F. Woods, J. McWirther, and J. McCanny (1989). Bit-level systolic architectures for high-performance IIR filtering. *Journal of VLSI Signal Processing*, 1(1):9–24.

Knuth, D. E. (1998). *The Art of Computer Programming: Seminumerical Algorithms*. Addison Wesley, Reading, Massachusetts, 3rd edition.

Kobayashi, H., and H. Ohara (1978). A synthesizing method for large parallel counters with a network of smaller ones. *IEEE Transactions on Computers*, C-27(8):753–57.

Kogge, P. M., and H. S. Stone (1973). A parallel algorithm for the efficient solution of a general class of recurrence equations. *IEEE Transactions on Computers*, C-22(8):783–91.

Kolagotla, R. K., H. R. Srinivas, and G. F. Burns (1997). VLSI implementation of a 200-MHz 16*16 left-to-right carry-free multiplier in 0.35 μm CMOS technology for next-generation DSPs. In *Proceedings of the IEEE 1997 Custom Integrated Circuits Conference*, pages 469–72.

König, D., and J. F. Böhme (1990). Optimizing the CORDIC algorithm for processors with pipeline architectures. In *Signal Processing V: Theories and Applications*. Elsevier Science, Amsterdam, The Netherlands.

Koren, I. (1993). *Computer Arithmetic Algorithms*. Prentice Hall, Englewood Cliffs, New Jersey.

Koren, I., and O. Zinaty (1990). Evaluating elementary functions in a numerical coprocessor based on rational approximations. *IEEE Transactions on Computers*, 39(8):1030–37.

Kornerup, P. (1994). Digit-set conversions: Generalizations and applications. *IEEE Transactions on Computers*, 43(5):622–29.

Kornerup, P. (1999). Necessary and sufficient conditions for parallel, constant time conversion and addition. In *Proceedings of the 14th IEEE Symposium on Computer Arithmetic*, pages 152–56.

Kota, K., and J. R. Cavallaro (1993). Numerical accuracy and hardware tradeoffs for CORDIC arithmetic for special-purpose processors. *IEEE Transactions on Computers*, 42(7):769–79.

Krieger, C., and B. Hosticka (1996). Inverse kinematics computations with modified CORDIC iterations. *IEE Proceedings Computer Digital Techniques*, 143(1):87–92.

Krishnamurthy, E. V. (1970). On optimal iterative schemes for high-speed division. *IEEE Transactions on Computers*, C-19(3):227–31.

Krishnamurthy, E. V. (1970). On range-transformation techniques for division. *IEEE Transactions on Computers*, C-19(2):157–60.

Kuck, D. J., S. Parker, and A. Sameh (1977). Analysis of rounding methods in floating-point arithmetic. *IEEE Transactions on Computers*, C-26:643–50.

Kuhlmann, M., and K. K. Parhi (1998). Power comparison of SRT and GST dividers. In *Proceedings of the SPIE—Advanced Signal Processing Algorithms, Architectures, and Implementations VIII*, volume 3461, pages 584–94.

Kuki, H., and W. J. Cody (1973). A statistical study of the accuracy of floating point number systems. *Communications of the ACM*, 16(14):223–30.

Kulisch, U. W. (1977). Mathematical foundation of computer arithmetic. *IEEE Transactions on Computers*, C-26(7):610–21.

Kulisch, U. W., and W. L. Miranker (1981). *Computer Arithmetic in Theory and Practice*. Academic Press, New York.

Kunemund, R., H. Soldner, S. Wohlleben, and T. Noll (1990). CORDIC processor with carry-save architecture. In *Proceedings of the 16th European Solid-State Circuit Conference*, pages 193–96.

Kuninobu, S., T. Nishiyama, H. Edamatsu, T. Taniguchi, and N. Takagi (1987). Design of high speed MOS multiplier and divider using redundant binary representation. In *Proceedings of the 8th IEEE Symposium on Computer Arithmetic*, pages 80–86.

Kutsuwa, T., M. Mun, and K. Ebata (1987). Configuration and evaluation of two's complement multiplication-division arrays. *IEEE Transactions Circuits and Systems.*, CAS-34:304–8.

Kwak, J.-H. (2000). *High Speed CORDIC Processor Design: Algorithms, Architectures, and Applications*. PhD thesis, University of Texas at Austin.

Kwak, J.-H., J. H. Choi, and E. E. Swartzlander (2000). High-speed CORDIC based on an overlapped architecture and a novel σ-prediction method. *Journal of VLSI Signal Processing Systems*, 25(2):167–78.

Kwon, O., K. Nowka, and E. E. Swartzlander (2000). A 16-bit by 16-bit MAC design using fast 5:2 compressors. In *Proceedings of the IEEE International Conference on Application-Specific Systems, Architectures, and Processors*, pages 235–43.

Ladner, R., and M. Fisher (1980). Parallel prefix computation. *Journal of the ACM*, 27(4):831–38.

Lai, H. C., and S. Muroga (1982). Logic networks of carry-save adders. *IEEE Transactions on Computers*, C-31:870–82.

Lang, T., and E. Antelo (1998). CORDIC vectoring with arbitrary target value. *IEEE Transactions on Computers*, 47(7):736–49.

Lang, T., and E. Antelo (2000). CORDIC-based computation of ArcCos and $\sqrt{1 - t^2}$. *Journal of VLSI Signal Processing*, 25:19–38.

Lang, T., and E. Antelo (2001). Correctly rounded reciprocal square root by digit recurrence and radix-4 implementation. In *Proceedings of the 15th IEEE Symposium on Computer Arithmetic*, pages 94–100.

Lang, T., and E. Antelo (2001). High-throughput 3D rotations and normalizations. In *Conference Record of the 35th Asilomar Conference on Signals, Systems and Computers*, volume 1, pages 846–51.

Lang, T., and P. Montuschi (1992). Higher radix square root with prescaling. *IEEE Transactions on Computers*, 41(8):996–1009.

Lang, T., and P. Montuschi (1999). Very high radix square root with prescaling and rounding and a combined division/square root unit. *IEEE Transactions on Computers*, 48(8):827–41.

Lang, T., and J.-M. Muller (2001). Bounds on runs of zeros and ones for algebraic functions. In *Proceedings of the 15th IEEE Symposium on Computer Arithmetic*, pages 13–20.

Lee, J.-A. (1990). *Redundant CORDIC: Theory and Its Application to Matrix Computations*. PhD thesis, University of California, Los Angeles.

Lee, J. A., and T. Lang (1992). Constant-factor redundant CORDIC for angle calculation and rotation. *IEEE Transactions on Computers*, 41(8):1016–25.

Lee, K., and K. Choi (1996). Self-timed divider based on RSD number system. *IEEE Transactions on Very Large Scale Integration (VLSI) Systems*, 4(2):292–95.

Leeser, M., and J. O'Leary (1995). Verification of a subtractive radix-2 square root algorithm and implementation. In *International Conference on Computer Design: VLSI in Computers and Processors*, pages 526–31.

Lefèvre, V., and J. Muller (1999). Table methods for the elementary functions. In *SPIE Symposium on Optical Science and Technology*, vol. 3807, pages 43–9.

Lefèvre, V., and J.-M. Muller (2001). Worst cases for correct rounding of the elementary functions in double precision. In *Proceedings of the 15th IEEE Symposium on Computer Arithmetic*, pages 111–18.

Lefèvre, V., J.-M. Muller, and A. Tisserand (1998). Toward correctly rounded transcendentals. *IEEE Transactions on Computers*, 47(11):1235–43.

Lehman, M. (1962). A comparative study of propagation speed-up circuits in binary arithmetic units. *Information Processing*, pages 671–77.

Lehman, M., and N. Burla (1961). Skip techniques for high-speed carry propagation in binary arithmetic units. *IRE Transactions on Electronic Computers*, EC-10:691–98.

Lewis, D. (1999). Complex logarithmic number system arithmetic using high-radix redundant CORDIC algorithms. In *Proceedings of the 14th IEEE Symposium on Computer Arithmetic*, pages 194–203.

Lewis, D. M. (1994). Interleaved memory function interpolators with application to an accurate lns arithmetic unit. *IEEE Transactions on Computers*, 43(8): 974–82.

Ligomenides, P. A. (1977). The skip-and-set fast division algorithm. *IEEE Transactions on Computers*, C-26:1030–32.

Lim, R. S. (1978). High-speed multiplication and multiple summand addition. In *Proceedings of the 4th IEEE Symposium on Computer Arithmetic*, pages 149–53.

Lim, Y. C. (1992). Single-precision multiplier with reduced circuit complexity for signal processing applications. *IEEE Transactions on Computers*, 41(10):1333–36.

Lin, H., and H. J. Sips (1987). A novel floating-point on-line division algorithm. In *Proceedings of the 8th IEEE Symposium on Computer Arithmetic*, pages 188–95.

Lin, H., and H. J. Sips (1990). On-line CORDIC algorithms. *IEEE Transactions on Computers*, 39(8):1038–52.

Ling, H. (1981). High-speed binary adder. *IBM Journal Research and Development*, 25(3):156–66.

Linhardt, R. J., and H. S. Miller (1969). Digit-by-digit transcendental function computation. *RCA Review*, 30:209–47.

Louie, M. E., and M. D. Ercegovac (1993). A digit-recurrence square root implementation for field programmable gate arrays. In *IEEE Workshop on FPGAs for Custom Computing Machines*.

Louie, M. E., and M. D. Ercegovac (1993). On digit-recurrence division implementation for field programmable gate arrays. In *Proceedings of the 11th IEEE Symposium on Computer Arithmetic*, pages 202–9.

Louie, M. E., and M. D. Ercegovac (1994). Implementing division with field programmable gate arrays. *Journal of VLSI Signal Processing*, 7(3):271–85.

Lu, F., and H. Samueli (1993). A 200-Mhz CMOS pipelined multiplier-accumulator using a quasi-domino dynamic full-adder cell design. *IEEE Journal of Solid-State Circuits*, 28:123–32.

Luk, W. K., and J. E. Vuillemin (1983). Recursive implementation of optimal time VLSI integer multipliers. In *VLSI '83. Proceedings of the IFIP TC WG 10.5 International Conference on Very Large Scale Integration*, pages 155–68. Elsevier Science Publishers (North-Holland).

Lutz, D. R., and D. N. Jayasimha (1996). Programmable modulo-k counters. *IEEE Transactions on Circuits and Systems I: Fundamental Theory and Applications*, 43(11):939–41.

Lutz, D. R., and D. N. Jayasimha (1997). The half-adder form and early branch condition resolution. In *Proceedings of the 13th Symposium on Computer Arithmetic*, pages 266–73.

Lynch, T., and E. E. Swartzlander (1992). A spanning tree carry lookahead adder. *IEEE Transactions on Computers*, C-41(8):931–39.

Lynch, T., A. Ahmed, M. J. Schulte, T. Callaway, and R. Tisdale (1995). The K5 transcendental functions. In *Proceedings of the 12th IEEE Symposium on Computer Arithmetic*, pages 163–70.

Lynch, T., and M. J. Schulte (1995). A high radix on-line arithmetic for credible and accurate computing. *Journal of Universal Computer Science*, 1(7):439–53.

Lyon, R. F. (1976). Two's complement pipeline multipliers. *IEEE Transactions on Communication*, pages 418–25.

Lyu, C. N., and D. W. Matula (1995). Redundant binary Booth recoding. In *Proceedings of the 12th IEEE Symposium on Computer Arithmetic*, pages 50–57.

MacSorley, O. L. (1961). High-speed arithmetic in binary computers. *IRE Proceedings*, 49:67–91.

Magenheimer, D. J., L. Peters, K. W. Pettis, and D. Zuras (1988). Integer multiplication and division on the HP precision architecture. *IEEE Transactions on Computers*, C-37:980–90.

Mahant-Shetti, S. S., P. T. Balsara, and C. Lemonds (1999). High performance low power array multiplier using temporal tiling. *IEEE Transactions on Very Large Scale Integration (VLSI) Systems*, 7(1):121–24.

Majerski, S. (1967). On determination of optimal distributions of carry skips in adders. *IEEE Transactions on Electronic Computers*, EC-16(1):45–58.

Majerski, S. (1985). Square-root algorithms for high-speed digital circuits. *IEEE Transactions on Computers*, C-34(8):724–33.

Majithia, J. C. (1972). Cellular array for extraction of squares and square roots of binary numbers. *IEEE Transactions on Computers*, C-21(9):1023–24.

Majithia, J. C., and R. Kitai (1971). A cellular array for the nonrestoring extraction of square roots. *IEEE Transactions on Computers*, C-20(12):1617–18.

Makino, H., Y. Nakase, H. Suzuki, H. Morinaka, H. Shinohara, and K. Mashiko (1996). An 8.8-ns 54 × 54-bit multiplier with high speed redundant binary architecture. *IEEE Journal of Solid-State Circuits*, 31(6):773–83.

Makino, H., H. Suzuki, H. Morinaka, Y. Nakase, H. Shinohara, K. Mashiko, T. Sumi, and Y. Horiba (1996). A design of high-speed 4-2 compressor for fast multiplier. *IEICE Transactions on Electronics*, E79-C(4):538–48.

Mandelbaum, D. M. (1990). A systematic method for division with high average bit skipping. *IEEE Transactions on Computers*, C-39(1):127–30.

Markstein, P. W. (1990). Computation of elementary functions on IBM RISC System/6000 processor. *IBM Journal of Research and Development*, pages 111–19.

Markstein, P. (2000). *IA-64 and Elementary Functions: Speed and Precision*. Hewlett-Packard Professional Books. Prentice Hall.

Martin, N. M., and S. P. Hufnagel (1980). Conditional-sum early completion adder logic. *IEEE Transactions on Computers*, C-29:753–56.

Mathews, J. H. (1992). *Numerical Methods for Mathematics, Science, and Engineering, Second Edition*. Prentice-Hall, Englewood Cliffs, New Jersey.

Matsubara, G., and N. Ide (1997). A low power zero-overhead self-timed division and square root unit combining a single-rail static circuit with a dual-rail dynamic circuit. In *Third International Symposium on Advanced Research in Asynchronous Circuits and Systems*, pages 198–209.

Matsubara, G., N. Ide, H. Tago, S. Suzuki, and N. Goto (1995). 30-ns 55-b shared radix-2 division and square root using a self-timed circuit. In *Proceedings of the 12th IEEE Symposium on Computer Arithmetic*, pages 98–105.

Matula, D. W. (1982). Basic digit sets for radix representation. *Journal of the ACM*, 29(4):1131–43.

Matula, D. W. (1991). Design of a highly parallel IEEE Standard floating point arithmetic unit. In *Proceedings of the Symposium on Combinatorial Optimization in Science and Technology at RUTCOR/DIMACS*.

Matula, D. W. (2001). Improved table lookup algorithms for postscaled division. In *Proceedings of the 15th IEEE Symposium on Computer Arithmetic*, pages 101–8.

Mazenc, C., X. Merrheim, and J.-M. Muller (1993). Computing functions $\cos^{-1}$ and $\sin^{-1}$ using CORDIC. *IEEE Transactions on Computers*, 42(1):118–22.

McIlhenny, R. D. (2002). *Complex Number On-line Arithmetic for Reconfigurable Hardware: Algorithms, Implementations, and Applications*. PhD thesis, University of California, Los Angeles.

McLeish, J. (1991). *Number*. Fawcett Columbine, New York.

McQuillan, S., and J. V. McCanny (1994). Fast algorithms for division and square root. *Journal of VLSI Signal Processing*, 8(2):151–68.

McQuillan, S., and J. V. McCanny (1995). A systematic methodology for the design of high performance recursive digital filters. *IEEE Transactions on Computers*, 44(8):971–82.

McQuillan, S. E. (1992). *Algorithms and Architectures for High Performance Arithmetic Processors*. PhD thesis, The Queen's University of Belfast.

McQuillan, S. E., and J. V. McCanny (1992). VLSI module for high-performance multiply, square root and divide. *IEE Proceedings E: Computers and Digital Techniques*, 139(6):505–10.

McQuillan, S. E., J. V. McCanny, and R. Hamill (1993). New algorithms and VLSI architectures for SRT division and square root. In *Proceedings of the 11th IEEE Symposium on Computer Arithmetic*, pages 80–86.

McQuillan, S. E., J. V. McCanny, and R. F. Woods (1991). High performance VLSI architecture for division and square root. *Electronics Letters*, V27(1): 19–21.

Mehlhorn, K., and F. P. Preparata (1987). Area-time optimal division for $t = \omega(\log n)^{1+\epsilon}$. *Information and Computation*, 72(3):270–82.

Mehta, M., V. Parmar, and E. E. Swartzlander (1991). High-speed multiplier design using multi-input counter and compressor circuits. In *Proceedings of the 10th IEEE Symposium on Computer Arithmetic*, pages 43–50.

Meier, P. C. H. (1999). *Analysis and Design of Low Power Digital Multipliers*. PhD thesis, Carnegie Mellon University.

Meier, P. C. H., R. A. Rutenbar, and L. R. Carley (1996). Exploring multiplier architecture and layout for low power. In *Proceedings of the IEEE 1996 Custom Integrated Circuits Conference*, pages 513–16.

Mencer, O., L. Semeria, M. Morf, and J.-M. Delosme (2000). Application of reconfigurable CORDIC architectures. *Journal of VLSI Signal Processing*, 24(2-3):211–21.

Meo, A. R. (1975). Arithmetic networks and their minimization using a new line of elementary units. *IEEE Transactions on Computers*, C-24(3):258–80.

Merrheim, X., J.-M. Muller, and H. J. Yeh (1993). Fast evaluation of polynomials and inverses of polynomials. In *Proceedings of the 11th IEEE Symposium on Computer Arithmetic*, pages 186–92.

Metafas, D. E., and C. E. Goutis (1995). A floating-point advanced CORDIC processor. *Journal of VLSI Signal Processing*, 10:53–65.

Metze, G. (1962). A class of binary divisions yielding minimally represented quotients. *IRE Transactions Electronic Computers*, EC-11(6):761–64.

Metze, G. (1967). Minimal square rooting. *IEEE Transactions on Computers*, EC-14(2):181–85.

Modiri, R., and T. Lang (1988). Alternative implementations of a radix-4 divider with scaling. Technical report CSD-880069, Computer Science Department, University of California, Los Angeles.

Moes, E. A. J., R. Nouta, and G. J. Hekstra (1993). Divider architectures for VLSI implementation. *International Journal of High Speed Electronics and Systems*, 4(1):1–33.

Montalvo, L. A., K. K. Parhi, and A. Guyot (1998). New Svoboda-Tung division. *IEEE Transactions on Computers*, 47(9):1014–20.

Montoye, R. K., E. Hokonek, and S. L. Runyan (1990). Design of the floating-point execution unit of the IBM RISC System/6000. *IBM Journal of Research and Development*, 34(1):59–70.

Montuschi, P. (1992). Parallel architectures for higher-radix division. *IEE Proceedings E: Computers and Digital Techniques*, 139(2):101–10.

Montuschi, P., and L. Ciminiera (1993). Reducing iteration time when result digit is zero for radix-2 SRT division and square root with redundant remainders. *IEEE Transactions on Computers*, 42(2):239–46.

Montuschi, P., and L. Ciminiera (1991). Algorithm and architectures for radix-4 division with over-redundant digit set and simple digit selection hardware.

In *Conference Record of the 25th Asilomar Conference on Signals, Systems and Computers*, pages 418–22.

Montuschi, P., and L. Ciminiera (1991). Simple radix 2 division and square root with skipping of some addition steps. In *Proceedings of the 10th IEEE Symposium on Computer Arithmetic*, pages 202–9.

Montuschi, P., and L. Ciminiera (1992). Design of a radix 4 division unit with simple selection table. *IEEE Transactions on Computers*, 41(12):1606–11.

Montuschi, P., and L. Ciminiera (1993). Reducing iteration time when result digit is zero for radix-2 SRT division and square root with redundant remainders. *IEEE Transactions on Computers*, 42(2):239–46.

Montuschi, P., and L. Ciminiera (1994). Over-redundant digit sets and the design of digit-by-digit division units. *IEEE Transactions on Computers*, 43(3): 269–77.

Montuschi, P., and L. Ciminiera (1994). Radix-8 division with over-redundant digit set. *Journal of VLSI Signal Processing*, 7(3):259–70.

Montuschi, P., and L. Ciminiera (1995). Quotient prediction without prescaling. *IEE Proceedings: Computers and Digital Techniques*, 142(1):15–22.

Montuschi, P., L. Ciminiera, and A. Giustina (1994). Division unit with Newton-Raphson approximation and digit-by-digit refinement of the quotient. *IEE Proceedings—Computers and Digital Techniques*, 141(6):317–24.

Montuschi, P., and T. Lang (2001). Boosting very-high radix division with prescaling and selection by rounding. *IEEE Transactions on Computers*, 50(1): 13–27.

Montuschi, P., and M. Mezzalama (1990). Survey of square rooting algorithms. *IEE Proceedings E: Computers and Digital Techniques*, 137(1):31–40.

Moore, J. S., T. W. Lynch, and M. Kaufmann (1998). A mechanically checked proof of the AMD5$_k$86 floating-point division program. *IEEE Transactions on Computers*, 47(9):913–26.

Morgan, C. P., and D. B. Jarvis (1959). Transistor logic using current switching routing techniques and its application to a fast carry-propagation adder. *Proceedings of the IEE*, 106B:467–68.

Muller, J.-M. (1994). Some characterizations of functions computable in on-line arithmetic. *IEEE Transactions on Computers*, 43(6):752–55.

Muller, J.-M. (1997). *Elementary Functions, Algorithms and Implementation*. Birkhauser, Boston.

Muller, J.-M. (1999). A few results on table-based methods. *Reliable Computing*, 5(3):279–88.

Nadler, M. (1956). A high speed electronic arithmetic unit for automatic computing machines. *Acta Technica* (6):464–78.

Naffziger, S. (1996). A sub-nanosecond 0.5 micron 64b adder design. *Digest of IEEE International Solid-State Circuits Conference*, pages 362–63.

Nagamatsu, M., S. Tanaka, J. Mori, K. Hirano, T. Noguchi, and K. Hatanaka (1990). A 15-ns 32 × 32-b CMOS multiplier with an improved parallel structure. *IEEE Journal of Solid-State Circuits*, 25(2):494–97.

Nagendra, C., M. J. Irwin, and R. M. Owens (1996). Area-time-power tradeoffs in parallel adders. *IEEE Transactions Circuits and Systems II: Analog and Digital Signal Processing*, 43(10):689–702.

Naini, A., A. Dhablania, W. James, and D. Das Sarma (2001). 1 GHz HAL SPARC64ʳ dual floating point unit with RAS features. In *Proceedings of the 15th IEEE Symposium on Computer Arithmetic*, pages 173–83.

Nannarelli, A. (1999). *Low Power Division and Square Root*. PhD thesis, University of California, Irvine.

Nannarelli, A., and T. Lang (1996). Low-power radix-4 divider. In *International Symposium on Low Power Electronics and Design*, pages 205–8.

Nannarelli, A., and T. Lang (1998). Low-power radix-8 divider. In *Proceedings International Conference on Computer Design. VLSI in Computers and Processors*, pages 420–26.

Nannarelli, A., and T. Lang (1998). Power-delay tradeoffs for radix-4 and radix-8 dividers. In *1998 International Symposium on Low Power Electronics and Design*, pages 109–11.

Nannarelli, A., and T. Lang (1999). Low-power radix-4 combined division and square root. In *Proceedings of the IEEE International Conference on Computer Design: VLSI in Computers and Processors (ICCD'99)*, pages 236–42.

Nannarelli, A., and T. Lang (1999). Low-power divider. *IEEE Transactions on Computers*, 48(1):2–14.

Nannarelli, A., and T. Lang (1999). Low-power division: Comparison among implementations of radix 4, 8 and 16. In *Proceedings of the 14th IEEE Symposium on Computer Arithmetic*, pages 60–67.

Naseem, A. (1984). *Implementation of Parallel Computational Algorithms on a Modified CORDIC Arithmetic Logic Unit*. PhD thesis, Michigan State University.

Naseem, A., and P. D. Fisher (1985). The modified CORDIC algorithm. In *Proceedings of the 7th IEEE Symposium on Computer Arithmetic*, pages 144–52.

Ngai, T. F., M. J. Irwin, and S. Rawat (1986). Regular, area-time efficient carry-lookahead adders. *Journal of Parallel and Distributed Computing*, 3(1):92–105.

Nicks, T. N., R. E. Fry, and P. E. Harvey (1994). POWER2 floating-point unit: Architecture and implementation. *IBM Journal of Research and Development*, 38(5):525–36.

Nielsen, A. M. (1997). *Number Systems and Digit Serial Arithmetic*. PhD thesis, Department of Mathematics and Computer Science, Odense University, Denmark.

Nielsen, A. M., D. W. Matula, C. N. Lyu, and G. Even (2000). An IEEE compliant floating-point adder that conforms with the pipelined packet-forwarding paradigm. *IEEE Transactions on Computers*, 49(1):33–47.

Noetzel, A. S. (1989). An interpolating memory unit for function evaluation: analysis and design. *IEEE Transactions on Computers*, 38(3):377–84.

Noll, T. (1991). Carry-save architectures for high-speed digital signal processing. *Journal of VLSI Signal Processing*, 3(1-2):121–40.

Noll, T., D. Schmitt-Landsiedel, H. Klar, and G. Enders (1986). A pipelined 330-MHz multiplier. *IEEE Journal of Solid-State Circuits*, SC-21(6):411–16.

Oberman, S. F. (1996). *Design Issues in High Performance Floating Point Arithmetic units*. PhD thesis, Department of Electrical Engineering, Stanford University.

Oberman, S. F. (1999). Floating-point division and square root algorithms and implementation in the AMD-K7 microprocessor. In *Proceedings of the 14th IEEE Symposium on Computer Arithmetic*, pages 106–15.

Oberman, S. F., G. Favor, and F. Weber (1999). AMD 3DNow technology: architecture and implementations. *IEEE Micro*, 19(2):37–48.

Oberman, S. F., and M. J. Flynn (1996). Fast IEEE rounding for division by functional iteration. Technical Report CSL-TR-96-700, Computer Systems Laboratory, Department of Electrical Engineering and Computer Science, Stanford University.

Oberman, S. F., and M. J. Flynn (1996). Implementing division and other floating-point operations: A system perspective. In *Scientific Computing and Validated Numerics (Proceedings of SCAN'95)*, pages 18–24.

Oberman, S. F., and M. J. Flynn (1997). Design issues in division and other floating-point operations. *IEEE Transactions on Computers*, 46(2):154–61.

Oberman, S. F., and M. J. Flynn (1998). Minimizing the complexity of SRT tables. *IEEE Transactions on Very Large Scale Integration (VLSI) Systems*, 6(1):141–49.

Oberman, S. F., and M. J. Flynn (1998). Reducing the mean latency of floating-point addition. *Theoretical Computer Science*, 196(1–2):201–14.

Oklobdzija, V. G. (1988). Simple and efficient CMOS circuit for fast VLSI adder realization. In *Proceedings of the IEEE Symposium on Circuits and Systems*, pages 235–38.

Oklobdzija, V. G. (1994). An algorithmic and novel design of a leading zero detector circuit: comparison with logic synthesis. *IEEE Transactions on Very Large Scale Integration (VLSI) Systems*, 2(1):124–28.

Oklobdzija, V. G. editor (1999). *High-Performance System Design: Circuits and Logic*. IEEE Press, Piscataway, New Jersey.

Oklobdzija, V. G., and E. R. Burnes (1985). Some optimal shemes for ALU implementation in VLSI technology. In *Proceedings of the 7th IEEE Symposium on Computer Arithmetic*, pages 2–8.

Oklobdzija, V. G., and M. D. Ercegovac (1982). An on-line square root algorithm. *IEEE Transactions on Computers*, C-31:70–75.

Oklobdzija, V. G., and D. Villeger (1995). Improving multiplier design by using improved column compression tree and optimized final adder in CMOS technology. *IEEE Transactions on VLSI*, 3(2):292–301.

Oklobdzija, V. G., D. Villeger, and S. S. Liu (1996). A method for speed optimized partial product reduction and generation of fast parallel multipliers using an algorithmic approach. *IEEE Transactions on Computers*, 45(3):294–306.

Omondi, A. R. (1994). *Computer Arithmetic Systems, Algorithms, Architecture and Implementations*. Prentice Hall International Series in Computer Science, Englewood Cliffs, New Jersey.

Oppenheim, A. V., R. W. Schafer, and J. R. Buck (1999). *Discrete-Time Signal Processing*. Prentice-Hall, Upper Saddle River, New Jersey.

Osorio, R. R., E. Antelo, J. D. Bruguera, J. Villalba, and E. L. Zapata (1995). Digit on-line large radix CORDIC rotator. In *Proceedings of ASAP-95* (Strasbourg, France), pages 246–57.

Overton, M. A. (2001). *Numerical Computing with IEEE Floating-Point Arithmetic*. SIAM.

Owens, R. M. (1980). *Digit On-line Algorithms for Pipeline Architectures*. PhD thesis, Department of Computer Science, Pennsylvania State University, University Park. (Technical Report CS-80-21.)

Owens, R. M., R. S. Bajwa, and M. J. Irwin (1995). Reducing the number of counters needed for integer multiplication. In *Proceedings of the 12th IEEE Symposium on Computer Arithmetic*, pages 38–41.

Owens, R. M., T. P. Kelliher, M. J. Irwin, M. Vishwanath, R. S. Bajwa, and W.-L. Yang (1993). The design and implementation of the Arithmetic Cube II, a VLSI signal processing system. *IEEE Transactions on Very Large Scale Integration (VLSI) Systems*, 1(4):491–502.

Paal, F. (1973). Implementation of truncated comparison and quotient prediction in the Q-P (quotient predictor) division algorithms. In *Proceedings of the 7th Asilomar Conference on Circuits, Systems and Computers*, pages 734–36.

Paliouras, V., K. Karagianni, and T. Stouraitis (2000). A floating-point processor for fast and accurate sine/cosine evaluation. *IEEE Transactions on Circuits and Systems II: Analog and Digital Signal Processing*, 47(5):441–51.

Parhami, B. (1987). On the complexity of table-lookup for iterative division. *IEEE Transactions on Computers*, C-36:1233–36.

Parhami, B. (1988). Carry-free addition of recoded binary signed-digit numbers. *IEEE Transactions on Computers*, C-37(11):1470–76.

Parhami, B. (1993). On the implementation of arithmetic support functions for generalized signed-digit number systems. *IEEE Transactions on Computers*, 42(3):379–84.

Parhami, B. (1996). Variations on multioperand addition for faster logarithmic-time tree multiplier. In *Proceedings of the 30th Asilomar Conference on Signals, Systems and Computers*, pages 899–903.

Parhami, B. (2000). *Computer Arithmetic: Algorithms and Hardware Designs*. Oxford University Press, New York.

Parhi, K. K. (1999). Low-energy CSMT carry generators and binary adders. *IEEE Transactions on VLSI Systems*, 7(12):450–62.

Park, W.-C., T.-D. Han, S.-D. Kim, and S.-B. Yang (1999). A floating point multiplier performing IEEE rounding and addition in parallel. *Journal of Systems Architecture*, 45(14):1195–1207.

Parker, A., and J. O. Hamblen (1992). Optimal value for the Newton-Raphson division algorithm. *Information Processing Letters*, 42(3):141–44.

Parks, M. (2000). Number-theoretic test generation for directed roundings. *IEEE Transactions on Computers*, 49(7):651–58.

Paterson, M., and U. Zwick (1993). Shallow circuits and concise formulae for multiple addition and multiplication. *Computational Complexity*, 3(3): 262–91.

Peng, V., S. Samudrala, and M. Gavrielov (1987). On the implementation of shifters, multipliers, and dividers in VLSI floating-point units. In *Proceedings of the 8th IEEE Symposium on Computer Arithmetic*, pages 95–102.

Pezaris, S. D. (1971). A 40 ns 17 bit by 17 bit array multiplier. *IEEE Transactions on Computers*, C-20(4):442–47.

Phatak, D. S. (1998). Comments on Duprat and Muller's branching CORDIC paper. *IEEE Transactions on Computers*, 47(9):1037–40.

Phatak, D. S. (1998). Double step branching CORDIC: a new algorithm for fast sine and cosine generation. *IEEE Transactions on Computers*, 47(5):587–602.

Phatak, D. S., T. Geoff, and I. Koren (2001). Constant-time addition and simultaneous format conversion based on redundant binary representation. *IEEE Transactions on Computers*, 50(11):1267–87.

Potkonjak, M., M. B. Srivastava, and A. P. Chandrakasan (1996). Multiple constant multiplications: Efficient and versatile framework and algorithms for exploring common subexpression elimination. *IEEE Transaction on Computer-Aided Design of Integrated Circuits and Systems*, 15(2):151–65.

Prabhu, J. A., and G. B. Zyner (1995). 167 MHz radix-8 divide and square root using overlapped radix-2 stages. In *Proceedings of the 12th IEEE Symposium on Computer Arithmetic*, pages 155–62.

Prasad, K., and K. K. Parhi (2001). Low-power 4-2 and 5-2 compressors. In *Proceedings of the 35th Asilomar Conference on Signals, Systems and Computers*, pages 129–33.

Priest, D. M. (1991). Algorithms for arbitrary precision floating point arithmetic. In *Proceedings of the 10th IEEE Symposium on Computer Arithmetic (Arith-10)*, pages 132–44.

Purdy, C. N., and G. B. Purdy (1987). Integer division in linear time with bounded fan-in. *IEEE Transactions on Computers*, C-36:640–44.

Quach, N. T., and M. J. Flynn (1992). High-speed addition in CMOS. *IEEE Transactions on Computers*, 41(12):1612–15.

Rabaey, J.-M., A. Chandrakasan, and B. Nikolić (2003). *Digital Integrated Circuits: A Design Perspective*. Prentice Hall, Englewood Cliffs, New Jersey, 2 edition.

Rajagopal, S., and J. R. Cavallaro (2001). On-line arithmetic for detection in digital communication receivers. In *Proceedings of the 15th IEEE Symposium on Computer Arithmetic*, pages 257–65.

Ramachandran, R., and S.-L. Lu (1996). Efficient arithmetic using self-timing. *IEEE Transactions on VLSI*, 4(4):445–54.

Ramamoorthy, C. V., J. R. Goodman, and K. H. Kim (1972). Some properties of iterative square-rooting methods using high-speed multiplication. *IEEE Transactions on Computers*, C-21(8):837–47.

Randell, B., editor (1975). *On the Mathematical Powers of the Calculating Engine (C. Babbage)*. Springer-Verlag, New York, 2nd edition.

Rauchwerger, L., and P. M. Farmwald (1990). A multiple floating point coprocessor architecture. In *Proceedings of the 23rd Annual Workshop and Symposium*, pages 216–22.

Reif, J. H., and S. R. Tate (1989). Optimal size integer division circuits. In *Proceedings of the 21st Annual ACM Symposium on Theory of Computing*, pages 264–73.

Reiser, J. F., and D. E. Knuth (1975). Evading the drift in floating-point addition. *Information Processing Letters*, 3(3):84–87.

Reitwiesner, G. W. (1960). Binary arithmetic. In *Advances in Computers*, volume 1, pages 232–308. Academic Press, New York.

Renaudin, M., B. E. Hassan, and A. Guyot (1996). A new asynchronous pipeline scheme: application to the design of a self-timed ring divider. *IEEE Journal of Solid-State Circuits*, 31(7):1001–13.

Robertson, J. E. (1955). Two's complement multiplication in binary parallel computers. *IEEE Transactions on Electronic Computers*, EC-34(3):118–19.

Robertson, J. E. (1957). Arithmetic unit (chapter 8). In *On the Design of Very High-Speed Computers*. Technical report no. 80, Computer Science Department, University of Illinois at Urbana-Champaign.

Robertson, J. E. (1958). A new class of digital division methods. *IRE Transactions Electronic Computers*, EC-7(3):88–92.

Robertson, J. E. (1960). *Theory of Computer Arithmetic Employed in the Design of the New Computer at the University of Illinois*. File no. 319, Computer Science Department, University of Illinois at Urbana-Champaign.

Robertson, J. E. (1964). *Introduction to Digital Computer Arithmetic*. File no. 599, Department of Computer Science, University of Illinois at Urbana-Champaign.

Robertson, J. E. (1965). *Methods of Selection of Quotient Digits during Digital Division*. File no. 663, Computer Science Department, University of Illinois at Urbana-Champaign.

Robertson, J. E. (1967). A deterministic procedure for the design of carry-save adders and borrow-save subtracters. Technical report No. 235, Dept. of Computer Science, University of Illinois, Urbana-Champaign.

Robertson, J. E. (1970). The correspondence between methods of digital division and multiplier recoding procedures. *IEEE Transactions on Computers*, C-19(8):692–701.

Rodrigues, M. R. D., J. H. P. Zurawski, and J. B. Gosling (1981). Hardware evaluation of mathematical functions. *IEE Proceedings E (Computers and Digital Techniques)*, 128(4):155–64.

Rohatsch, F. A. (1967). *A Study of Transformations Applicable to the Development of Limited Carry-Borrow Propagation Adders*. PhD thesis, Department of Computer Science, University of Illinois, Urbana-Champaign.

Rubinfield, L. P. (1975). A proof of the modified Booth's algorithm for multiplication. *IEEE Transactions on Computers*, C-24(4):1014–15.

Ruess, H., N. Shankar, and M. K. Srivas (1999). Modular verification of SRT division. *Formal Methods in System Design*, 14(1):45–73.

Rusinoff, D. (1998). A mechanically checked proof of IEEE compliance of a register-transfer-level specification of the AMD-k7 floating-point multiplication, division, and square root instructions. *LMS Journal of Computation and Mathematics*, 1:148–200.

Salomon, D. (1987). A design for an efficient NOR-gate only, binary ripple adder with carry-completion detection logic. *Computer Journal*, 30(3):283–85.

Sam, H., and A. Gupta (1990). A generalized multibit recoding of the two's complement binary numbers and its proof with application in multiplier implementations. *IEEE Transactions on Computers*, C-39(8):1006–15.

Santoro, M. R. (1989). *Design and Clocking of VLSI Multipliers*. PhD thesis, Stanford University.

Santoro, M. R., G. Bewick, and M. A. Horowitz (1989). Rounding algorithms for IEEE multipliers. In *Proceedings of the 9th Symposium on Computer Arithmetic*, pages 176–83.

Santoro, M. R., and M. A. Horowitz (1989). SPIM: a pipelined 64 × 64-bit iterative multiplier. *IEEE Journal of Solid-State Circuits*, 24:487–93.

Schmid, H. (1974). *Decimal Computation*. John Wiley & Sons, New York.

Schulte, M. J., P. I. Balzola, A. Akkas, and R. W. Brocato (2000). Integer multiplication with overflow detection or saturation. *IEEE Transactions on Computers*, 49(7):681–91.

Schulte, M. J., J. Omar, and E. Swartzlander, Jr. (1994). Optimal initial approximations for the Newton-Raphson division algorithm. *Computing*, 53(3–4): 233–42.

Schulte, M. J., and J. E. Stine (1997). Symmetric bipartite tables for accurate function approximation. In *Proceedings of the 13th IEEE Symposium on Computer Arithmetic*, pages 175–83.

Schulte, M. J., and J. E. Stine (1997). Accurate function evaluation by symmetric table lookup and addition. In *Proceedings of the IEEE International Conference on Application-Specific Systems, Architectures and Processors*, pages 144–53.

Schulte, M. J., and J. E. Stine (1999). Approximating elementary functions with symmetric bipartite tables. *IEEE Transactions on Computers*, 48(8):842–47.

Schulte, M. J., J. E. Stine, and J. G. Jansen (1999). Reduced power dissipation through truncated multiplication. In *Proceedings of the IEEE Alessandro Volta Memorial Workshop on Low-Power Design*, pages 61–69.

Schulte, M. J., and E. Swartzlander (1993). Truncated multiplication with correction constant (for DSP). In *Proceedings of the IEEE Workshop on VLSI Signal Processing*, pages 388–96.

Schulte, M. J., and E. E. Swartzlander (1993). Exact rounding of certain elementary functions. In *Proceedings of the 11th IEEE Symposium on Computer Arithmetic*, pages 138–45.

Schulte, M. J., and E. E. Swartzlander (1994). Hardware designs for exactly rounded elementary functions. *IEEE Transactions on Computers*, 43(8):964–73.

Schwarz, E. M. (1995). Rounding for quadratically converging algorithms for division and square root. In *Conference Record of the 29th Asilomar Conference on Signals, Systems and Computers*, volume 1, pages 600–3.

Schwarz, E. M., R. Averill, III, and L. Sigal (1997). A radix-8 CMOS S/390 multiplier. In *Proceedings of the 13th IEEE Symposium on Computer Arithmetic*, pages 2–9.

Schwarz, E. M., and M. J. Flynn (1996). Hardware starting approximation method and its application to the square root operation. *IEEE Transactions on Computers*, 45(12):1356–69.

Schwarz, E. M., R. M. Smith, and C. A. Krygowski (1999). The S/390 G5 floating-point unit supporting hex and binary architectures. In *Proceedings of the 14th IEEE Symposium on Computer Arithmetic*, pages 258–65.

Scott, N. R. (1985). *Computer Number Systems & Arithmetic*. Prentice Hall, Englewood Cliffs, New Jersey.

Seidel, P.-M., and G. Even (2001). On the design of fast IEEE floating-point adders. In *Proceedings of the 15th IEEE Symposium on Computer Arithmetic*, pages 184–94.

Seidel, P.-M., L. D. McFearin, and D. W. Matula (2001). Binary multiplication radix-32 and radix-256. In *Proceedings of the 15th IEEE Symposium on Computer Arithmetic (Arith-15)*, pages 23–32.

Shaham, Z., and Z. Riesel (1972). A note on division algorithms based on multiplication. *IEEE Transcations on Computers*, C-21(5):513–14.

Sharangpani, H., and K. Arora (2000). Itanium processor microarchitecture. *IEEE Micro*, 20(5):24–43.

Shedletsky, J. J. (1977). Comment on the sequential and indeterminate behaviour of an end-around-carry adder. *IEEE Transactions on Computers*, C-26(3):271–72.

Shively, R. R. (1963). *Stationary Distributions of Partial Remainders in SRT Digital Division*. PhD thesis, University of Illinois.

Singh, M., and S. M. Nowick (2000). Fine-grain pipelined asynchronous adders for high-speed DSP applications. In *Proceedings of the IEEE Computer Society Workshop on VLSI 2000, System Design for a System-on-Chip Era*, pages 111–18.

Sips, H. J. (1984). Bit-sequential arithmetic for parallel processors. *IEEE Transactions on Computers*, C-33(1):7–20.

Sklansky, J. (1960). Conditional-sum addition logic. *IRE Transactions on Electronic Computers*, EC-9:226–31.

Sklansky, J. (1960). An evaluation of several two-summand binary adders. *IRE Transactions on Electronic Computers*, EC-9:213–26.

Smith, S. G., and P. Denyer (1988). *Serial-Data Computation*. Kluwer Academic Publishers.

Soceneantu, A., and C. I. Toma (1972). Cellular logic array for redundant binary division. In *Proceedings of the IEE*, volume 119, pages 1452–56.

Soderquist, P., and M. Leeser (1996). Area and performance tradeoffs in floating-point division and square root implementations. *ACM Computing Surveys*, 28(3):518–64.

Soderstrand, M., W. Jenkins, G. Jullien, and F. Taylor (1986). *Residue Number System Arithmetic: Modern Applications in Digital Signal Processing*. IEEE Press, New York.

Song, P. J., and G. D. Micheli (1991). Circuit and architecture trade-offs for high-speed multiplication. *IEEE Journal of Solid-State Circuits*, 26(9):1184–98.

Spaniol, O. (1981). *Computer Arithmetic: Logic and Design*. John Wiley & Sons, Inc., New York.

Specker, W. H. (1965). A class of algorithms for $\ln(x)$, $\exp(x)$, $\sin(x)$, $\cos(x)$, $\tan^{-1}(x)$ and $\cot^{-1}(x)$. *IEEE Transactions on Electronic Computers*, EC-14: 85–86.

Spira, P. M. (1973). Computation times of arithmetic and Boolean functions in (d,r) circuits. *IEEE Transactions on Computers*, C-22(6):552–55.

Srinivas, H. R., and K. K. Parhi (1995). A fast radix-4 division algorithm and its architecture. *IEEE Transactions on Computers*, 44(6):826–31.

Srinivas, H. R., and K. K. Parhi (1999). A radix 2 shared division/square root algorithm and its VLSI architecture. *Journal of VLSI Signal Processing Systems for Signal, Image, and Video Technology*, 21(1):37–60.

Stan, M. R., A. F. Tenca, and M. D. Ercegovac (1998). Long and fast up/down counters. *IEEE Transactions on Computers*, 47(7):722–35.

Stelling, P. F. (1995). *Application of Combinatorics to Parallel Multiplier Design, Tree Reconstruction, and the Analysis of Strings*. PhD thesis, University of California, Davis.

Stelling, P. F., C. U. Martel, V. G. Oklobdžija, and R. Ravi (1998). Optimal circuits for parallel multipliers. *IEEE Transactions on Computers*, 47(3):273–85.

Stelling, P. F., and V. G. Oklobdzija (1996). Design strategies for optimal hybrid final adders in a parallel multiplier. *Journal of VLSI Signal Processing*, 14(3):321–31.

Stelling, P. F., and V. G. Oklobdzija (1997). Implementing multiply-accumulate operation in multiplication time. In *Proceedings of the 13th IEEE Symposium on Computer Arithmetic*, pages 99–106.

Stenzel, W. J., W. J. Kubitz, and G. H. Garcia (1977). A compact high-speed parallel multiplication scheme. *IEEE Transactions on Computers*, C-26(10):948–57.

Sterbenz, P. H. (1974). *Floating Point Computation*. Prentice-Hall, Englewood Cliffs, New Jersey.

Stine, J. E., and M. J. Schulte (1999). The symmetric table addition method for accurate function approximation. *Journal of VLSI Signal Processing*, 21:167–77.

Story, S., and P. T. P. Tang (1999). New algorithms for improved transcendental functions on IA-64. In *Proceedings of the 14th IEEE Symposium on Computer Arithmetic*, pages 4–11.

Strandberg, R. H., L. G. Bustamante, V. G. Oklobdzija, M. Soderstrand, and J. C. LeDuc (1996). Efficient realizations of squaring and reciprocal used in adaptive sample rate notch filter. *Journal of VLSI Signal Processing Systems*, 14(3):303–9.

Sung, T. Y., Y. H. Hu, and H. J. Yu (1986). Doubly pipelined CORDIC array for digital signal processing. In *IEEE International Conference on ASSP*, pages 1169–72.

Suzuki, H., H. Makino, K. Mashiko, and H. Hamano (1997). A floating-point divider using redundant binary circuits and an asynchronous clock scheme. In *Proceedings of the International Conference on Computer Design: VLSI in Computers and Processors*, pages 685–89.

Suzuki, H., Y. Nakase, H. Makino, H. Morinaka, and K. Mashiko (1995). Leading-zero anticipatory logic for high-speed floating point addition. In *Proceedings of the IEEE 1995 Custom Integrated Circuits*, pages 589–92.

Svoboda, A. (1963). An algorithm for division. *Information Processing Machines (Stroje na Zpracovani Informaci)*, 9:25–34.

Svoboda, A. (1970). Adder with distributed control. *IEEE Transactions on Computers*, C-19(8):749–51.

Swartzlander, E. (1999). Truncated multiplication with approximate rounding. In *Proceedings of the 33rd Asilomar Conference on Signals, Systems, and Computers*, volume 2, pages 1480–83.

Swartzlander, E. E. (1973). Parallel counters. *IEEE Transactions on Computers*, C-22:1021–24.

Swartzlander, E. E. editor. (1990). *Computer Arithmetic, Vol. 1 and Vol. 2*. IEEE Computer Society Press, Los Alamitos, California.

Sweeney, D. W. (1965). An analysis of floating-point addition. *IBM Systems Journal*, 4:31–42.

Sweitz, F. J. (1987). *Capitalism & Arithmetic: The New Math of the 15th Century*. Open Court, La Salle, Illinois.

Szabo, N. S., and R. I. Tanaka (1967). *Residue Arithmetic and Its Applications to Computer Technology*. McGraw-Hill, New York.

Takagi, N. (1987). *Studies on Hardware Algorithms for Arithmetic Operations with a Redundant Binary Representation*. PhD thesis, Department of Information Science, Kyoto University.

Takagi, N. (2001). A hardware algorithm for computing reciprocal square root. In *Proceedings of the 15th IEEE Symposium on Computer Arithmetic*, pages 94–100.

Takagi, N., T. Asada, and S. Yajima (1991). Redundant CORDIC methods with a constant scale factor. *IEEE Transactions on Computers*, 40(9):989–95.

Takagi, N., and T. Horiyama (1999). A high-speed reduced-size adder under left-to-right input arrival. *IEEE Transactions on Computers*, 48(1):76–80.

Takagi, N., H. Yasukura, and S. Yajima (1985). High speed multiplication algorithm with a redundant binary addition tree. *IEEE Transactions on Computers*, C-34(9):789–96.

Tan, K. G. (1978). The theory and implementation of high-radix division. In *Proceedings of the 4th IEEE Symposium on Computer Arithmetic*, pages 154–63.

Tang, P. T. P. (1989). Table-driven implementation of the exponential function in IEEE floating-point arithmetic. *ACM Transactions on Mathematical Software*, 15(2):144–57.

Tang, P. T. P. (1990). Table-driven implementation of the logarithm function in IEEE floating-point arithmetic. *ACM Transactions on Mathematical Software*, 16(4):378–400.

Tang, P. T. P. (1991). Table lookup algorithms for elementary functions and their error analysis. In *Proceedings of the 10th IEEE Symposium on Computer Arithmetic*, pages 232–36.

Tang, P. T. P. (1992). Table-driven implementation of the expm1 function in IEEE floating-point arithmetic. *ACM Transactions on Mathematical Software*, 18(2):211–22.

Taylor, F. (1984). Residue arithmetic: A tutorial with examples. *IEEE Computer Magazine*, 17(5):50–62.

Taylor, G. S. (1981). Compatible hardware for division and square root. In *Proceedings of the 5th IEEE Symposium on Computer Arithmetic*, pages 127–34.

Taylor, G. S. (1985). Radix-16 SRT dividers with overlapped quotient-selection stages. In *Proceedings of the 7th IEEE Symposium on Computer Arithmetic*, pages 64–71.

Taylor, G. S., and D. A. Patterson (1981). VAX hardware for the proposed IEEE Floating-Point Standard. In *Proceedings of the 5th IEEE Symposium on Computer Arithmetic*, pages 190–96.

Tenca, A. F. (1998). *Variable Long-Precision Arithmetic (VLPA) for Reconfigurable Coprocessor Architectures*. PhD thesis, University of California, Los Angeles.

Tenca, A. F., and M. D. Ercegovac (1999). On the design of high-radix on-line division for long precision. In *Proceedings of the 14th IEEE Symposium on Computer Arithmetic*, pages 44–51.

Tenca, A. F., M. D. Ercegovac, and M. E. Louie (1999). Fast on-line multiplication units using LSA organization. In *Proceedings of the SPIE—Advanced Signal Processing Algorithms, Architectures, and Implementations IX*, volume 3807, pages 74–83.

Tenca, A. F., and S. U. Hussaini (2001). A design of radix-2 on-line division using LSA organization. In *Proceedings of the 15th IEEE Symposium on Computer Arithmetic*, pages 266–73.

Timmermann, D., H. Hahn, and B. J. Hosticka (1992). Low latency time CORDIC algorithms. *IEEE Transactions on Computers*, 41(8):1010–15.

Timmermann, D., H. Hahn, B. J. Hosticka, and B. Rix (1991a). A new addition scheme and fast scaling factor compensation methods for CORDIC algorithms. *INTEGRATION, The VLSI Journal*, 11:85–100.

Timmermann, D., H. Hahn, B. J. Hosticka and G. Schmidt (1991). A programmable CORDIC chip for digital signal processing applications. *IEEE Journal of Solid-State Circuits*, 26(9):1317–21.

Timmermann, D., B. Rix, H. Hahn, and B. J. Hosticka (1994). A CMOS floating-point vector-arithmetic unit. *IEEE Journal of Solid-State Circuits*, 29(5):634–39.

Tisserand, A., P. Marchal, and C. Piguet (1999). FPOP: field-programmable on-line operators. In *Proceedings of the SPIE—Advanced Signal Processing Algorithms, Architectures, and Implementations IX*, volume 3807, pages 31–42.

Tocher, K. D. (1958). Techniques of multiplication and division for automatic binary computers. *Quart. J. Mech. Appl. Math.*, XI(Pt. 3):364–84.

Trivedi, K. S., and M. D. Ercegovac (1977). On-line algorithms for division and multiplication. *IEEE Transactions on Computers*, C-26(7):161–67.

Trivedi, K. S., and J. G. Rusnak (1978). Higher radix on-line division. In *Proceedings of the 4th IEEE Symposium on Computer Arithmetic*, pages 164–74.

Tsunekava, Y., M. Hinosugi, and M. Miura (1998). Design and VLSI evaluation of a high-speed cellular array divider with a selection function. *Electrical Engineering in Japan*, 124(4):760–97.

Tu, P. K.-G. (1990). *On-line Arithmetic Algorithms for Efficient Implementations*. PhD thesis, University of California, Los Angeles.

Tu, P. K.-G., and M. D. Ercegovac (1989). Design of on-line division unit. In *Proceedings of the 9th IEEE Symposium on Computer Arithmetic*, pages 42–49.

Tu, P. K.-G., and M. D. Ercegovac (1991). Gate array implementation of on-line algorithms for floating-point operations. *Journal of VLSI Signal Processing*, 3(4):307–17.

Tullsen, D. M., and M. D. Ercegovac (1986). Design and implementation of an on-line algorithm. In *Proceedings of the SPIE Real Time Signal Processing IX*, volume 698, pages 92–99.

Tung, C. (1968). A division algorithm for signed-digit arithmetic. *IEEE Transactions on Computers*, C-17(9):887–89.

Tung, C. (1970). Signed-digit division using combinational arithmetic nets. *IEEE Transactions on Computers*, C-19(8):746–48.

Tung, C. (1972). Arithmetic (Chapter 3). In *Computer Science*. Wiley-Interscience, New York.

Tung, C., and A. Avizienis (1970). Combinational arithmetic systems for the approximation of functions. In *AFIPS Conference Proceedings 1970 Spring Joint Computer Conference*, pages 95–107.

Turrini, S. (1989). Optimal group distribution in carry-skip adders. In *Proceedings of the 9th IEEE Symposium on Computer Arithmetic*, pages 96–103.

Tyagi, A. (1993). A reduced area scheme for carry-select adders. *IEEE Transactions on Computers*, C-42(10):1163–70.

Unwala, I. H., and E. E. Swartzlander (1993). Superpipelined adder designs. In *Proceedings of the International Symposium on Circuits and Systems (ISCAS)*, volume 3, pages 1841–44.

Uya, M., K. Kaneko, and J. Yasui (1984). A CMOS floating-point multiplier. *IEEE Journal of Solid-State Circuits*, SC-19(5):697–702.

Van, L. D., S.-S. Wang, and W.-S. Feng (2000). Design of the lower error fixed-width multiplier and its application. *IEEE Transactions Circuits and Systems II: Analog and Digital Signal Processing*, 47(10):1112–18.

Vanmeulebroecke, A., E. Vanzieleghem, T. Denyer, and P. G. A. Jespers (1990). A new carry-free division algorithm and its application to a single-chip 1024-b RSA processor. *IEEE Journal of Solid-State Circuits*, SC-25(3): 748–65.

Vassiliadis, S., D. S. Lemon, and M. Putrino (1989). S/370 sign-magnitude floating-point adder. *IEEE Journal of Solid-State Circuits*, 24:1062–70.

Vassiliadis, S., E. M. Schwarz, and D. J. Harahan (1989). A general proof for overlapped multiple-bit scanning multiplication. *IEEE Transactions on Computers*, 38:172–73.

Vassiliadis, S., J. Philips, and B. Blaner (1993). Condition code predictor for fixed-point arithemtic units. *IEEE Transactions on Computers*, 42(7):825–39.

Verdonk, B., A. Cuyt, and D. Verschaeren (2001). A precision- and range-independent tool for testing floating-point arithmetic. i. Basic operations, square root, and remainder. *ACM Transactions on Mathematical Software*, 27(1):92–118.

Villalba, J., T. Lang, and E. L. Zapata (1998). Parallel compensation of scale factor for the CORDIC algorithm. *Journal of VLSISignal Processing*, 19:227–41.

Villalba, J., E. L. Zapata, E. Antelo., and J. D. Bruguera (1998). Radix-4 vectoring CORDIC. *Journal of VLSI Signal Processing*, 19:127–47.

Volder, J. (1959). The CORDIC computing techniquie. *IRE Transactions on Electronic Computers*, EC-8(3):330–34.

Volder, J. E. (2000). The birth of CORDIC. *Journal of VLSI Signal Processing Systems*, 25(2):101–5.

Vuillemin, J. E. (1991). Constant time arbitrary length synchronous binary counters. In *Proceedings of the 10th IEEE Symposium on Computer Arithmetic*, pages 180–83.

Wakerly, J. F. (2001). *Digital Design Principles & Practices*. Prentice Hall, Englewood Cliffs, New Jersey.

Wallace, C. S. (1964). A suggestion for a fast multiplier. *IEEE Transactions on Electronic Computers*, EC-13(2):14–17.

Walter, C. D. (1995). Verification of hardware combining multiplication, division and square root. *Microprocessors and Microsystems*, 19(5):243–45.

Walther, J. S. (1971). A unified algorithm for elementary functions. In *Proceedings of the Spring Joint Computer Conference*, pages 379–85.

Walther, J. S. (2000). The story of unified CORDIC. *Journal of VLSI Signal Processing Systems*, 25(2):107–12.

Wang, C.-C., C.-J. Huang, and G.-C. Lin (2000). Cell-based implementation of radix-4/2 64b dividend 32b divisor signed integer divider using the COMPASS cell library. *IEE Proceedings—Computers and Digital Techniques*, 147(2):109–15.

Wang, S. (1998). *A CORDIC Arithmetic Processor*. PhD thesis, University of Texas at Austin.

Wang, S., V. Piuri, and E. E. Swartzlander (1997). Hybrid CORDIC algorithms. *IEEE Transactions on Computers*, 46(11):1202–7.

Wang, Z., G. A. Jullien, and W. C. Miller (1995). A new design technique for column compression multipliers. *IEEE Transactions on Computers*, 44(8):962–70.

Ware, F. A., W. McAllister, J. R. Carlson, D. K. Sun, and R. J. Vlach (1982). 64 bit monolithic floating-point processors. *IEEE Journal of Solid-State Circuits*, SC-17(5):898–907.

Wasser, S., and M. J. Flynn (1982). *Introduction to Arithmetic for Digital Computers*. Holt, Rinehart, Winston, New York.

Watanuki, O. (1981). *Floating-Point On-line Arithmetic for Highly Concurrent Digit-Serial Computation: Application to Mesh Problems*. PhD thesis, University of California, Los Angeles.

Watanuki, O., and M. D. Ercegovac (1983). Error analysis of certain floating-point on-line algorithms. *IEEE Transactions on Computers*, C-32:352–58.

Wei, B. W. Y., and C. D. Thompson (1990). Area-time optimal adder design. *IEEE Transactions on Computers*, 39(5):666–75.

Weinberger, A. (1978). High-speed zero-sum detection. In *Proceedings of the 4th IEEE Symposium on Computer Arithmetic*, pages 200–207.

Weinberger, A. (1981). 4:2 carry-save adder module. *IBM Technical Disclosure Bulletin*, 23.

Weinberger, A., and J. L. Smith (1958). A logic for high-speed addition. *Nat. Bur. Stand. Circ.*, 591:3–12.

Wells, D. (1997). *Curious and Interesting Numbers*. Penguin Books, New York.

Weste, N. H. E., and K. Eshragian (1993). *Principles of CMOS VLSI Design: A System Perspective*. Addison-Wesley Publishing Co., Reading, Massachusetts, 2nd edition.

Wey, C. L. (2000). Design of fast high-radix SRT dividers and their VLSI implementation. *IEE Proceedings—Computers and Digital Techniques*, 147(4):275–81.

Wey, C. L., and C.-P. Wang (1999). Design of a fast radix-4 SRT divider and its VLSI implementation. *IEE Proceedings—Computers and Digital Techniques*, 146(4):205–10.

Wilkinson, J. H. (1963). *Rounding Errors in Algebraic Processes*. Prentice-Hall, Englewood Cliffs, New Jersey.

Williams, J., and V. C. Hamacher (1981). A linear-time divider array. *Canadian Electr. Engineering Journal*, 6:14–20.

Williams, T., N. Patkar, and G. Shen (1995). SPARC64: a 64-b 64-active-instruction out-of-order-execution MCM processor. *IEEE Journal of Solid-State Circuits*, 30(11):1215–26.

Williams, T. E. (1991). *Self-Timed Rings and Their Application to Division*. PhD thesis, Stanford University. Computer Systems Laboratory technical report no. CSL-TR-91-482.

Williams, T. E., and M. Horowitz (1991). A 160ns 54-bit CMOS division implementation using self-timed and symmetrically overlapped SRT stages. In *Proceedings of the 10th IEEE Symposium on Computer Arithmetic*, pages 210–17.

Williams, T. E., and M. A. Horowitz (1991). A zero-overhead self-timed 160-ns 54-b CMOS divider. *IEEE Journal of Solid-State Circuits*, 26(11):1651–61.

Wilson, J. B., and R. S. Ledley (1961). An algorithm for rapid binary division. *IRE Transactions Electronic Computers*, EC-10(4):662–70.

Winograd, S. (1965). On the time required to perform addition. *Journal of the ACM*, 12(2):277–85.

Winograd, S. (1967). On the time required to perform multiplication. *Journal of the ACM*, 14(4):793–802.

Wires, K. E., M. J. Schulte, L. P. Marquette, and P. I. Balzola (1999). Combined unsigned and 2's complement squarers. In *Proceedings of the 33rd Asilomar Conference on Signals, Systems, and Computers*, pages 1215–19.

Wires, K. E., M. J. Schulte, and J. E. Stine (2000). Variable-correction truncated floating point multipliers. In *Proceedings of the 34th Asilomar Conference on Signals, Systems, and Computers*, pages 1344–48.

Wires, K. E., M. J. Schulte, and J. E. Stine (2001). Combined IEEE compliant and truncated floating point multipliers for reduced power dissipation. In *Proceedings of the IEEE International Conference on Computer Design: VLSI in Computers and Processors (ICCD'01)*, pages 497–500.

Wold, E. H. and A. M. Despain (1984). Pipeline and parallel-pipeline FFT processors for VLSI implementations. *IEEE Transactions on Computers*, C-33(5):414–26.

Won, J.-H., and K. Choi (2000). Low power self-timed radix-2 division. In *ISLPED'00: Proceedings of the 2000 International Symposium on Low Power Electronics and Design*, pages 210–12.

Wong, D., and M. J. Flynn (1992). Fast division using accurate quotient approximations to reduce the number of iterations. *IEEE Transactions on Computers*, 41(8):981–95.

Wong, D. C., G. De Micheli, and M. J. Flynn (1993). Designing high-performance digital circuits using wave pipelining: Algorithms and practical experiences. *IEEE Transactions Computer-Aided Design of Integrated Circuits and Systems*, 12(1):25–46.

Wong, W. F., and E. Goto (1994). Fast hardware-based algorithms for elementary function computations using rectangular multipliers. *IEEE Transactions on Computers*, 43(3):278–94.

Yang, B., D. Timmermann, J. F. Bome, H. Hahn, B. J. Hosticka, G. Schmidt, and G. Zimmer (1987). Special computers: graphics, robotics. In *Proceedings of VLSI Computer, COMPEURO*, pages 727–30.

Yeager, K. C. (1996). The Mips R10000 superscalar microprocessor. *IEEE Micro*, 16(2):28–41.

Yeh, C.-H., and B. Parhami (1996). Efficient pipelined multi-operand adders with high throughput and low latency: Designs and applications. In *Proceedings of the 30th Asilomar Conference on Signals, Systems and Computers*, pages 894–98.

Yohe, J. M. (1973). Roundings in floating-point arithmetic. *IEEE Transactions on Computers*, C-22(6):577–86.

Yoshida, N., E. Goto, and S. Ichikawa (1991). Pseudorandom rounding for truncated multipliers. *IEEE Transactions on Computers*, 40(9):1065–67.

Yoshimura, H., T. Nakanishi, and H. Yamauchi (1989). A 50-MHz CMOS geometrical mapping processor. *IEEE Transactions on Circuits and Systems*, 36(10):1360–63.

Yu, R. Y., and G. B. Zyner (1995). 167 MHz radix-4 floating point multiplier. In *Proceedings of the 12th IEEE Symposium on Computer Arithmetic*, pages 149–54.

Zimmerman, R. (1998). *Binary Adder Architectures for Cell-Based VLSI and Their Synthesis (Ph.D. dissertation)*. Series in Microelectronics, Vol. 37. Hartung-Gore, Konstanz, Switzerland.

Zou, F., and P. Kornerup (1995). High speed DCT/IDCT using a pipelined CORDIC algorithm. In *Proceedings of the 12th IEEE Symposium on Computer Arithmetic*, pages 180–87.

Zuras, D. (1993). On squaring and multiplying large integers. In *Proceedings of the 11th IEEE Symposium on Computer Arithmetic*, pages 260–71.

Zuras, D. (1994). More on squaring and multiplying large integers. *IEEE Transactions on Computers*, 43(8):899–908.

Zuras, D., and W. H. McAllister (1986). Balanced delays trees and combinatorial division in VLSI. *IEEE Journal of Solid-State Circuits*, SC-21:814–19.

Zurawski, J. H. P. (1980). *High Performance Evaluation of Division and Other Elementary Functions*. PhD thesis, University of Manchester, England.

Zurawski, J. H. P., and J. B. Gosling (1981). Design of high-speed digital divider units. *IEEE Transactions on Computers*, C-30(9):691–99.

Zurawski, J. H. P., and J. B. Gosling (1987). Design of a high-speed square root, multiply, and divide unit. *IEEE Transactions on Computers*, C-36(9):13–23.

Index